Netcentric System of Systems Engineering with DEVS Unified Process

System of Systems Engineering Series

Series Editor: Mo Jamshidi

Discrete-Time Inverse Optimal Control for Nonlinear Systems,
Edgar N. Sanchez and Fernando Ornelas-Tellez

Netcentric System of Systems Engineering with DEVS Unified Process,
Saurabh Mittal and José L. Risco Martín

Netcentric System of Systems Engineering with DEVS Unified Process

Saurabh Mittal
José Luis Risco Martín

With contributions by:
Deniz Cetinkaya • Alexander Verbraeck • Mamadou D. Seck

CRC Press is an imprint of the
Taylor & Francis Group, an **informa** business

CRC Press
Taylor & Francis Group
6000 Broken Sound Parkway NW, Suite 300
Boca Raton, FL 33487-2742

First issued in paperback 2017

Version Date: 20121214

ISBN 13: 978-1-138-07659-4 (pbk)
ISBN 13: 978-1-4398-2706-2 (hbk)

Visit the Taylor & Francis Web site at
http://www.taylorandfrancis.com

and the CRC Press Web site at
http://www.crcpress.com

Dedication

This book is dedicated to our families, who have never doubted our dreams, no matter how crazy they might be. Specially dedicated to our kids, who are at the age of asking the question "Why?" We love our kids as they are, and we love that question as scientists. The answer is usually reached in simulation,

... in our minds and ... beyond.

Contents

SECTION I The Basics

SECTION III Netcentric System of Systems

SECTION IV Case Studies

SECTION V Next Steps

Preface

The foundation for this book was laid during Dr. Saurabh Mittal's doctoral study (2005 through 2007) at the Arizona Center for Integrative Modeling and Simulation (ACIMS), earlier known as the Artificial Intelligence Simulation Laboratory, in the Department of Electrical and Computer Engineering of the University of Arizona, Tucson, under the advisement of Professor Bernard P. Zeigler, the pioneer of Discrete Event Systems Specification (DEVS) formalism. Dr. Mittal was involved in multiple projects such as the modernization of the High Frequency Global Communications System (HFGCS) simulation model for the U.S. Air Force, the Automated Test-Case Generator (ATC-Gen) at the Joint Interoperability Test Command (JITC), Ft. Huachuca, and various other efforts. These projects were accomplished at ACIMS, which acted as a subcontractor to Northrop Grumman Information Technology (NGIT), JITC, Ft. Huachuca. The methodology adopted for all the projects had recurring themes of model specification, transformation, coding, testing, and execution. Most of the time, specifications directly led to executable code in DEVSJAVA (©). Sometime in 2006, Dr. Mittal coauthored an eXtensible Markup Language (XML) representation based on Finite Deterministic DEVS (XFD-DEVS) that allowed automated transformation of models into XML and executable code. Integration of XML with DEVS was a synergistic development as the ATC-Gen project also pushed the XML-based System Entity Structure (SES) ontology development. The research on platform-independent DEVS modeling was just beginning.

During the summer of 2006, Dr. José Luis Risco Martín arrived at the ACIMS Laboratory as a visiting research scholar. A strong collaboration was forged during his two-month stay in Tucson. Prior to coming at ACIMS lab, Dr. Martin had already developed the XML version of DEVS coupled models that acted as the common ground for the ensuing collaboration. Drs. Mittal and Risco Martín exchanged a lot of ideas, and the first version of the DEVS Modeling Language (DEVSML) stack was conceived at that point. They developed the execution of DEVS using Web Services technology and also developed the DEVS/Service Oriented Architecture (SOA) framework. They now had an architecture that takes in platform-independent XML-based DEVS model and executes them on a netcentric platform. The seeds of a platform-neutral DEVS Virtual Machine (DEVSVM) were laid.

During his doctoral study, Dr. Mittal developed many transformations for sources like XML-based state machines, structured natural language message-based systems, the Department of Defense Architecture Framework (DoDAF) (Version 1.0), Business Process Modeling Notation (BPMN), Unified Modeling Language (UML), XFD-DEVS, and so on, that all led to executable DEVS. Based on the bifurcated model-continuity-based life cycle methodology developed during the ATC-Gen project and various efforts in developing the DEVS back-end, the DEVS Unified Process (DUNIP) was conceptualized in 2007.

As a research scientist at L-3 Communications and a contractor to the U.S. Air Force Research Laboratory (AFRL), at Wright-Patterson Air Force Base, Ohio,

Dr. Mittal is now implementing the DEVSML stack at various AFRL initiatives. A new version, DEVSML 2.0, was proposed as a part of the Large-scale Cognitive Modeling (LSCM) initiative that integrates domain-specific languages (DSLs) with the prior version of DEVSML stack. A DEVS DSL called DEVSML, a platform-independent language in the DEVSML 2.0 stack, was also proposed that redefines the way DEVS models are written in a structured ecosystem such as Eclipse.

XML was not the only means to DEVS transformation, and Drs. Mittal and Risco Martín ventured into the domain of model transformations and model-to-DEVS transformation through a platform-independent DEVS DSL. Various tools were explored, most notably the Eclipse Modeling Framework (EMF), Eclipse Graphical Modeling Framework (GMF), Xtext, Xpand, Generic Modeling Environment, Groovy, Scala, Ruby, and so on that allow the development of DSLs. As DEVSML is an integral part of DUNIP, DUNIP was updated to include DSLs as a means to incorporate other domains that want to interface with a netcentric DEVSVM.

While Dr. Mittal was making advances at the modeling level, Dr. Risco Martín was making advances in the DEVS to non-DEVS interoperability using XML message communication and using the DEVS netcentric platform to run large-scale performance studies. Based on the advances and the capabilities of netcentric DEVS, their work was a strong force in the development of a DEVS Modeling and Simulation (M&S) Standard.

In 2011, Drs. Mittal and Risco Martín updated their server-side capabilities to the latest Web Service specification standards such as JAX-WS and developed advanced concepts for session replication and platform scalability. They also developed various domain-authoring tools in Eclipse and NetBeans ecosystems. This book is an outcome of the application of a lot of foundational ideas by the DEVS community and the latest in netcentric DEVS systems engineering by the authors. This book presents the latest on DUNIP and how it makes model-driven engineering (MDE) concepts applicable to the DEVS and at-large communities. DUNIP creates a possibility for wider adoption of DEVS systems engineering methodologies by integrating various DSLs, existing or non-existing, to a formal computational complex dynamical systems framework. This book is largely based on the authors' experience with the DEVS formalism and application of MDE in the context of netcentric system of systems.

HOW TO USE THIS BOOK

The authors expect the reader to be aware of the basic principles of object-oriented programming. The book is divided into five sections with each section covering a particular community of interest. A summary of each chapter is provided below.

SECTION I: BASICS

Section I is designed for the undergraduate students and novices who want to venture into the DEVS world. The students can refer to some of the case studies in Section IV.

Chapter 1: Introduction to Systems Modeling and Simulation

Chapter 1 provides the basics of computational M&S. This chapter also provides an overview of UML and how various concepts of Object-oriented modeling are applied to M&S.

Chapter 2: System of Systems Modeling and Simulation with DEVS

Chapter 2 provides definitions of system of systems (SoS), complex systems, complex adaptive systems (CAS), and netcentric SoS. This chapter also provides the DEVS framework for M&S.

Chapter 3: DEVS Formalism and Variants

Chapter 3 describes DEVS formalism and its computational representation in Java. Various examples are provided that aid understanding of DEVS systems specification concepts.

Chapter 4: DEVS Software: Model and Simulator

Chapter 4 provides more examples and the computational representation of DEVS simulators. This chapter gives in-depth descriptions on how to write your own DEVS simulators. While this chapter and Chapter 3 give computational representations, Chapter 5 provides details on DEVS modeling without any programming language constructs.

Chapter 5: DEVS Modeling Language

Chapter 5 describes a DEVS DSL. This chapter is designed for early DEVS adopters who want to delve into the DEVS world at a much coarser level. This chapter provides an Eclipse editor to allow creation of platform-independent DEVS models. Various design issues of DEVS grammar are explained and its XML representation is also presented.

Chapter 6: DEVS Unified Process

Chapter 6 provides an overview of DUNIP and how it can be realized in a systems M&S effort. This chapter also describes the DEVSML 2.0 stack, which is the linchpin of netcentric DEVS systems engineering. This chapter binds the foundational concepts and presents a vision of an integrated spiral methodology using DUNIP.

Section II: Modeling and Simulation-Based Systems Engineering

Section II is for graduate students and advanced practitioners of DEVS. It is also for industry professionals who would like to learn the advanced capabilities of DEVS-based systems engineering methodology and to use MDE in their efforts. This section provides methodologies to apply M&S principles to SoS design and also provides an overview on the development of executable architectures based on architecture framework like DoDAF.

Chapter 7: Reconfigurable DEVS

Chapter 7 describes the variable structure capability in the DEVS component-based M&S framework. This chapter also presents the dynamic structure capability through the use of experimental frames and enhanced Model View Simulator Controller (MSVC) paradigm.

Chapter 8: Real-Time and Virtual-Time DEVS

Chapter 8 describes real-time and virtual-time DEVS and provides more details on the development of real-time systems.

Chapter 9: Model-Driven Engineering and Its Application to Modeling and Simulation

Chapter 9 is a collaborative effort with Deniz Cetinkaya, Dr. Alexander Verbraeck, and Dr. Mamadou D. Seck, all at the Delft University of Technology, The Netherlands. Deniz is a PhD student, Dr. Seck is an assistant professor, and Dr. Verbraek is a professor, and they have all contributed their expertise in sharing the authors' vision. This chapter formally presents a framework for model-driven development (MDD) as applicable to M&S domain.

Chapter 10: System Entity Structures and Contingency-Based Systems

Chapter 10 presents SES formalism, its computational representation, a constraint-based language, and a larger Knowledge-Based Contingency Driven Generative Systems (KCGS) framework for designing contingency-based artificial systems with DUNIP. The foundation of KCGS was laid at Cognitive Models and Agents Branch, AFRL, where a Cognitive Systems Specification Framework (CS2F) was developed, a realization of KCGS.

Chapter 11: Department of Defense Architecture Framework (Version 1.0)

Chapter 11 dives into the area of architecture frameworks and describes the U.S. DoDAF Version 1.0. This chapter analyzes the DoDAF specification document, finds gaps in the specification, and suggests augmenting the DoDAF 1.0 specification with two new documents that allow the M&S community to build executable architectures.

Chapter 12: Modeling and Simulation-Based Testing and DoDAF Compliance

Chapter 12 gives an overview on various model-based testing methodologies at various levels of system specifications and on how architecture frameworks such as DoDAF can be subjected to test and evaluation studies.

Section III: Netcentric System of Systems

Section III is for graduate students, advanced users of DEVS, and industry professionals who are interested in building DEVS virtual machines and netcentric SoS. This section can be read directly after Section I.

Chapter 13: DEVS Standard

Chapter 13 gives an overview of the complexities in standardization efforts at the modeling level as well as at the simulation level. This chapter also describes the integration of DEVS and non-DEVS (MATLAB®) systems. Examples are presented in sufficient detail.

Chapter 14: Architecture for DEVS/SOA

Chapter 14 describes architecture for DEVS/SOA and the DEVS Virtual Machine. This chapter provides details about the server and client designs and how the DEVSML 2.0 stack is implemented in a netcentric infrastructure.

Chapter 15: Model and Simulator Deployment in a Netcentric Environment

After an understanding of the netcentric DEVS is given in Chapter 14, Chapter 15 provides practical examples on how to deploy the models at the client side and the simulators at the server side. Various netcentric architectural constructs are described in sufficient detail.

Chapter 16: Netcentric System of Systems with DEVS-Based Event-Driven Architectures

Chapter 16 integrates the emerging paradigm of Event Driven Architecture (EDA) with the netcentric DEVSML 2.0, the DEVSVM, and MDE toward the development of agile netcentric DEVS-based EDAs. This chapter also provides details about integrating the Web Service Description Language (WSDL), a DSL, into the DEVSML 2.0 stack.

Chapter 17: Metamodeling in Department of Defense Architecture Framework (Version 2.0) Metamodel

Chapter 17 revisits the DoDAF with its latest version, which is oriented toward data engineering, sharing, and interoperability. This chapter represents DoDAF 2.0 viewpoints with the SES ontology and presents an argument that DoDAF 2.0 is now in a state where it can effectively act as a DSL that can be anchored to a formal M&S framework. This chapter addresses issues of interoperability and integration of DoDAF DSL based on SES ontology to the netcentric DEVSVM.

Section IV: Case Studies

Section IV presents the case studies that realize many of the concepts defined in Sections I–III.

Chapter 18: Joint Close Air Support: Building from Informal Scenarios

Chapter 18 provides a complete example using DEVSML. Joint Close Air Support (JCAS) is a U.S. Joint Measure Thread (JMT) that describes the JCAS scenario. The user is led from informal natural language specification through a requirements refinement process into a DEVSML specification that is executable. This chapter can be read in conjunction with Chapter 5.

Chapter 19: DEVS Simulation Framework for Multiple Unmanned Aerial Vehicles in Realistic Scenarios

Chapter 19 brings a realistic context in the specification of scenarios. This chapter presents a simulation framework executable on DEVS/SOA in the context of the modeling of multiple unmanned aerial vehicles. Instead of DEVSML, the model is directly built into the implementation programming language. Various performance studies are conducted in a netcentric DEVS/SOA infrastructure. This chapter can be read in conjunction with Chapters 4 and 15.

Chapter 20: Generic Network Systems Capable of Planned Expansion: From Monolithic to Netcentric Systems

Chapter 20 presents an overview of the state-of-the-art HFGCS simulation model, also known as the Generic Network System Capable of Planned Expansion (GENETSCOPE). A methodology is presented that made a monolithic system, a modular system that is executable in a DEVS component-based framework. Further ideas are presented to make it netcentric. This chapter can be read in conjunction with Chapters 7 and 16.

Chapter 21: Executable UML

Chapter 21 takes a hard look at the UML framework and develops various transformations that take UML elements to the DEVS world and vice versa. This chapter can be read in conjunction with Chapters 5, 6, and 10.

Chapter 22: BPMN to DEVS: Application of MDD4MS Framework in Discrete Event Simulation

Chapter 22 takes BPMN to the DEVS executable using the MDD4MS framework developed in collaboration with Centikaya, Dr. Verbraeck, and Dr. Seck. This chapter describes a BPMN metamodel and the transformations to a DEVS metamodel. The metamodel is also integrated with the DEVSML metamodel. This chapter can be read in conjunction with Chapters 5 and 9.

Section V: Next Steps

Section V presents the application of DEVS extensions in a netcentric environment to the discipline of CAS.

Chapter 23: Netcentric Complex Adaptive Systems

Chapter 23 provides an overview of CAS and how modeling of CAS using DEVS concepts can be attempted. This chapter also conceptualizes a netcentric CAS and argues that a netcentric SoS becomes a CAS. Finally, this chapter provides various features and requirements to achieve integration and interoperability in a netcentric CAS.

Online

Various other supporting materials are available online at:
http://www.duniptechnologies.com/book/sos.

Exercises

Some of the chapters come with exercises that advance the idea and concept described earlier. A more serious student is encouraged to attempt these. Partial solutions to the exercises are available online at:
http://www.duniptechnologies.com/book/sos/exercises.

Software

Instructions to use and install the software used in this book are available online at:
http://www.duniptechnologies.com/book/sos/software.

WHAT THIS BOOK IS NOT

This book is not about learning a programming language or a specific tool. It is about the application of MDE concepts and how to build models of complex dynamical that are executable on a DEVS platform. Specifically, this book will not help if you are looking to learn the following:

- Advanced programming skills in Java or associated implementation languages
- DSL development tools like Xtext, Xpand, EMF, Eclipse GMF, or eCore
- Server administration (Glassfish 3.x or Tomcat 6.x)
- Graphical user interface development

QUESTIONS AND CONTACT INFORMATION

For any questions, concerns, or errors in this book, please send an e-mail to:
sosbook@duniptechnologies.com.

Acknowledgments

In this arduous journey of two years, we have to express many thanks to our families and our young kids. Both of us became parents during the manuscript preparation phase. We especially thank our wives for those moments in which they took care of the home and especially the kids, who are now toddlers, while we stayed away working on the book. We wholeheartedly thank our families for their unconditional love, patience, kindness, understanding, and encouragement.

We are grateful toward our series editor, Professor Mo Jamshidi, and the publisher, Taylor & Francis Group/CRC Press, for finding merit in our work and giving us an opportunity to develop it further.

We sincerely thank Professor Bernard Zeigler for his support in conceptualizing the ideas in the early stages of our work at the ACIMS Laboratory. We came to know many people at NGIT who helped create the foundation of DUNIP. We thank all our friends and colleagues at NGIT for their support.

We would like to thank Deniz Cetinkaya, Alexander Verbraeck, and Mamadou Seck at Delft University of Technology for collaborating on two chapters for this book. Their important contributions are very much appreciated.

We especially thank our friends and colleagues at L-3 Communications, AFRL, and the Department of Computer Architecture and Automation (Universidad Complutense de Madrid [UCM], Spain).

Finally, we thank our friends who collaborated on the development of certain ideas and experiments given in this book. We thank Jesús Manuel de la Cruz and Eva Besada (UCM), and Joaquín Aranda and Alejandro Moreno (Universidad Nacional de Educación a Distancia, Spain).

Saurabh Mittal
José Luis Risco Martín

Authors

Saurabh Mittal is the founder and principal scientist at Dunip Technologies, which he manages in his free time. He is currently a full-time research scientist at L-3 Communications and is a contractor to the U.S. Air Force Research Laboratory, Wright-Patterson Air Force Base, Ohio. Here in this capacity, he is working on large-scale cognitive M&S using DEVS systems engineering, cognitive domain ontologies extending SES theory and various other cross-directorate M&S integration and interoperability efforts using architecture frameworks such as Department of Defense Architecture Framework. He is a recipient of the highest civilian contractor recognition "Golden Eagle" award by Joint Interoperability Test Command, Defense Information Systems Agency, U.S. Department of Defense. He serves on various conference program committees and is a reviewer for many prestigious international journals. His areas of interest include web-based M&S using SOA, executable architectures, netcentricity, DUNIP, domain ontologies, platform independent modeling, domain specific languages, SES theory and applications, parallel-distributed simulation, complex systems, complex adaptive systems, and system of systems engineering using DoDAF.

José L. Risco Martín is an associate professor at the Computer Architecture and Automation Department of UCM, Spain. His research interests focus on the design methodologies for integrated systems and high-performance embedded systems, including new modeling frameworks to explore thermal management techniques for multiprocessor system-on-chip, novel architectures for logic and memories in forthcoming nanoscale electronics, dynamic memory management and memory hierarchy optimizations for embedded systems, networks-on-chip interconnection design, and low-power design of embedded systems. He is also interested in theory of M&S, with an emphasis on DEVS, and the application of bioinspired optimization techniques in computer-aided design problems.

Section I

The Basics

1 Introduction to Systems Modeling and Simulation

1.1 THE NATURE OF SIMULATION

A simulation is an imitation of some real thing, state of affairs, or process in action. The act of simulating something generally entails representing certain key characteristics or dynamic behaviors of a selected physical or abstract system. Simulation is used in many contexts, including the modeling of natural systems or human systems to gain insight into their functioning. Other contexts include simulation of technology for performance optimization, safety engineering, testing, training, and education. Simulation can also be used as a prediction tool to show the eventual real effects of alternative conditions and courses of action.

Key issues in developing a simulation technology include the acquisition of valid source information about the referent, selection of key characteristics and behaviors, use of simplifying approximations and assumptions within the simulation model, and fidelity and validity of the simulation outcomes.

The definitions taken from WordNet (2012) are as follows:

1. *Simulation*: The act of imitating the behavior of some situation or some process by means of something suitably analogous (especially for the purpose of study or personnel training).
2. ***Simulation, computer simulation****(computer science)*: The technique of representing the real world by a computer program. "A simulation should imitate the internal processes and not merely the results of the thing being simulated."
3. *Model, simulation*: Representation of something (sometimes on a smaller scale).
4. *Pretense, pretence, pretending, simulation, feigning*: The act of giving a false appearance. "His conformity was only pretending."

This book is focused on the computer simulation concept, as per the highlighted definition above. A computer simulation attempts to simulate an abstract model (that is computationally represented) of a particular system. Computer simulations have become useful parts of mathematical modeling of many natural systems in physics, chemistry, biology, human systems, economics, psychology, and social science, and in the process of engineering new technology, to gain insights into the operation of those systems.

Computer simulations vary from computer programs that run a few minutes, to network-based groups of computers running for hours, to ongoing simulations that run for days. The scale of events being simulated by computer simulations has far exceeded anything possible (or perhaps even imaginable) using the traditional paper-and-pencil mathematical modeling over 30 years ago. For example, a desert-battle simulation of one force invading another involved the modeling of 66,239 tanks, trucks, and other vehicles on simulated terrain around Kuwait using multiple supercomputers in the U.S. Department of Defense (DoD High Performance Computing Modernization Program, 1992). Another simulation ran a 1-billion-atom model, where previously a 2.64-million-atom model of a ribosome, in 2005, had been considered a massive computer simulation (Schwede & Peitsch, 2008), and the Blue Brain project at EPFL (Switzerland) began in May 2005 to create the first computer simulation of the entire human brain, right down to the molecular level (Blue Brain Project, 2005).

1.2 SYSTEMS, MODELS, AND MODELING

1.2.1 Systems

A system is a part of the real world under study that can be identified from the rest of its environment for a specific purpose. Such a system is called a *real* system because it is physically part of the real world. The state of a system is defined as the collection of variables necessary to describe a system at a particular time, relative to the objectives of a study.

1.2.1.1 Components of a System

A system is composed of a set of entities (or components) that interact among themselves and with the environment to accomplish the system's goal. This interaction influences the behavior of the system. An entity has a structure and a behavior. The behavior is the definition of states and state transitions generated by actions the system takes during the time interval in which it is studied. We may also define the behavior as a set of actions or events that change the system's state. The structure, to a large part, captures the aspects of the system that stay invariant during the time interval in which it is studied, if the system does not modify its structure internally, as in many organic and natural systems.

1.2.1.2 Discrete and Continuous Systems

Systems may be categorized into two types, discrete and continuous. "Few systems in practice are wholly discrete or wholly continuous, but since one type of change predominates for most systems, it will be usually possible to classify a system as being either discrete or continuous" (Law & Kelton, 2000). A discrete system is one for which the state variables may change only at discrete values of time. It is also known as a discrete-time system.

A bank is an example of a *discrete* system because the state variable, the number of customers in the bank, changes only when a customer arrives or when service

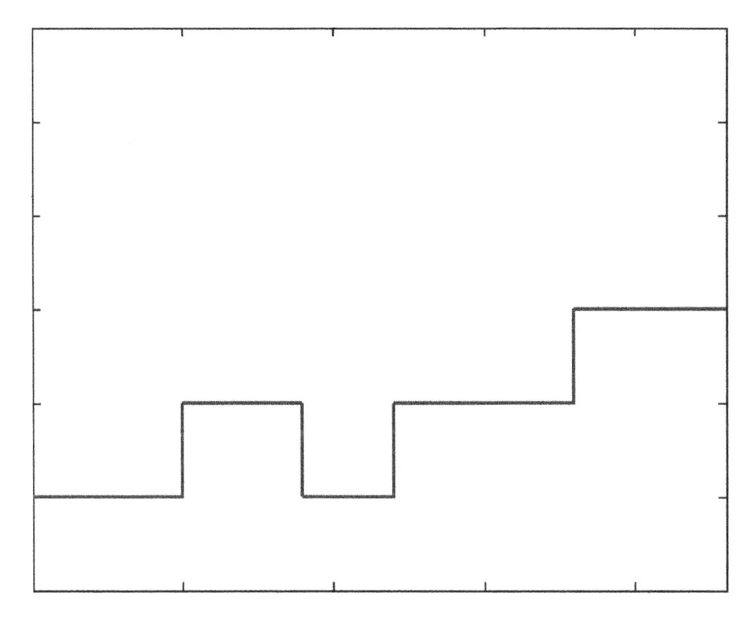

FIGURE 1.1 Discrete system state variable.

provided for a customer is completed. Figure 1.1 shows how the number of customers changes only at discrete points of time.

A *continuous* system is one whose state is capable of changing at any instant of time. It is also known as continuous-time signal system. An example is the head of water behind a dam. During a rain storm and after the rain storm has passed, water continues to flow into the lake behind the dam. Water is drawn from the dam for flood control and to make electricity. Evaporation also decreases the water level. Figure 1.2 shows how the sate variable, head of water behind the dam, changes for this continuous system.

1.2.2 Model of a System

A *model* is an abstract representation of a real system. The model is simpler than the real system, but it should be equivalent to the real system in all relevant aspects. The act of developing a model of the system under study is called modeling. Abstraction is very useful in modeling large and complex systems, such as computer operating systems and real-life natural systems.

Every model has a specific purpose and goal. A model only includes those aspects of the real system that were decided as being important, according to the initial requirements of the model. This implies that the limitations of the model have to be clearly understood and documented. Just as systems are represented by entities, states, structures, and behavior, models are represented in the same manner as well. However, the model contains only those elements that are relevant to be studied.

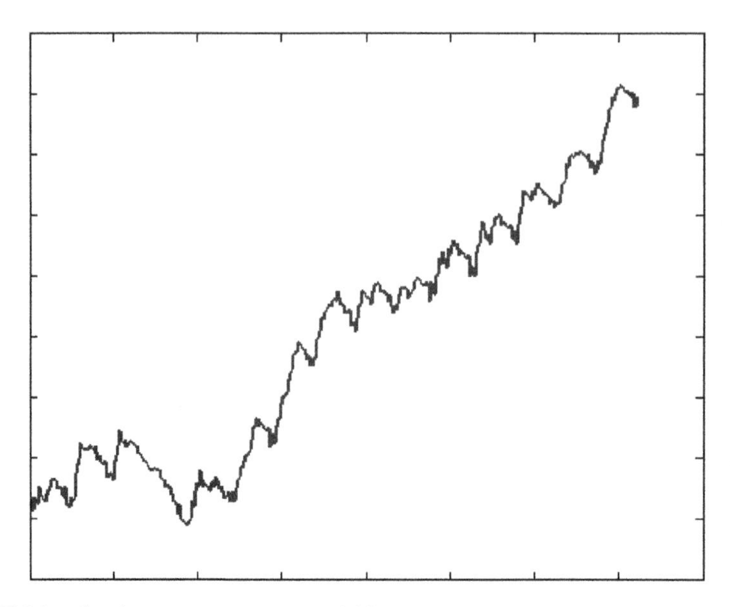

FIGURE 1.2 Continuous system state variable.

Models can be categorized as being mathematical or physical. A physical model is used in various contexts to imply a physical representation of something. That thing may be a single item or object (e.g., a bolt) or a large system (e.g., a scale model of the Solar System), and are not typical of the kinds of models that are usually of interest in operations research and system analysis. A mathematical model represents a system in terms of logical and quantitative relationships that are manipulated and changed to see how the model reacts to changing environment conditions. A simulation model is a particular type of mathematical model of a system.

1.2.3 Modeling

Modeling, especially scientific modeling, refers to the process of developing a model as a conceptual representation of some phenomenon as discussed above. Typically, a model will refer only to some aspects of the phenomenon in question, and two models of the same phenomenon may be essentially different, that is, in which the difference is more than just a simple renaming. This may be due to differing requirements of the model's end users or conceptual or aesthetic differences by the modelers and decisions made during the modeling process. Aesthetic considerations that may influence the structure of a model might be the modeler's preference for a reduced ontology, preferences regarding probabilistic models vis-à-vis deterministic ones, discrete versus continuous time, and so on. For this reason, model developers need to understand the model's original purpose and the assumptions of its validity. Having found a model for some desired aspect of reality, it can serve as the basis for simulation, the only way for non-invasive examination of physical reality besides real-world experiments.

1.3 PHASES IN MODEL DEVELOPMENT

The development process of a simulation model is a sequence of phases that starts with the specification of modeling goals. The sequence of phases is possibly carried out in an iterative manner. The development process for a simulation model is defined by the following general steps:

1. *Problem formulation and setting of objectives*: This statement must provide the description of the purpose for building the model, the questions it must help to answer, and the performance measures relevant to these questions. This is a description of what is to be accomplished with the simulation model to be constructed.
2. *Model conceptualization*: It is best to start with a simple model and build toward greater complexity. This description includes the list of relevant components, the interaction and the relationships among them, and the dynamic behavior of the model. It is not necessary to have one-to-one mapping between the model and the real system, to begin with. Abstraction plays a crucial role here.
3. *Model design*: This stage describes the details of the computational data structures necessary to implement the components and the details of the algorithms for the relationships and the dynamic behavior of the model.
4. *Model translation or model implementation*: Most of complex models require a great deal of information storage and computation. The model must be entered into a computer-recognizable format. The implementation of the model can be in a simulation language or in general-purpose high-level programming languages such as C#, Java, C++, and so on. The simulation package (or tool) to use is also an important practical decision. The main tasks in this phase are coding, debugging, and testing the simulation model.
5. *Verification of the model*: From different runs of the implementation of the model program, this stage compares the output results with those that would have been produced by a correct implementation of the conceptual model. Verification refers to the correctness of the algorithm or the simulation package in running the model.
6. *Data collection*: The objective of the model development process defines indirectly the kind of data to be collected. Some of this data will be used to validate the simulation model.
7. *Validation of the model*: This stage compares the outputs of the verified model with the output of the real system. This stage also compares the model with reality.

1.4 TYPES OF SIMULATION MODELS

It is useful to categorize simulation models among four different dimensions.

1.4.1 Stochastic or Deterministic

Simulation models that contain no random variables are classified as deterministic. Deterministic models have a known set of inputs, which will result in a unique set of outputs. A stochastic simulation model has one or more random variables as inputs. Random inputs lead to random outputs. Since the outputs are random, they can be considered only as estimates of the true characteristics of a model.

1.4.2 Steady State or Dynamic

Steady-state models use equations that define the relationships between elements of the modeled system and attempt to find the system's equilibrium state. Such models are often used in simulating physical systems as a simpler modeling case before dynamic simulation is attempted. A dynamic simulation model studies the changes in a system in response to (usually changing) input signals.

1.4.3 Continuous or Discrete

A continuous dynamic simulation performs numerical solution of differential-algebraic equations or differential equations (either partial or ordinary). Periodically, the simulation program solves all the equations, and uses the numbers to change the state and output of the simulation. Applications include flight simulators, simulation games, chemical process modeling, and simulations of electrical circuits. A discrete event simulation (DES) manages events in time. Most computer, logic-test, and fault-tree simulations are of this type. In this type of simulation, the simulator maintains a queue of time-ordered events and processes the imminent event. The simulator reads the queue and triggers new events as each event is processed. New events can be placed at the head of the queue.

1.4.4 Local or Distributed

Local simulation models run on a single computer or processor. Distributed models run on a network of interconnected computers, possibly through the Internet. Simulations dispersed across multiple host computers are often referred to as distributed simulations. There are several standards for distributed simulation, including the aggregate level simulation protocol (ALSP) (Seidel, 1993), distributed interactive simulation (DIS),* high level architecture (HLA) (U.S. Department of Defense, 2001), and test and training enabling architecture (TENA) (TENA, 2012).

1.5 EXAMPLES OF MODELS

1.5.1 Population Growth

An approximate model of population growth is the Malthusian growth model. The Malthusian growth model, sometimes called the simple exponential growth model,

* The standard was developed over a series of "DIS Workshops" at the Interactive Networked Simulation for Training symposium, held by the University of Central Florida's Institute for Simulation and Training.

is essentially exponential growth based on a constant rate of compound interest. The model is named after the Reverend Thomas Malthus, who authored *An Essay on the Principle of Population* (1798), one of the earliest and most influential books on population.

$$P(t) = P_0 \cdot e^{r \cdot t}$$

where P_0 is the initial population, r is the growth rate, and t is the time.

1.5.2 Particle in a Potential Field

We consider a particle as being a point of mass m that describes a trajectory, which is modeled by a function $x : \mathbb{R} \to \mathbb{R}^3$ given its coordinates in space as a function of time. The potential field is given by a function $V : \mathbb{R} \to \mathbb{R}^3$, and the trajectory is a solution of the differential equation:

$$m \frac{d^2}{dt^2} x(t) = -\nabla\left(V\left(x(t)\right)\right)$$

Note that this model assumes the particle is a point mass, which is certainly known to be false in many cases we use this model; for example, as a model of planetary motion.

1.5.3 Amortization Process Model

If we purchase a house and take a loan of d dollars with a fixed interest rate of R percent per year ($r = R/12$ per month), then the loan is paid back through the process known in economics as amortization. Using simple logic, it is not hard to conclude that the outstanding principal, $y[k]$, at $k + 1$ discrete-time instant (month) is given by the following recursive formula:

$$y[k+1] = y[k] + ry[k] - f[k+1] = (1+r)\, y[k] - f[k+1]$$

where $f[k + 1]$ is the payment made in $(k+1)$ at discrete-time instant (month).

1.6 SOFTWARE SYSTEMS ENGINEERING WITH OBJECTS, CLASSES, AND UML

Object-oriented modeling is a general modeling approach for dealing with large and complex systems. In this paradigm, everything is abstracted to an "object" that has some properties and displays some behavior. We present an overview of object-oriented modeling in this section as it is a practical and widely accepted paradigm for modeling. We provide an overview on Unified Modeling Language

(UML), which is used in several case study descriptions throughout the book. UML has been accepted as the standard notation for object-oriented modeling.

As stated earlier, the modeling process starts with the construction of a conceptual model for a real-world system under study. Next, the model is implemented as a target operational system that consists of software and hardware subsystems. The conceptual model is simpler, easier to understand, and easier to manipulate than the real system. Conceptual models can be described using UML.

In object-oriented modeling, objects are the main components of a system. An object represents a real-world entity and is the first step of developing abstractions. Every object has data (a set of attributes that constitute the state of the object) and behavior (a set of methods or member functions).

When modeling with UML, an object is conceptualized as a class and is graphically represented as a rectangle divided in three parts. The first one contains the name of the object. In the second part, the set of attributes is included. Finally, the third part includes the set of methods.

A class defines the structure and the behavior for all possible attributes of that class. The class also defines a set of objects that share common characteristics. An object is a particular instance of a class. Then, in an object-oriented paradigm, a class represents a group of objects with common characteristics and behavior. *Classification* is the task of identifying real objects, and then the common characteristics and behavior of those objects, grouping them into classes. In a simulation, these instance objects send messages to other instance objects, changing their state and attribute values.

A class is a *static* notion, whereas an instance object is a *dynamic* notion. Thus, an object-oriented programming has two views: (1) the static view as a group of classes in the main decomposition unit in an object-oriented program and (2) a dynamic view, where the program is executed and a set of objects with individual behavior interacts among them.

In the following, we describe how we can define or conceptualize systems using UML. Next, we introduce some features of Java as it is the main programming language used in this book.

1.6.1 UML for Object-Oriented Modeling

In the field of software engineering, UML is a standardized visual specification language for object modeling (The Object Management Group [OMG], 2012). UML is a general-purpose modeling language that includes a graphical notation used to create an abstract model of a system, referred to as a UML model.

UML is officially defined at the OMG (2012) by the UML metamodel. A metamodel is a model that describes the model. The UML metamodel and UML models may be serialized in eXtensible Markup Language (XML). UML was designed to specify, visualize, construct, and document software-intensive systems. However, UML is not restricted to modeling software. UML is also used for business process modeling, systems engineering modeling and representing organizational structures. UML has been a catalyst for the evolution of model-driven technologies. By establishing an industry consensus on a graphic notation to represent common concepts like classes, components, generalization, aggregation, and behaviors, UML has allowed software developers to concentrate more on design and architecture.

UML models may be automatically transformed to other representations (e.g., Java) by means of XSLT or QVT-like transformation languages, supported by the OMG. In addition, UML is extensible, offering the following mechanisms for customization: profiles and stereotype. The semantics of extension by profiles have been improved since the UML 2.0 major revision.

UML 2.0 has 13 types of diagrams, which can be categorized hierarchically as shown in Figure 1.3. Structure diagrams emphasize what constitutes the system structurally and include a class diagram, component diagram, composite structure diagram, deployment diagram, object diagram, and package diagram. Behavior diagrams emphasize what must happen in the system being modeled. They include an activity diagram, state machine diagram, and use case diagram. UML also includes interaction diagrams, a subset of behavior diagrams, used to define the flow of control and data among the entities in the system being modeled. Interaction diagrams are communication diagrams, interaction overview diagrams, sequence diagrams, and timing diagrams.

Not all diagrams are necessary to completely model an application; these different types of modeling diagrams allow modelers to select the best diagram to describe the modeled system. The most relevant UML diagrams are described in this chapter.

1.6.1.1 Use Case Diagrams

Use case diagrams describe behavior in terms of the high-level functionality and usage of a system by the stakeholders and other members such as developers who build the system. The use case diagram describes the usage of a system (subject) by its actors (environment) to achieve a goal, which is realized by the subject providing a set of services to the selected actors. The use case can also be viewed as functionality and/or capabilities that are accomplished through the interaction between the subject and its actors. Use case diagrams include the use case and actors and the associated

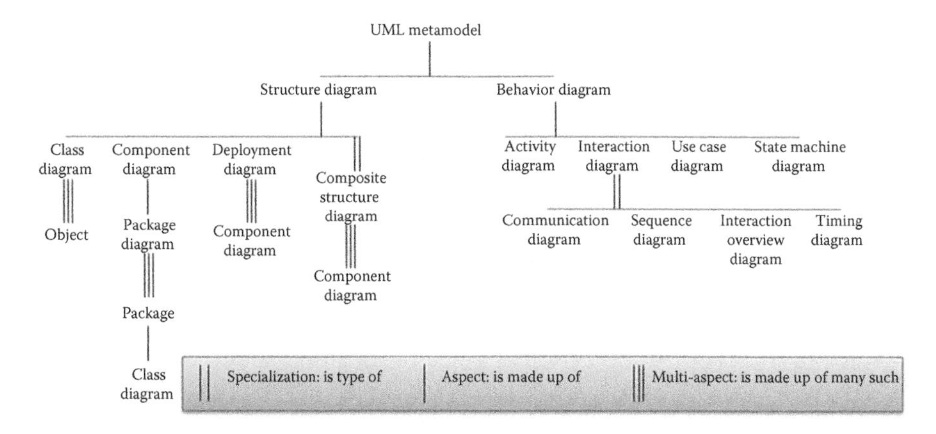

FIGURE 1.3 Unified Modeling Language (UML) metamodel.

communication between them. Actors represent classifier roles that are external to the system that may correspond to users, systems, and/or other environmental entities. They may interact either directly or indirectly with the system.

The use case relationships are *communication*, *include*, *extend*, and *generalization*. Actors are connected to use cases via communication paths, which are represented by an association relationship. The *include* relationship provides a mechanism for factoring out common functionality that is shared among multiple use cases and is always performed as part of the base use case. The *extend* relationship provides optional functionality, which extends the base use case at defined extension points under specified conditions. The *generalization* relationship provides a mechanism to specify variants of the base use case.

The use cases are often organized into packages with the corresponding dependencies between them.

Figure 1.4 depicts how use cases help delineate specific goals associated with driving and parking a vehicle. In Figure 1.4, the *extend* relationship specifies that the behavior of a use case may be extended by the behavior of another (usually supplementary) use case. The "Start the vehicle" use case is modeled as an extension of "Drive the vehicle." This means that there are conditions that may exist that require the execution of an instance of "Start the vehicle" before an instance of "Drive the vehicle" is executed.

The use cases "Accelerate," "Steer," and "Brake" are modeled using the *include* relationship. *Include* is a directed relationship between two use cases, implying that the behavior of the included use case is inserted into the behavior of the *including* use case. The *including* use case may only depend on the result (value) of the *included* use case. This value is obtained as a result of the execution of the *included* use case. This means that "Accelerate," "Steer," and "Brake" are all parts of the normal process of executing an instance of "Drive the car."

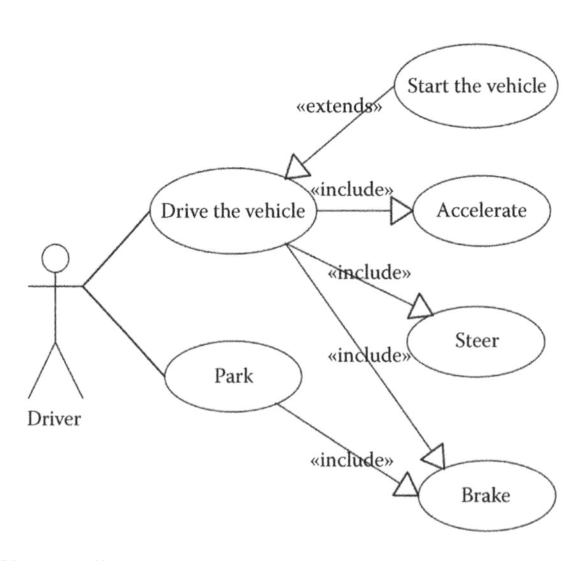

FIGURE 1.4 Use case diagram.

In many situations, the usage of *include* and *extend* relationships is subjective and may be reversed, based on the approach of an individual modeler.

With respect to UML-based modeling and simulation (M&S), use case diagrams are used to establish the system context, defining system boundaries and multilevel resolution capabilities at appropriate hierarchical level. They should be used like a starting point in the model development. However, there are no special rules to design the best use cases. The resolution depends mainly on the model context under development.

1.6.1.2 Class Diagrams

UML 2 class diagrams are the mainstay of object-oriented analysis and design. UML 2 class diagrams show the classes of the system, their interrelationships: inheritance, aggregation, and association, and the operations and attributes of the classes. Class diagrams are used for a wide variety of purposes, including both conceptual/domain modeling and detailed design modeling.

As stated earlier, a class is graphically represented as a rectangle divided in three parts. The first part contains the name of the class. The second part includes the set of attributes. The methods or operations in the class are included in the third part. Figure 1.5 shows the complete graphical representation of a class Car and a corresponding instance.

An *association* describes the relationship between two or more classes. The most basic type of association is the binary association, which is represented by a solid line connecting two classes. Any association may include an association name (above the line), roles and multiplicity (at the end of the line), and a solid small triangle to indicate the direction in which to read the association name.

Figure 1.6 shows a binary relation between the *Car* class and the *Driver* class. The association name is hidden. The roles are person and vehicle. Multiplicity is denoted with a star and a number 1. The star in the side of *Car* denotes that there can be zero or many objects of this class in association with the *Driver* class. There is only one object of the *Driver* class.

An *aggregation* relationship is stronger than the association. It means that a class is contained within another class. The larger class is called the owner, and the smaller class is called the component. In UML, this relation is denoted with a connecting line with a diamond at the owner class side. When the diamond is solid, UML denotes a *composition*, which means that the owner class has exclusive ownership of the contained class. Figure 1.7 shows an owner class in association with three component classes.

Generalization is the association between a general class and a more specialized class or extended class. This association is also called inheritance, being an important relationship between classes. In UML, an arrow points from the derived class to the parent class.

Figure 1.8 illustrates a simple class hierarchy using inheritance. The most general class is *Vehicle*, which is the parent class. The other three classes (*Car*, *Truck*, and *Motorcycle*) inherit the features from the parent class.

1.6.1.3 Sequence Diagrams

The sequence diagram describes the flow of control between actors and systems or between parts of a system. This diagram represents the sending and receiving of messages between the interacting entities called lifelines, where time is represented

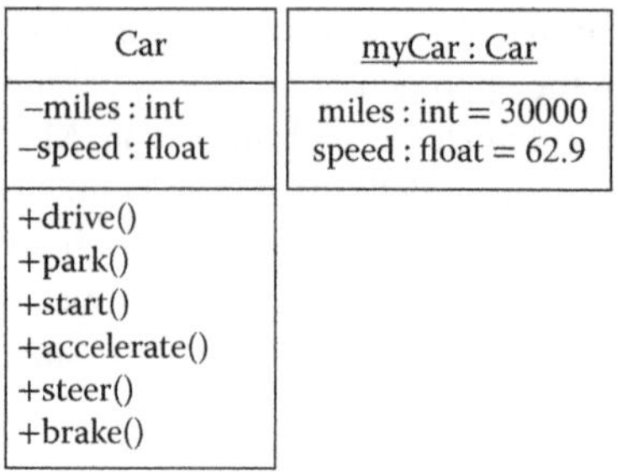

Car
–miles : int
–speed : float
+drive()
+park()
+start()
+accelerate()
+steer()
+brake()
myCar : Car
miles : int = 30000
speed : float = 62.9

FIGURE 1.5 UML class diagram and object diagram.

–Person
1
*
–Vehicle
Driver
Car
–miles : int
–speed : float
+drive()
+park()
+start()
+accelerate()
+steer()
+brake()

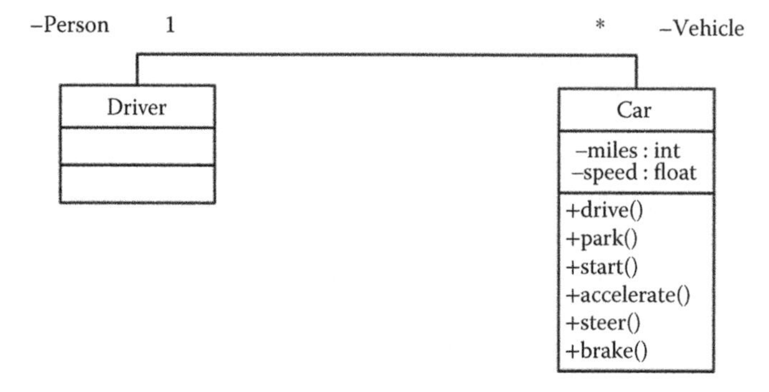

FIGURE 1.6 A binary association between *Driver* and *Car.*

1
OwnerClass
–Rol5
–Rol1
1
1
–Rol3
*
–Rol2
*
–Rol4
*
–Rol6
ComponentClassA
ComponentClassB
ComponentClassC

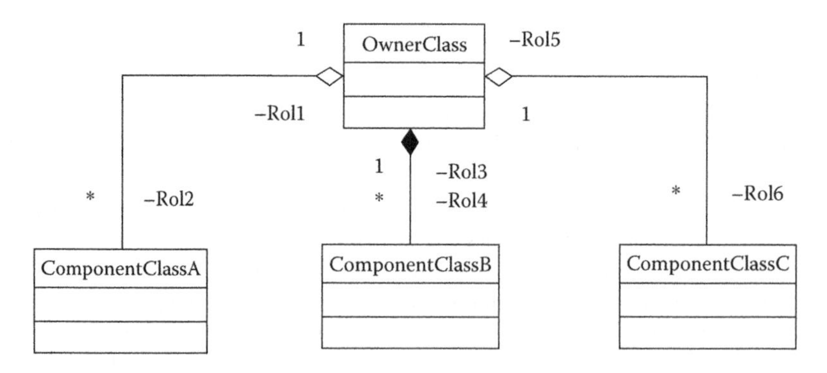

FIGURE 1.7 An aggregation relationship with three component classes.

along the vertical axis. The sequence diagrams can represent highly complex interactions with special constructs to represent various types of control logic, reference interactions on other sequence diagrams, and decomposition of lifelines into their constituent parts.

We make use of state diagrams defined in Choi et al. (2006), which uses three operators (seq, alt, and loop) and adds the parameter "sigma" as information to explicitly specify the timing constraint among components of a system. We define

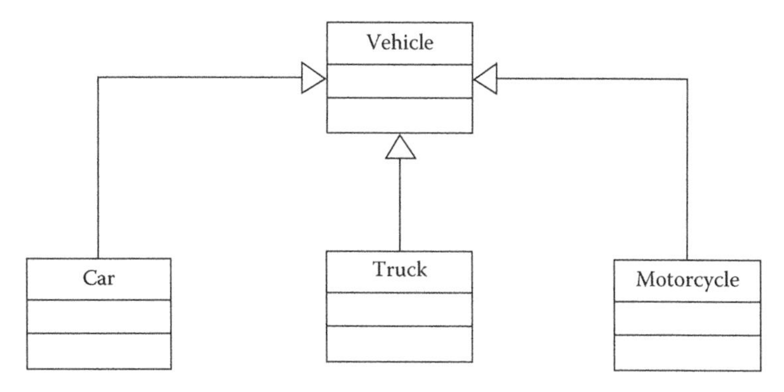

FIGURE 1.8 An inheritance relationship.

both the phase and sigma of the model by means of constraints in the sequence diagram. Sigma specifies a point in time and the event occurs at the specified time, and the phase provides information about the model's global state. For example, the constraint "active, 5s" means that an event occurs after 5 seconds while the global state of the model is "active." All the sending and receiving events should have sigma and phase, except those lifelines that represent models without state. If sigma is infinity, it is denoted as "inf" and implies that the object waits for the incoming event indefinitely until any message arrives. Finally, if sigma is not defined explicitly, it is supposed that sigma is updated using the elapsed time of the coming event; that is

$$\sigma = \sigma - \varepsilon$$

where ε is the elapsed time since the last state transition.

Figure 1.9 depicts an illustrative example. From the model M1 point of view, there is an initialization message, which initializes phase to "active" and sigma to 5 seconds. If no message is received, the next event (message) is sent after 5 seconds followed by a state transition, which sets the phase to "passive" and sigma to infinity. After 9 seconds, M1 receives a message from M2 and a state transition happens. This transition sets phase of M1 to passive and sigma to infinity. Note that in the case of M2, its sigma is updated to 4 seconds after the M1 transition (9 seconds minus elapsed time, 5 seconds) implicitly.

1.6.1.4 Timing Diagrams

Timing diagrams are one of the new artifacts added to UML 2. They are used to explore the behaviors of one or more objects throughout a given period of time. There are two basic flavors of timing diagram: the concise notation and the robust notation.

Figure 1.10 depicts an example of both concise and robust notations of timing diagrams. M1 starts in "active" state for 5 seconds. After that, M1 sends a timeout message called "M1OutputMessage" and makes an internal transition changing its state to "passive." The message changes the state of M2 from "active, 9s" to "active" (sigma = 4s implicitly) through a state transition in M2. Four seconds later, M2 sends a timeout message and makes a state transition, changing its state from "active" to

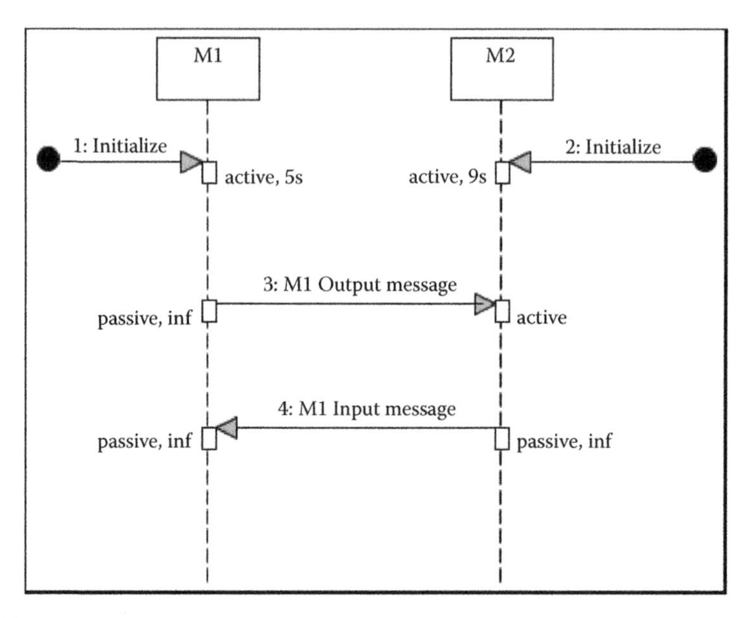

FIGURE 1.9 Example sequence diagram.

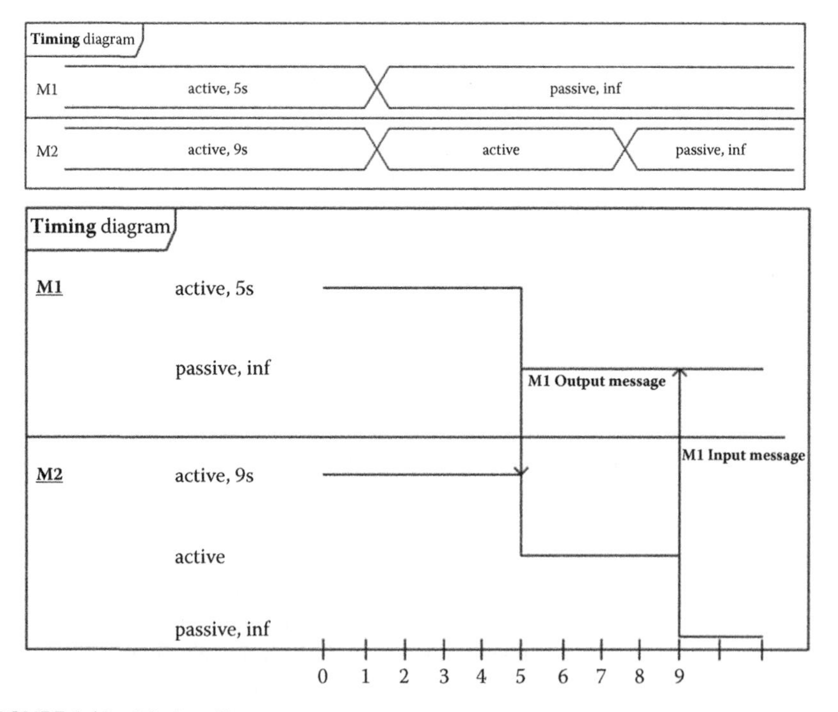

FIGURE 1.10 Timing diagram.

"passive." The message is called "M1InputMessage," which executes a state transition in M1 without effects.

1.6.1.5 State Diagrams

UML state machine diagrams define a set of concepts that can be used for modeling discrete behavior through finite state transition systems. The state machine represents behavior as the state history of an object in terms of its transitions and states. The activities that are invoked during the transition of the states are specified along with the associated event and guard conditions following the format "event[guard]/activity."

Figure 1.11 shows the state machine diagram for M1. M1 starts in state "active" for 5 seconds. After that, the message "M1OutpurMessage" is sent and a state transition happens, changing the state to "(passive, inf)." If M1 receives a message "M1InputMessage," a new state transition is executed, without changes in the state of M1.

1.6.2 Introduction to Java

Java is a general-purpose high-level programming language that is object oriented. The main characteristics of Java are as follows:

- Java is portable. The source code is compiled to an intermediate code called bytecode. This code can be ported from one machine to any other, where a platform-specific interpreter (the Java Virtual Machine or JVM) can interpret and execute the bytecode.
- Java supports multithreading. A program can have multiple execution paths called threads. Simultaneous activities in a real system can be modeled as simultaneous tasks.
- Java includes a large number of standard libraries, which facilitates the use of graphical user interfaces, web applications, data structures, and so on.

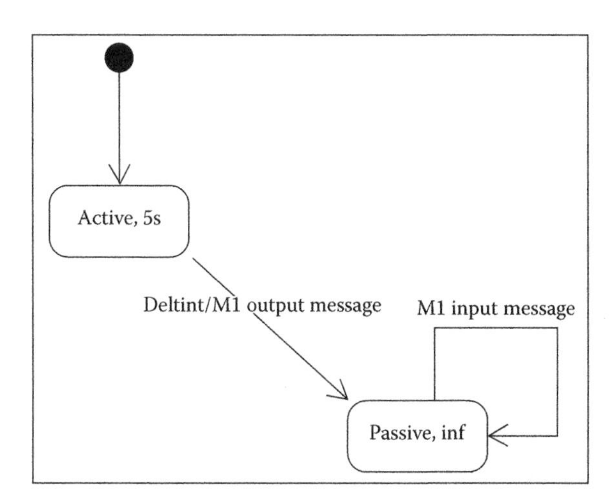

FIGURE 1.11 M1 state machine diagram.

A Java program is a collection of classes. The main method must be included in one of these classes. There are no global variables or objects allowed to the programmer. All the members (functions or attributes) in a class are private, protected, or public. These constraints are called the access modifiers. All members of a base class can be redefined or overridden in a derived class (when using inheritance). All class definitions must include one or more constructors. In Java, all the variables are references (with the exception of primitive types like boolean, char, int, double, etc.) and are automatically destroyed by the garbage collector mechanism.

1.6.2.1 Java Classes

Java classes are organized in packages. A class is a group of attributes or variables and member functions (also called methods) (Listing 1.1). These methods operate with the attributes. The keyword **public** is optional. All the methods and attributes must be defined inside the block {…}.

LISTING 1.1 CLASS DEFINITION

```
[public] class Classname {

//Definition of attributes and methods
…
}
```

An *instance* (or object) is a concrete class sample (Listing 1.2). Classes are like data types, whereas objects are like variables belonging to a particular data type.

LISTING 1.2 CLASS INSTANTATION

```
ClassName anObject = new ClassName();
ClassName anotherObject = new ClassName();
```

Every attribute or method must belong to a class. If a class extends another class, the extended class inherits all the base attributes and methods. A class can only extend one preexisting class. If the base class is not specified, then the *Object* class is extended, which is the base of all the class hierarchy defined in Java.

A class to define a circle with a certain coordinate (x,y) and a radius r can be implemented, as shown in Listing 1.3.

LISTING 1.3 CLASS DEFINING A CIRCLE

```
package main;

//File: Circle.java
public class Circle extends Shape {

  static int numCircles = 0;
  public static final double PI = 3.14159265358979323846;
  public double x, y, r;

  public Circle(double x, double y, double r) {
    this.x = x;
    this.y = y;
    this.r = r;
    numCircles++;
  }

  public Circle(double r) {
    this(0.0, 0.0, r);
  }

  public Circle(Circle c) {
    this(c.x, c.y, c.r);
  }

  public Circle() {
    this(0.0, 0.0, 1.0);
  }

  public double perimeter() {
    return 2.0 * PI * r;
  }

  public double area() {
    return PI * r * r;
  }
  //method designed to compare circles

  public Circle theGreatest(Circle c) {
    if (this.r >= c.r) {
      return this;
    } else {
      return c;
  }
  }
  //class method designed to compare circles
```

```
public static Circle theGreatest(Circle c, Circle d) {
    if (c.r >= d.r) {
        return c;
    } else {
        return d;
    }
}
}
```

In this example, we can see how both attributes and methods are defined in the *Circle* class. Both can be declared for the object or for the class (static). The file name matches the name of the class with a ".java" extension.

1.6.2.2 Attributes

To the contrary of the structured programming, centered on functions, object-oriented programming is centered on data. A class is a set of data and several methods that operate on them.

Every object has its own copy of the attributes. For example, every instance of the *Circle* class has its own coordinates (x, y) and its radius r. A method can be called by writing the object name followed by a dot and the method name. For example, to compute the area of a circle called c1 we write *c1.area()*. An attribute can belong to a primitive type (boolean, int, long, double, ...) or can reference to other classes (called *composition*). All the data must be initialized. Usually, these attributes are initialized in a special method called *constructor*. However, as in the *Circle* example, these attributes can be initialized even before calling the constructor. Attributes and methods have one of the access modifiers: *public*, *private*, and *protected*.

An attribute can belong to a class instead of the object. These types of attributes are called *static* attributes. In this case, there is one copy of a static attribute per class and all the objects share the same static attributes. In the *Circle* class, there are two static variables: the constant PI and a counter *numCircles* to compute the number of circles instantiated. To have access to static attributes, we write the name of the class, followed by a point and the attribute name: *Circle.numCircles*.

There are also *final* attributes in Java, which imply that the value of the attribute cannot be changed at runtime. A final attribute can be considered as a constant. However, declaring a reference to an object as final makes the reference a constant, not the object. Thus, final attributes are usually declared with basic data types.

1.6.2.3 Methods

Methods are functions defined inside a class. They are invoked (avoiding static methods) by writing the object name plus a point and the method name. The object that invokes the function is like an implicit argument.

See the following method (Listing 1.4), taken from the *Circle* class:

LISTING 1.4 A CLASS METHOD

```
public Circle theGreatest(Circle c) {
    if (this.r >= c.r) {
        return this;
    } else {
        return c;
    }
}
```

The declaration starts with the access qualifier (*public* in this case), the data type returned (*Circle* in this case, *void* if the method returns nothing), the method name, and the list of arguments. All the methods have direct access to the attributes of the object that invoke the method. The attributes can be used without any identifier, although the object is referenced by a pointer called **this**, which can be used if, for example, there are local variables or arguments with the same name than an attribute.

Java allows the programmer to overload methods. Overloaded methods are different methods with the same name, but different in the number or arguments type. The *Circle* class, for example, presents two overloaded methods: the *constructor* (four versions) and the method *theGreatest* (two versions).

Methods can be *overridden*. A class can override an inherited method of a superclass. Overriding a method consist of giving a new implementation in the derived class. In this case, the method must exactly have the same arguments in type and number. Methods can also be static. Static methods cannot use **this** because these methods are called through the class, without a calling object. For example, we can use **Math.sin(angle)** to compute the sine of an angle.

The *constructor* class is one of the most important methods in object-oriented programming. A constructor is automatically called every time we create an object. The principal purpose of a constructor is to initialize the attributes in memory. A constructor does not have return data type and its name is the same as the class name. A constructor can call another constructor of the same class using **this**. In this case, **this** must appear in the first line. A subclass constructor can call the corresponding superclass constructor using **super**, followed by the list of arguments. In this way, a constructor must initialize only the attributes that are not inherited. A constructor has also an access modifier. If we declare a constructor as **private**, no other class can instantiate this class (it must be done by static methods, called factory methods).

1.6.2.4 INHERITANCE

We can build a class derived from another class using inheritance. To declare that a class is inherited, the keyword **extends** is used (Listing 1.5).

LISTING 1.5 CLASS CIRCLE EXTENDS CLASS SHAPE

```
public class Circle extends Shape {…}
```

When a class extends another class, it inherits all its attributes and methods. These attributes and methods can be overridden in the extended class, which also can add new attributes and methods. In Java, a class cannot extend multiple classes. There is no multiple inheritance.

Every single class has a superclass. When we define a class that does not extend other class, Java uses the *Object* class, which is the root of the class hierarchy in Java.

1.6.2.5 Abstract Classes and Methods

A class declared as *abstract* cannot create objects. An abstract class forces the use of derived class as special implementations of the properties of the abstract class (Listing 1.6).

LISTING 1.6 ABSTRACT CLASS

```
public abstract class Shape {
 public abstract double area();
 ...
}
```

An abstract class can contain methods also declared as abstract. In this case, the method definition is not given. If a class has an abstract method, it is mandatory for the class to be abstract. An abstract class, on the contrary, can have no abstract methods. An abstract method cannot be static.

1.6.2.6 Access Modifiers

Inside a class, all the members (attributes and method) are accessible. Private members are only accessible inside the class. If a constructor is declared as private, only a static method of that class can call the constructor.

From a subclass, the class inherits all the superclass private members, but they can be only accessed using public or protected methods of the superclass.

In the same package, the class has access to all the members belonging to other classes in the same package and not declared as private.

Regarding classes in different packages, a class A can access to a member of another class B (in a different package) if the class B is public and the member of B is public. However, if class A extends B, the class A also has access to all the protected members.

Table 1.1 shows a summary of the access modifiers in Java.

1.6.2.7 Polymorphism

Polymorphism is the capability to do different things by an action or a method, based on the object that it is acting upon. Overloading, overriding, and dynamic method binding are three types of polymorphism.

Overloaded methods are methods with the same name signature but with either a different number of parameters or different types in the parameter list. For example, "spinning" a number may mean increase it, and "spinning" an image may mean rotate it by 90°. By defining a method for handling each type of parameter we control the desired effect.

TABLE 1.1
Summary of Access Modifiers in Java

Visibility	Public	Protected	Private	Default
Same class	Yes	Yes	Yes	Yes
Same package	Yes	Yes	No	Yes
Different package	Yes	No	No	No
Same package, subclass	Yes	Yes	No	Yes
Different package, subclass	Yes	Yes	No	No

Overridden methods are methods that are redefined within an inherited or subclass. They have the same signature, and the subclass definition is used.

Dynamic (or late) method binding is the ability of a program to resolve references to subclass methods at runtime. As an example (Listing 1.7) assume that three subclasses (*Circle*, *Triangle*, and *Rectangle*) have been created based on the abstract *Shape* class, each having their own *area()* method. Although each method reference is to a shape (but no shape objects exist), the program will resolve the correct method reference at runtime:

LISTING 1.7 POLYMORPHISM

```
double area;
Shape shape;
Triangle triangle = new Triangle();
Circle circle = new Circle();
Rectangle rectangle = new Rectangle();
shape = triangle; area = shape.area();
shape = circle; area = shape.area();
shape = rectangle; area = shape.area();
```

1.6.2.8 Java Generics

Generics add stability to the code, managing attributes of "any type" instead of a particular one. To show the utility of Generics, we follow the same example given in (Lesson: Generics, 2012). This example shows the necessary syntax and terminology to use generics.

The example consists of a non-generic *Box* class that operates on objects of any type (Listing 1.8). It provides two methods: *add*, which adds an object to the box, and *get*, which retrieves the object back. Since its methods accept or return the object, we are free to pass in whatever we want, provided that it is not one of the primitive types. However, if we need to restrict the contained type to something specific (like Integer), our option would be to specify the requirement in documentation (or in this case, a comment), which of course, the compiler knows nothing about.

LISTING 1.8 NON-GENERIC CLASS

```
public class Box {

  private Object object;

  public void add(Object object) {
    this.object = object;
  }

  public Object get() {
    return object;
  }

  public static void main(String[] args) {
    // ONLY place Integer objects into this box!
    Box integerBox = new Box();

    integerBox.add(new Integer(10));
    Integer someInteger = (Integer) integerBox.get();
    System.out.println(someInteger);
  }
}
```

In the previous example, we performed the *right* casting approach. By casting the object returned by the *get* method, we perform the right transformation because the variable stored is an integer. However, we can find the situation (Listing 1.9) in which we are typecasting incorrectly.

LISTING 1.9 NON-GENERIC CLASS WITH ERROR

```
public class Box {
...
  public static void main(String[] args) {
    // ONLY place Integer objects into this box!
    Box integerBox = new Box();

    // Oops, another programmer replaced this line:
    integerBox.add("10");
    Integer someInteger = (Integer) integerBox.get();
    System.out.println(someInteger);
  }
}
```

The cast from object to integer has mistakenly been overlooked. This is clearly a bug, but because the code still compiles, we would not know anything is wrong until runtime when the application crashes with a *ClassCastException*.

Now we update the *Box* class to use generics. We first create a generic type declaration by changing the code `public class Box` to `public class Box<T>`; this introduces one *type* variable, named *T*, that can be used anywhere inside the class. There is nothing particularly complex about this concept. *T* is as a special kind of variable, whose "value" will be whatever type we pass in; this can be any class type or another type variable. It just cannot be any of the primitive data types. In this context, we also say that *T* is a formal type parameter of the *Box* class (Listing 1.10).

LISTING 1.10 GENERIC CLASS

```
public class Box<T> {

  //T stands for "Type"
  private T t;

  public void add(T t) {
    this.t = t;
  }

  public T get() {
    return t;
  }

  public static void main(String[] args) {
    Box<Integer> integerBox = new Box<Integer>();

    integerBox.add(new Integer(10));
    Integer someInteger = integerBox.get();
    System.out.println(someInteger);
  }
}
```

In this case, no cast is needed. Indeed, if we perform the same mistake adding a *String* object, then compilation will fail, alerting us to what previously would have been a runtime bug (Listing 1.11).

LISTING 1.11 USING *GENERICS* IN BOX CLASS

```
public class Box<T> {
...
  public static void main(String[] args) {
    Box<Integer> integerBox = new Box<Integer>();

    integerBox.add("10");
    Integer someInteger = integerBox.get();
    System.out.println(someInteger);
  }
}
```

```
Box.java:15: add(java.lang.Integer) in
    Box<java.lang.Integer> cannot be applied to
    (java.lang.String)integerBox.add("10");
                                    ^
1 error
```

1.6.2.9 Arrays and Collection Classes

1.6.2.9.1 Arrays

Java arrays can be treated as objects that belong to a predefined class. Arrays are objects, but with some particular characteristics:

- Arrays are instantiated with the **new** operator followed by the type and number of elements.
- We can obtain the number of elements in an array with the implicit attribute **length**.
- We can access to one element with square brackets and one index varying from **0** to **length-1**.
- We can create arrays of objects of any type.
- Elements in an array are initialized with the default value of the corresponding data type (0 for numeric arrays, empty strings for Strings, false for Boolean, null for references).

Arrays can be initialized with values enclosed by braces and with commas between elements. We can also use the same format but with several **new** operators inside. If two arrays are equal, we have two array names pointing to a single array object. An array reference can be defined in two ways, whereas the space for the elements is reserved using the **new** operator (Listing 1.12).

LISTING 1.12 ARRAYS DEFINITION

```
double[] x;//Array reference (1)
double x[];//Array reference (2)
x = new double[100];//Memory for 100 elements
double[] x = new double[100];//Two in one, reference and
memory for 100 elements
```

We can see in the following code some examples of array creation (Listing 1.13).

LISTING 1.13 EXAMPLE OF ARRAYS

```
// Array of 10 integers
int v1[] = new int[10];
// Array initialized with some values
int v2[] = {0, 1, 2, 3, 4, 5, 6, 7, 8, 9};
String days[] = {"monday", "tuesday", "wednesday",
       "thursday", "friday", "saturday", "sunday"};
// Array of 5 object
Box boxes[] = new Box[5];//5 null references
for (int i = 0; i < 5; i++) {
  boxes[i] = new Box();
}
```

A bidimensional array is created with dynamic memory. In Java, a matrix is a vector of references to row vectors. With this scheme, each row could have a different number of elements. A matrix can be directly created in two different ways (Listing 1.14).

LISTING 1.14 MATRIX DEFINITION

```
int nrows = 100, ncols = 100;
int[][] matrix1 = new int[3][4];//(1)
int[][] matrix2; // Begin (2)
// 2.1 We fist create the references array:
matrix2 = new int[nrows][];
// 2.2 We allocate memory for each vector in a row:
for (int i = 0; i < nrows; i++) {
  matrix2[i] = new int[ncols];
}
// See some examples of creation
double matrix3[][] = new double[3][3];
int[][] matrix4 = {{1, 2, 3}, {4, 5, 6}};
int[][] matrix5 = new int[3][];
matrix5[0] = new int[5];
matrix5[1] = new int[4];
matrix5[2] = new int[8];
```

1.6.2.9.2 Collections

Java also has an extensive library to work with object collections. We enumerate two examples here. For a complete reference to Java collections, we recommend the reader to refer to Trail: Collections (2012).

A linked list (*LinkedList* class) implements all optional list operations and permits all elements, including *null*. The *LinkedList* class provides uniformly named methods to *get*, *remove*, and *insert* an element at the beginning and end of the list. These operations allow linked lists to be used as a stack, queue, or double-ended queue. All the operations perform, as could be expected, for a doubly linked list as well. Operations that index into the list will traverse the list from the beginning or the end, whichever is closer to the specified index. Table 1.2 shows a summary of the main methods we can find in the *LinkedList* class.

An *ArrayList* class is a resizable-array implementation. It implements all optional list operations and permits all elements, including null. In addition to implementing list operations, this class provides methods to manipulate the size of the array that is used internally to store the list. The *size*, *isEmpty*, *get*, *set*, and other operations run in constant time. The *add* operation runs in amortized constant time; that is, adding *n* elements requires *O(n)* time. All the other operations run in linear time (roughly speaking). Each ArrayList instance has a *capacity*. The capacity is the size of the array used to store the elements in the list. It is always at least as large as the list size. As elements are added to an *ArrayList*, its capacity grows automatically. The details of the growth policy are not specified beyond the fact that adding an element has constant amortized time cost. An application can increase the capacity of an *ArrayList* instance before adding a large number of elements using the *ensureCapacity* operation. This may reduce the amount of incremental reallocation. Table 1.3 lists the operations available for the *ArrayList* class.

In Listing 1.15, we can see an example of the use of both collections.

TABLE 1.2
Some LinkedList Member Functions

Method	Function
LinkedList<E>()	Constructs an empty list
boolean add(E e)	Appends the specified element to the end of this list
void clear()	Removes all the elements from this list
E element()	Retrieves, but does not remove, the head (first element) of this list
E remove(), E removeFirst()	Retrieves and removes the head (first element) of this list
E removeLast()	Removes and returns the last element from this list
int size()	Returns the number of elements in this list

TABLE 1.3
ArrayList Member Functions

Method	Function
ArrayList<E>()	Constructs an empty list with an initial capacity of 10
boolean add(E e)	Appends the specified element to the end of this list
void add(int index, E element)	Inserts the specified element at the specified position in this list

TABLE 1.3 (*Continued*)
ArrayList Member Functions

Method	Function
void clear()	Removes all the elements from this list
E get(int index)	Returns the element at the specified position in this list
boolean isEmpty()	Returns true if this list contains no elements
E remove(int index)	Removes the element at the specified position in this list
int size()	Returns the number of elements in this list

LISTING 1.15 USING COLLECTIONS

```
ArrayList<Box> vector = new ArrayList<Box>();
for (int i = 0; i < 10; ++i) {
  Box box = new Box();
  Integer element = new Integer(i);
  box.add(element);
  vector.add(box);
}
for (Box box : vector) {
  System.out.println(box.get().toString());
}
LinkedList<Box> list = new LinkedList<Box>();
for (int i = 0; i < 10; ++i) {
  Box box = new Box();
  Integer element = new Integer(i);
  box.add(element);
  list.add(box);
}
while (list.size() > 0) {
  System.out.println(list.removeFirst().get().toString());
}
```

REFERENCES

Blue Brain Project. (2005). Last accessed Oct. 2012 at: http://bluebrain.epfl.ch/page-52658-en.html.

Choi, K., Jung, S., Kim, H., Bae, D.-H., & Lee, D. (2006). UML-based modeling and simulation method for mission-critical real-time embedded system development. *IASTED Conf. on Software Engineering* (pp. 160–165). Retrieved from http://www.scopus.com/inward/record.url?eid = 2-s2.0-34047162457&partnerID = 40&md5 = a4f662c1f5f336095df2889c2a71e4d3.

Seidel, D. (1993). *Aggregate Level Simulation Protocol (ALSP): Program Status and History*. The MITRE Corporation.

DoD High Performance Computing Modernization Program. (1992) Retrieved from http://www.hpcmo.hpc.mil/cms2/index.php.

Law, A. M., & Kelton, W. D. (2000). *Simulation Modeling and Analysis. McGraw-Hill Series in Industrial Engineering and Management Science* (3rd ed.). Boston, MA: McGraw-Hill.

Lesson: Generics. (2012). Retrieved from http://docs.oracle.com/javase/tutorial/java/generics/index.html.

The Object Management Group (OMG)—The Unified Modeling Language (UML). (2012). Retrieved from http://www.uml.org.

Schwede, T., & Peitsch, M. C. (2008). *Computational Structural Biology: Methods and Applications* (p. 792). World Scientific Hackensack, NJ.

Test and Training Enabling Architecture (TENA). (2012). Retrieved from https://www.tena-sda.org/display/intro/Home.

Trail: Collections. (2012). Retrieved from http://docs.oracle.com/javase/tutorial/collections/.

U.S. Department of Defense. (2001). *RTI 1.3-Next Generation Programmer's Guide Version 4.*

WordNet. (2012). Retrieved from http://wordnet.princeton.edu.

2 System of Systems Modeling and Simulation with DEVS

2.1 DEFINITIONS

In the first chapter, the concepts of systems modeling and simulation (M&S) were introduced. Various types of systems at their fundamental level were briefly described at the computational level. Generally, a system is a collection of components working together to achieve a result not achievable by any single component constituting a system. There are many implicit questions within this statement, such as

- How many components?
- How are components organized?
- Is the organization flat or hierarchical?
- What are the interactions between the components?
- Do the interactions define the system or vice versa?
- Are the interactions local or displaced in space and time?
- What is the system's boundary?
- Is it a natural system or an artificial system?
- What result is being achieved?
- Is an objective necessary for a system to exit?
- How does a system come into being in the first place?

Answers to each of the questions above will start a new journey of discovery into the study of the system, system of systems (SoSs), complex systems, and lastly, complex adaptive systems (CASs). We shall now look at each of these different types of systems.

Beginning with the notion of system, a few things are universally accepted at the systems level:

- System is composed of interacting components and components are identified at multiple levels of resolution.
- Hierarchy is used to manage the large number of components and complexities.
- Interactions among components play a vital role in system's behavior.
- A system has a boundary.
- A system manifests outward behavior and internal states.
- A system interacts with the environment.

At the SoS level, the focus is more on the integration and interoperability to achieve a common goal or mission. According to Jamshidi (2008: pp. 38), "system of systems are large-scale integrated systems that are heterogeneous and independently operable on their own, but are networked together for a common goal." There are numerous contextual definitions of SoS, and a survey and discussion is available in Jamshidi (2008). As a summary of various concepts, an SoS has the following characteristics:

- *Operational independence of individual systems*: This implies that if decomposed into constituent systems, these constituent systems are capable of performing independently of one another, but not the overall SoS goal.
- *Managerial independence of the systems*: The component systems are independently acquired and have their specific purpose for the organization that acquired them. Each of the constituent systems may be managed by a different authority or organization.
- *Geographic distribution*: An SoS is never a monolithic system and is distributed geographically. Knowledge and information sharing is mostly exchanged with each other through well-defined communication channels.
- *Emergent behavior*: An SoS performs functions and purposes that may not reside in any constituent systems. The principal purpose behind engineering an SoS and the constituent systems is the collective emergent behavior.
- *Evolutionary and adaptive development*: An SoS is always in a state of evolution and development with constituent systems having their own life cycles. Constituent systems may be restructured, updated, or removed during the life cycle of the SoS.
- *Integration of constituent systems*: Engineering an SoS involves integration at multiple levels such as organizational, technological, national, functional, budget, and so on.
- *Interoperability*: Architecting an SoS where the constituent systems interoperate toward a common goal is a technological as well as a managerial challenge.
- *Open system*: An SoS is an open system. There is no closed-form engineering solution for architecting an SoS. An open system has no boundaries. A closed system has defined boundaries. The constituent systems may or may not be closed systems either.
- *Evolving requirements*: An SoS comes into being to achieve a purpose and is dissolved as soon as the goal is achieved. The same constituent systems may be engaged again toward a different configuration of a new SoS to achieve a different objective.
- *Optimization of systems*: An SoS may be designed to optimize the operation of constituent systems working together toward a common goal.

A netcentric SoS is an SoS that utilizes *standards* to integrate and operate in a network-centric environment. The network, most likely the World Wide Web or Internet, is the underlying communication mechanism. Usage of widely adopted standards facilitates integration and interoperability. A new class of netcentric SoSs, called event-driven architectures, that are built on netcentric service-oriented architecture, is an example of dynamic interoperable SoS.

At the complex systems level, things become a bit more complicated in the sense that the complex system is inherently dynamic due to a lot of moving parts. An SoS may transform into a complex system. A complex system exhibits the following (Levy, 2000):

- *Nonlinear behavior*: The constituent systems have nonlinear relationships with other systems, and therefore long-term planning is impossible. Metasystem behavior cannot be derived by analyzing the behavior of component systems.
- *Dramatic changes that occur unexpectedly*: Small changes can cause huge perturbations. Cascading failures are the Achille's heel of complex systems.
- *Patterns and short-term predictability*: Long-term behavior forecasting is impossible. Complex SoSs are used for short-term predictable dynamic behavior. This is a characteristic of complex dynamical systems, where only the next state is predictable from the current state.

Then, there are the CASs that add more dynamism to the complex systems. In addition to the properties displayed by complex SoSs, the CASs portray the following properties (Mittal, 2012):

- *Network topology*: The interconnectedness of the constituent systems defines CAS behavior. Presence of hubs and clusters (systems with high degree of connectivity) redefine CAS's overall purpose. The overall network is dynamic and portrays scale-free topology (Barabasi, 2003).
- *Self-organization and emergence*: Self-organization is the ability of a system to organize itself toward a common objective. It is entirely possible that a CAS self-organizes to achieve an unintended objective as it is impossible to predict long-term behaviors. A CAS displays strong and weak emergence. In strong emergence, the systems take proactive action to avoid a future state. A CAS involving human-in-the-loop may result in a well-intended proactive action leading to cascaded failures. Examples are stock markets, brain, economy, and so on. Weak emergence is displayed when the emergent behavior can be traced back to the constituent systems such as Game of Life and connectionist networks.
- *Upward and downward causation*: CAS spans multiple levels of hierarchy due to strong interconnectedness and weak connectivity to remote systems. This adds more unpredictability to the entire system behavior.

Discrete Event Systems Specification (DEVS) formalism is based on dynamical systems theory and has been used to model complex dynamical systems for the last 40 years (Zeigler et al. 2000). However, CAS is an emerging computational paradigm. Among the many contributors to CAS, the works of Rouse (2007), Sage (2002), and Sheard (2008, 2009) have strong relevance to systems engineering. DEVS for CAS has been considered by Zeigler (2004) through the quantization principle (see Section 2.2), and recent work by Mittal (2012) formally analyzes CAS with respect to scale-free topologies and how DEVS is positioned to address CAS M&S.

In this chapter, we take a close look at the framework for M&S of complex systems underlying the DEVS formalism. Chapter 3 will discuss the complex dynamical behavior of DEVS components.

2.2 DEVS HIERARCHY OF SYSTEMS SPECIFICATION

The DEVS theory as proposed by Zeigler in 1976 is made up of two orthogonal concepts:

1. *Levels of systems specification*: Describes how systems behave.
2. *System specification formalisms*: Incorporates various modeling styles, such as continuous or discrete.

Systems theory distinguishes between system structure (how the system is constituted internally) and system behavior (how the system manifests externally). Understanding the system structure allows us to deduce its behavior. The internal structure of a system is laden with many concepts, such as

- *State representation*: Different states the system may exist in
- *Transition functions*: Mechanisms that allow moving from one state to another
- *State-to-output functions*: Mechanisms that make system a producer in an environment
- *Composition*: Capacity to form a larger system by coupling smaller systems
- *Decomposition*: Capacity to decompose into smaller systems from a larger system
- *Hierarchical construction*: Capacity to continue to portray composition
- *Modular*: Capacity to have defined input and output interfaces to enable composition

Systems theory is *closed under composition* in that the structure and behavior of a composition of systems can be expressed in original systems theory terms. This is the foundation of modular systems that have defined input and output interfaces through which all interaction with the environment occurs. Such modular systems are coupled together to form larger ones, leading to a hierarchical construction.

Descriptions of alternate system specification formalisms such Differential Equation System Specification (DESS), Discrete Time System Specification (DTSS), and Quantized Systems are briefly described in Chapter 3 and can be found in much greater detail in Zeigler et al. (2000). Although DESS and DTSS as their names suggest are self-explanatory, quantized DEVS warrants a definition. Quantization is a process for representing and simulating continuous systems as an alternative to the more conventional time axis. It is built on the *threshold crossing* model. While

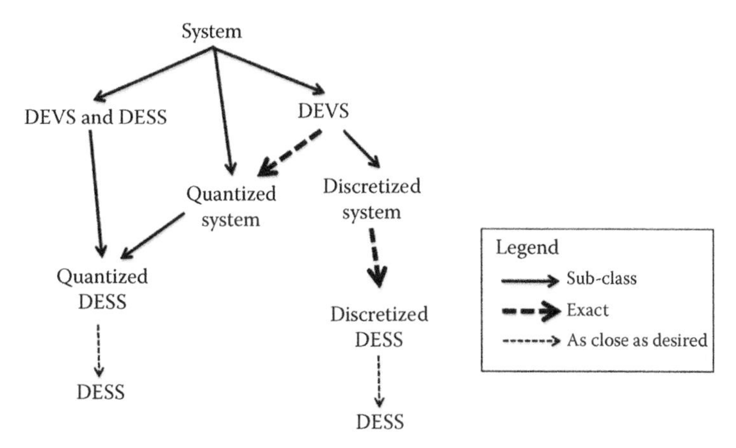

FIGURE 2.1 Discrete Event Systems Specification (DEVS) formalism and quantized systems. (From Zeigler, B. P. et al., *Theory of Modeling and Simulation*, Academic Press, New York, NY, 2000. With permission.)

discretization leads to DTSS, *quantization* leads to discrete-event systems. Figure 2.1 shows a mapping of various formalisms. As can be seen, the universality of DEVS formalism allows specification of hybrid systems. Having seen the scope of the DEVS formalism, we focus our attention on the DEVS formalism and levels of systems specifications.

2.2.1 Hierarchy of Systems Specification

The hierarchy of systems specification has five levels. At the most basic level (Level 0) is the observation frame that defines which inputs stimulate the system, what variables to measure, and how to observe them over a time base. At this level, we also think about the possible range of values the inputs may take. The observation also correlates the input trajectory with specific outputs the system produces. Such correlation between inputs and outputs is called an *I/O frame* linked over a time base at Level 0. A collection of such I/O frames is called an *I/O behavior* at Level 1. It is entirely possible that two or more input trajectories may lead to the same output trajectory. At Level 1 we have all those trajectories collected, and at Level 2 we distinguish them based on the initial state of the system when input is injected. The initial state determines a "unique" response to any input and is represented as an *I/O function*. We can define not only the initial state but also the state transitions when the system responds to input trajectories. We have a system that has a state-space and characteristic functions that map specific input trajectories to specific output trajectories. The black box system at Level 3 is an "atomic" component in DEVS parlance, capable of dealing with external inputs and that undergoes state transitions to produce external outputs. At Level 4, we have coupled and interactive systems that are connected using "coupling" relationships. The systems are coupled using ports and outputs of one system and

TABLE 2.1
Levels of Systems Specifications

Level	Name	System Specification at This Level
4	Coupled systems	Systems built from component systems with a coupling recipe
3	I/O system structure	System with state space and transitions to generate the behavior
2	I/O function	Collection of I/O pairs constituting the allowed behavior partitioned according to initial state of the system
1	I/O behavior	Collection of I/O pairs constituting the allowed behavior of the system from external black box view
0	I/O frame	I/O variables and port together with values over a time base

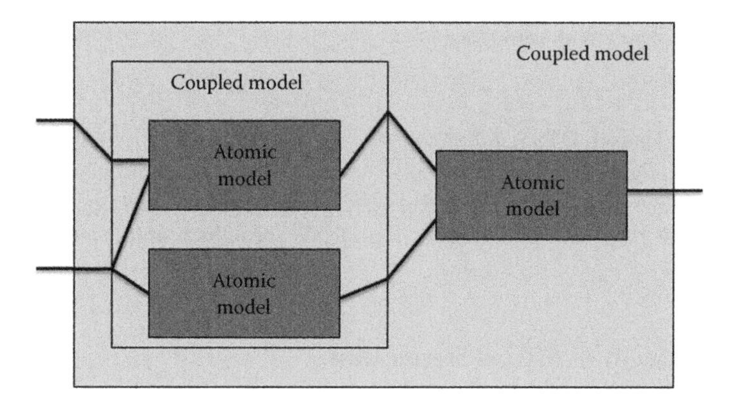

FIGURE 2.2 DEVS hierarchical system with atomic and coupled components.

are connected to inputs of another system. Such coupling allows composition and hierarchical construction. Table 2.1 summarizes these levels.

2.2.2 DEVS System Components

Structurally, hierarchical DEVS system is composed of three elements—atomic, coupled, and associated couplings—between the atomic and coupled components. The atomic components are defined at Level 3, and coupled systems are specified at Level 4 of the *systems hierarchy specifications*. This allows construction of modular and hierarchical systems. Figure 2.2 shows the concept. The atomic and coupled components are connected by well-defined couplings at well-defined ports. The atomic component describes the dynamic behavior with a finite state machine, and the coupled component describes the connectivity and the hierarchical composition of both atomic and coupled models. Formal specification of atomic and coupled models is described in Chapter 3.

2.3 FRAMEWORK FOR M&S

The framework for M&S defines four entities and the relationships between these entities (Figure 2.3):

1. *Source system (real world or artificial system under study)*: The source system is the time-indexed observed behavior of the system. Not every observed behavior is of interest. The subset of this behavior database is bounded by the experiments, motivations, or objectives of the system under study.
2. *Experimental frame (the scope of the system under study)*: This defines the scope of the modeling exercise and is based on the objectives and scope of the entire M&S effort. It acts as a data provider to experiment with the constructed model as well as a data gatherer from the model. The validation of the model occurs in the experimental frame.
3. *Model (the abstract description of the system under study)*: A model is an abstraction of the source system and is represented by a set of instructions, rules, equations, or constraints for generating I/O behavior. Alternatively, a model is a state transition and output generation mechanism that accepts input trajectories and generates output trajectories depending on its initial state.
4. *Simulator (an algorithm that runs model over a time base)*: A computational system that executes model on a time base to generate model's behavior.

Table 2.2 relates the entities to DEVS levels of systems specification. There are two relationships defined in the preceding framework:

1. *Modeling relation (validity)*: The modeling relation relates an experimental frame, a model, and a system. This relation defines *validity*. In Zeigler's formulation, validity refers to the condition where it is impossible to distinguish the model and system in the experimental frame of interest. The most basic is the *replicative validity* at Level 1 of system specifications, where the model and system agree on I/O behavior within acceptable tolerance. Next is the

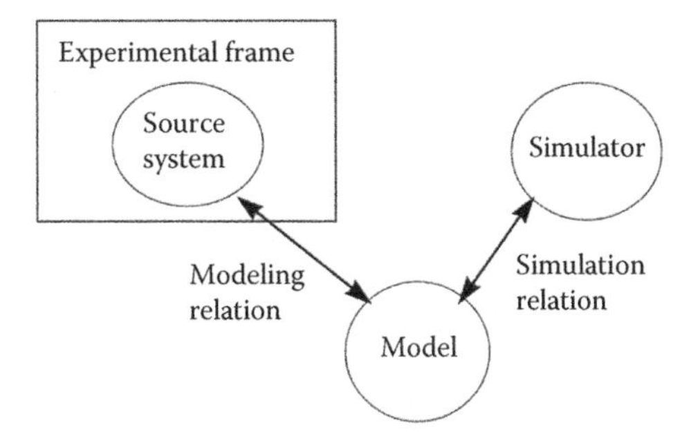

FIGURE 2.3 Framework for modeling and simulation (M&S). (From Zeigler, B. P. et al., *Theory of Modeling and Simulation*, Academic Press, New York, NY, 2000. With permission.)

TABLE 2.2
Framework Entities Constructed at Various DEVS Levels of Systems Specifications

Level	Name	Framework for M&S
4	Coupled systems	Model, simulator, experimental frame
3	I/O system structure	Model, simulator, experimental frame
2	I/O function	Model, source system
1	I/O behavior	Source system
0	I/O frame	Source system

predictive validity at Level 2 of systems specification that allows the model to predict unseen system behavior, provided the model is set to the same state as the real system. Finally, there is *structural validity* at Level 3 or higher of systems specification, which implies that the model is capable of not only replicating the system's behavior but also predicting step-by-step component-by-component the way the system does its transitions (Zeigler et al., 2000). Validation answers the question: "Did you build the right thing?"

2. *Simulation relation (verification)*: The *simulation relation* relates the model and the simulator that executes the model. This relation ensures simulator correctness; that is, it displays model's behavior the way it is meant to be. Alternatively, verification is an attempt to establish that the simulation relation holds between a simulator and a model. Verification answers the question, "Did you build it right?"

The framework for M&S categorically separates various entities. This has manifold advantages when we are building complex systems. Some of the advantages are listed as follows:

- *Repository*: All the entities can be a part of the repository. Multiple experimental frames can be applicable to a single model, or a single experimental frame is applicable to the same model at a different level of resolution or fidelity. For example, in case of a lumped model derived from a base model, the same experimental frame can easily evaluate the validity of the lumped model.
- *Separating the phenomenon and the experiment*: The model is separate from the scope of the system under study. Model is built on abstractions. Abstractions can be specified in various other formalisms, for example, UML, SysML, BPMN, and so on. Model describes a system's behavior "phenomenon" and not just a snapshot of system state.
- *Language independence*: All the entities can be language agnostic provided both the M&S relations hold. This is the basis of interoperability and the starting point toward netcentric M&S.
- *Independent research areas*: M&S have emerged as complete research areas, and separation of these two is critical to achieve model-based

systems engineering (MBSE). The simulator can be implemented on a local machine, a distributed system, a grid, or a netcentric platform. The simulation relation makes the simulator transparent.

- *Different time base of simulators*: Because simulators are abstracted behind the simulation relation, the simulators can implement time in a variety of ways. Time can be logical or real wall clock time. In a monolithic system, the simulator can either run as fast as it can or as slow as it configured (relative to real time) when the time base is *logical*. For example, assuming real wall clock time proceeds in seconds, a translation factor of 0.1 implies that the simulation proceeds 10 times faster than real time, provided the computational time is less than the logical time. It is entirely possible that the computational time overrides the logical time and the simulation becomes dependent on the computational processing power, that is, "as fast as it can" and still not achieve the desired translation factor. In a heterogeneous, distributed system, the simulator is distributed on different clocks and possibly in different time zones. Logical time in such a heterogeneous system is hard to manage as there is no synchronism between the simulators. These networked systems have message latency, so simulators are designed to account for lost real time. With today's high-speed networks, real-time heterogeneous simulations are a reality where message latency could be minimized to a great extent. This is the basis for event-driven architectures and netcentric M&S.

The reader is encouraged to refer to the work of Zeigler et al. (2000) for in-depth description of the preceding concepts.

2.3.1 Computational Representation

Using Unified Modeling Language (UML), we can represent the framework as a set of classes and relations, as illustrated in Figures 2.4 and 2.5 (Mittal et al., 2008). Various implementations support different subsets of the classes and relations.

2.4 SUMMARY

In the highly interconnected world of today, systems have transformed into SoSs. Monolithic systems are a thing of the past. Today's systems are built from component systems designed by third parties, whether it is a software application or a mission critical system. The success of integrated development environments, like Eclipse, that are built on modular components is a clear example of scalability and integration that we can relate to as a modeler. Engineering an SoS is a challenge. M&S of an SoS is equally difficult, as there is an additional step of developing abstractions. The task is equally harder at the testing and evaluation stages. In this chapter, we briefly discussed the definitions and concepts behind systems, SoS, netcentric SoS, complex systems, and CAS. We introduced the DEVS hierarchical levels of systems

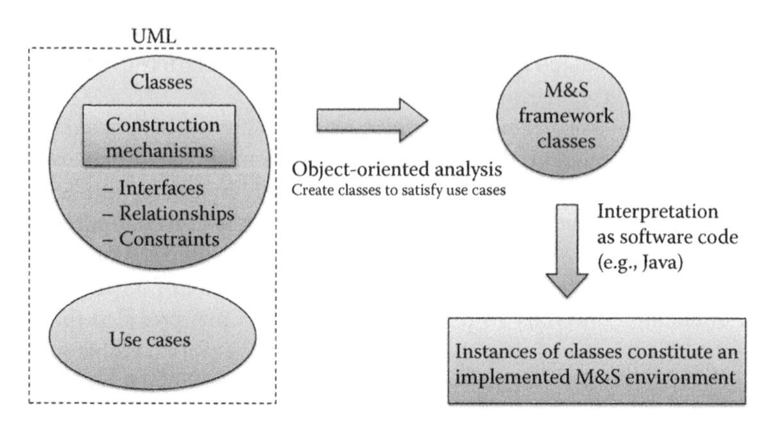

FIGURE 2.4 M&S framework formulated within Unified Modeling Language (UML).

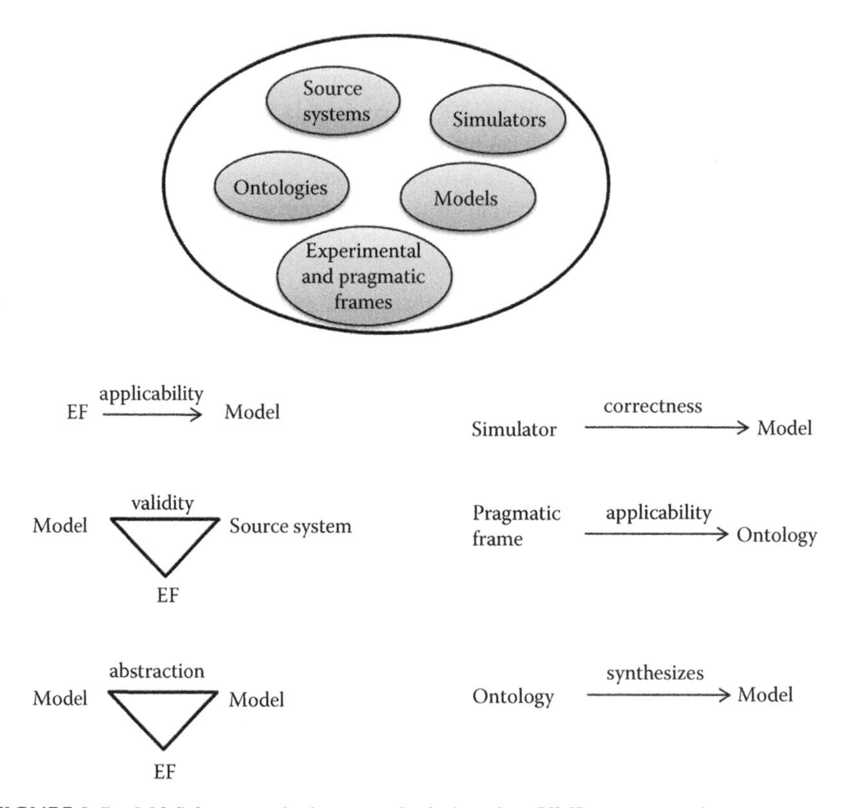

FIGURE 2.5 M&S framework classes and relations in a UML representation.

engineering and it has been used in designing complex dynamical systems for the last four decades. We provided an overview of basic DEVS systems engineering concepts and how systems are analyzed at various levels of specifications. Let us move forward in understanding the formal specification of complex dynamical systems.

REFERENCES

Barabasi, A.-L. (2003). *Linked: How Everything is Connected to Everything Else and What it Means for Business, Science and Everyday Life*. New York, NY: Penguin Books.

Jamshidi, M. (2008). *System of Systems Engineering—Principles and Applications*. Boca Raton, FL: Taylor and Francis, CRC Publishers.

Levy, D. (2000). Applications and limitations of complexity theory in organizational theory and strategy. In Rabin, J., Miller, G. J., & Hildreth, W. B. (Eds.) *Handbook of Strategic Management*. New York: Marcel Dekker.

Mittal, S., Zeigler, B. P., Martin, J. L. R., Sahin, F., & Jamshidi, M. (2008). *Modeling and Simulation for System of Systems Engineering, in System of Systems Engineering for 21st Century*. Jamshidi (Ed.), San Francisco, CA: Wiley & Sons.

Mittal, S. (2012). Emergence in stigmergic and complex adaptive systems: A formal discrete events perspective. *Journal of Cognitive Systems Research*, Special Issue on Stigmergy http://dx.doi.org/10.1016/j.cogsys.2012.06.003.

Rouse, W. (2007). Complex engineered, organizational and natural systems. *Systems Engineering*, 10(3), 260–271.

Sage, A. (2002). Complex adaptive systems. In *Yearbook of Science and Technology* (pp. 51–53). New York: New York, McGraw-Hill.

Sheard, S. (2008). Complex adaptive systems in systems engineering and management. In A. R. Sage, *Handbook of Systems Engineering and Management*. Hoboken, NJ: John Wiley & Sons.

Sheard, S. (2009). Principles of complex systems for systems engineering. *Systems Engineering*, 12(2): 295–311.

Zeigler, B. (2004). Discrete event abstraction: An emerging paradigm for modeling complex adaptive systems. In *Perspectives on Adaptation in Natural and Artificial Systems, Essays in Honor of John Holland*. Oxford: Oxford University Press.

Zeigler, B. P., Praehofer, H., & Kim, T. G. (2000). *Theory of Modeling and Simulation*. New York, NY: Academic Press.

3 DEVS Formalism and Variants

The Discrete Event System Specification (DEVS) formalism was first introduced by Zeigler in 1976 to provide a rigorous common basis for discrete event modeling and simulation (M&S) (Zeigler et al., 2000). A "common" basis means that it is possible to express popular discrete event formalisms such as event scheduling, activity scanning, and process interaction using the DEVS formalism.

The class of formalisms denoted as discrete event is characterized by a continuous time base, where only a finite number of events can occur during a finite time span. This contrasts with Discrete Time System Specification (DTSS) formalisms where the time base is isomorphic to $\mathbb{N}$ and with Differential Equation System Specification (DESS or continuous-time) formalisms in which the state of the system may change continuously over time.

The Formalism Transformation Graph (FTG) published by H. Vangheluwe in 2000 and shown in Figure 3.1 depicts behavior-conserving transformations between some important formalisms. The graph distinguishes between continuous-time formalisms on the left-hand side and discrete formalisms (both discrete time and discrete event) on the right-hand side. Although the graph suggests that formalisms can be mapped onto a common formalism on their respective sides, very few transformations allow crossing the middle line; this illustrates why hybrid systems (those that bring together both discrete and continuous systems) are difficult to solve.

The traditional approach to solving continuous-time problems is based on discretization, which approximates a continuous-time model by a discrete-time system (difference equations). However, a partitioning of the time axis, as is the case in discretization, is hard to harmonize with a partitioning of the state space, as is performed in discrete event systems. In this regard, mapping continuous-time formalisms (Ordinary Differential Equations [ODEs] and semiexplicit Differential Algebraic Equation [DAEs]) onto the DEVS formalism (this corresponds to the arrow going from "scheduling-hybrid-DAE to "DEVS" on the FTG) may be performed through quantization.

The closure property (under composition or coupling) of systems such as DEVS offers the possibility to describe a model as a hierarchical composition of simpler subcomponents. Apart from the obvious advantages associated with modularity (conceptual level, component reusability), a significant gain in the efficiency of simulating large, complex dynamic systems can also be achieved by using multirate integration (employing different integration frame rates for the simulation of fast and slow subcomponents), either on a single or on multiple processors.

Although some continuous-time formalisms (e.g., causal block diagram simulation tools) allow model hierarchization, multirate integration mixes poorly with

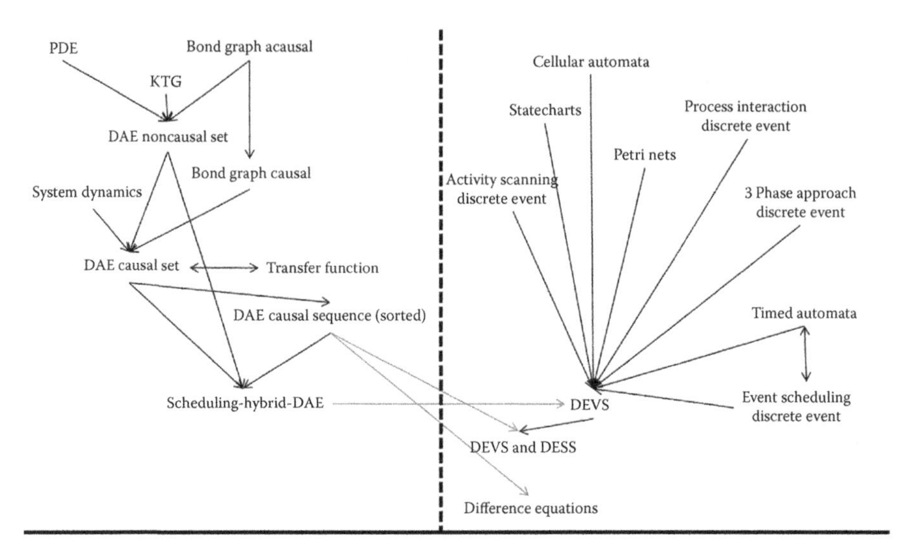

FIGURE 3.1 Formalism transformation graph.

traditional approaches where discretization along the time axis forces the simulator to work explicitly with the global time base. This is in contrast to discrete event formalisms where the simulator is concerned with local state space changes, and the time base is dealt with implicitly. Discrete event concepts are thus better suited for parallel-distributed simulation, and much effort has been devoted to the development of conservative (e.g., Chandy–Misra approach), optimistic (e.g., time-warp) and real-time (e.g., DARPA's Distributed Interactive Simulation [DIS]) parallel discrete event simulation techniques. The relevance of DEVS in that context is illustrated by the concept of the DEVS bus, which concerns the use of DEVS models as "wrappers" to enable a variety of models to interoperate in a networked simulation. The DEVS bus has been implemented on top of the Department of Defense's High Level Architecture (HLA) standard, itself based on the Run-Time Infrastructure (RTI) protocol.

Taking into account that DEVS is able to describe the behavior of other formalisms, our goal here is to introduce DEVS, focused on one particular form of this formalism: parallel DEVS (P-DEVS), and giving some details about other variants. The reader should refer to the book *Theory of Modeling and Simulation* by Zeigler et al. (2000) to understand the details behind the mathematical background of these techniques.

According to the DEVS theory, the system of interest is seen as a source system and a model uses abstractions to represent the source system. The model represents a simplified version of reality and its structure. The model is built considering the conditions of experimentation of the system of interest, including the work conditions of the real system and its application domain. Thus, the model is restricted to the experimental framework under which it was developed.

This model is subsequently used to build a simulator. The simulator is able to change the state of the model by running all the necessary state transitions already defined in the model. All the transitions are executed in an appropriate order, according to the model definition.

DEVS was created for M&S of discrete event dynamic systems. As a result, it defines a formal way to define systems whose states change either upon the reception of an input event or due to the expiration of a time delay. To deal with the system under study, the model can be organized hierarchically in such a way that higher level components in a system are decomposed into simpler elements.

The formal separation between model and simulator and the hierarchical and modular nature of the DEVS formalism have enabled the carrying out of formal proofs on the different entities under study. One of them is the proof of composability of the subcomponents (including legitimacy and equivalence between multicomponent models). The second is the ability to conduct proofs of correctness of the simulation algorithms, which results in rigorously verified simulators. All the proofs are based on formal transformations between each of the representations, trying to prove the equivalence between the entities under study at different levels of abstraction and preserving *homomorphism* (Zeigler et al., 2000). For instance, we can prove that the mathematical entity *simulator* is able to execute correctly the behavior described by the mathematical entity *model*, which represents the system.

Different mathematical mechanisms are used to prove these points, including the mathematical manipulation of the abstraction hierarchy, observation of input and output (I/O) trajectories (to ensure that different levels of specification correctly describe the system's structure), and decomposition concepts (DEVS is closed under composition, which means that a composite model integrated by multiple components is equivalent to an atomic component).

3.1 DEVS FORMALISM

We first introduce the original DEVS formalism known as classic DEVS. The question whether the formalism describes a "system" (i.e., under which conditions it is well behaved is a system-theory sense) is also covered. It turns out that even a well-behaved DEVS model can behave in a counterintuitive manner. Finally, the P-DEVS formalism, which removes some deficiencies of the original DEVS, is presented.

3.1.1 Classic DEVS Formalism

Classic DEVS is an intrinsically sequential formalism that allows for the description of system behavior at two levels: (1) at the lowest level, an atomic DEVS describes the autonomous behavior of a discrete event system as a sequence of deterministic transitions between states as well as how it reacts to external inputs and (2) at the highest level, a coupled DEVS describes a discrete event system in terms of a network of coupled components, each an atomic-DEVS model or a coupled DEVS in its own right, as we will see in Section 3.1.1.2.

3.1.1.1 Atomic DEVS

An atomic-DEVS M is specified by a 7-tuple

$$M = \langle X, Y, S, \delta_{int}, \delta_{ext}, \lambda, ta \rangle$$

where

- X is the input set
- Y is the output set
- S is the state set
- $\delta_{int}: S \rightarrow S$ is the internal transition function
- $\delta_{ext}: Q \times X \rightarrow S$ is the external transition function
- $Q = \{(s,e): s \in S, e \in [0, ta(s)]\}$ is the total state set, and e is the elapsed time since the last transition
- $\lambda: S \rightarrow Y$ is the output function
- $ta: S \rightarrow R_0^+ \cup \infty$ is the time advance function

There are no restrictions on the sizes of the sets, which typically are product sets; that is, $S = S_1 \times S_2 \times \cdots S_n$. In the case of the state set S, this formalizes multiple concurrent parts of a system while it formalizes multiple I/O ports in the case of sets X and Y. The time base T is not mentioned explicitly and is continuous. For a discrete event model described by an atomic DEVS M, the behavior is uniquely determined by the initial total state $(s_0, e_0) \in Q$ and is obtained by means of the following iterative simulation procedure (Figure 3.2).

At any given moment, a DEVS model is in state $s \in S$. In the absence of external events, it remains in that state for a period of time defined by ta(s). When ta(s) expires, the model outputs the value $\lambda(s)$ through a port and then it changes to a new state s_1 given by $\delta_{int}(s)$. This transition is called an *internal* transition. Then, the process starts again (see bottom gray arrow in Figure 3.2). On the contrary, an external transition may occur due to the reception of external events through input ports. In this case, the external transition function determines the new state s_2 given by $\delta_{ext}(s, e, x)$, where s is the current state, e is the time elapsed since the last transition (external or internal), and x is the external event received. After an external transition, the model is rescheduled and the process starts again (see left gray arrow), setting the elapsed time e to 0.

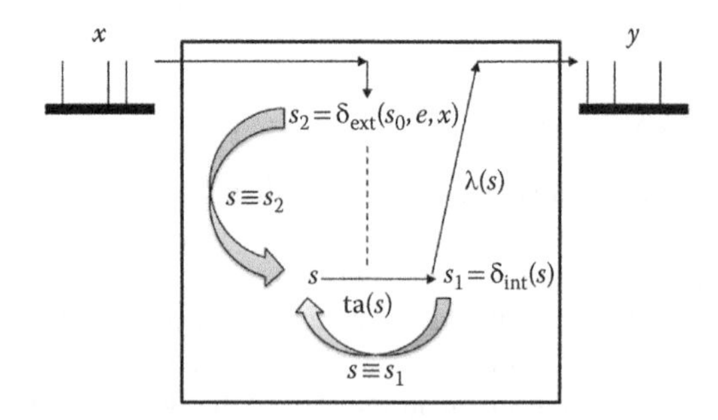

FIGURE 3.2 State transitions of an atomic Discrete Event System Specification (DEVS) model (1/2).

Following the previous definition, an atomic model has structure and behavior. Regarding the structure, we can find:

- The set of input ports through which external events are received. The set of input events X is composed by a set of pairs of input port and valid data: $X = \{(p, v) \mid p \in \text{InPorts}, v \in X_p\}$, where InPorts represents the set of input ports and X_p represents the set of values for the input port p.
- The set of output ports through which external events are sent. The set of output events Y is composed by a set of pairs of output port and valid data: $Y = \{(p, v) \mid p \in \text{OutPorts}, v \in Y_p\}$, where OutPorts represents the set of output ports and Y_p represents the set of values for the output port p.
- The set of state variables and parameters. One state variable is always present, *sigma* (in the absence of external events, the system stays in the current state for the time given by *sigma*: σ).

With respect to the behavior, we can find:

- The time advance function, which controls the timing of internal transitions—usually, this function just returns the value of *sigma*.
- The internal transition function, which specifies to which next state the system will transit after the time given by the time advance function (*sigma*) has elapsed.
- The external transition function, which specifies how the system changes the state when an input is received—the effect is to place the system in a new state and *sigma* thus scheduling it for a next internal transition; the next state is computed on the basis of the present state, the input port and value of the external event, and the time that has elapsed in the current state.
- The output function, which generates an external output just before an internal transition takes place.

In summary, *sigma* holds the time remaining to the next internal transition. This is precisely the time advance value to be produced by the time advance function. In the absence of external events, the system stays in the current state for the time given by *sigma*.

The time advance function can take any real number between 0 and ∞. A state for which $ta(s) = 0$ is called transient state. In contrast, if $ta(s) = \infty$, then s is said to be a passive state, in which the system will remain perpetually unless an external event is received.

Example

Consider the following timing diagrams in Figure 3.3.

At any time t, the system is in state $s_1 \in S$. No external event occurs, so the system will stay in state s_1 until the elapsed time e reaches $ta(s_1)$. The time left, $\sigma = ta(s_1) - e$, is often introduced as an alternate way to check for the time until the next (internal) transition. The system first produces the output value $\lambda(s_1)$ and makes a transition to state $s_3 = \delta_{int}(s_1)$. Next, an external event $x \in X$ occurs before

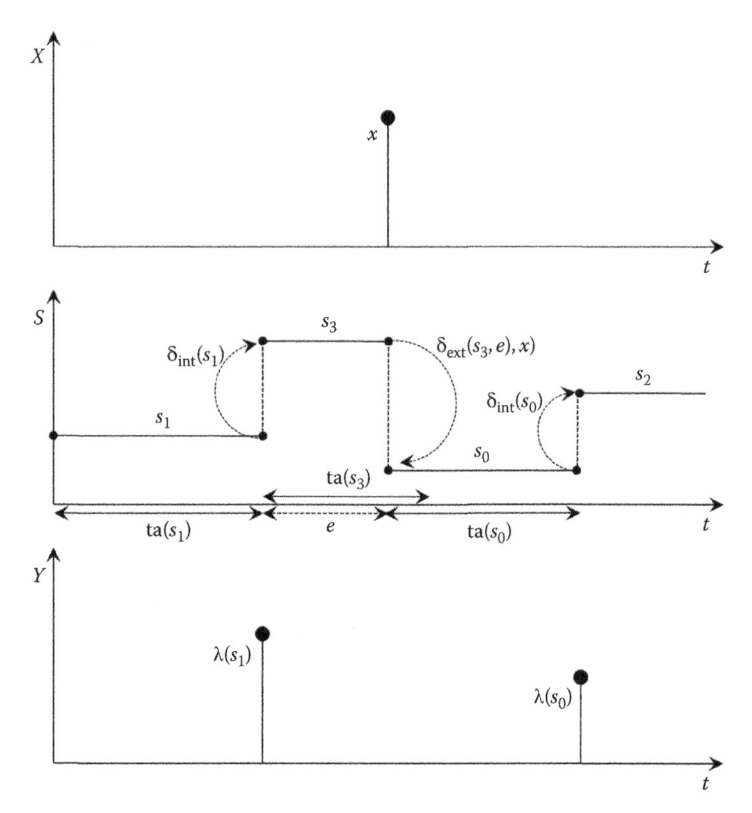

FIGURE 3.3 States transition of an atomic-DEVS model (2/2).

e reaches ta(s_3) and the system interrupts its autonomous behavior and instantaneously goes to state $s_0 = \delta_{ext}((s_3, e), x)$. Thus, the internal transition function dictates the system's new state based on its old state in the absence of external events. The external transition function dictates the system's new state whenever an external event occurs, based on this event *x*, the current state *s*, and how long the system has been in this state, *e*. After both types of transitions, the elapsed time e is reset to 0.

Example

A *processor* atomic model consumes a job *j*. When the processor receives a job through an input port, the processor remains busy until the processing time j_p is finished. Then it sends the job through an output port.

The processor model can be formally described as

$$\text{Processor} = \langle X, S, Y, \delta_{int}, \delta_{ext}, \lambda, \text{ta} \rangle$$

where,

$$X = \{(\text{in}, j \in \boldsymbol{J})\}, \text{ where } J \text{ is a set of jobs.}$$

$$S = (\text{phase} = \{\text{"busy"}, \text{"passive"}\}) \times \sigma \in R_0^+ \times j \in \boldsymbol{J}$$

$$Y = \{(\text{out}, j \in \boldsymbol{J})\}$$

$$\text{ta}(\text{phase}, \sigma, j) = \sigma$$

$$\lambda(\text{phase}, \sigma, j) = j$$

$$\delta_{\text{int}}(\text{phase}, \sigma, j) = (\text{"passive"}, \infty, \varnothing)$$

$$\delta_{\text{ext}}(\text{phase}, \sigma, j, e, (\text{in}, j')) = \begin{cases} (\text{"busy"}, j'_p, j') & \text{if phase} = \text{"passive"} \\ (\text{"busy"}, \sigma - e, j) & \text{if phase} = \text{"busy"} \end{cases}$$

The term *collision* refers to the situation where an external transition occurs at the same time as an internal transition. When such a collision occurs, the atomic-DEVS formalism specifies that the tie between the two transition functions shall be solved by first carrying out the internal, then the external transition function with $e = 0$.

Outputs are associated only with internal transitions to impose a delay on the propagation of events.

3.1.1.2 Coupled DEVS

A coupled-DEVS N is specified by a 7-tuple:

$$N = \langle X, Y, D, \{M_i\}, \{I_j\}, \{Z_{j,k}\}, \gamma \rangle$$

where

X is the input set

Y is the output set

D is the set of component indexes

$\{M_i | i \in D\}$ is the set of components, each M_i being an atomic DEVS:

$$M_i = \langle X_i, Y_i, S_i, \delta_{\text{int},i}, \delta_{\text{ext},i}, \lambda_i, \text{ta}_i \rangle$$

$\{I_j | j \in D \cup \{\text{self}\}\}$ is the set of all influencer sets, where $I_j \subseteq D \cup \{\text{self}\}, j \notin D$ is the influencer set of j.

$\{Z_{j,k} | j \in D \cup \{\text{self}\}, k \in I_j\}$ is the set of output-to-input translation functions, where

$$Z_{j,k} : X \to X_k, \text{ if } j = \text{self}$$

$$Z_{j,k} : Y_j \to Y, \text{ if } k = \text{self}$$

$$Z_{j,k} : Y_j \to X_k, \text{ otherwise}$$

$\gamma : 2^D \to D$ is the select function

The sets X and Y typically are product sets, which formalize multiple I/O ports. Each atomic DEVS in the network is assigned a unique identifier in the set D. This corresponds to model names or references in a modeling language. The coupled-DEVS N itself is referred to by means of self $\notin D$. This provides a natural way of indexing the components in the set $\{M_i\}$ and to describe the sets $\{I_j\}$, which explicitly describes the network structure, and $\{Z_{j,k}\}$.

Figure 3.4 shows an example of a coupled DEVS. In this case, I_A = {self}, I_B = {self,A}, and I_{self} = {B}. For modularity reasons, a component may not be influenced by components outside its enclosing scope, defined as $D \cup$ {self}. The condition $j \notin I_j$ forbids a component to directly influence itself to prevent instantaneous dependency cycles. The functions $Z_{j,k}$ describe how an influencer's output is mapped onto an influencer's input. The set of output-to-input transition functions implicitly describes the coupling network structure, which is sometimes divided into external input couplings (EICs, from the coupled-DEVS' input to a component's input), external output couplings (EOCs, from a component's output to the coupled-DEVS' output), and internal couplings (ICs, from a component's output to a component's input).

As a result of coupling concurrent components, multiple internal transitions may occur at the same simulation time t. Since in sequential simulation systems only one component can be activated at a given time, a tie-breaking mechanism to select which of the components should be handled first is required. The classic coupled-DEVS formalism uses the select function γ to choose a unique component from the set of imminent components, defined as

$$\Pi_t = \{i \mid i \in D, \sigma_i = 0\}$$

that is, those components that have an internal transition scheduled at time t. The component returned by $\gamma(\Pi_t)$ will thus be activated first. For the other components

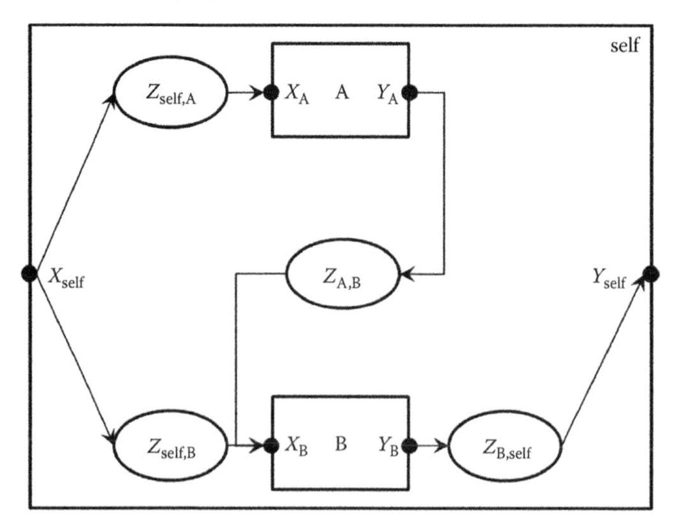

FIGURE 3.4 A coupled DEVS.

in the imminent set, we are left with the following ambiguity: when an external event is received by a model at the same time as its scheduled internal transition, which elapsed time should be used by the external transition: $e = 0$ of the new state or $e = \mathrm{ta}(s)$ of the old state? These collisions are resolved by letting $e = 0$ for the unique activated component and $e = \mathrm{ta}(s)$ for all the others.

3.1.2 P-DEVS Formalism

Because of the inherent sequential nature of classic DEVS, modeling using this formalism requires extra care. As a matter of fact, resolving collisions by means of the select function γ might result in counterintuitive behaviors.

The P-DEVS formalism (to distinguish it from parallel implementations of both classic DEVS and P-DEVS) was introduced to solve these problems by properly handling collisions between simultaneous internal and external events. As the name indicates, P-DEVS is a formalism whose semantics successfully describes (irrespective of sequential or parallel implementations) concurrent transitions of imminent components, without the need for a priority scheme.

Just as in the case of classic DEVS, P-DEVS allows for the description of system behavior at the atomic and coupled levels. The formalism is closed under coupling, which leads to hierarchical model construction. Other concepts like legitimacy introduced later also apply to P-DEVS.

The formalism uses a bag as the message structure: a bag X^b of elements in X is similar to a set except that multiple occurrences of elements are allowed $\left(\text{e.g., } X^b = \{a,b,a\}\right)$. As with sets, bags are unordered. Note that this is the only difference between a set and a bag. Thus, either using sets or bags (i.e., classic DEVS or P-DEVS) to collect inputs sent to a component, we recognize that inputs can arrive from multiple sources and that more than one input with the same identity may arrive simultaneously.

The atomic formalism for P-DEVS M is specified by an 8-tuple:

$$M = \langle X, Y, S, \delta_{\mathrm{int}}, \delta_{\mathrm{ext}}, \delta_{\mathrm{con}}, \lambda, \mathrm{ta} \rangle$$

The definition is almost identical to that of the classic version, except that we introduce the concept of a bag in the external transition and output functions:

$$\delta_{\mathrm{ext}} : Q \times X^b \rightarrow S$$

$$\lambda : S \rightarrow Y^b$$

This reflects the idea that more than one input can be received simultaneously, and similarly for the generation of outputs. P-DEVS also introduces the confluent transition function

$$\delta_{\mathrm{con}} : S \times X^b \rightarrow S$$

which gives the modeler complete control over the collision behavior when a component receives external events at the time of its internal transition. Rather than serializing model behavior at collision times through the select function γ at the coupled level, P-DEVS leaves the decision of what serialization to use to the individual component. The default definition of the confluent function simply applies the internal transition function before applying the external transition function to the resulting state.

Example

Our processor atomic model can be defined using P-DEVS as

$$\text{Processor} = \langle X, S, Y, \delta_{\text{int}}, \delta_{\text{ext}}, \delta_{\text{con}}, \lambda, \text{ta} \rangle$$

where

$X = \{(\text{in}, j \in \boldsymbol{J})\}$, where J is a set of jobs.

$S = (\text{phase} = \{\text{"busy"}, \text{"passive"}\}) \times \sigma \in R_0^+ \times j \in \boldsymbol{J}$

$Y = \{(\text{out}, j \in \boldsymbol{J})\}$

$\text{ta}(\text{phase}, \sigma, j) = \sigma$

$\lambda(\text{phase}, \sigma, j) = j$

$\delta_{\text{int}}(\text{phase}, \sigma, j) = (\text{"passive"}, \infty, \varnothing)$

$$\delta_{\text{ext}}(\text{phase}, \sigma, j, e, (\text{in}, j')) = \begin{cases} (\text{"busy"}, j'_p, j') & \text{if phase} = \text{"passive"} \\ (\text{"busy"}, \sigma - e, j) & \text{if phase} = \text{"busy"} \end{cases}$$

$\delta_{\text{con}}(\text{phase}, \sigma, j, (\text{in}, j')) = \delta_{\text{ext}}(\delta_{\text{int}}(\text{phase}, \sigma, j), 0, (\text{in}, j'))$

The coupled formalism for P-DEVS N is specified by a 6-tuple:

$$N = \langle X, Y, D, \{M_i\}, \{I_j\}, \{Z_{j,k}\} \rangle$$

We note the absence of the select function γ. All the remaining elements have the same interpretation as in the classic version, except that here again the bag concept must be introduced in the output-to-input translation functions $\{Z_{j,k}\}$.

The semantics of the formalism is simple: at any event time t, all components in the imminent set Π_t first generate their output, which get assembled into bags at the proper inputs. Then, to each component in Π_t is applied either the confluent or the internal transition function, depending whether it has received inputs or not. The external transition function is applied to those components that have received inputs and are outside the imminent set.

A different definition of coupled models (that we use in the following) is

$$N = \langle X, Y, D, \{M_d \mid d \in D\}, \text{EIC}, \text{EOC}, \text{IC} \rangle$$

where

X is the set of input events
Y is the set of output events
D is the set of component names (atomic or coupled)
M_d is a DEVS model for each $d \in D$
EIC is the set of the external input couplings
EOC is the set of the external output couplings
IC is the set of the internal couplings

Figure 3.5 shows an example of a DEVS-coupled model with three components, M_1, M_2, and M_3, as well as their couplings. These models are interconnected through the corresponding I/O ports presented in Figure 3.5.

The models are connected to the external coupled models through the EIC and EOC connectors. M_1, M_2, and M_3 can be atomic or coupled models.

Following the previous coupled model definition, the model in Figure 3.5 can be formally defined as

$$N = \left\langle X, Y, D, \left\{ M_d | d \in D \right\}, \mathrm{EIC}, \mathrm{EOC}, \mathrm{IC} \right\rangle$$

where

X = the set of input events
Y = the set of output events

$$D = \{M_1, M_2, M_3\}$$

$$M_d = \{M_{M_1}, M_{M_2}, M_{M_3}\}$$

$$\mathrm{EIC} = \{(N, \mathrm{in}) \rightarrow (M_1, \mathrm{in})\}$$

$$\mathrm{EOC} = \{(M_3, \mathrm{out}) \rightarrow (N, \mathrm{out})\}$$

$$\mathrm{IC} = \{(M_1, \mathrm{out}) \rightarrow (M_2, \mathrm{in}), (M_2, \mathrm{out}) \rightarrow (M_3, \mathrm{in})\}$$

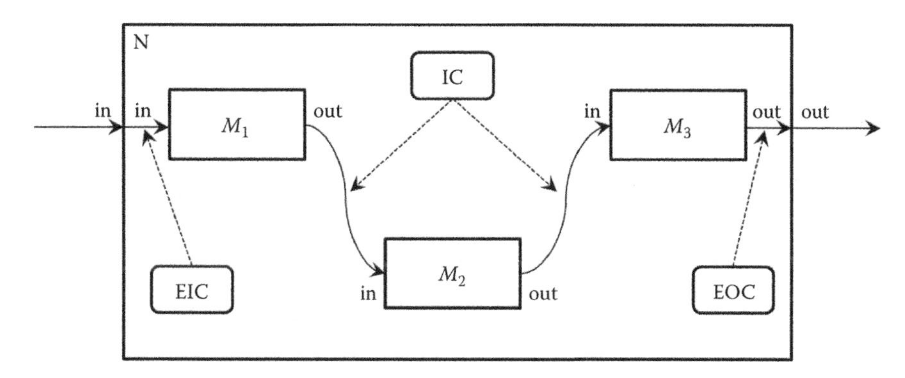

FIGURE 3.5 A DEVS-coupled model.

Exercise 3.1

Write a P-DEVS specification for a system that does nothing. It does not respond with outputs no matter what the input trajectory happens to be.

Exercise 3.2

Write a P-DEVS specification for a system that responds to its input and stores it forever or until the next input comes along.

3.2 WELL-DEFINED SYSTEMS AND LEGITIMACY

The DEVS formalism is closed under coupling: given a coupled model N with atomic-DEVS components, we can construct an equivalent atomic-DEVS M. The construction procedure is compliant with our intuition about concurrent behavior and resembles the implementation of event-scheduling simulators. At its core is the total time order of all events in the system. By induction, *closure under coupling* leads to hierarchical model construction, where the components in a coupled model can themselves be coupled DEVS. This means that the results developed for atomic DEVS in this section also apply to coupled models.

In a modular construct, zero-time propagation could result in infinite instantaneous loops. Such ill-behaved systems can of course still be constructed using transitory states, despite only associating outputs with internal transitions. Thus, transitory states in a DEVS model could result in an ill-behaved system when zero-time advance cycles are present. Legitimacy is the property of DEVS that formalizes these notions.

For an atomic-DEVS M, legitimacy is defined by first introducing an iterative internal transition function $\delta_{\text{int}}^{+} : S \times \mathbb{N} \rightarrow S$, which returns the state reached after n iterations starting at state $s \in S$ when no external event intervenes. It is recursively defined as

$$\delta_{\text{int}}^{+}(s,n) = \delta_{\text{int}}\left(\delta_{\text{int}}^{+}(s,n-1)\right)$$

$$\delta_{\text{int}}^{+}(s,0) = 0$$

Next, we introduce a function $\Gamma : S \times \mathbb{Z} \rightarrow \mathbb{R}_0^{+}$ that accumulates the time the system takes to make these n transitions:

$$\Gamma(s,n) = \Gamma(s,n-1) + \text{ta}\left(\delta_{\text{int}}^{+}(s,n-1)\right) = \sum_{i=0}^{n-1} \text{ta}\left(\delta_{\text{int}}^{+}(s,i)\right)$$

$$\Gamma(s,0) = 0$$

With these definitions, we say that a DEVS is legitimate if for each $s \in S$:

$$\lim_{n \to \infty} \Gamma(s,n) \rightarrow \infty$$

Equivalently, legitimacy can be interpreted as a requirement that there are only a finite number of events in a finite time span. It can be shown that the structure specified by a DEVS is a well-defined system if and only if the DEVS is legitimate.

For atomic-DEVS M with S finite, a necessary and sufficient condition for legitimacy is that every cycle in the state diagram of δ_{int} contains at least one nontransitory state. For the case where S is infinite, however, there exists only a stronger-than-necessary sufficient condition, namely, that there is a positive lower bound to the time advances; that is, $\forall s \in S, \text{ta}(s) > b$.

Actually, instantaneous loops are at the heart of the legitimacy issue. Since outputs are only generated in the absence of external events, the atomic-DEVS formalism is a Moore machine. From an implementation point of view, it is easy to emulate the effect of generating an output upon entering a state by using $\lambda(\delta_{int}(s))$.

3.3 DEVS MODEL EXAMPLE

The Experimental Frame-Processor (EFP) model is usually presented as one of the initial examples to start to practice with DEVS M&S. It is a DEVS-coupled model consisting of three atomic models and one coupled model (see Figure 3.6).

The *Generator* atomic model generates job-messages at fixed time intervals and sends them via the "out" port. The *Transducer* atomic model accepts job-messages from the Generator at its "arrived" port and remembers their arrival time instances. It also accepts job-messages at the "solved" port. When a message

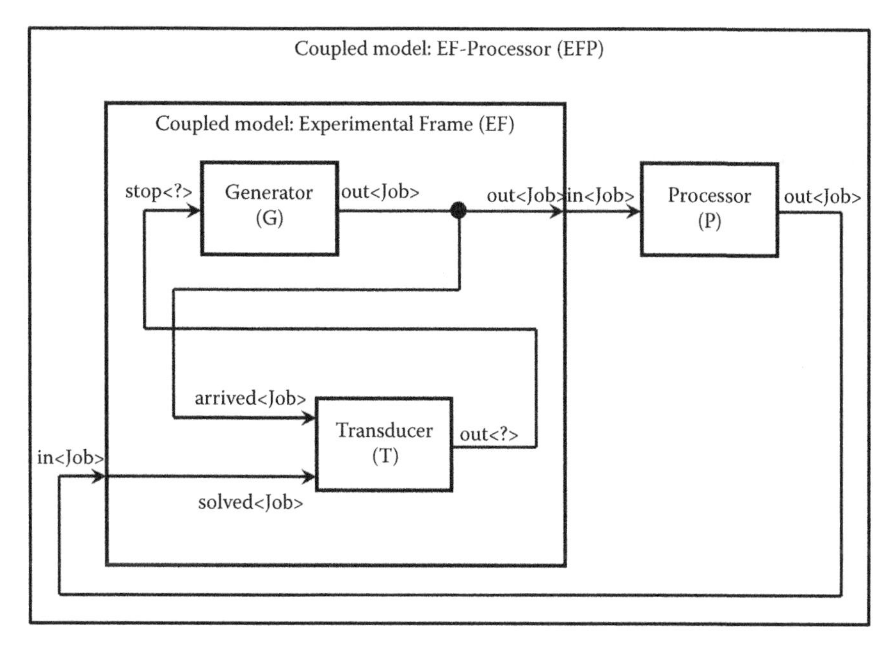

FIGURE 3.6 Experimental Frame-Processor (EFP) model; boxes: models; arrows: couplings; arrow labels: input/output port names.

arrives at the "solved" port, Transducer matches this job with the previous job that had arrived on the "arrived" port earlier and calculates their time difference. Together, these two atomic models form an EF-coupled model. The EF sends Generator's job-messages on the "out" port and forwards the messages received on its "in" port to Transducer's "solved" port. The Transducer observes the response (in this case, the turnaround time) of messages that are injected into an observed system. The observed system in this case is the Processor atomic model. The Processor accepts jobs at its "in" port and sends them via "out" port again after some finite, but nonzero time period. If the processor is busy when a new job arrives, it discards it. Finally, the Transducer stops the generation of jobs by sending any event from its "out" port to the "stop" port at the generator, after a given simulation time interval.

Based on Figure 3.6, we can define the coupled model for this example as

$$N_{\mathrm{EFP}} = \left\langle X, Y, D, \left\{ M_d \middle| d \in D \right\}, \mathrm{EIC}, \mathrm{EOC}, \mathrm{IC} \right\rangle$$

where

$X = \varnothing$

$Y = \varnothing$

$D = \{\mathrm{EF}, \mathrm{P}\}$

$M_d = \{M_{\mathrm{EF}}, M_{\mathrm{P}}\}$

$\mathrm{EIC} = \varnothing$

$\mathrm{EOC} = \varnothing$

$\mathrm{IC} = \{(\mathrm{EF, out}) \rightarrow (\mathrm{P, in}), (\mathrm{P, out}) \rightarrow (\mathrm{EF, in})\}$

The EF-coupled model can be defined as

$$N_{\mathrm{EF}} = \left\langle X, Y, D, \left\{ M_d \middle| d \in D \right\}, \mathrm{EIC}, \mathrm{EOC}, \mathrm{IC} \right\rangle$$

where

$\mathrm{X} = \{(\mathrm{in}, j \in \boldsymbol{J})\}$, where $\boldsymbol{J}$ is a set of jobs

$Y = \{(\mathrm{out}, j \in \boldsymbol{J})\}$, where $\boldsymbol{J}$ is a set of jobs

$D = \{\mathrm{G}, \mathrm{T}\}$

$M_d = \{M_{\mathrm{G}}, M_{\mathrm{T}}\}$

$\mathrm{EIC} = \{(\mathrm{EF, in}) \rightarrow (\mathrm{T, solved})\}$

$$\text{EOC} = \{(\text{G, out}) \rightarrow (\text{EF, out})\}$$

$$\text{IC} = \{(\text{G, out}) \rightarrow (\text{T, arrived}), (\text{T, out}) \rightarrow (\text{G stop})\}$$

We have defined the behavior of the processor model in a previous example. Now, we describe the functionality of both the generator and transducer models. The generator model can be formally described as

$$\text{Generator} = \langle X, S, Y, \delta_{\text{int}}, \delta_{\text{ext}}, \delta_{\text{con}}, \lambda, \text{ta} \rangle$$

$X = \{(\text{stop}, \nu)\}$, where ν is any event.

$S = (\text{phase} = \{\text{"active"}, \text{"passive"}\}) \times \sigma \in R_0^+ \times i = 1, 2, \ldots, N : j_i \in \boldsymbol{J}$

$Y = \{(\text{out}, j_i \in \boldsymbol{J})\}$

$\text{ta}(\text{phase}, \sigma, i) = \sigma$

$\lambda(\text{phase}, \sigma, i) = j_i$

$\delta_{\text{int}}(\text{phase}, \sigma, i) = (\text{"active"}, \sigma, i + 1)$

$\delta_{\text{ext}}(\text{phase}, \sigma, i, e, (\text{in}, \nu)) = (\text{"passive"}, \infty, i)$

$\delta_{\text{con}}(\text{phase}, \sigma, i, (\text{in}, \nu)) = \delta_{\text{ext}}(\delta_{\text{int}}(\text{phase}, \sigma, i), 0, (\text{in}, \nu))$

The transducer model can be formally described as

$$\text{Transducer} = \langle X, S, Y, \delta_{\text{int}}, \delta_{\text{ext}}, \delta_{\text{con}}, \lambda, \text{ta} \rangle$$

$X = \{(\text{arrived}, j \in \boldsymbol{J}), (\text{solved}, j \in \boldsymbol{J})\}$, where $\boldsymbol{J}$ is a set of jobs.

$S = (\text{phase} = \{\text{"active"}, \text{"passive"}\}) \times \sigma \in R_0^+ \times \text{clock} \in R_0^+ \times J_A \in \boldsymbol{J} \times J_S \in \boldsymbol{J}$

where J_A and J_S are sets of arrived and solved jobs, respectively.

$Y = \{(\text{stop}, \nu)\}$, where ν is any event.

$\text{ta}(\text{phase}, \sigma, \text{clock}, J_A, J_S) = \sigma$

$\lambda(\text{phase}, \sigma, \text{clock}, J_A, J_S) = \nu$

$\delta_{\text{int}}(\text{phase}, \sigma, \text{clock}, J_A, J_S) = (\text{"passive"}, \infty, \text{clock} + \sigma, J_A, J_S)$

$$\delta_{\text{ext}}\left(\text{phase},\sigma,\text{clock},J_A,J_S,e,\left(\text{arrived},j^a\right),\left(\text{solved},j^s\right)\right)=\left(\text{active},\ \sigma-e,\text{clock}+e,J_A\right.$$
$$=\left\{j^a,J_A\right\}\ \mathit{if}\ j^a\neq\varnothing,J_s$$
$$\left.=\left\{j^s,J_S\right\}:j_t^s=\text{clock}\ \mathit{if}\ j^s\neq\varnothing\right)$$

where the time in which the job is solved is set to clock with j_t^s = clock.

$$\delta_{\text{con}}\left(\text{phase},\sigma,\text{clock},J_A,J_S,\left(\text{arrived},j^a\right),\left(\text{solved},j^s\right)\right)$$
$$=\delta_{\text{ext}}\left(\delta_{\text{int}}\left(\text{phase},\sigma,\text{clock},J_A,J_S\right),0,\left(\text{arrived},j^a\right),\left(\text{solved},j^s\right)\right)$$

3.4 DEVS REPRESENTATION OF QUANTIZED SYSTEMS

Numerical analysis is concerned with the study of convergence and stability, and a suitable choice of the step-size h. For a difference approximation to be usable for a class of functions $\boldsymbol{f}(y,t)$, it is necessary that any function in this class satisfies three requirements:

1. The existence and uniqueness of a solution. This is satisfied by explicit schemes and can usually be ascertained for implicit schemes.
2. For sufficiently small h, $\boldsymbol{y}_i$ should be close in some sense to $\boldsymbol{y}(t_i)$. Since the scheme we use is an approximation of the original problem, we expect it to introduce an error upon each iteration: assuming infinite precision arithmetic, we call this approximation error the local truncation error τ_i (from the truncation of the Taylor expansion). If we can prove for a given scheme that

$$\lim_{h\to 0}\tau_i=0$$

then the method is said to be consistent (or accurate). However, we are interested in the accumulation of these errors: we write $\boldsymbol{y}_i=\boldsymbol{y}(t_i)+\boldsymbol{e}_i$, where $\boldsymbol{e}_i$ is the global truncation error (equivalent to summing t_i under the assumption that $\boldsymbol{e}_0=0$). If we can show for a given system that

$$\lim_{h\to 0}\boldsymbol{e}_i=0$$

then the method is said to be convergent. For instance, we can find for the Euler–Cauchy method that $|\tau_i|=O(h^2)$ (consistent of order 2) and $|e_i|=O(h)$ (convergent of order 1).
3. The solution should be *effectively computable*. This concerns, on the one hand, the computational efficiency of the implemented method; on the other hand, since we cannot assume infinite precision arithmetic in practice, we want to estimate the growth of round-off errors in the solution. This is related to the stability of the method, which is actually a much more general concept: a method is said to be unstable if, during the solution procedure, the result becomes unbounded. This phenomenon arises when

the difference equations themselves tend to amplify errors to the point that they obliterate the solution itself. A method is said to be stable (or 0-stable) if the corresponding difference equation is stable.

As an alternative to the traditional discretization approach to the solution of ODEs, Zeigler proposed an approach based on partitioning of the state space rather than of the time domain. This *quantization* approach requires a change in viewpoint. The question "at which point in the future is the system going to be in a given state" is now asked instead of "in which state is the system going to be at a given future time." In both questions, a numerical procedure to produce the answer is derived from the ODEs model.

When applied to a continuous signal, both quantization and discretization approaches yield an exact representation of the original signal only in the limit case where the partition size goes to zero (assuming a well-posed problem). Whereas DTSS seem to match discretized signals well, it turns out that DEVS is an appropriate formalism for quantized systems.

A simple quantization of an interval Y over $\mathbb{R}$ can be defined as follows: we first introduce the sets $d_i = \left\{ y \mid y \in Y, \frac{q}{2}(2i-1) \leq y < \frac{q}{2}(2i+1) \right\}, i \in \mathbb{Z}$. Each denotes a quantum (or cell, block) of Y, where q is the quantum size. In general, the sets d_i represent a tessellation of the space Y; that is, $\cup_i d_i = Y$ and $\forall i \neq j, d_i \cap d_j = \varnothing$. This can be extended to higher dimensions, defining tiles of arbitrary shapes or of nonuniform sizes.

In each quantum, a representative item y_i is designated. Usually the middle element of the quantum is chosen, $y_i = q \cdot i$.

For a time base $T = \mathbb{R}$, a function f defined in an (open or closed) interval $f : [t_a, t_b] \rightarrow Y$ is called a segment over Y and T. Using the simple quantization scheme introduced above, we define the quantization of a segment $f^{[t_0,t_n]}$ as the piecewise-continuous segment:

$$f_*^{[t_0,t_n]} = f_1^{[t_0,t_1]} \cdot f_2^{[t_1,t_2]} \cdots f_n^{[t_{n-1},t_n]}$$

where each $f_i^{[t_{i-1},t_i]}$ is a constant segment of value y_j, such that the range of the corresponding segment $f^{[t_{i-1},t_i]}$ lies entirely in quantum d_j (see Figure 3.7).

Quantization suggests a new approach to solving ODEs, where a system updates its output only when a "sufficiently important" change has occurred in its state.

Quantization of systems is a general concept that imposes no constraints on the internal system. We will assume for our present purpose that it represents a continuous-time system. The quantized system is equivalent to the internal system only in the limit case where the quantum tends to 0.

It turns out that every quantized system can be simulated, without error, by a DEVS model. To represent a quantized system by a DEVS model, we allow the model to remember its last (quantized) input. The time advance function ta is then the time to the next change in output produced by the quantized output. The output function λ outputs the representative of the new quanta, whereas the internal transition function δ_{int} updates the state accordingly. If a new input x' is received, δ_{ext} updates the DEVS state as specified in the system.

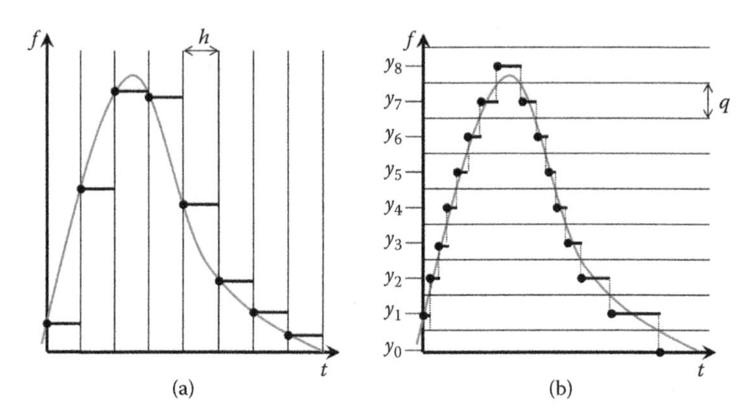

FIGURE 3.7 (a) Discretization and (b) quantization of the same segment.

There is a first consequence of this example: a quantized ODE can be simulated by a DEVS model. We derive some interesting perspectives. Since DEVS is closed under coupling, a quantized ODE can be coupled with purely discrete event components. However, some care must be taken to avoid sending a quantized signal to a quantizer with a different quantum size, which could result in unexpected results. This requirement is called *partition refinement.*

Example

The autonomous, first-order form of an ODE is

$$\dot{x} = f(x)$$
$$x(t_0) = x_0$$

Integrating both sides of the ODE, it can be rewritten as

$$x(t_1) = x(t_0) + \int_{t_0}^{t_1} f(x(t))dt$$

In causal block diagram simulation systems, this system can be implemented as an integrator block with feedback, as Figure 3.8 depicts.

The Euler–Cauchy method can be obtained by discretizing the equation, after approximating the integral function by $(t_1 - t_0) \cdot f(x(t_0))$. Using a DEVS-quantized integrator instead of the Euler–Cauchy approximation, we approximate the integral by $e \cdot r$, where e is the elapsed time and r is the last input. It follows that when the system enters into a new state (either after an external or internal transition), the time of residency in that state, that is, the time advance function, is obtained by solving the equation for the time until the current quantum is departed. As a result, the DEVS-quantized integrator is defined as follows:

- The same quantum size q is used for both the I/O of the integrator.
- The state of the integrator is defined as $s = (x, r, y)$, where x is the state itself, r stores the last input received, and y is the representative item for the current quantum.

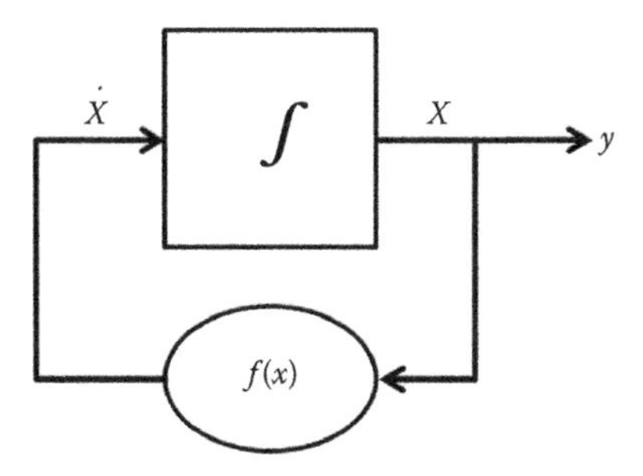

FIGURE 3.8 Causal block diagram of an ordinary differential equation.

- The time advance function returns the time to the next output and internal transitions, that is, the time till the current quantum is departed:

$$\mathrm{ta}(x,r,y)=\begin{cases}+\infty & \text{if } r=0\\ \dfrac{\left|x-\left(y+\dfrac{q}{2}\cdot\mathrm{sign}(r)\right)\right|}{|r|} & \text{otherwise,}\end{cases}$$

where the numerator is the distance between the state x and the relevant quanta interface.

- The internal transition function brings the state component x to the quanta interface "above" or "below," depending on the sign of the slope r:

$$\delta_{\mathrm{int}}(x,r,y)=\left(y+\frac{q}{2}\cdot\mathrm{sign}(r),r,y+q\cdot\mathrm{sign}(r)\right)$$

- The external transition function applies the Euler–Cauchy approximation of the integral function and stores the input received:

$$\delta_{\mathrm{ext}}((x,r,y),e,r')=(x+er,r',y)$$

- The output function outputs the representative of the new quanta:

$$\lambda(x,r,y)=y+q\cdot\mathrm{sign}(r)$$

Figure 3.9 depicts an example of the behavior defined by the DEVS-quantized integrator. Suppose that at a certain instant the input r is greater than 0. In this case, both the state x and the output y increase their values in time. However, if at a time instant t_5 the integrator receives an input less than 0, the new state is computed and both the state and the output decrease in time.

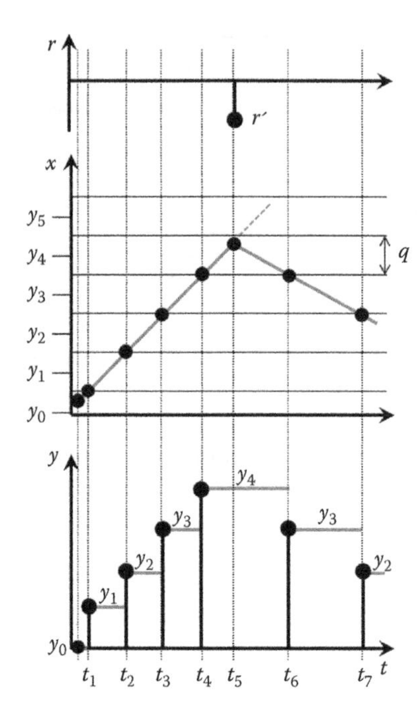

FIGURE 3.9 DEVS quantized integrator.

3.5 DEVS REPRESENTATION OF SYSTEMS

In this section, we provide the DEVS formulation of two other discrete systems: DTSS and DESS.

3.5.1 DTSS Models

Here, we define the DTSS formalism. A DTSS model is a structure:

$$\text{DTSS } M = \langle X_M, Y_M, S_M, \delta_M, \lambda_M, h \rangle$$

where

- X_M is the input set
- Y_M is the output set
- S_M is the state set
- $\delta_M : S_M \times X_M \to S_M$ is the transition function
- λ_M is the output function, and there are two possibilities:
- $\lambda_M : S_M \to Y_M$ Moore-type
- $\lambda_M : S_M \times X_M \to Y_M$ Mealy type
- h is a constant employed for the specification of the time base, where $t = k \cdot h$, with k integer

Regarding the structure of a DTSS-coupled model, there are four types of DTSS components to consider:

1. *Input-free Moore DTSS*: These components drive the simulation forward as the ultimate generators of the inputs in a closed coupled model.
2. *Multiported memoryless FuNction Specified Systems (FNSSs)*: These collect outputs from Moore components and transform them into inputs without using any state information.
3. *Moore DTSS with input*: The outputs are generated for the next cycle based on the current input.
4. *Mealy DTSS*: They include memoryless FNSSs. In a well-defined coupled, they form a directed acyclic graph of computations, taking zero time and propagating from Moore outputs back to Moore inputs.

We examine each of them in turn in the following sections.

3.5.1.1 Input-Free Moore DTSS

These input-free systems are sometimes called *autonomous systems*. There is no input, so we could name these systems as subclass of *source* systems. They have the following substructure of a DTSS:

$$\text{DTSS } M = \langle Y_M, S_M, \delta_M, \lambda_M, h \rangle$$

A step signal with given values for the initial and final step values (y_i, y_f) and the instant of change t_c can be built as an asynchronous signal generator with the output given the initial value and changing only when the step time arrives and as a synchronous signal generator given the value of the output at every time step kh.

A DEVS representation of an input-free Moore system is straightforward since the input-free system acts as an uninterruptable generator with a fixed generation period. Thus, given the previous DTSS system, define:

$$\text{DEVS } M = \langle Y, S, \delta_{\text{int}}, \lambda, \text{ta} \rangle$$

where

$$Y = Y_M$$

$$S = S_M \times \mathbb{R}^+_\infty = (s_M, \sigma)$$

$$\delta_{\text{int}}(s_M, \sigma) = (\delta_M(s_M, \varnothing), h)$$

$$\lambda(s_M, \sigma) = \lambda_M(s_M)$$

$$\text{ta}(s_M, \sigma) = \sigma$$

The reason to retain the *sigma* variable even though it is apparently always set to the predictable step time, *h*, is that initialization of coupled model DEVS simulations requires setting *sigma* initially to zero in order to output the initial value.

3.5.1.2 Multiported FNSS

Another special type of DTSSs is the memoryless or FNSSs. These systems do not have a state transition function, and the output is a function of input only. We can consider them as Mealy-type systems with only one state that can be omitted entirely.

They have the following substructure of a DTSS:

$$\text{DTSS } M = \langle X_M, Y_M, \lambda_M \rangle$$

DEVS simulation of a single-input FNSS is straightforward. When receiving the input value, the model goes into a transitory state and outputs the computed output value. However, if there are multiple input ports, the system must wait until one of the inputs changes its value before going into the output mode. Since inputs may be generated at different points in the processing, we cannot assume that all inputs come in the same message. We formulate an FNSS with multiple input ports and define a DEVS model to simulate it:

$$\text{DTSS } M = \langle X_M, Y_M, \lambda_M \rangle$$

where

$X_M = \{(p,v) \mid p \in \text{InPorts}, v \in V\}$ is the set of input ports and values (we assume that all ports accept the same value set, for simplicity)

$Y_M = \{(p,v) \mid p \in \text{OutPorts}, \ v \in V_{\text{out}}\}$ is the single output port and its values

We define

$$\text{DEVS } M = \langle X, Y, S, \delta_{\text{ext}}, \delta_{\text{int}}, \delta_{\text{con}}, \lambda, \text{ta} \rangle$$

where

$X = X_M$

$Y = Y_M$

$S = \{x_p : p \in \text{InPorts}\} \times \mathbb{R}^+_\infty$, initially: $s_0 = (\{x_p = \varnothing : p \in \text{InPorts}\}, \sigma = \infty)$

$$\delta_{\text{ext}}(\{x_p\}, \infty, e, (p_1, x_1'), (p_2, x_2'), \ldots, (p_n, x_n')) = \textit{if } x_i \neq x_i' \Rightarrow x_i = x_i' \wedge \sigma = 0, i = 1, \ldots, n$$

$$\delta_{\text{int}}(\{x_p\}, 0) = (\{x_p\}, \infty)$$

$$\lambda(\{x_p\}, 0) = \lambda_M(\varnothing, \{x_p\}) \Leftrightarrow \forall p \in \text{InPorts}, x_p \neq \varnothing$$

$$\text{ta}(\{x_p\}, \sigma) = \sigma$$

Since inputs might not all arrive together, we cache each arriving input. When one of the inputs changes its state and each input has been received at least once, then the output is computed. The initial transition function resets the *sigma* state variable to infinity, waiting for the change in one of the inputs.

Exercise 3.3

Define an FNSS specification for an AND gate with two inputs. Translate it to the equivalent DEVS.

3.5.1.3 Moore DTSS with Input

The DEVS representation of a Moore DTSS with input combines features of the memoryless and input-free representations. The DEVS waits for a change in an input port to be heard from before computing the next state of the DTSS and scheduling its next output by setting σ to h. This guarantees an output departure to trigger the next cycle. In the subsequent internal transition, the DEVS continues holding in the new DTSS state waiting for another change in one of the inputs. The initial output must be given (or the initial state or both) to provide outputs even when not all the inputs have been received. These systems have the full structure of the general DTSS described above; that is, DTSS $M = X_M, Y_M, S_M, \delta_M, \lambda_M, h$. A possible DEVS formulation is

$$\text{DEVS } M = \langle X, Y, S, \delta_{\text{ext}}, \delta_{\text{int}}, \delta_{\text{con}}, \lambda, \text{ta} \rangle$$

where

$X = X_M$

$Y = Y_M$

$S = S_M \times \{x_p : p \in \text{InPorts}\} \times \mathbb{R}^+_\infty = (s_M, \{x_p\}, \sigma)$, the initial state s_0 (usually with $\sigma = 0$) is provided.

$$\delta_{\text{ext}}\left(s_M, \{x_p\}, \sigma, e, (p_1, x_1'), (p_2, x_2'), \ldots, (p_n, x_n')\right) = \begin{cases} \textit{if } x_i \neq x_i' \Rightarrow x_i = x_i', i = 1, \ldots, n \\ \quad s_M = \delta(s_M, \{x_p\}) \\ \quad \sigma = \sigma - e \end{cases}$$

$$\delta_{\text{int}}\left(s_M, \{x_p\}, \sigma\right) = \left(s_M, \{x_p\}, h\right)$$

$$\lambda\left(s_M, \{x_p\}, \sigma\right) = \lambda_M\left(s_M\right)$$

$$\text{ta}\left(s_M, \{x_p\}, \sigma\right) = \sigma$$

Exercise 3.4

Write a DTSS specification for a D flip-flop. The D flip-flop propagates the input ($Q = D$) when the clock input changes its state from 0 to 1. Write the equivalent DEVS specification. Set the clock period to 0.2 and draw a timing diagram to show the system behavior.

Exercise 3.5

Describe a DTSS specification for a pulse generator. The output of this system is a square wave with a given amplitude and period. Try to write directly an equivalent DEVS model.

Exercise 3.6

Write a DTSS specification for a ramp (slope = 1, h = 0.1). Translate this specification to a DEVS model. Finally, approximate the ramp using quantization and DEVS.

3.5.1.4 Mealy DTSS

A mealy DTSS is represented as a DEVS in a manner similar to a memoryless function with the difference that the state is updated when one of the inputs change. However, the output cannot be prescheduled for the next cycle. So the Mealy DEVS passivates (remains in a state with $\sigma = \infty$) after a state update just as if it were a memoryless element. Thus, given a Mealy DTSS $M = X_M, Y_M, S_M, \delta_M, \lambda_M, h$, one possible DEVS representation is

$$\text{DEVS } M = \langle X, Y, S, \delta_{\text{ext}}, \delta_{\text{int}}, \delta_{\text{con}}, \lambda, \text{ta} \rangle$$

where

$$X = X_M$$

$$Y = Y_M$$

$$S = S_M \times \{x_p : p \in \text{InPorts}\} \times \mathbb{R}^+_\infty, \text{ initially: } s_0 = \left(s_{M,0}, \{x_p = \varnothing : p \in \text{InPorts}\}, \sigma = \infty\right)$$

$$\delta_{\text{ext}}\left(s_M, \{x_p\}, \infty, e, (p_1, x_1'), (p_2, x_2'), \ldots, (p_n, x_n')\right) = \begin{cases} \textit{if } x_i \neq x_i' \Rightarrow x_i = x_i', i = 1, \ldots, n \\ s_M = \delta(s_M, \{x_p\}) \\ \sigma = 0 \end{cases}$$

$$\delta_{\text{int}}\left(s_M, \{x_p\}, \sigma\right) = \left(s_M, \{x_p\}, \infty\right)$$

$$\lambda\left(s_M, \{x_p\}, 0\right) = \lambda_M\left(s_M, \{x_p\}\right)$$

$$\text{ta}\left(\{x_p\}, \sigma\right) = \sigma$$

3.5.2 DESS Models

A DESS is an M&S formalism described using mathematical set theory. A DESS specification is a structure:

$$\text{DESS } M = \langle X_M, Y_M, S_M, \delta_M, \lambda_M \rangle$$

where

X_M is the set of inputs
Y_M is the set of outputs
S_M is the set of states
$\delta_M : S_M \times X_M \rightarrow S_M$ is the transition function

and the output function is

$\lambda_M : S_M \rightarrow Y_M$ (Moore type)

$\lambda_M : S_M \times X_M \rightarrow Y_M$ (Mealy type)

The transition function can be defined for every state s and bounded continuous input segment as the solution of the state differential equation obtained integrating from the initial time to the final time, with the given initial state and given input segment from initial time to final time, $[t_i, t_f]$:

$\frac{ds}{dt} = \delta_M\left(s, x^{[t_i,t_f]}\right)$, with known $s(t_i), x^{[t_i,t_f]}$, and being $s(t_f)$ the solution to the differential equation.

There are two approaches to DEVS representation of DESS. The first is to employ standard numerical methods that result in a DTSS simulation of the DESS. To this end, the DESS transition function is approximated using (for example) the Euler–Cauchy method and then simulated as a Moore DTSS with inputs. This DTSS system can be easily formulated as a DEVS model. The second approach is quite similar, but in this case the differential equation is approximated using quantization. As shown in Section 3.4, a quantized system can be modeled by a DEVS system.

Exercise 3.7

Write a DESS specification for the simple harmonic oscillator: $F = -kx$. Write the same system using DEVS and discrete time (DESS $\rightarrow$ DTSS $\rightarrow$ DEVS) and quantization and DEVS (DESS $\rightarrow$ DEVS).

REFERENCES

Vangheluwe, H. L. M. (2000). DEVS as a common denominator for multi-formalism hybrid systems modelling. Proceedings of the 2000 IEEE International Symposium on Computer-Aided Control System Design, 129–134, Anchorage, Alaska, USA.

Zeigler, B. P., Praehofer, H., & Kim, T. G., (2000). *Theory of modeling and simulation*. New York, NY: Academic Press.

4 DEVS Software
Model and Simulator

4.1 INTRODUCTION

In this chapter, we present how to develop a Discrete Event Systems Specification Modeling and Simulation (DEVS M&S) framework. To this end, we have selected NetBeans 7.0.1 as the Integrated Development Environment (IDE). It is free and open-source. Besides, it includes all the tools needed to create professional desktop, enterprise, web, and mobile applications with the Java platform, as well as with languages like C/C++, PHP, JavaScript, and Groovy.

The Java SE Development Kit 6 (JDK6) Update 26 or later is required to install NetBeans IDE. The 7.0.1 version of the IDE cannot be installed using JDK 5.0.* On the NetBeans IDE Download page,† one of several installers can be obtained, each of which contains the base IDE and additional tools. We recommend one of the following two:

1. Java SE: Supports all standard Java SE development features and provides support for the NetBeans Rich Client Platform (RCP) development environment.
2. Java EE: Provides tools for developing Java SE and Java EE applications as well as support for the NetBeans RCP development platform. This download option also includes GlassFish Server Open Source Edition 3.1 and Apache Tomcat 7.0.14 software.

Once NetBeans is installed, we start with the development of the DEVS M&S framework. To this end, open NetBeans and select File→NewProject.... In the new window, select Java Application and click Next (Figure 4.1).

In the next window, introduce MicroSim as the name of the new project, as well as the target directory for the project (Figure 4.2). Uncheck the Create Main Class option, and then press Finish.

* The latest update of JDK 6 can be downloaded at www.oracle.com/technetwork/java/javase/downloads.
† http://netbeans.org/downloads.

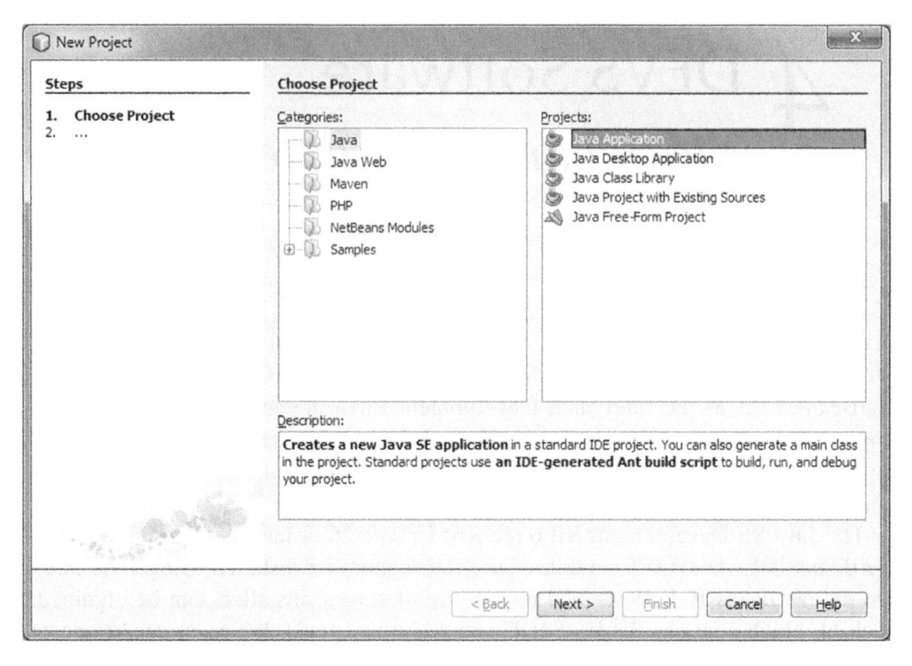

FIGURE 4.1 New Java project in NetBeans.

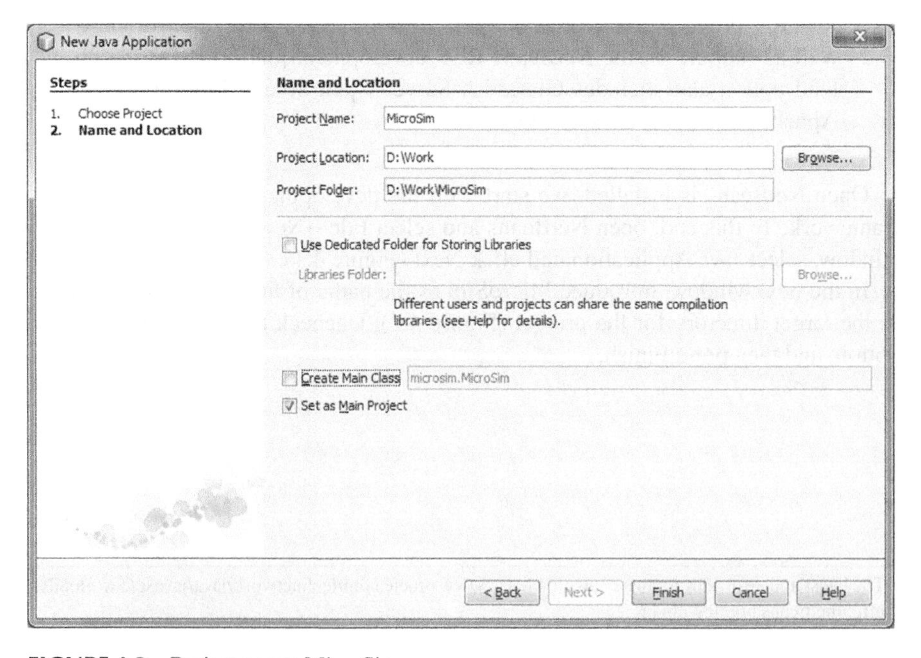

FIGURE 4.2 Project name: MicroSim.

4.2 DEVS MODELING METAMODEL

We first describe the DEVS metamodel by means of a Unified Modeling Language (UML) class diagram. As seen in previous chapters, DEVS is formed basically by coupled and atomic models, connections, ports, and states. We can start with the atomic model definition:

$$M = < X,\ Y,\ S,\ \delta_{\text{int}},\ \delta_{\text{ext}},\ \delta_{\text{con}},\ \lambda, \text{ta} >$$

Both inputs and outputs are defined in terms of pairs (port, value). Thus, we need a *Port* class to store those values. Besides, the atomic model itself needs its own *Atomic Port* class. We also need five functions: three transition functions $\left(\delta_{\text{int}}, \delta_{\text{ext}}, \delta_{\text{con}}\right)$, the output function ($\lambda$), and the time advance function (ta). Regarding the state, we may consider that the set of attributes in the *Atomic* class will represent its state. Thus, we do not need a class representation for the state.

With respect to the coupled model definition,

$$N = < X,\ Y,\ D,\ \{M_d \mid d \in D\},\ \text{EIC},\ \text{EOC},\ \text{IC} >$$

As in the atomic model, both X and Y are implicitly defined by the port class. The coupled model also needs its class *Coupled*. We also need a *Coupling* class to define a single connection. Finally, the coupled model is a container; that is, the coupled model contains connections, as well as atomic and coupled models. We will use a superclass *Component* to define a component in the model that can be atomic or coupled. Thus, M_d will be a set of components. To store both connections and components, we use the doubly linked list data structure, which is implemented in the *LinkedList<E>* Java class. Figure 4.3 depicts the DEVS modeling metamodel, containing the five classes mentioned above. Next, we describe all the classes and show their implementation.

4.2.1 PORT CLASS

First of all, we divide MicroSim into three parts or packages: modeling, simulation, and other packages related to non-DEVS classes. Thus, we start creating a package called modeling. In the Projects tab, click the right mouse button on the Source Packages subfolder, and select New→Java Package… (Figure 4.4).

Write *microsim.kernel.modeling* as the new name for the package and press Finish (Figure 4.5).

Next, we start creating the *Port* class (Figure 4.6). Press the right mouse button in the *microsim.kernel.modeling* package and select New→Java Class….

In the next window (Figure 4.7), we name the class *Port*, and press Finish. Now we are ready to write the code of the *Port* class, as Figure 4.8 shows.

A *Port* object will belong to a connection (see Figure 4.3). Furthermore, when we create an atomic or coupled model, it will have ports as part of their attributes. In our case, a DEVS port contains a name and a bag of values. There are multiple choices

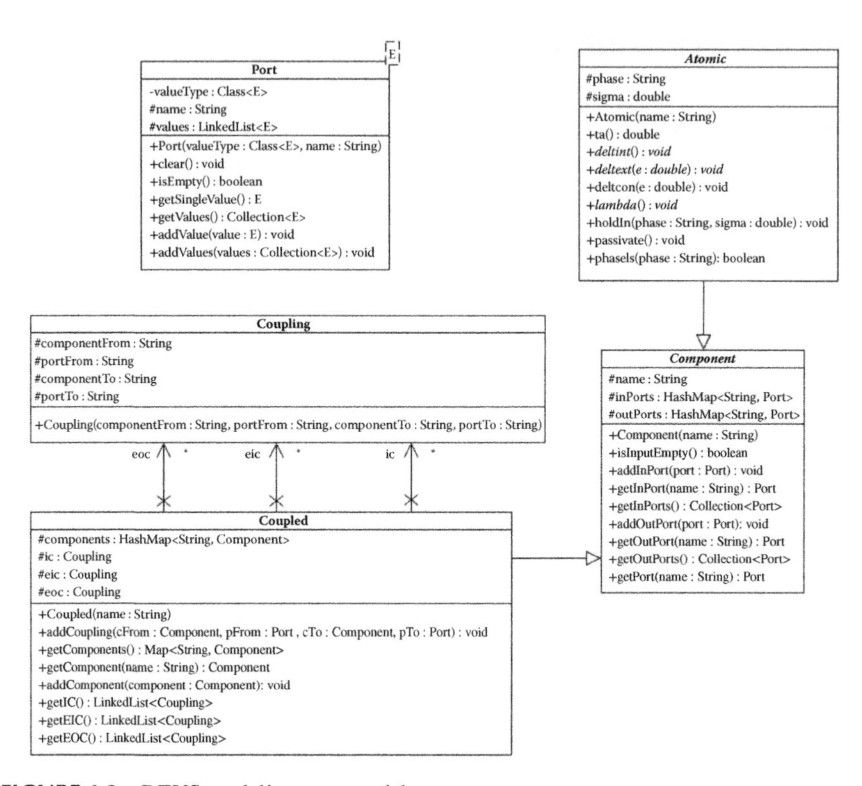

FIGURE 4.3 DEVS modeling metamodel.

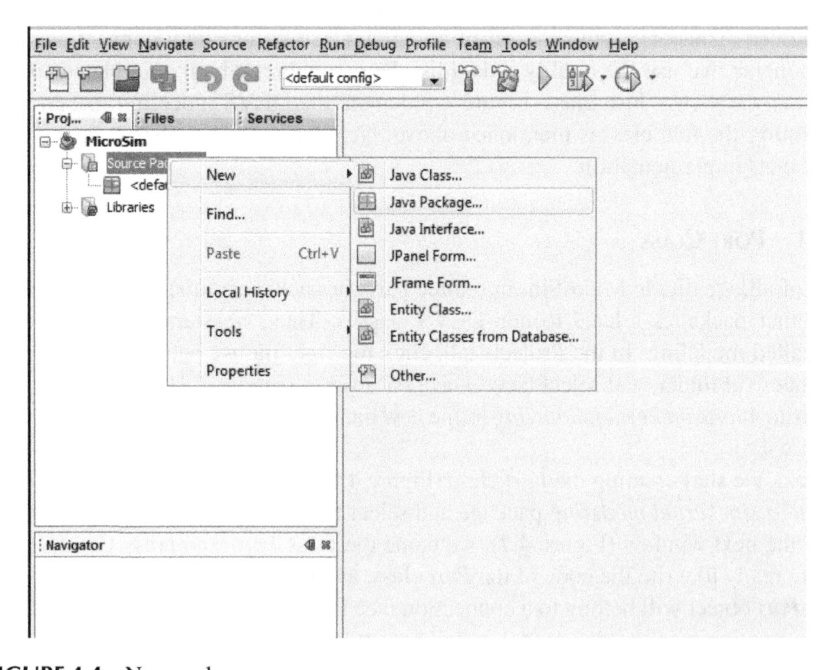

FIGURE 4.4 New package.

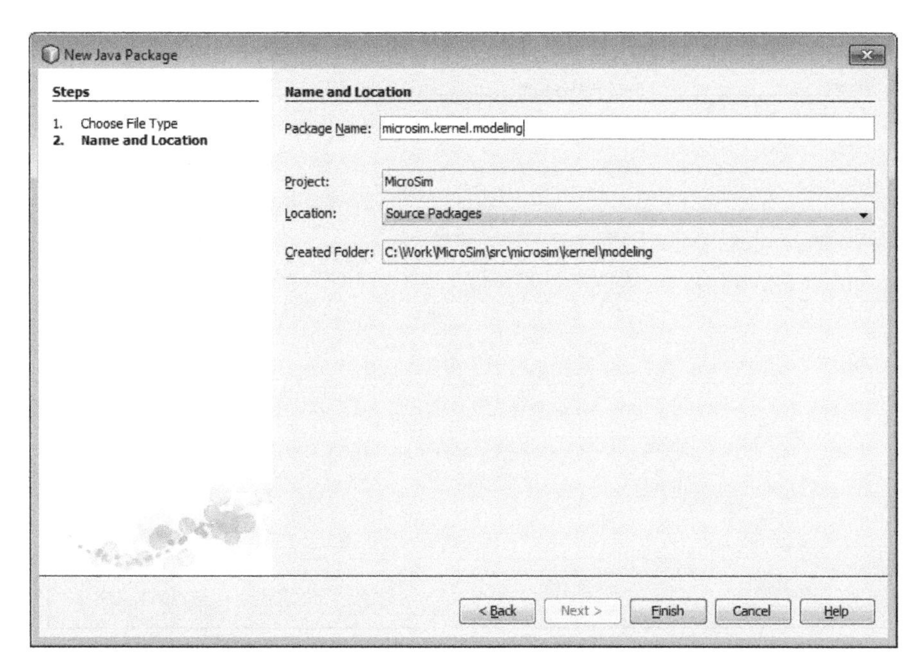

FIGURE 4.5 Microsim.kernel.modeling package.

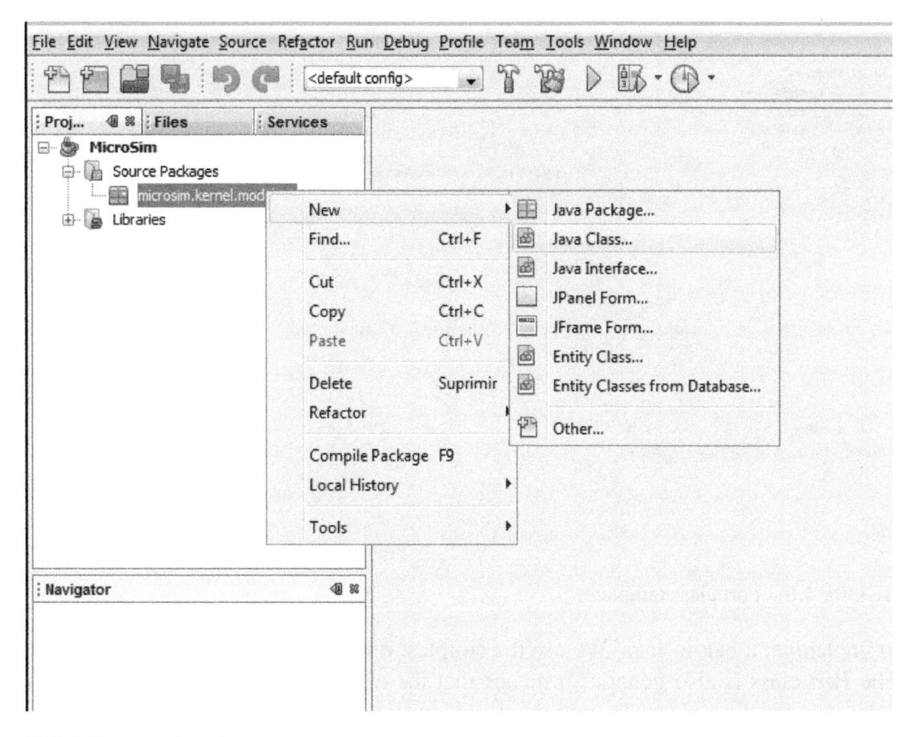

FIGURE 4.6 Creating a new Java class.

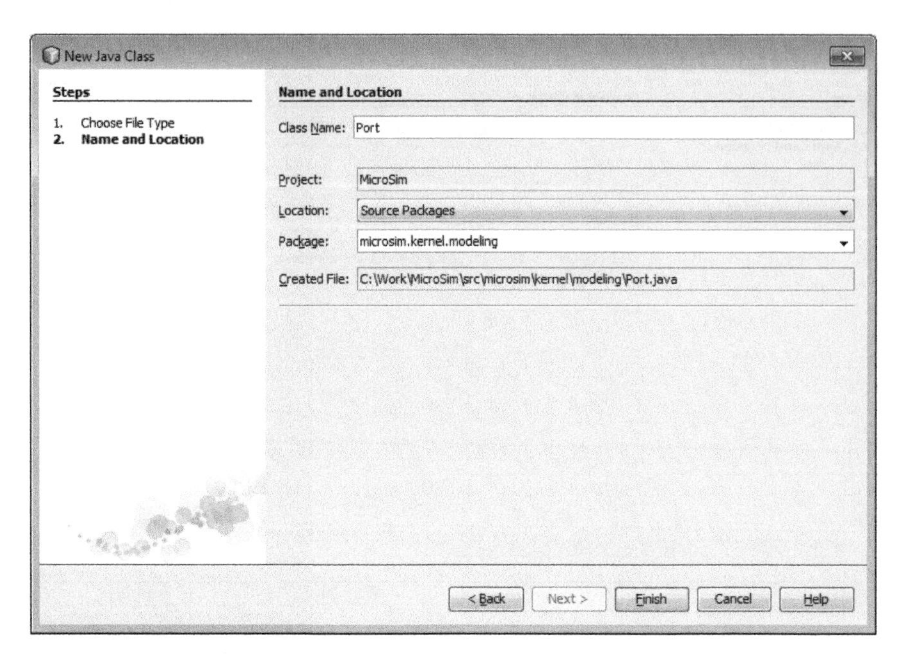

FIGURE 4.7 Class name: Port.

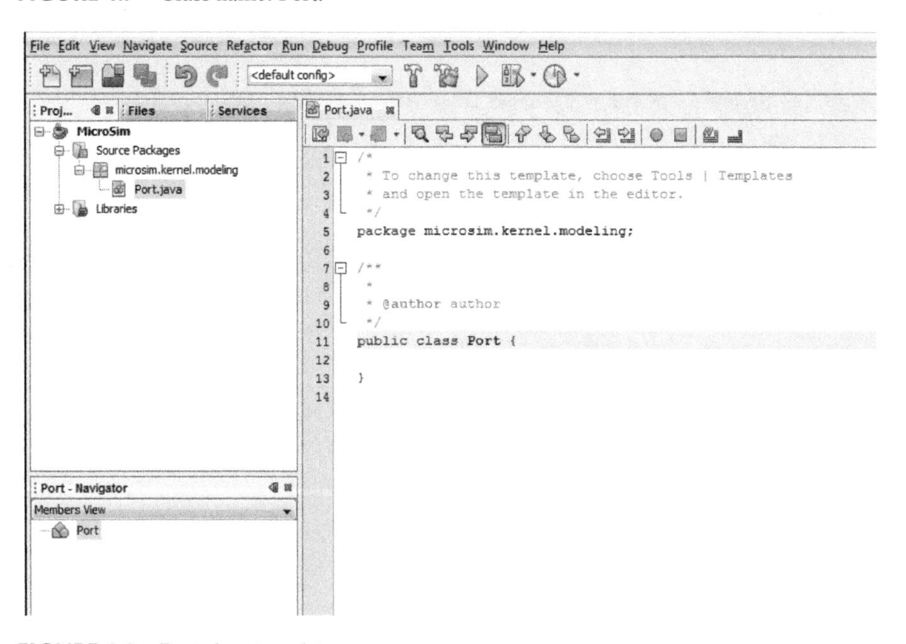

FIGURE 4.8 Port class template.

to implement a bag in Java. We use the simplest one: a doubly linked list of values. The *Port* class is also generic; it means that the elements in the bag of values can belong to any Java class. Around these two attributes, there is a set of methods that operates over them. The source code for the *Port* class is shown in Listing 4.1.

LISTING 4.1 PORT SOURCE CODE

```
public class Port<E> {

  private Class<E> valueType;
  protected String name;
  protected LinkedList<E> values = new LinkedList<E>();

  public Port(Class<E> valueType, String name) {
    this.valueType = valueType;
    this.name = name;
  }

  public void clear() {
    values.clear();
  }

  public boolean isEmpty() {
    return values.isEmpty();
  }

  public E getSingleValue() {
    return values.element();
  }

  public String getName() {
  return name;
  }

  public void setName(String name) {
    this.name = name;
  }

  public Collection<E> getValues() {
    return values;
  }

  public void addValue(E value) {
    values.add(value);
  }

  public void addValues(Collection<E> values) {
    this.values.addAll(values);
  }

  public Class<E> getValueType() {
    return valueType;
  }
}
```

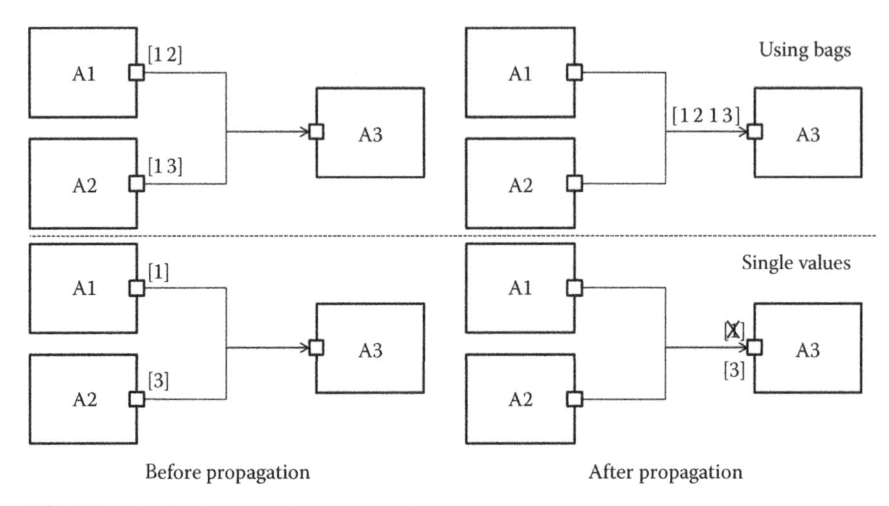

FIGURE 4.9 Difference between ports that use bags and ports that use single values.

Most of the functions in the *Port* class are quite intuitive and do not need an explanation, with the exception of *getValueType*. Although this class' return type is generic, we are using an argument to define the type of events managed by this port. We must use this redundant constructor by now. This *valueType* parameter will be very useful in distributed simulation using web services (DEVS/SOA). *Collection<T>* is a Java interface compatible with any Java data structure. It means, for example, that the method *addValues* can receive any data structure when it is called. The function *getSingleValue* will be very useful because sometimes an atomic model sends just a single value through its port, and thus the list of values will contain only one element.

The parallel DEVS (P-DEVS) formalism uses a bag as the message structure: a bag X^b of elements in X is similar to a set except that multiple occurrences of elements are allowed $\left(\text{e.g., } X^b = \{1, 2, 1\}\right)$. Note that this is the only difference between a set and a bag. As with sets, bags are unordered. Thus, using either sets or bags (i.e., classic DEVS or P-DEVS) to collect inputs sent to a component, we recognize that inputs can arrive from multiple sources and that more than one input with the same identity may arrive simultaneously. This fact is shown in Figure 4.9. When we use bags, multiple values can be propagated between atomic or coupled modes. On the contrary, when we use ports with single values, only one value will survive in the propagation.

4.2.2 Component Class

The *Component* class contains the name of the component, as well as input and output ports. In addition to usual functions like getters, setters, and adders, there is also a member function called *isInputEmpty*, which will be very useful in the simulation package. Note that both atomic and coupled models are also components. Thus, both *Atomic* and *Coupled* classes extend component. The source code for the component class is shown in Listing 4.2.

LISTING 4.2 COMPONENT SOURCE CODE

```
public abstract class Component {

  protected String name;
  protected HashMap<String, Port> inPorts =
     new HashMap<String, Port>();
  protected HashMap<String, Port> outPorts =
     new HashMap<String, Port>();

  public Component(String name) {
    this.name = name;
  }

  public String getName() {
    return name;
  }

  public void setName(String name) {
    this.name = name;
  }

  public boolean isInputEmpty() {
    Collection<Port> ports = inPorts.values();
    for (Port port : ports) {
      if (!port.isEmpty()) {
        return false;
      }
    }
    return true;
  }

  public void addInPort(Port port) {
    inPorts.put(port.getName(), port);
  }

  public Port getInPort(String name) {
    return inPorts.get(name);
  }

  public Collection<Port> getInPorts() {
    return inPorts.values();
  }

  public void addOutPort(Port port) {
    outPorts.put(port.getName(), port);
  }

  public Port getOutPort(String name) {
    return outPorts.get(name);
  }
```

```
  public Collection<Port> getOutPorts() {
    return outPorts.values();
  }
}
```

4.2.3 Atomic Class

Before defining the *Atomic* class, we first show an auxiliary class that will be useful to manage constants. It is not contained in the metamodel of Figure 4.3 for simplicity, as this class is directly placed in the *microsim.kernel* package (Listing 4.3).

LISTING 4.3 CONSTANTS.JAVA

```
public class Constants {
   public static final double INFINITY =
      Double.POSITIVE_INFINITY;
   public static final String PHASE_PASSIVE = "passive";
}
```

We have defined two constants. The first one is *INFINITY*, which is frequently used. The second, *PHASE_PASSIVE*, is a common DEVS phase name: "passive," which is the phase when an atomic model enters into a passive state (with sigma being infinity).

Thus, the atomic class (Listing 4.4) is an abstract class that contains the basic P-DEVS functionality. It contains no ports and the minimal state possible, a phase and a sigma. As the atomic class also extends components, it inherits attributes *name* and *isInputEmpty*.

LISTING 4.4 ATOMIC.JAVA

```
public abstract class Atomic extends Component {

    protected String phase;

    public String getPhase() {
        return phase;
    }

    public void setPhase(String phase) {
        this.phase = phase;
    }
    protected double sigma;

    public double getSigma() {
```

```
        return sigma;
    }

    public void setSigma(double sigma) {
        this.sigma = sigma;
    }

    public Atomic(String name) {
        super(name);
        // Passivate the model
        phase = Constants.PHASE_PASSIVE;
        sigma = Constants.INFINITY;
    }

    public double ta() {
        return sigma;
    }

    abstract public void deltint();
    abstract public void deltext(double e);

    public void deltcon(double e) {
        deltint();
        deltext(0);
    }

    abstract public void lambda();

    public void holdIn(String phase, double sigma) {
        this.phase = phase;
        this.sigma = sigma;
    }

    public void passivate() {
        this.phase = Constants.PHASE_PASSIVE;
        this.sigma = Constants.INFINITY;
    }

    public boolean phaseIs(String phase) {
        return this.phase.equals(phase);
    }
}
```

Around these attributes, we can also find the basic DEVS functions: the transition functions $(\delta_{int}, \delta_{ext}, \delta_{con})$, the output function ($\lambda$), and the time advance function (ta). Both ta and δ_{con} have their default implementations. The time advance function just returns sigma, whereas the δ_{con} implements the default behavior:

$$S = \delta_{con}\left(S, X^b\right) = \delta_{ext}\left(\delta_{int}(S), 0, X^b\right)$$

The first question is: where is the state *S* in this implementation? We have mentioned above that the state is defined through the attributes of an atomic model. Thus, all the methods in the atomic model will have access to these attributes and thus to the state. As a result, we do not need to pass the state to the transition functions since they have access to it. The second question is: where is the input X^b in the implementation? The same answer applies. When we define an atomic model (that will extend the *Atomic* class), it will have a set of attributes such as ports, and because of that, the input will be present in the atomic model; more precisely, the input will be contained in the set of input ports. We will see this point in detail below.

There are also other interesting methods in the *Atomic* class: *holdIn* keeps the atomic model in a given phase and sigma, *passivate* returns the model to the idle state, and *phaseIs* asks whether the model is in a given phase. With port, component, and atomic classes defined, we are ready to see some examples.

4.2.3.1 Step

The Step atomic model provides a step between two definable levels at a specified time. If the simulation time is less than the step time parameter value (*stepTime*), the block's output is the initial value parameter value (*initialValue*). For simulation time greater than or equal to the step time, the output is the final value parameter value (*finalValue*). Step time specifies the time when the output jumps from the initial value parameter to the final value parameter. Initial value specifies the block output until the simulation time reaches the step time parameter. Final value defines the block output when the simulation time reaches and exceeds the step time parameter.

Thus, we have the following structure and behavior for the step model:

Output ports:
  portOut
Primary states:
  Phases: "initialValue," "finalValue"
  Sigma: Any non-negative number
Secondary states:
  initialValue: Any real number
  finalValue: Any real number
Parameters:
  stepTime: Any positive number
Initialization:
  Initial state is ("initialValue," 0.0, initialValue, finalValue)
  stepTime is also initialized
Output function:
  If (phase = = "initialValue") send initialValue to output port portOut
  Else if (phase = = "finalValue") send finalValue to the same port.
Internal transition function:
  If (phase = = "initialValue") hold-in "finalValue" for stepTime
  Else if (phase = = "finalValue") hold-in "passive" for infinity.

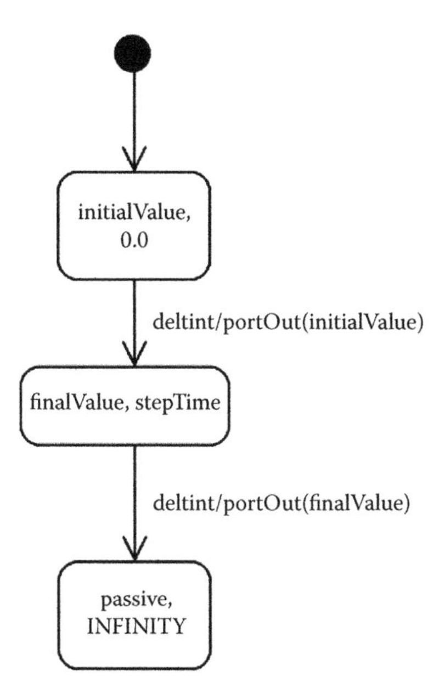

FIGURE 4.10 Step model machine specification.

The behavior of the step atomic model is specified as a UML state machine diagram in Figure 4.10.

Exercise 4.1

Write a P-DEVS formal specification for the step atomic model.

Given the structure and behavior of the step atomic model, the implementation is straightforward (Listing 4.5).

LISTING 4.5 STEP.JAVA

```
public class Step extends Atomic {

  public Port<Double> portOut = new
       Port<Double>(Double.class, "out");
  protected double initialValue;
  protected double stepTime;
  protected double finalValue;

  public Step(String name, double initialValue,
       double stepTime, double finalValue) {
    super(name);
    super.addOutPort(portOut);
    this.initialValue = initialValue;
```

```
    this.stepTime = stepTime;
    this.finalValue = finalValue;
    super.holdIn("initialValue", 0.0);
  }

  @Override
  public void deltint() {
    if (super.phaseIs("initialValue")) {
      super.holdIn("finalValue", stepTime);
    } else if (super.phaseIs("finalValue")) {
      super.passivate();
    }
  }

  @Override
  public void deltext(double e) {
  }
  @Override
  public void lambda() {
    if (super.phaseIs("initialValue")) {
      portOut.addValue(initialValue);
    } else if (super.phaseIs("finalValue")) {
      portOut.addValue(finalValue);
    }
  }
}
```

We can see that the external transition function is empty. This is because this atomic model will never receive an external event since it does not have input ports.

4.2.3.2 Pulse Generator

The Pulse generator block generates square-wave pulses at regular intervals. The block's waveform parameters, *amplitude*, pulse width (state variable *pulseWidth*), *period*, and phase delay (state variable *phaseDelay*), determine the shape of the output waveform. Figure 4.11 shows how each parameter affects the waveform.

Amplitude is the pulse amplitude. *Period* is the pulse period. Pulse width is the duty cycle. Phase delay is the delay before the pulse is generated.

Figure 4.12 shows the behavior of the pulse model as a state machine. The model stays in phase "delay" for 0.0 seconds. Immediately, the output is 0.0 and an internal transition happens. Then, the primary state of the model goes into "high" for *phaseDelay* seconds. After that, the output is amplitude and the state changes again into "low" for *pulseWidth* seconds. Then, the model enters into a loop where the pulse takes its periodic form.

When we follow the state machine diagram in Figure 4.12, the implementation is simple (Listing 4.6).

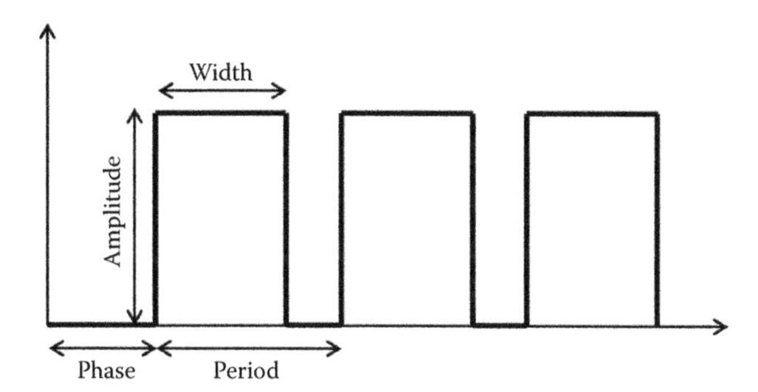

FIGURE 4.11 Pulse generator waveform.

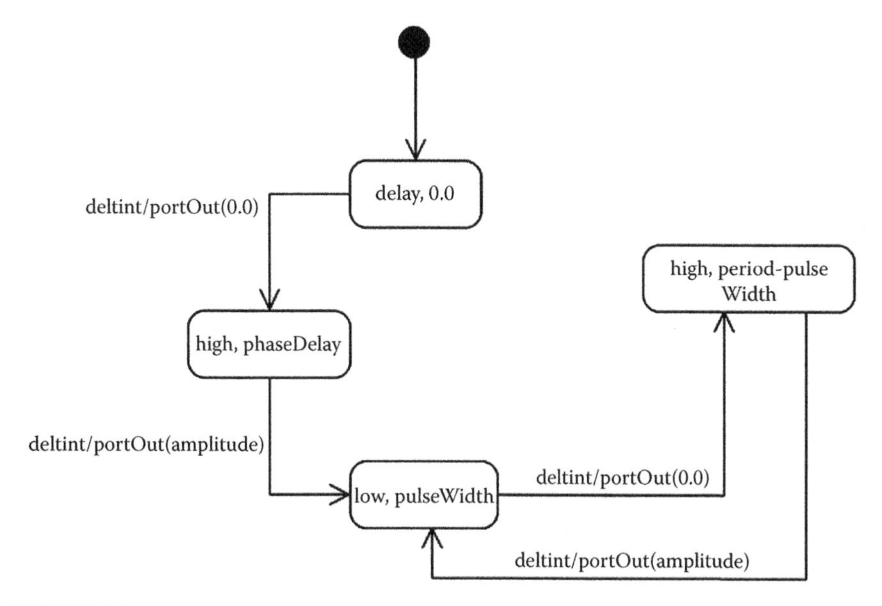

FIGURE 4.12 Pulse-state machine diagram.

LISTING 4.6 PULSEGENERATOR.JAVA

```
public class PulseGenerator extends Atomic {

  public Port<Double> portOut =
       new Port<Double>(Double.class, "out");
  protected double amplitude;
  protected double pulseWidth;
  protected double period;
  protected double phaseDelay;
```

```
  public PulseGenerator(String name, double amplitude,
       double pulseWidth, double period, double phaseDelay) {
    super(name);
    super.addOutPort(portOut);
    this.amplitude = amplitude;
    this.pulseWidth = pulseWidth;
    this.period = period;
    this.phaseDelay = phaseDelay;
    super.holdIn("delay", 0);
  }

  @Override
  public void deltint() {
    if (super.phaseIs("delay")) {
      super.holdIn("high", phaseDelay);
    } else if (super.phaseIs("high")) {
      super.holdIn("low", pulseWidth);
    } else if (super.phaseIs("low")) {
      super.holdIn("high", period - pulseWidth);
    }
  }

  @Override
  public void deltext(double e) {
  }

  @Override
  public void lambda() {
    if  (super.phaseIs("delay")) {
      portOut.addValue(0.0);
    } else if (super.phaseIs("high")) {
      portOut.addValue(amplitude);
    } else if (super.phaseIs("low")) {
      portOut.addValue(0.0);
    }
  }
}
```

4.2.3.3 Ramp

The Ramp atomic model generates a signal that starts at a specified time and value, and changes by a specified rate. The block's *slope*, start time (*startTime* attribute), and initial output (*initialOutput* attribute) parameters determine the characteristics of the output signal. *Slope* specifies the rate of change of the generated signal. *Start time* defines the time at which the block begins generating the signal. *Initial output* specifies the initial value of the output signal. Although we could use a quantified ramp, for the sack of clarity in this example, we use a sample time parameter (*sampleTime*) to

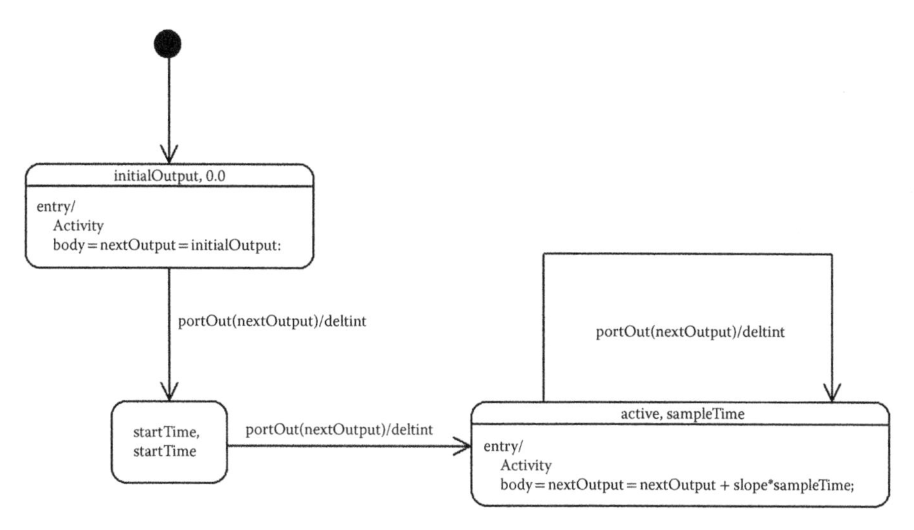

FIGURE 4.13 Ramp-state machine diagram.

generate the output after the start time parameter. This is then a Discrete Time System Specification (DTSS) atomic model, which of course can be defined using P-DEVS.

Figure 4.13 depicts the ramp-state machine diagram. The model output two values before entering into a periodic state. The first is the *initialOutput*, when the model is initialized. The second is the same value, just before the ramp starts. After that, a point in the ramp is generated every *sampleTime* seconds. The implementation follows the state machine diagram (Listing 4.7).

LISTING 4.7 RAMP.JAVA

```
public class Ramp extends Atomic {

  public Port<Double> portOut = new Port<Double>
      (Double.class, "out");
  protected double startTime;
  protected double slope;
  protected double sampleTime;
  protected double nextOutput;

  public Ramp(String name, double initialOutput,
      double startTime, double slope, double sampleTime) {
    super(name);
    super.addOutPort(portOut);
    this.nextOutput = initialOutput;
    this.startTime = startTime;
    this.slope = slope;
    this.sampleTime = sampleTime;
    super.holdIn("initialOutput", 0.0);
  }
```

```
    @Override
    public void deltint() {
      if (super.phaseIs("initialOutput")) {
        super.holdIn("startTime", startTime);
      } else {
        nextOutput += slope * sampleTime;
        super.holdIn("active", sampleTime);
      }

    @Override
    public void deltext(double e) {
    }

    @Override
    public void lambda() {
      portOut.addValue(nextOutput);
    }
}
```

4.2.4 Coupling Class

As the DEVS metamodel shown in Figure 4.3, the *Coupling* class is only responsible for representing one connection in the DEVS model. This class contains, as attributes, the source port of the connection, the destination port, and their respective components. The implementation is quite simple (Listing 4.8).

LISTING 4.8 COUPLING.JAVA

```
public class Coupling {

  protected String componentFrom;
  protected String portFrom;
  protected String componentTo;
  protected String portTo;

  public Coupling(String componentFrom, String portFrom,
      String componentTo, String portTo) {
    this.componentFrom = componentFrom;
    this.portFrom = portFrom;
    this.componentTo = componentTo;
    this.portTo = portTo;
  }
  public String getComponentFrom() {
    return componentFrom;
  }
```

```
  public String getComponentTo() {
    return componentTo;
  }

  public String getPortFrom() {
    return portFrom;
  }

  public String getPortTo() {
    return portTo;
  }
}
```

4.2.5 Coupled Model

A coupled model is basically a *Container* class. It contains other components that can be atomic or coupled, and three additional sets (the internal, external input, and external output connections). Correspondingly, the attributes responsible for storing these elements are *components, ic, eic*, and *eoc*. Thus, according to Figure 4.3, a basic implementation follows in Listing 4.9.

LISTING 4.9 COUPLED.JAVA

```
public class Coupled extends Component {

  protected HashMap<String, Component> components =
       new HashMap<String, Component>();

  // Internal connections
  protected LinkedList<Coupling> ic =
       new LinkedList<Coupling>();
  // External input connections
  protected LinkedList<Coupling> eic =
       new LinkedList<Coupling>();
  // External output connections
  protected LinkedList<Coupling> eoc =
       new LinkedList<Coupling>();

  public Coupled(String name) {
    super(name);
  }

  public void addCoupling(Component cFrom, Port pFrom,
       Component cTo, Port pTo) {
    Coupling coupling = new Coupling(
```

```
                cFrom.getName(), pFrom.getName(),
                cTo.getName(), pTo.getName());
    // Add to connections
    if (cFrom == this) {
      eic.add(coupling);
    } else if (cTo == this) {
      eoc.add(coupling);
    } else {
      ic.add(coupling);
    }
  }

  public Map<String, Component> getComponents() {
    return components;
  }

  public Component getComponent(String name) {
    return components.get(name);
  }

  public void addComponent(Component component) {
    components.put(component.getName(), component);
  }

  public LinkedList<Coupling> getIC() {
    return ic;
  }

  public LinkedList<Coupling> getEIC() {
    return eic;
  }

  public LinkedList<Coupling> getEOC() {
    return eoc;
  }
}
```

As can be seen in the source code (Listing 4.9), the *addCoupling* method recognizes if the connection is ic, eic, or eoc by just examining *this*, that is, the coupled model instance.

4.3 DEVS SIMULATION METAMODEL

DEVS treats a model and its simulator as two distinct elements. The DEVS simulation protocol describes how a DEVS model should be simulated: whether in a standalone fashion or in a coupled model. Such a protocol is implemented by a processor, which can be a simulator or a coordinator. As illustrated in Figure 4.14, the DEVS protocol is executed as follows:

1. First, the hierarchy is built. Note that in Figure 4.14, the simulation is performed over a root-coupled model. Besides, one of the child components can be a coupled model; as a consequence, a simulator in Figure 4.14 should be a coordinator.
2. A cycle is then entered in which the coordinator requests that each simulator provide its time of the next event and determines the minimum of the returned values to obtain the global time of the next event. Current global (and virtual) time is fixed to this value: $t = \min(\text{tN}_i)$. In Figure 4.14, $i = 2$.
3. These simulators with $t = \text{tN}_i$ apply the λ_i method in the corresponding model to produce an output.
4. In this step, output propagation is carried out. The propagation is performed in all ports in which the corresponding model has left one or more values as an output. Thus, after a simulator i has executed λ_i, if there is output, then this output is propagated.
5. All the simulators execute a special transition function that tries to determine the combined effect of the propagated output and internal scheduling on its state.
6. A side effect of the execution of this transition function is to produce the time of next event, tN—for DEVS simulators, this state change is computed according to the DEVS formalism and tN is updated using the time advance of its model.

Finally, the coordinator obtains the next global time of the next event and the cycle repeats at point 2.

It should be noted that although the *Coordinator* class will be unique, its actual implementations could be quite different. This is not only because of the software design, but also because of the coordinator itself. The coordinator does not impose

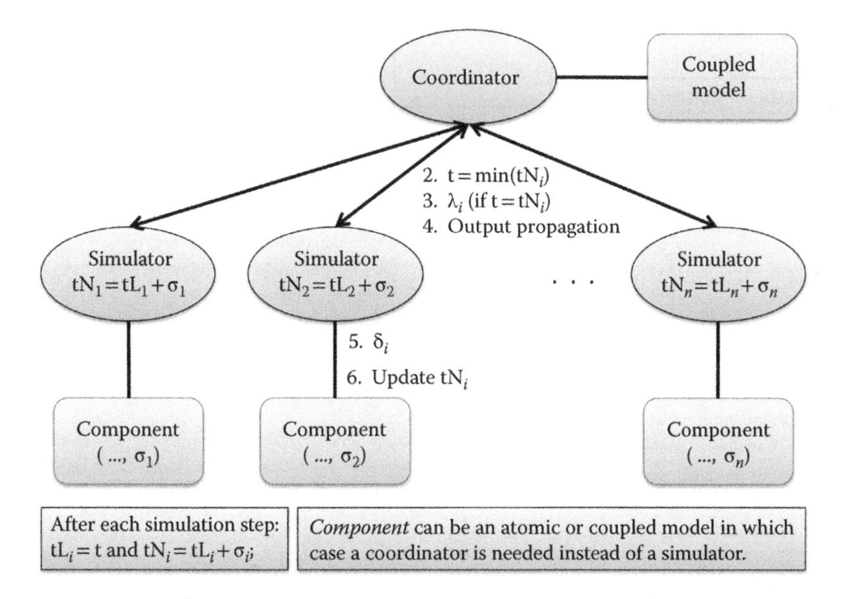

FIGURE 4.14 DEVS simulation protocol.

any strict ordering for the messages sent/received when multiple components are scheduled to receive inputs at the same time. For example, when the coordinator requests are sent to two or more simulators, the order in which the λ_i responses are received can be arbitrary. This is expected since the P-DEVS formalism is defined to assume no dependency between two messages received from one or more components. Therefore, there cannot be any dependencies between the two coordinators (or simulators) that are used together in a distributed fashion.

According to the simulation process in Figure 4.14, one coordinator contains many simulators and coordinators. Thus, as in the DEVS metamodel, there is a base *Component* class; we would need an abstract base class, *DevsSimulator*. Both *Simulator* (for atomic models) and *Coordinator* (for coupled models) classes would extend the *DevsSimulator* class. Regarding attributes, *DevsSimulator* should contain a set of attributes that are common to both *Simulator* and *Coordinator*. These are *tL* (time of the last event) and *tN* (time to the next event). With respect to the methods implemented in *DevsSimulator*, we should follow the process described in Figure 4.14. Propagation of events is only performed in the coordinator. Thus, we have *initialization, output function, transition function* and *time advance function*. Note that the protocol is quite equivalent to the DEVS modeling formalism. Indeed, we try to simulate a DEVS model using a DEVS structure.

Figure 4.15 shows the proposed DEVS simulation metamodel. The *Simulator* class just adds an atomic model to the list of attributes. The coordinator instead has, in addition to the associated coupled model, a set of simulators (each one being either a

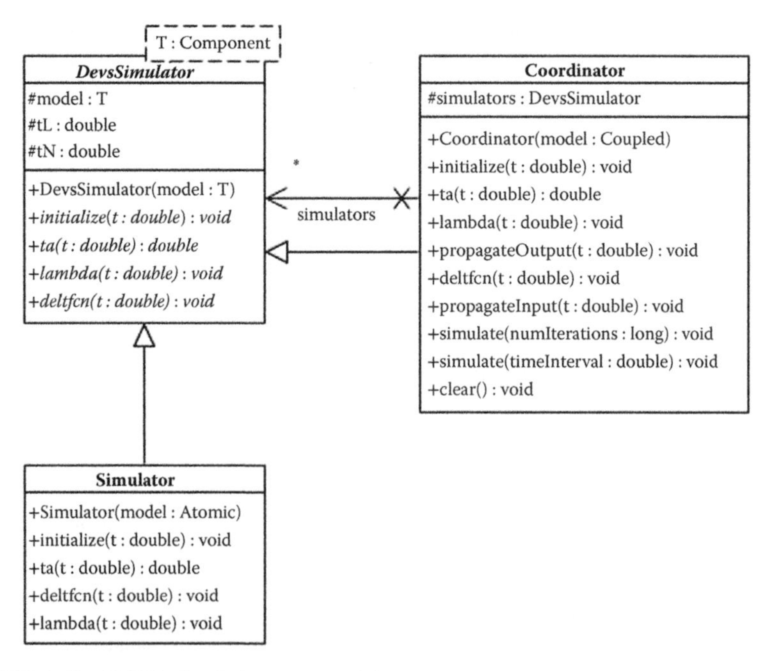

FIGURE 4.15 DEVS simulation metamodel.

Simulator or *Coordinator*). The coordinator has also more methods: *propagateOutput* propagates the output in internal connections and external output connections, *propagateInput* propagates the events in the external input connections, *clear* empties all the ports in the connections of the associated coupled model, and *simulate* simulates the coupled model. There are two versions of the *simulate* method. The first one just runs the points 2 to 6 in Figure 4.14 for *numIterations* times. The second one simulates the coupled model for a prescribed time interval.

4.3.1 DevsSimulator Class

The implementation of the abstract *DevsSimulator* class is straightforward (Listing 4.10).

LISTING 4.10 DEVSSIMULATOR.JAVA

```
public abstract class DevsSimulator {

    protected double tL; // Time of last event

    public double getTL() {
        return tL;
    }

    public void setTL(double tL) {
        this.tL = tL;
    }
    protected double tN; // Time of next event

    public double getTN() {
        return tN;
    }

    public void setTN(double tN) {
        this.tN = tN;
    }

    public DevsSimulator() {
    }

    abstract public void initialize(double t);

    abstract public void lambda(double t);

    abstract public void deltfcn(double t);
}
```

4.3.2 Simulator Class

In the *Simulator* class, the constructor takes the atomic model to be simulated. The method *initialize* just initializes both *tL* (to an initial global time *t*) and *tN* attributes. The time advance function *ta* just invokes the same function in the associated atomic model returning *sigma*. The same thing happens with the output function *lambda*, but as the coordinator calls all the output functions, we first check whether current global time is equal to the next event time. The transition function *deltfcn* will be called after the output function. Thus, this function must check if a transition does not exist (*input* empty and *t* not equal to *tN*), or if there exists an internal transition function (*input* empty and *t* equal to *tN*), or an external transition function (*input* not empty and *t* is not equal to *tN*), or a confluent function (*input* not empty and *t* equal to *tN*) (Listing 4.11).

LISTING 4.11 SIMULATOR.JAVA

```
public class Simulator extends DevsSimulator {

    protected Atomic model; // The atomic model

    public Atomic getModel() {
        return model;
    }

    public void setModel(Atomic model) {
        this.model = model;
    }

    public Simulator(Atomic model) {
        super();
        this.model = model;
    }

    @Override
    public void initialize(double t) {
        tL = t;
        tN = t + ta(t);
    }

    @Override
    public double ta(double t) {
        return model.ta();
    }

    @Override
    public void deltfcn(double t) {
        boolean isInputEmpty = model.isInputEmpty();
        if (isInputEmpty && t != tN) {
            return;
        } else if (!isInputEmpty && t == tN) {
            double e = t - tL;
            model.setSigma(model.getSigma()-e);
```

```
                model.deltcon(e);
            } else if (isInputEmpty && t == tN) {
                model.deltint();
            } else if (!isInputEmpty && t != tN) {
                double e = t - tL;
                model.setSigma(model.getSigma()-e);
                model.deltext(e);
            }
            tL = t;
            tN = tL + model.ta();
        }

        @Override
        public void lambda(double t) {
            if (t == tN) {
                model.lambda();
            }
        }
}
```

4.3.3 Coordinator Class

The *Coordinator* class has more functionality than the *Simulator* class, mainly because the *Coordinator* must manage itself and all the contained simulators or coordinators. As mentioned, a *Coordinator* has the derived attributes *tL* and *tN*, as well as the associated coupled model (*model*). It also contains the set of abstract simulators associated with the components in the root-coupled model (*simulators*).

The constructor not only assigns the associated coupled model but also initializes the hierarchy. To this end, all the components in the coupled model are visited, and a simulator is created for them. If the component is an atomic model, a *simulator* object is created. However, if the component is a coupled model, a *coordinator* is instantiated and the process starts again in a recursive manner. The method *initialize* not only sets *tL* and *tN* to their initial values but also calls the *initialize* method of all the children.

The time advance function *ta* must return the minimal *sigma* of all the components. To this end, this method searches the component simulators for the minimal difference between the next event time and the current global time. This is the time remaining until the next event.

The output function *lambda* is quite simple. It calls the output function in all the simulators (that will check if it is time for the next event). If imminent, the coordinator propagates the output in the connections of the associated coupled model (more precisely in the internal connections and external output connections; see *propagateOutput* in the source code).

The transition function *deltfcn* propagates inputs in the external input connections (see *propagateInput*) and tells all the simulators to execute their transition functions. At the end, *tL* and *tN* events are updated. Listing 4.12 shows the resulting source code in the *Coordinator* class.

LISTING 4.12 COORDINATOR.JAVA SOURCE CODE

```
public class Coordinator extends DevsSimulator<Coupled> {

   /** Simulators stored in the coupled model. */
   protected LinkedList<DevsSimulator> simulators =
        new LinkedList<DevsSimulator>();

   public Coordinator(Coupled model) {
     super(model);
     // Build hierarchy
     Collection<Component> components = model.
        getComponents().values();
     for (Component component : components) {
        if (component instanceof Coupled) {
          DevsSimulator simulator =
            new Coordinator((Coupled) component);
          simulators.add(simulator);
        } else if (component instanceof Atomic) {
          DevsSimulator simulator =
            new Simulator((Atomic) component);
          simulators.add(simulator);
        }
     }
   }

   @Override
   public void initialize(double t) {
     for (DevsSimulator simulator : simulators){
       simulator.initialize(t);
     }
     tL = t;
     tN = t + ta(t);
   }
   @Override
   public double ta(double t) {
     double tn = Constants.INFINITY;
     for (DevsSimulator simulator : simulators){
       if (simulator.tN < tn) {
         tn = simulator.tN;
       }
     }
     return tn - t;
   }
   @Override
   public void lambda(double t) {
     for (DevsSimulator simulator : simulators) {
        simulator.lambda(t);
     }
     propagateOutput(t);
   }
```

```
    public void propagateOutput(double t) {
      LinkedList<Coupling> ic = model.getIC();
      for (Coupling c : ic) {
        model.getComponent(c.getComponentTo()).
          getInPort(c.getPortTo()).addValues(
          model.getComponent(c.getComponentFrom()).
          getOutPort(c.getPortFrom()).getValues());
      }

      LinkedList<Coupling> eoc = model.getEOC();
      for (Coupling c : eoc) {
        model.getOutPort(c.getPortTo()).
        addValues(model.getComponent(c.getComponentFrom()).
         getOutPort(c.getPortFrom()).getValues());
      }
    }

    @Override
    public void deltfcn(double t) {
      propagateInput(t);
      for (DevsSimulator simulator : simulators) {
        simulator.deltfcn(t);
      }
      tL = t;
      tN = tL + ta(t);
    }

    public void propagateInput(double t) {
      LinkedList<Coupling> eic = model.getEIC();
      for (Coupling c : eic) {
        model.getComponent(c.getComponentTo()).
          getInPort(c.getPortTo()).addValues(model.
          getInPort(c.getPortFrom()).getValues());
      }
    }

    public void simulate(long numIterations) {
      double t = tN;
      long counter;
      for (counter = 1; counter < numIterations &&
          t < Constants.INFINITY; counter++) {
        lambda(t);
        deltfcn(t);
        clear();
        t = tN;
      }
    }

    public void simulate(double timeInterval) {
      double t = tN;
      double tF = t + timeInterval;
```

```
    while (t < Constants.INFINITY && t < tF) {
      lambda(t);
      deltfcn(t);
      clear();
      t = tN;
    }
  }

  public void clear() {
    for (DevsSimulator simulator : simulators) {
      if (simulator instanceof Coordinator) {
        ((Coordinator) simulator).clear();
      }
    }

    LinkedList<Coupling> eic = model.getEIC();
    for (Coupling c : eic) {
      model.getInPort(c.getPortFrom()).clear();
      model.getComponent(c.getComponentTo()).
        getInPort(c.getPortTo()).clear();
    }

    LinkedList<Coupling> eoc = model.getEOC();
    for (Coupling c : eoc) {
      model.getComponent(c.getComponentFrom()).
        getOutPort(c.getPortFrom()).clear();
      model.getOutPort(c.getPortTo()).clear();
    }

    LinkedList<Coupling> ic = model.getIC();
    for (Coupling c : ic) {
      model.getComponent(c.getComponentFrom()).
        getOutPort(c.getPortFrom()).clear();
      model.getComponent(c.getComponentTo()).
        getInPort(c.getPortTo()).clear();
    }
  }
}
```

4.4 SIMULATION OF COUPLED MODELS

A simulation execution always needs to synthesize model outputs in a meaningful way to the end user. In this section, we first show how to quickly build a *Scope*, an atomic model that will represent numerical values. Currently, we are using JFreeChart (www.jfreechart.org), which is a Java chart library that makes it easy for developers to display professional quality charts in their applications. Go to the website and download both JFreeChart and JCommon (an external dependency). Once downloaded, JCommon contains a JAR library in its main directory (*jcommon-1.0.17.jar* in our case), and JFreeChart contains a JAR library in the *lib* subdirectory (*jfreechart-1.0.14.jar* in our case). Using NetBeans in our MicroSim project (Figure 4.16), press the right mouse button in the Libraries subfolder and select Add JAR/Folder....

Just add the two libraries mentioned above (*jcommon-1.0.17.jar* and *jfreechart-1.0.14.jar*) (Figure 4.17).

Next, we implement a *Scope* (Listing 4.13). Write the following class. Refer to the JFreeChart documentation to understand the set of classes and methods used.

An instance of the *Scope* class first creates a window and an empty step chart with the main title, and titles for the *x*- and *y*-axis. Next, the *Scope* object is able to receive the discrete points through the method *addPoint* that are plotted in the step chart.

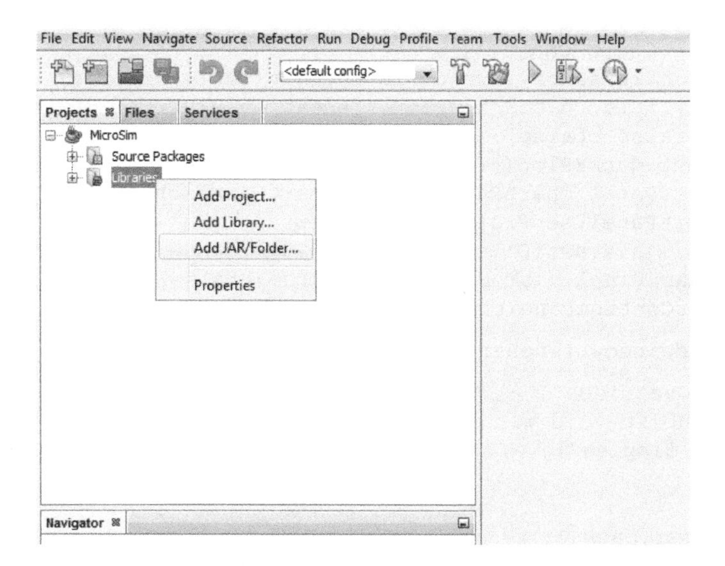

FIGURE 4.16 Adding external libraries.

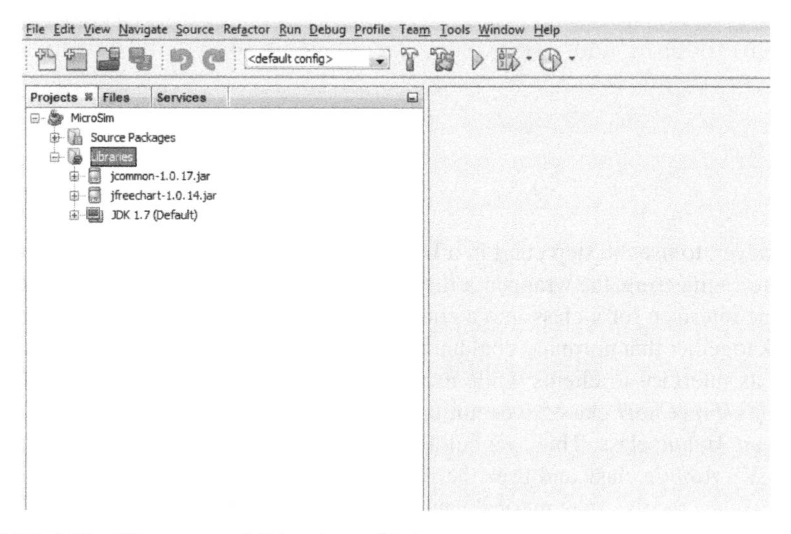

FIGURE 4.17 JCommon and JFreechart added.

LISTING 4.13 SCOPEJFREECHART.JAVA

```
public class ScopeJFreeChart extends Jframe {
  protected XYSeries serie;
  public ScopeJFreeChart(String windowsTitle,
      String title, String xTitle, String yTitle) {
    super(windowsTitle);
    XYSeriesCollection dataSet = new XYSeriesCollection();
    serie = new XYSeries(yTitle);
    dataSet.addSeries(serie);
    JfreeChart chart =
       ChartFactory.createXYStepChart(title, xTitle,
       yTitle, dataSet, PlotOrientation.VERTICAL, true,
       false, false);
    chart.getXYPlot().setDomainAxis(new NumberAxis());
    ChartPanel chartPanel = new ChartPanel(chart);
    chartPanel.setPreferredSize(new
         java.awt.Dimension(500, 270));
    chartPanel.setMouseZoomable(true, false);
    setContentPane(chartPanel);
    addWindowListener(new WindowAdapter() {
     @Override
     public void windowClosing(WindowEvent e) {
       dispose();
     }
    });
    super.pack();
    RefineryUtilities.centerFrameOnScreen(this);
    this.setVisible(true);
  }
  public void addPoint(double x, double y) {
    serie.add(x, y);
  }
}
```

However, to use the step chart in a DEVS model, we need to build a *wrapper*. In software engineering, the wrapper or the adapter pattern is a design pattern that translates one interface for a class into a compatible interface. A wrapper allows classes to work together that normally could not because of incompatible interfaces, by providing its interface to clients while using the original interface. We can think like the *ScopeJFreeChart* class "does not talk the DEVS language", that is, it does not extend an *Atomic* class. Thus, we build an auxiliary class called *Scope* that extends our DEVS *Atomic* class and uses the previous *ScopeJFreeChart* class. We assume that the *Scope* receives a numerical value by one port, and represents the global time versus the numerical values received by its input port (Listing 4.14).

LISTING 4.14 SCOPE.JAVA

```
public class Scope extends Atomic {

  public Port<Number> portIn =
     new Port<Number>(Number.class, "in");
  protected double time;
  protected ScopeJFreeChart chart;

  public Scope(String yTitle) {
    super("Scope");
    super.addInPort(portIn);
    chart = new ScopeJFreeChart("Scope", "Scope",
                               "time", yTitle);
    this.time = 0.0;
  }

  @Override
  public void deltint() {
    time += super.getSigma();
    super.passivate();
  }

  @Override
  public void deltext(double e) {
    time += e;
    if (!portIn.isEmpty()) {
      chart.addPoint(time,
         portIn.getSingleValue().doubleValue());
    }
  }

  @Override
  public void lambda() {
  }
}
```

As can be seen, the scope atomic model is in an idle state (the superclass always *passivates* an object in its constructor). The output function does nothing, since there is no output port. The internal transition function just updates the global clock and *passivates* the model. The external transition function updates the global clock and sends the computed time and the number received to the step chart.

4.4.1 Pulse

We now show how to simulate the *Pulse* atomic model with the *Scope* atomic model. To simulate a DEVS model, we first need to build a root-coupled model containing all the components involved in the simulation. Figure 4.18 depicts a scheme of the

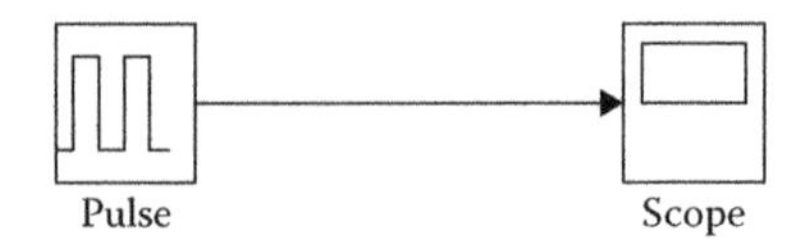

FIGURE 4.18 *PulseGeneratorExample* coupled model.

LISTING 4.15 PULSEGENERATOREXAMPLE.JAVA

```
public class PulseGeneratorExample extends Coupled {

  public PulseGeneratorExample() {
    super("PulseGeneratorExample");
    PulseGenerator pulse =
      new PulseGenerator("Pulse", 10, 3, 5, 5);
    super.addComponent(pulse);
    Scope scope = new Scope(pulse.getName() + "::" +
      pulse.portOut.getName());
    super.addComponent(scope);
    super.addCoupling(pulse, pulse.portOut, scope,
      scope.portIn);
  }

  public static void main(String[] args) {
    PulseGeneratorExample pulseExample =
      new PulseGeneratorExample();
    Coordinator coordinator = new Coordinator(
      pulseExample);
    coordinator.initialize(0.0);
    coordinator.simulate(30.0);
  }
}
```

root-coupled model. It can be developed with a class extending the coupled model. The implementation is straightforward (Listing 4.15).

The constructor creates both atomic models as well as their connections. The *Scope* only needs the title of the *y*-axis in the step chart. Furthermore, we have created a pulse atomic model with an amplitude equal to 10, pulse width of 3, period equal to 5, and a phase delay of 5.

There is also a main function that creates an instance of the coupled model and run the simulation using the coordinator previously developed. In NetBeans, just click with the right mouse button over the *PulseGeneratorExample* class and select Run …. It will show the step chart showing the simulation results (Figure 4.19).

4.4.2 Ramp

An example to simulate the Ramp atomic model (Figure 4.20) is performed in an identical way (Listing 4.16).

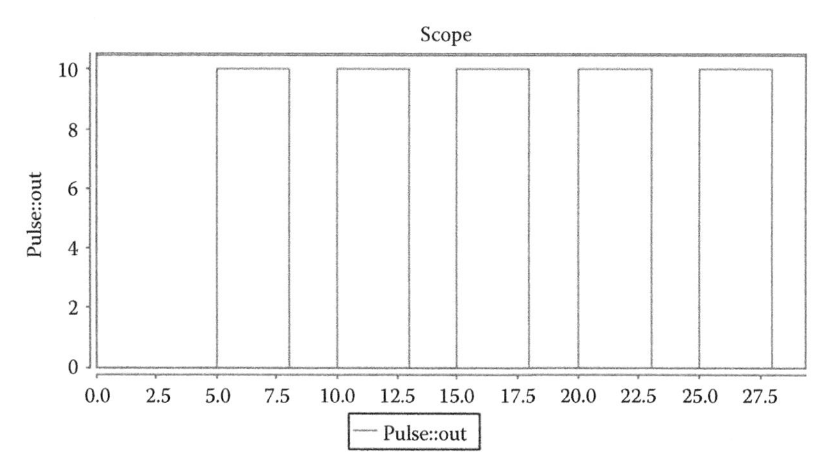

FIGURE 4.19 Pulse generator simulation output.

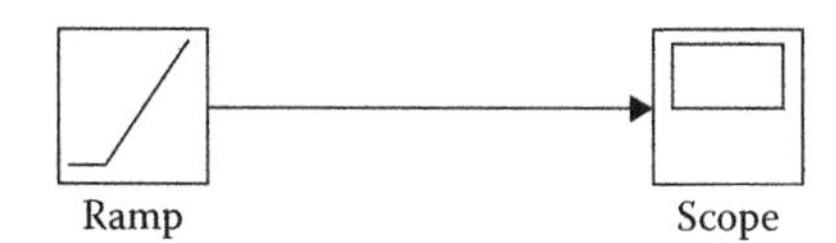

FIGURE 4.20 *RampExample* coupled model.

LISTING 4.16 RAMPEXAMPLE.JAVA

```
public class RampExample extends Coupled {

  public RampExample() {
    super("RampExample");
    Ramp ramp = new Ramp("Ramp", 2, 10, 2, 0.1);
    super.addComponent(ramp);
    Scope scope = new Scope(ramp.getName() + "::" +
       ramp.portOut.getName());
    super.addComponent(scope);
    super.addCoupling(ramp, ramp.portOut, scope,
        scope.portIn);
  }

  public static void main(String[] args) {
    RampExample example = new RampExample();
    Coordinator coordinator = new Coordinator(example);
    coordinator.initialize(0.0);
    coordinator.simulate(30.0);
  }
}
```

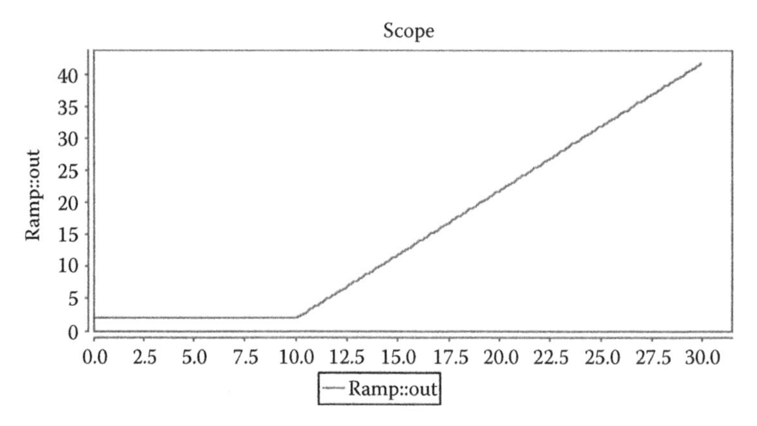

FIGURE 4.21 Ramp simulation output.

In this case, the Ramp has an initial output equal to 2, the start time is 10, the slope is 2, and the sample time is set to 0.1. Figure 4.21 shows the output of the simulation.

The Ramp atomic model is implemented using DTSS. In the following, we show how to create a quantized ramp (discretizing the *y*-axis instead of the time axis; see Chapter 3). We first need to select a quantum size, *qSize*. When selected, the ramp grows with the rate of *slope*. Thus, the time needed for the ramp to vary in *qSize* will be

$$\frac{qSize}{|slope|}$$

The implementation is shown in Listing 4.17.

LISTING 4.17 QRAMP.JAVA

```
public class QRamp extends Atomic {

  public Port<Double> portOut = new Port<Double>(
      Double.class, "out");
  protected double startTime;
  protected double slope;
  protected double nextOutput;
  protected double qOutput;

  public QRamp(String name, double initialOutput,
      double startTime, double slope, double qOutput) {
    super(name);
    super.addOutPort(portOut);
    this.nextOutput = initialOutput;
    this.startTime = startTime;
    this.slope = slope;
    this.qOutput = qOutput;
    super.holdIn("initialOutput", 0.0);
  }
```

```
    @Override
    public void deltint() {
      if (super.phaseIs("initialOutput")) {
        super.holdIn("startTime", startTime);
      } else {
        double sampleTime = qOutput / Math.abs(slope);
        nextOutput += slope * sampleTime;
        super.holdIn("active", sampleTime);
      }
    }

    @Override
    public void deltext(double e) {
    }

    @Override
    public void lambda() {
      portOut.addValue(nextOutput);
    }
  }
```

The *QRampExample* is built exactly like the *RampExample* (Listing 4.18). The output is shown in Figure 4.22.

LISTING 4.18 QRAMPEXAMPLE.JAVA

```
public class QrampExample extends Coupled {

  public QrampExample() {
    super("QrampExample");
    Qramp qramp = new Qramp("Ramp", 2, 10, 2, 0.1);
    super.addComponent(qramp);
    Scope scope = new Scope(qramp.getName() + "::" +
      qramp.portOut.getName());
    super.addComponent(scope);
    super.addCoupling(qramp, qramp.portOut, scope,
      scope.portIn);
  }

    public static void main(String[] args) {
    QrampExample example = new QrampExample();
    Coordinator coordinator = new Coordinator(example);
    coordinator.initialize(0.0);
    coordinator.simulate(30.0);
  }
}
```

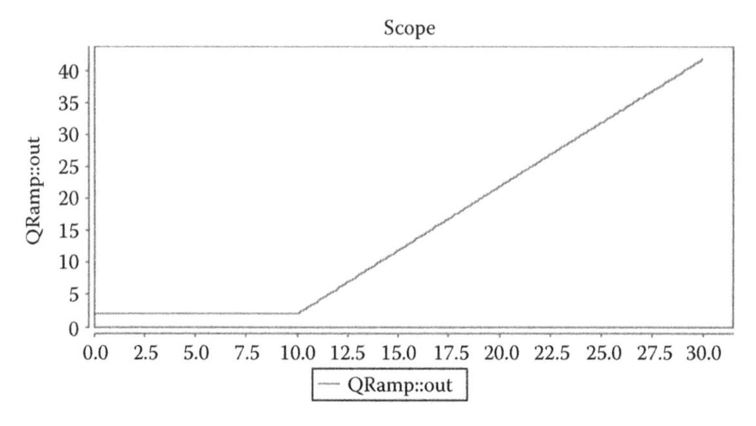

FIGURE 4.22 *QRampExample* simulation output.

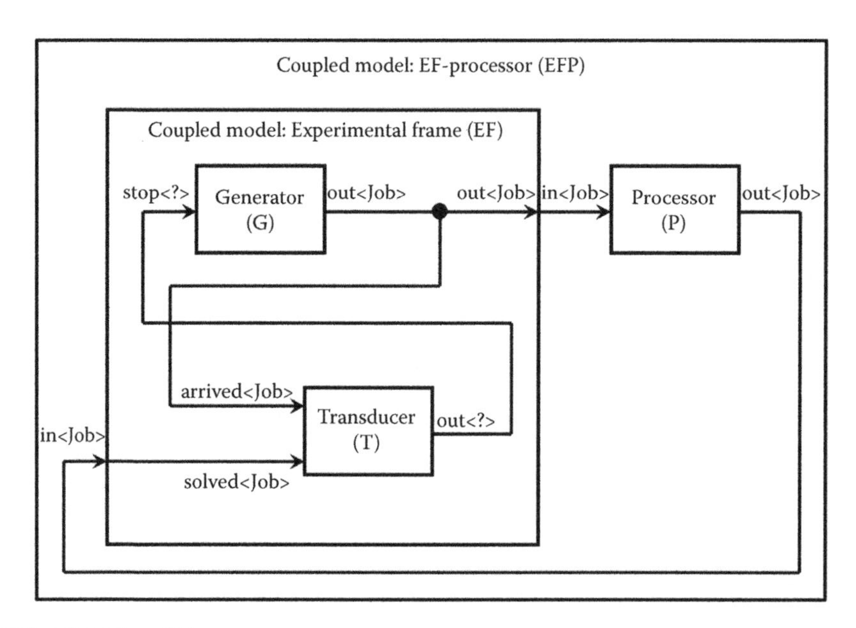

FIGURE 4.23 EFP structure.

4.4.3 Experimental Frame and Processor Model

We have already presented the Experimental Frame-Processor (EFP) model in Chapter 3. Indeed, we have written the DEVS specification for all the components in the model. In this section, we build and simulate the EFP model using MicroSim.

Figure 4.23 shows the structure of the EFP model. From the implementation point of view, we need two coupled models (the root-coupled model EFP and the experimental frame EF), three atomic models (Generator, Processor, and Transducer) and one *Job* class.

Figure 4.24 shows the UML class diagram for the EFP model.

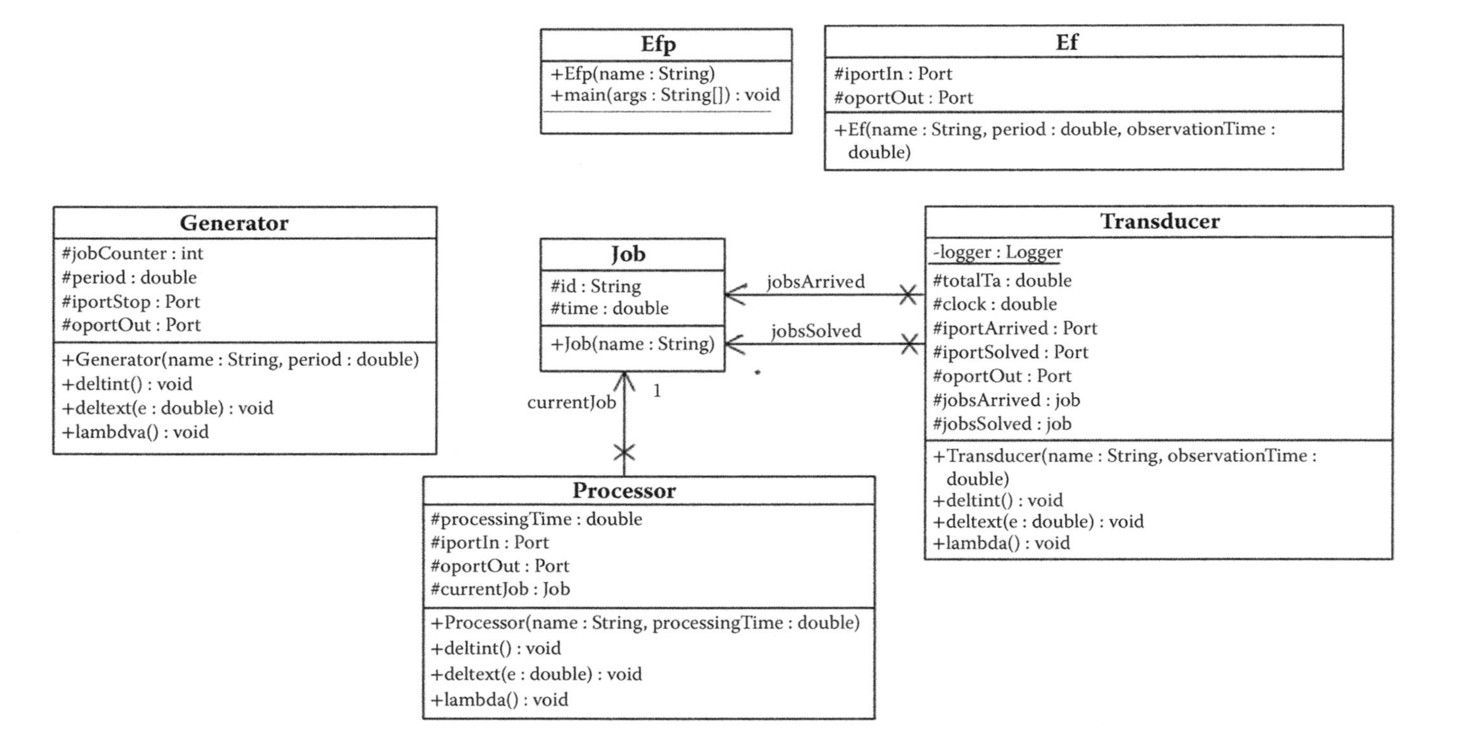

FIGURE 4.24 EFP UML class diagram.

4.4.3.1 Job

The *Job* class (Listing 4.19) contains a string to uniquely name the job and an attribute called time to store the time in which the job was generated and the time in which the job was solved (overwriting the first value):

LISTING 4.19 JOB.JAVA

```
public class Job {

  protected String id;
  protected double time;

  public Job(String name) {
    this.id = name;
    this.time = 0.0;
  }
}
```

4.4.3.2 Generator Atomic Model

The Generator has a period for job generation. To stop the generator, we add an input port called *stop*. If an event is received at the inport *stop*, the generator will change from the *active* state to the *idle* state. We also add a counter to generate a different name for each job. The implementation just follows the DEVS description already given in this book (Listing 4.20).

LISTING 4.20 GENERATOR.JAVA

```
public class Generator extends Atomic {

  protected Port<Job> iportStop = new Port<Job>
                                    (Job.class, stop");
  protected Port<Job> oportOut = new Port<Job>
                                   (Job.class, "out");
  protected int jobCounter;
  protected double period;

  public Generator(String name, double period) {
    super(name);
    super.addInPort(iportStop);
    super.addOutPort(oportOut);
    this.period = period;
    jobCounter = 1;
    this.holdIn("active", period);
  }
```

```
  @Override
  public void deltint() {
    jobCounter++;
    this.holdIn("active", period);
  }

  @Override
  public void deltext(double e) {
    super.passivate();
  }

  @Override
  public void lambda() {
    Job job = new Job("" + jobCounter + "");
    oportOut.addValue(job);
  }
}
```

4.4.3.3 Processor Atomic Model

The Processor has two ports, one to receive jobs and the other to send processed jobs. For simplicity, every job is solved in the same interval *processingTime*. The Processor does not accept jobs if there is a current job (*currentJob*) being processed (Listing 4.21).

LISTING 4.21 PROCESSOR.JAVA

```
public class Processor extends Atomic {

  protected Port<Job> iportIn = new Port<Job>
                              (Job.class, "in");
  protected Port<Job> oportOut = new Port<Job>
                               (Job.class, "out");
  protected Job currentJob = null;
  protected double processingTime;

  public Processor(String name, double processingTime) {
    super(name);
    super.addInPort(iportIn);
    super.addOutPort(oportOut);
    this.processingTime = processingTime;
  }
```

```
  @Override
  public void deltint() {
    super.passivate();
  }

  @Override
  public void deltext(double e) {
    if (super.phaseIs("passive")) {
      Job job = iportIn.getSingleValue();
      currentJob = job;
      super.holdIn("active", processingTime);
    }
  }

  @Override
  public void lambda() {
    oportOut.addValue(currentJob);
  }
}
```

4.4.3.4 Transducer Atomic Model

The Transducer has two input ports, one for jobs generated and the other for the set of jobs solved. It also stores these jobs (*arrived* and *solved*) in two separated linked lists. Additionally, there is a global clock (*clock*) to store the global instant of time (and thus updated in both transition functions). The *totalTa* parameter accumulates the time needed to solve each job. When the transducer operates for a given observation time, it sends an output to the generator to stop generation. Note that in the implementation, we use a Java *Logger* object to see the result (Listing 4.22).

LISTING 4.22 TRANSDUCER.JAVA

```
public class Transducer extends Atomic {

  private static final Logger logger =
      Logger.getLogger(Transducer.class.getName());
  protected Port<Job> iportArrived =
      new Port<Job>(Job.class, "arrived");
  protected Port<Job> iportSolved =
      new Port<Job>(Job.class, "solved");
  protected Port<Job> oportOut =
      new Port<Job> (Job.class, "out");
  protected LinkedList<Job> jobsArrived =
      new LinkedList<Job>();
  protected LinkedList<Job> jobsSolved = new
    LinkedList<Job>();
```

```
protected double totalTa;
protected double clock;

public Transducer(String name, double observationTime) {
  super(name);
  super.addInPort(iportArrived);
  super.addInPort(iportSolved);
  super.addOutPort(oportOut);
  totalTa = 0;
  clock = 0;
  super.holdIn("active", observationTime);
}

@Override
public void deltint() {
  clock = clock + getSigma();
  double throughput;
  double avgTaTime;
  if (!jobsSolved.isEmpty()) {
    avgTaTime = totalTa / jobsSolved.size();
    if (clock > 0.0) {
      throughput = jobsSolved.size() / clock;
    } else {
      throughput = 0.0;
    }
  } else {
    avgTaTime = 0.0;
    throughput = 0.0;
  }
  logger.info("End time: " + clock);
  logger.info("Jobs arrived : " + jobsArrived.size());
  logger.info("Jobs solved : " + jobsSolved.size());
  logger.info("Average TA = " + avgTaTime);
  logger.info("Throughput = " + throughput);
  super.passivate();
}

@Override
public void deltext(double e) {
  clock = clock + e;
  Job job = null;
  if (!iportArrived.isEmpty()) {
    job = iportArrived.getSingleValue();
    logger.info("Start job " + job.id + " @ t = " +
        clock);
    job.time = clock;
    jobsArrived.add(job);
  }
```

```
    if (!iportSolved.isEmpty()) {
      job = iportSolved.getSingleValue();
      totalTa += (clock - job.time);
      logger.info("Finish job " + job.id + " @ t = " +
        clock);
      job.time = clock;
      jobsSolved.add(job);
    }
  }

  @Override
  public void lambda() {
    Job job = new Job("null");
    oportOut.addValue(job);
  }
}
```

4.4.3.5 Experimental Frame Coupled Model

According to Figure 4.23, the EF-coupled model contains the generator and the transducer as well as one input port and one output port. The implementation is straightforward, since we must follow the graphical description and add components and connections as they appear in Figure 4.23 (Listing 4.23).

LISTING 4.23 EF.JAVA

```
public class Ef extends Coupled {

  protected Port<Job> iportIn = new Port<Job>
                                (Job.class, "in");
  protected Port<Job> oportOut = new Port<Job>
                                  (Job.class, "out");

  public Ef(String name, double period, double
      observationTime) {
    super(name);
    super.addInPort(iportIn);
    super.addOutPort(oportOut);
    Generator generator = new Generator
      ("Generator", period);
    addComponent(generator);
    Transducer transducer = new Transducer
      ("Transducer", observationTime);
    addComponent(transducer);
    addCoupling(this, this.iportIn, transducer,
      transducer.iportSolved);
```

```
        addCoupling(generator, generator.oportOut, this,
            this.oportOut);
        addCoupling(generator, generator.oportOut,
            transducer, transducer.iportArrived);
        addCoupling(transducer, transducer.oportOut,
            generator, generator.iportStop);
    }
}
```

4.4.3.6 Experimental Frame-Processor Coupled Model

As with the previous coupled model, we follow Figure 4.23 to develop this class. Additionally, we include a main static function to simulate this root coupled model. Note that the simulation is configured for an infinite number of cycles (Listing 4.24). This is not an issue, as the Transducer will passivate the generator, which is the only Moore DTSS model in the system.

LISTING 4.24 EXPERIMENTAL FRAME—PROCESSOR (EFP.JAVA)

```
public class Efp extends Coupled {
    public Efp(String name) {
        super(name);
        Ef ef = new Ef("ExpFrame", 1, 20);
        addComponent(ef);
        Processor processor = new Processor("Processor", 2);
        addComponent(processor);
        addCoupling(ef, ef.oportOut, processor,
            processor.iportIn);
        addCoupling(processor, processor.oportOut, ef,
            ef.iportIn);
    }

    public static void main(String args[]) {
        MicroSimLogger.setup(Level.INFO);
        Efp efp = new Efp("Coordinator");
        Coordinator coordinator = new Coordinator(efp);
        coordinator.initialize(0);
        coordinator.simulate(Long.MAX_VALUE);
    }
}
```

We can see the simulation output in the NetBeans *Output* tab. We have formatted the Java Logger to see the wall clock time of the simulation (Listing 4.25).

LISTING 4.25 SIMULATION OUTPUT

```
[INFO|00:00:00.047]: Start job 1 @ t = 1.0
[INFO|00:00:00.125]: Start job 2 @ t = 2.0
[INFO|00:00:00.125]: Start job 3 @ t = 3.0
[INFO|00:00:00.125]: Finish job 1 @ t = 3.0
[INFO|00:00:00.125]: Start job 4 @ t = 4.0
[INFO|00:00:00.125]: Start job 5 @ t = 5.0
[INFO|00:00:00.125]: Finish job 3 @ t = 5.0
[INFO|00:00:00.125]: Start job 6 @ t = 6.0
[INFO|00:00:00.125]: Start job 7 @ t = 7.0
[INFO|00:00:00.125]: Finish job 5 @ t = 7.0
[INFO|00:00:00.141]: Start job 8 @ t = 8.0
[INFO|00:00:00.141]: Start job 9 @ t = 9.0
[INFO|00:00:00.141]: Finish job 7 @ t = 9.0
[INFO|00:00:00.141]: Start job 10 @ t = 10.0
[INFO|00:00:00.141]: Start job 11 @ t = 11.0
[INFO|00:00:00.156]: Finish job 9 @ t = 11.0
[INFO|00:00:00.156]: Start job 12 @ t = 12.0
[INFO|00:00:00.156]: Start job 13 @ t = 13.0
[INFO|00:00:00.156]: Finish job 11 @ t = 13.0
[INFO|00:00:00.156]: Start job 14 @ t = 14.0
[INFO|00:00:00.156]: Start job 15 @ t = 15.0
[INFO|00:00:00.156]: Finish job 13 @ t = 15.0
[INFO|00:00:00.156]: Start job 16 @ t = 16.0
[INFO|00:00:00.156]: Start job 17 @ t = 17.0
[INFO|00:00:00.156]: Finish job 15 @ t = 17.0
[INFO|00:00:00.172]: Start job 18 @ t = 18.0
[INFO|00:00:00.172]: Start job 19 @ t = 19.0
[INFO|00:00:00.172]: Finish job 17 @ t = 19.0
[INFO|00:00:00.172]: End time: 19.0
[INFO|00:00:00.172]: Jobs arrived : 19
[INFO|00:00:00.172]: Jobs solved : 9
[INFO|00:00:00.172]: Average TA = 2.0
[INFO|00:00:00.188]: Throughput = 0.47368421052631576
[INFO|00:00:00.188]: Start job 20 @ t = 19.0
[INFO|00:00:00.188]: Finish job 19 @ t = 20.0
```

We can validate that the jobs are generated and solved according to the specification of the Generator (period of 1 second) and the Processor (processing time of 2 seconds). However, note that we are working using virtual time. The wall clock time generates *Jobs*, as the real processor in the PC is able to execute all the Java sentences in the MicroSim simulator (in the order of milliseconds). We will attend to the issue of real-time simulation in Chapter 8.

REFERENCES

JFreeChart. Retrieved from www.jfree.org/jfreechart/. Last accessed October 2012.
NetBeans. Retrieved from www.netbeans.org. Last accessed October 2012.

5 DEVS Modeling Language

A Domain-Specific Language (DSL) is a dedicated language for a specific problem domain and is not intended to solve problems outside it. For example, HTML, Verilog, and VHDL are DSLs for a very specific domain. A DSL can be textual, graphical, or hybrid. A DSL builds abstractions so that the respective domain experts could specify their problem well suited to their domain understanding without paying much attention to the general-purpose computational programming languages such as C, C++, and Java, which have their own learning curve. The notion of domain-specific modeling arises from this concept, and the DSL designers are tasked with creating a domain-specific modeling language. If a DSL is also meant for simulation purposes, then one more task of mapping a specific DSL to a general-purpose computational language is also in the cards. There are many DEVS DSLs that implement a subset of rigorous DEVS formalism. One example of DEVS DSL is DEVSpecL (Hong & Kim, 2006), built on the BNF grammar. DSL writing tools such as Xtext and Ruby focusing directly on the Extended Backus–Naur Form (EBNF) grammar provide a much easier foundation to develop the Abstract Syntax Tree (AST) for model-to-model transformations. The rich integration and code generation capabilities with open-source tools such as Eclipse give them strong acceptance in the software modeling community.

This chapter presents another DEVS DSL, called DEVS modeling language (DEVSML) (Mittal & Douglass, 2012), based on finite deterministic DEVS (Hwang & Zeigler, 2007) and an earlier developed XML-based XFD-DEVS (Mittal, 2007; Mittal et al., 2012).

5.1 LANGUAGE

Although XFD-DEVS was a good start toward platform-independent DEVS modeling, it had many shortcomings, such as no confluent function, no multiple inputs, no multiple outputs, no complex message types, and no state variables. These shortcomings are removed in the proposed language that was done as a part of the Large-Scale Cognitive Modeling initiative at the Air Force Research Laboratory (Mittal & Douglass, 2011). This DEVS DSL is closer to the true DEVS formalism with some necessary abstractions. Another example of a DEVS DSL is DEVSpecL. It provides a platform-independent way to specify DEVS models that are transformed to platform-specific language implementation in C++. Like any language, the DEVS Modeling Language (DEVSML) has keywords, as shown in Table 5.1.

TABLE 5.1
DEVSML Keywords

package	**import**	**entity**
extends	coupled	models
interfaceIO	couplings	atomic
ic	eoc	eic
vars	state-time-advance	state-machine
start in	confluent	deltint
deltext	outfn	sigma
continue	reschedule	ignore-input
input-only	input-first	input-later
infinity	int	double
String	boolean	input
output	S:	S":
this	X:[]	Y:[]

A DEVSML file has the extension *.fds*, and the specification language contains three primary element types: entity, atomic and coupled.

```
Type:Atomic | Coupled | Entity;
```

While the atomic DEVS formalism has a notion of ports (input and output), the DEVSML has a notion of messages specified as entity structures that are eventually transformed to port definitions. The DEVSML grammar is specified using the Xtext EBNF notation. For more details on Xtext EBNF capabilities, refer the Xtext manual. In the following subsections, we will look at each of the elements.

5.1.1 Entity

DEVS is a component-based framework where each of the components communicates using messages. When DEVS is tied to a platform-specific implementation, these messages are object instances defined in the implemented language. These message objects are exchanged according to the port-value pairs specified in the atomic model structure. Consequently, the message structure is declared and defined in an atomic model, to begin with. Using the object-oriented principles, this message object structure is then reused in other atomic models. In DEVSML, as the port-value assignment is abstracted and automated, the entities are defined not as a part of the component but as a first-class citizen. These entities are then declared in atomic or coupled components for their reuse. The components exchange these entities through automated assignments as port-entity pairs. Such a framework allows these entities to be used in other standardized message-exchange frameworks such as WSDL-based Service Oriented Architecture (SOA).

The EBNF specification of entity is as follows:

```
Entity:
'entity' name = ID
('extends' superType = [Entity|QualifiedName])?
('{' (pairs + = Variable)* '}')?;

Variable: Type = VarType name = ID;
VarType:
simple = ('int'|'double'|'String'|'boolean') | complex =
   [Entity | QualifiedName];
```

An entity is specified by a name, `name = ID`. It may or may not extend another entity. The expression `[Entity|QualifiedName]` means that the supertype is to be specified as a `QualifiedName`, which is an Xtext construct and is of type `Entity`. It provides the full package path of the entity defined in the project. For more details on `QualifiedName`, refer the Xtext manual. Each entity contains a set of key-value pairs as `Variable` that contains a `VarType`, which may be primitive `('int'|'double'|'String'|'boolean')` or complex `[Entity | QualifiedName]`.

5.1.2 Atomic

While designing the atomic DEVSML grammar, we have tried to stay as close as possible to the parallel DEVS formalism. However, some abstractions are necessary. The primary abstraction as laid out in Section 5.1.1 is the port-value to message-entity abstraction. The other omission is the notion of elapsed time in the external event transition. The handling of elapsed time is limited to the implementation of features such as `'continue'` and `'reschedule'` keywords, which are described later in the section. The other last piece of omission is the state transition based on the message content. We believe that once the message content is copied over to the locally scoped variables, all the operations can be executed on the message content. However, any state change based on the message content is where the limitations of finite deterministic DEVS come into the picture.

The atomic type is specified in EBNF grammar as follows:

```
Atomic:      'atomic' name = ID
   ('extends' superType = [Atomic|QualifiedName])?'{'
   'vars''{'(variables += Variable)*'}'
   'interfaceIO' '{'(msgs += Msg)*'}'
   'state-time-advance' '{'(stas += STA)*'}'
   'state-machine''{'
       start in ' init = InitState
       (atBeh += AtomicBeh)* '}'
    '}';

Msg: type = ('input'|'output') ref = [Entity|QualifiedName]
  name = ID;
STA: name = ID timeAdv = TimeAdv;
TimeAdv: tav = FLOAT | inf = "infinity" | var = [Variable];
```

```
InitState: state = [STA] (code = Code)?;
AtomicBeh: deltext = Deltext | outfn = Outfn | deltint =
  Deltint | confluent = Confluent;
Code: '{'str = STRING '}';
```

The atomic type can extend from another *Atomic* type and is composed of the following:

- Set of variables of the type `Variable`: These are the locally scoped variables for the defined atomic model.
- interfaceIO specification that has a set of messages of the type `Msg`: An `Msg` is either an input or an output message and is referenced as an `Entity` type. Note that the keywords `'input' and 'output'` are used to automate the port-entity assignments and map them to the port-value definitions in the canonical DEVS formalism.
- Set of states and the associated time-advances: Each state-time-advance pair is defined as `STA`. The time-advance `TimeAdv` can have values of `FLOAT`, infinity, or a `Variable` declared above in the atomic model. Using the declared variable allows the user to specify the value at execution time. This value gets assigned in the model initialization phase.
- State machine that contains the initial state `InitState` and the atomic behavior `AtomicBeh`.

Now let us look at the atomic-state machine definition in more detail. The `InitState` specifies the initial state of the atomic model. The expression `state = [STA]` implies that the model references the state already defined in the construct *STA* defined earlier in the model. An element is said to have been defined when there is an ID associated with it. The next expression `(code = Code)?` implies that there may be a code snippet associated with setting up the initial state. As we shall see in a later section on code generation, the code expressed as a `STRING` is syntactically checked at runtime for any compilation error.

The atomic DEVS formalism has *deltint*, *deltext*, *deltcon*, and *lamda* functions to specify the atomic behavior and is implemented accordingly as functions in DEVSJAVA. Consequently, the DEVSML has a set of atomic behaviors, and the behavior `AtomicBeh` is of the form of

```
AtomicBeh:
deltext = Deltext | outfn = Outfn | deltint = Deltint |
  confluent = Confluent;
```

Let us look at each of these:

```
Deltint: 'deltint' '(''S:'state = [STA] ')'' =>' 'S":'
  target = [STA] (code = Code)?;
```

The `Deltint` specification refers to a source state `'(''S:'state = [STA]')'`and always transitions `'=>'` to the target state `'S":'target = [STA]`. Transition to the same source state is allowed.

```
Deltext: 'deltext' '(''S:'state = [STA]','X:' '['(in +=
   [Msg])+ '])' ' =>' 'S":'(target = [STA])? (res = Resched |
   cont = 'continue')? (code = Code)?;
Resched: 'reschedule''(' setSig = SetSigma ')';
```

The Deltext specification refers a source state and an input message reference set '(''S:'state = [STA]',''X:' '['(in += [Msg])+ '])' that may have transitions to a target state. The 'continue' keyword allows the model to stay in the same source state, but advances the elapsed time and redefines the sigma or time-advance as *sigma-e*. The 'reschedule' keyword allows the resetting of time-advance of the target state. This feature overrides the time-advance of the referenced state defined earlier in STA. This rescheduling is a characteristic of the FDDEVS specification.

```
Confluent: 'confluent' con = ('ignore-input'|'input-only'
   |'input-first'| 'input-later');
```

The Confluent specification has four implementations as the options suggest above. During code generation, the option translates to making calls to either *deltint* or *deltext* or both in a specific order as dictated by the selection.

```
Outfn: 'outfn' '(''S:'state = [STA]')' ' =>' 'Y:' '['(out +=
   [Msg])+ ']' (code = Code)?;
```

Finally, the Outfn refers a source state '(''S:'state = [STA]')' and maps it to an output message set 'Y:' '['(out += [Msg])+ ']'.

All the above behavior specifications are code-assisted and validated, as behavior is specified in the editor. Let us look at some of those.

5.1.2.1 Code-Assist Features for Atomic DEVSML

Xtext is seamlessly integrated with the Eclipse Modeling Framework (EMF) and the designed EBNF grammar is transformed into the native ecore format for AST manipulations and code generation. As the DEVSML editor is text-based to begin with, the code assist feature in Eclipse provides recommendations and code completion capabilities by pressing Ctrl+Space. All the references in the grammar specified as [elementX] are available during the model design phase. As a result, once the element is defined in early stages of the design and is identified by name = ID, it is available as a reference if the scope permits it. By default, the element is visible in the package scope, but it is highly customizable. As we shall see in the later section, we use this capability to expedite the couplings design in a coupled model.

To reap the benefits of the code completion features, the sequential process to design an atomic DEVSML model is as follows:

1. Define entities. They may be in the same package or in a different file with a different package.

2. Declare an atomic model type with a name and import the package containing entities. Now all the entities are available as references for their reuse as message types with Ctrl+Space.
3. Start with defining atomic model variables. Here also, entities are available for complex variable types.
4. Define interface inputs and outputs. Entities are available as message types.
5. Define states and their time-advance. All the states will be available in the state machine from this point onward. This is important as the state machine must not use any state for which time-advance has not been defined. The time-advance can be a double, a string "infinity," or a variable that may be assigned a value in the initialize code snippet. If there is no value specified, the default value for a double is assigned, which is zero.
6. Define the state machine.
   a. Begin with an initial state. Select any state from the stateset defined in the previous step. Ctrl+Space gives you all the available states.
   b. Now you are in a position to specify the `Deltint`, `Deltext`, `Outfn`, or `Confluent` behavior. Type the appropriate keyword, that is, `'deltint'`, `'deltext'`, `'outfn'`, or `'confluent'`, and press Ctrl+Space. Appropriate hints and options will appear that speed up the behavior specification process.

5.1.2.2 Runtime Model Validation of Atomic DEVSML

The Xtext framework provides a rich validation extension mechanism that allows writing customized validators for the defined EBNF grammar. We have currently defined the following validations that execute at runtime when the atomic behavior is being designed in the DEVSML Eclipse editor:

1. Unique Model names across packages.
2. No same source state for internal transitions; that is, there should be no two internal transitions with the same source state.
3. Output message in Outfn must be defined in interfaceIO; that is, the message being sent in Outfn must be defined as an "output" in the atomic models' interfaceIO definition.
4. Input message in Deltext must be referenced in interfaceIO; that is, the message being received in Deltext must be defined as an "input" in the atomic models' interfaceIO definition.
5. No same source state and input message set for external input transitions; that is, there should be no two external input transitions with the same source state and the identical input message set.
6. No same source state for output functions; that is, there should be only one output function associated with a specific state.

These validations provide the next level of model checking. The first level is provided by managing the entity scope in the editor, as described in Section 5.1.2.1. The transformed code after these validations is DEVS-correct by construction.

5.1.3 Coupled

The DEVSML specification of a coupled DEVS is as follows:

```
Coupled: 'coupled' name = ID
  ('extends' superType = [Coupled|QualifiedName])?'{'
  'models' '{'(components += Component)*'}'
  'interfaceIO' '{'(msgs += Msg)*'}'
  'couplings' '{' (couplings += Coupling)*'}'
  '}';
Component: AtomicComp | CoupledComp;
AtomicComp: 'atomic' at = [Atomic | QualifiedName] name = ID;
CoupledComp: 'coupled' cp = [Coupled | QualifiedName] name =
  ID;
Coupling: ic = IC | eoc = EOC | eic = EIC;
EIC: 'eic''this'':' msgtype = [Msg] '->' dest = [Component]':'
  destMsgType = [Msg];
IC:'ic'
  src = [Component | QualifiedName] ':'msgtype =[Msg] '->'
  dest = [Component | QualifiedName] ':' destMsgType = [Msg];
EOC: 'eoc'
  src = [Component] ':' msgtype = [Msg] '->' 'this'':'
  destMsgType = [Msg];
```

The `Coupled` DEVSML type can extend from another DEVSML `Coupled` component. It is composed of the following:

- Set of models of the type `Component`: The `Component` can be of the type `AtomicComp` or `CoupledComp`. Note that each has a name. This allows multiple components of the same type within a coupled model. For example, multiple *processor* components with different names, all of type *processor.*
- Interface specification: Please see atomic above for detailed description.
- Set of couplings of type `Coupling`: In the original DEVS formalism, the couplings are specified by connecting port specifications from one component to another. In DEVSML, as we abstracted away from port specifications, we use the message `Msg` entity that flows between the components. This abstraction is transformed to port specification in the code generation phase. Each `Coupling` specification is of any of the following types:
  - External Input Coupling, `EIC`: This type of coupling is defined for connections originating from the input interface of the coupled model to its subcomponents. Consequently, the source is identified as `'this'` keyword and the coupling allows routing of the input message defined in its `'interfaceIO'` to the destination submodel specified through `Component`.
  - Internal Coupling, `IC`: This type of coupling is specified between the subcomponents as enumerated in `Component`.
  - External Output Coupling, `EOC`: This type of coupling is specified from the contained `Component` to the outside interface of the coupled model. Consequently, the destination is specified as `'this'`.

5.1.3.1 Code-Assist Features of Coupled DEVSML

Having designed the atomic DEVSML models, the Eclipse Xtext editor is further used to design the coupled DEVSML models. The coupled model may be in the same package as an atomic model or in a different package or in a different file altogether. Usage of import statements provides dependencies from other components. To define a coupled model, the following steps are executed. The code assist features are made available incrementally.

1. Specify the package and imports.
2. Specify the coupled model name and any supertype of type `Coupled`.
3. Specify components. The components are referenced on `'atomic'` or `'coupled'` keyword.
4. Specify the `'interfaceIO'` using the `'input'` or `'output'` keyword. Based on the package scope or imports, the message entities are made available.
5. Specify couplings. The coupling statements begin with `'eic'`, `'ic'`, or `'eoc'` keywords.
   a. IC coupling: Pressing Ctrl+Space will show the available subcomponents defined in the component section above. Pressing Ctrl+Space for the message-outgoing source will show the referenced message name defined in interfaceIO of the selected subcomponent. If there is no output interface message, then no message is available as an option. Consequently, this subcomponent cannot have any output couplings. Here, the scope of the output source message has been customized. Pressing Ctrl+Space again, select the destination subcomponent model. Again, using scopes, only those components that can receive the same entity will show up in content assist. Moving along, all the input messages of the selected destination subcomponent that are of the same entity type are made available using Ctrl+Space.
   b. EIC coupling: All the operations remain the same as above, and the only change is that the source is always available as `'this'` and the outgoing source message is scoped from the interfaceIO defined for the current coupled model.
   c. EOC coupling: All the operations remain the same as above, and the change is visible in the destination component as “this” and input destination message is scoped from the interfaceIO defined for the current coupled model.

5.1.3.2 Runtime Model Validation of Coupled DEVSML

We have currently defined the following validations for a coupled DEVSML model:

1. Unique model names across packages.
2. Model types for supertype and component set are already filtered based on the keyword `'atomic'` or `'coupled'`.

3. IC coupling:
   a. The source output message must be defined in the interfaceIO of the source component as `'output'`.
   b. The destination input message must be defined in the destination interfaceIO as `'input'` and of the same type `Entity` as source message type.
4. EIC coupling:
   a. The source output message must be defined in the current coupled models' interfaceIO as `'input'`.
   b. The destination input message must be defined in the destination interfaceIO as `'input'` and of the same type `Entity` as source message type.
5. EOC coupling:
   a. The source output message must be defined in the component's interfaceIO as `'output'`.
   b. The destination input message must be defined in the current coupled model's interfaceIO as `'output'`.

When the coupled models are constructed using code-assist features and validated by the rules above, the coupled DEVSML models are DEVS-correct. The next section will describe the code generation mechanisms within the Xtext Eclipse framework.

5.2 DYNAMIC CODE GENERATION

Once the models are created using the Xtext Eclipse-based editor, the next step is to get an executable DEVS code. The DEVSML is semantically anchored in DEVS formalism, and the abstractions in DEVSML are unpacked during the code generation phase. The most prominent abstraction is the port-value to message-entity mapping.

After the specification of the fds grammar, the Xtext framework generates a bunch of artifacts to specify the scope providers, validators, and code generators. These are provided as functions that can be overridden to support the grammar under design. The Xtext framework is seamlessly integrated with the Eclipse Java framework; any *save* process (by invoking Ctrl+S) in the editor invokes the *doGenerate()* function that generates the entire codebase. As a result, one can dynamically view the generated code from the abstract DEVSML.fds specification. This capability is one of the very important features in our current selection of the Xtext framework, and the logic code specified in the CODE element in atomic files is readily checked for syntactical errors. Since the generated code is already in an Eclipse java project, the modeler is informed of any compilation errors that the DEVSML model might introduce.

The template to generate any platform-specific code is specified in the `FdsGenerator.xtend`. Let us unpack the FdsGenerator.xtend to better understand the transformation from the platform-independent DEVS specification to the platform-specific DEVS specification such as DEVSJAVA or MicroSim. This specific file is a holder for the class called FdsGenerator that implements the org.eclipse.xtext.generator.IGenerator interface. It provides only one method, doGenerate(Resource resource, IFileSystemAccess fss){}, that needs to be overridden to provide our customized code generation snippets. The entire fds AST is available to us for manipulation and code generation in the .xtend file. The first step is extracting elements of specific types (Listing 5.1). Recall from the previous section that there are three types in the fds grammar: entity, atomic, and coupled. Listing 5.1 shows the library function doGenerate() and how the occurrence of specific elements can be extracted and further processed. The FdsGenerator is equipped with a code generator for three engines: DEVSJAVA, MicroSim, and MicroNet. While the first two are in Java, the last one, MicroNet, is in .NET language. In addition, the AST is also realized in XML for each particular engine for any further use. For each of the top-level types in the DEVSML grammar, all the platform-specific code is generated in their own specific folders. For brevity, Listing 5.1 shows only the DEVSJAVA engine and the accompanying xml mode.

LISTING 5.1 EXTRACTION OF SPECIFIC ELEMENTS OF THE DEVSML GRAMMAR

```
for(e: resource.allContentsIterable.filter(typeof
  (Entity))){
      fsa.generateFile(
            "devsjava/"+e.fullyQualifiedName.
            toString.replace(".","/")+".java",
                  e.compile("devsjava")
      )
      fsa.generateFile(
                  "devsjava/"+"xml/"+e.eContainer.
                  fullyQualifiedName.
                  toString.replace(".","/")+"/"+
                  e.name+".xml",
                  e.compileXml("devsjava")
      )
}
for(e: resource.allContentsIterable.filter(typeof
  (Coupled))){
      coupledPackageName = e.eContainer.
        fullyQualifiedName.toString

      fsa.generateFile(
                  "devsjava/"+"xml/"+e.eContainer.
                  fullyQualifiedName.
                  toString.replace(".","/")+"/"+
                  e.name+".xml",
                  e.compileXml("devsjava")
      )
      fsa.generateFile(
                  "devsjava/"+e.fullyQualifiedName.
                  toString.replace(".","/")+".java",
                  e.compile("devsjava")
      )
      ...
}
for(e: resource.allContentsIterable.filter(typeof
  (Atomic))){
```

```
        packageName = e.eContainer.fullyQualifiedName.
          toString
        observerPackageName = packageName.concat
          (".observers")

        ...

        fsa.generateFile(
                    "devsjava/"+"xml/"+e.eContainer.
                      fullyQualifiedName.
                    toString.replace(".","/")+"/"+
                    e.name+"_xml",
                    e.compileXml("devsjava")
        )
        fsa.generateFile(
                    "devsjava/"+e.fullyQualifiedName.
                    toString.replace(".","/")+".java",
                    e.compile("devsjava")
        )
        ...
    }
    for(e: resource.allContentsIterable.filter(typeof
      (Model))){
        fsa.generateFile(
                   "devsjava/"+"run/Main.java",
                   e.compile("devsjava")
        )
    }
```

The Xtend language (www.eclipse.org/xtend) is a Java DSL in many aspects with the addition of Xtext constructs and runtime access of the AST specified by Xtext. As can be seen in Listing 5.1, the atomic model's code generation is a bit more complex than the entity or coupled models. For more information regarding the Xtend language, refer to the online manual.

5.2.1 Entity

When the entity element in a DSL is encountered, a file with EntityName.xxx is created, where .xxx is either .java or .net, and the control is passed on to the `compile(Entity e, String engineType)` function and `compileXml (Entity e, String engine Type)` for EntityName.xml output. The content of the entity element is specified in Listing 5.2. A class is created with the specified name in a package specified under the "engineType" folder. The libraries corresponding to the specific `engineType` are imported. The code below specifies a constructor, and a *toString()* function. It also unpacks the `Variable` element in a DSL and delegates each variable-type definition to `compile(VarType type)`.

LISTING 5.2 GENERATE PSM BASED ON ENGINETYPEFOR ENTITY IN FDSGENERATOR.XTEND

```
def compile(VarType type)'''
      «IF type.simple ! = null»
      «type.simple»«ELSE»«type.complex.
        fullyQualifiedName.toString»
      «ENDIF»
'''

def compile(Entity e, String engineType)'''
      «IF e.eContainer ! = null »
            package «engineType».«e.eContainer.
              fullyQualifiedName»;
      «ENDIF»

      «FOR f: e.getImports(engineType)»
      import «f»;
      «ENDFOR»

      public class «e.name» extends
            «IF e.superType ! =
            null»«e.superType.name.
              toFirstUpper»«ELSE»«e.superBaseType
              (engineType)»
            «ENDIF»{

            «FOR f: e.pairs»
                  public «f.type.compile» «f.name»;
            «ENDFOR»

            public «e.name»(){
                  this("«e.name.toFirstUpper»");
            }
            public «e.name»(String name){
                  super(name);
            }
            public String toString(){
                  StringBuilder sb = new StringBuilder();
                  sb.append(getName()+"{");
                  «FOR f: e.pairs»

                  sb.append("(«f.name.toFirstUpper» "+
                  «IF f.type.simple ! = null&&
                    f.type.simple.equals("double")»
                  util.doubleFormat.niceDouble«ENDIF»
                    («f.name»)+")");
                  «ENDFOR»
                  sb.append("}");
                  return sb.toString();
            }
      }
'''
```

```
def compileXml(Entity e, String engineType){'''
      <entity name = "«e.name»" package = "
      «IF e.eContainer ! = null »
      «engineType».
         «e.eContainer.fullyQualifiedName»«ENDIF»">
      <extends type = "
             «IF e.superType ! = null»«e.superType.name.
               toFirstUpper»
             «ELSE»«e.superBaseType(engineType)»«ENDIF»"/>
             <imports>
                    «FOR f: e.getImports(engineType)»
                    <import>«f»</import>
                    «ENDFOR»
             </imports>
             <vars>
                   «FOR f: e.pairs»
                   <var type = "«f.type.compile»" name =
                     "«f.name»"/>
                   «ENDFOR»
             </vars>
      </entity>
'''
```

For the sake of brevity, only code generation with respect to the DEVSJAVA engine will be described ahead. The reader is encouraged to look at the SubVersioN (SVN) repository for detailed multi-engine code generation.

5.2.2 Atomic

On encountering an atomic element in an .fds file, the appropriate block in the *doGenerate()* function is invoked. An atomic DSL is a complex one, and we need to store information for efficient code generation of DEVS elements. Consequently, local variables in the xtend file are used to manipulate the available AST. The control is delegated to `compile (Atomic e, String engineType)`. The pseudocode (Listing 5.3) is as follows:

LISTING 5.3 THE PSEUDOCODE FOR THE ATOMIC PLATFORM-SPECIFIC CODE GEN

```
def compile (Atomic e, String engineType)'''
      //1. specify package
      //2. add DEVS modeling imports for specific
         engineType
      //3. create class and extend from DEVS atomic{
      //4.  define variables
      //5.  define empty hashtable for State-time-advances
```

```
      //6.  define state strings
      //7.  Create constructor with name{
                  //initialize hashtable for
                   state-time-advances
                  //unpack messages
                        //add inports
                        //add outports
                  //add test inputs
      //8.  override initialize()
                    //unpack initiatlization code
                    //unpack state-time-advances
                    //hold in initial state
      //9.  override deltint()
                  //unpack AtomicBeh.deltint
      //10.  override deltext()
                  //advance elapsed time
                  //unpack AtomicBeh.deltext
      //11.  override deltcon()
                  //unpack AtomicBeh.confluent
      //12.  override out()
                 //unpack AtomicBeh.outfn
      //13.  Override logging function
      //14.  write processing function
      //15.  End class
'''
```

Steps 1 through 4 are identical to the steps used for entity PSM generation. Let us see Steps 5 and 6 (Listing 5.4).

LISTING 5.4 HANDING STATE SPECIFICATION

```
//5. define empty hashtable for State-time-advances
     private Hashtable<String,Double> sta;

//6. define state strings
     «FOR f: e.stas»
           private final String «f.name.toUpperCase»
             = "«f.name»";
     «ENDFOR»
```

These two steps allow declaring a hashtable to store the states of the atomic model and associate with the corresponding time-advance of the state. Step 5 declares a hashtable. Step 6 unpacks the defined states in a DSL and makes them available as static strings in the atomic model. The complete hashtable is built in the *initialize* function described ahead.

Step 7 creates a constructor and unpacks the entity messages and classifies them as DEVS constructs for inports and outports (Listing 5.5). The inports are added with a prefix "in," and the outports with a prefix "out." The `name = ID` concept in the Xtext grammar matches perfectly with the port name concept in DEVS specifications. It also creates test inputs based on the defined inports. For each input message, two test inputs are available: one at elapsed time = 0 and the other at elapsed time = 1 from the time the input is injected manually. The test-input construct is provided to test piecewise the atomic model behavior and aligns itself to the test-driven methodologies in agile programming techniques.

LISTING 5.5 CODE-GEN FOR THE ATOMIC CONSTRUCTOR

```
//7. Create constructor with name
public «e.name»(String name){
      super(name);

      //initialize hashtable for state-time-advances
      sta = new Hashtable<String, Double>();

      //unpack messages
      //add inports
      //add outports
      «FOR f: e.msgs»
            «IF f.type.equals("input")»
            addInport("in«f.name.toFirstUpper»");
            «ENDIF»

            «IF f.type.equals("output")»
            addOutport("out«f.name.toFirstUpper»");
            «ENDIF»
      «ENDFOR»

      //add test inputs
      «FOR f: e.msgs»
      «IF  f.type.equals('input')»
       addTestInput("in«f.name.
       toFirstUpper»", new «f.ref.fullyQualifiedName.
       toString»());
      «ENDIF»
      «IF f.type.equals('input')»
       addTestInput("in«f.name.
       toFirstUpper»", new «f.ref.fullyQualifiedName.
       toString»(),1);
      «ENDIF»
      «ENDFOR»
}
```

Step 8 specifies the initialize function (Listing 5.6). This function is called every time the model is reset and/or constructed the first time after executing the constructor. It unpacks the code segment from the AST for `Atomic.init` element and builds the hashtable for state-time-advance mapping. It unpacks the double value of time-advance, whether it is infinity or non-zero or a declared variable of the type double. Finally, it executes the DEVS library function *holdIn (state, timeAdvance)* that sets the state of the atomic model to the initial state for time-advance stored in the hashtable.

LISTING 5.6 CODE-GEN FOR THE INITIALIZE FUNCTION

```
//8. override initialize()
    @Override
    public void initialize(){
            //unpack initialization code
            «IF e.init.code ! = null»«e.init.code.str»
              «ENDIF»;

            //unpack state-time-advances
            «FOR f: e.stas»
            sta.put(«f.name.toUpperCase»,
              «IF f.timeAdv.inf ! =null» INFINITY
                «ELSE»
                «IF f.timeAdv.^var ! =null»
                  «f.timeAdv.^var.name»
                «ELSE» «f.timeAdv.tav»
            «ENDIF» «ENDIF»);
            «ENDFOR»

            //hold in initial state
            holdIn(«e.init.state.name.toUpperCase»,
            sta.get(«e.init.state.name.toUpperCase»));
    }
```

Step 9 defines the *deltint* behavior in a DSL specified by a collection of `Deltint` elements (Listing 5.7).

LISTING 5.7 CODE-GEN FOR THE DELTINT STRUCTURE

```
//9. override deltint()
    @Override
    public void deltint(){
    «FOR atBeh: e.atBeh»
        «IF atBeh.deltint ! = null»
                //unpack AtomicBeh.deltint
        «atBeh.deltint.compile(engineType)»
        «ENDIF»
    «ENDFOR»
    }
```

The `Deltint` element is unpacked in the `compile (Deltint e)` function specified below in Listing 5.8. There are three elements in the `Deltint` element: source state, target state, and code element. The function *compile* loops through all the `Deltint` elements and put them in an if-else tree for every source state within the `Deltint` elements. Before the library function *holdIn* is called for the target state, the generated code puts a guard condition in between these two steps. The guard condition returns true or false based on the execution of function of the form "processDeltintSrcTarget()." The function by default returns true. The code element specified in `Deltint.code` is put as a string within this defined function for runtime execution. This is an important implementation aspect of runtime code generation and interfacing of a platform-independent DSL with platform-dependent implementation such as Java. Since the Java code is generated at runtime while composing the DSL, it is compiled by the JVM within the integrated Xtext framework, and the compile errors aid writing the procedural logic code. Remember that the DSL does not have the grammar for the procedural logic but relies on the runtime compilation of a DSL to platform-specific implementation to identify errors at compile-time for the platform-specific implementation. Although we succeed in generating a correction compile-time platform-specific code, the runtime platform-specific behavior may still contain the logical errors in our behavior specification. On returning true, the execution proceeds to the target state by holding in the time-advance extracted from the hashtable. On returning false, the execution does not proceed to the target state. The concept will get clearer through an example in the later section of the chapter.

LISTING 5.8 CODE-GEN FOR THE DELTINT FUNCTION

```
def compile(Deltint deltint, String engineType)'''
      «getElsePrefixDeltint»
      if(phaseIs(«deltint.state.name.toUpperCase»)){
            if(«createProcessingInstr("Deltint",deltint.
              state.name,
            deltint.target.name, deltint. code)»){
            holdIn(«deltint.target.name.toUpperCase»,
            sta.get(«deltint.target.name.toUpperCase»));
            }
      }
      «incrementDeltint»
'''

def String createProcessingInstr(String type, String src,
  String target, Code code){
      var processStr = type + src.toFirstUpper +
        target.toFirstUpper
      createProcessingFunction(processStr.toString,
        "process", "", code)
      return "process" + processStr + LB + RB
}
```

Step 10 defines the deltext behavior in DSL specification by a collection of `Deltext` elements (Listing 5.9). The code below also semantically anchors to the DEVS *Continue(e)* function that implies the advancement of elapsed time whenever an external event occurs and is received by the corresponding component.

LISTING 5.9 CODE-GEN FOR THE DELTEXT STRUCTURE

```
//10. override deltext()
      @Override
      public void deltext(double e, message x){
             //advance elapsed time
             Continue(e);
             «FOR atBeh: e.atBeh»
                    «IF atBeh.deltext ! = null»
                           //unpack AtomicBeh.deltext
                           «atBeh.deltext.compile
                             (engineType)»
                    «ENDIF»
             «ENDFOR»
      }
```

There are a couple of elements in the `Deltext` specification. Just to refresh your memory, the EBNF for deltext is provided below again.

```
Deltext:
'deltext' '(''S:'state = [STA]',''X:' '['(in += [Msg])+
  ']')' ' = >'
'S":'(target = [STA])? (res = Resched | cont = 'continue')?
(code = Code)?
;
```

So, we have a source state, **S**; a nonempty set of input messages, **X**; a target state, **S**′; a reschedule component, `Resched`, which reschedules the time-advance of the target state other than the time-advance stored in the hashtable; a continue component, `'continue'`, which keeps the source state as a target state along with resetting the time-advance to `(sigma - e);` and, finally, the logic code component as String in `Code`. The `Deltext` element is unpacked in the `compile (Deltext e)` function specified in Listing 5.10. The grammar allows for multiple inputs in a source state. The semantics 'and's all the input messages with the source state. It implies that all the messages of the input set must arrive at the same time instant. The combination of $S + X_1 + X_2 + \ldots + X_n$ is unique; accordingly, one has to design the state machine for specific cases of $S + X_1$ or $S + X_2$ or $S + X_1 + S_2$. Next, it gathers all the inputs at the respective inports and delegates them to the processing function of the form `processDeltextSourceInputX`. The `Code` element places the String `code` in the generated function that returns a Boolean. On successful compilation and a run-time `true` value, the function holds in the target state. The absence of the target state or the target state being null implies that the feature `'continue'` is being used and the target state is same as the source state. In the case of a non-null value, the time-advance is calculated by the function `deltextTimeAdvance(Deltext deltext)`. The function returns a non-negative value based on `'reschedule'`, which is user specified, or `'continue'`, which is `(sigma - e)`, or the default time-advance from the state-time-advance hashtable. The time-advance string is also shown below. It may be infinity, a positive double, or a variable name declared in the atomic component earlier.

LISTING 5.10 CODE-GEN FOR THE DELTEXT FUNCTION

```
def compile(Deltext deltext, String engineType)'''
    «getElsePrefixDeltext»
        if(phaseIs(«deltext.state.name.toUpperCase»)
        «FOR in: deltext.in»
            && somethingOnPort(x, "in«in.name.
              toFirstUpper»")
        «ENDFOR»){

        «FOR in: deltext.in»
            «in.ref.fullyQualifiedName.toString»
            «in.name.toFirstLower» =
            («in.ref.fullyQualifiedName»)
              getEntityOnPort (x,"in«in.name.
              toFirstUpper»");
        «ENDFOR»

        «IF deltext.target = = null»
            «createDeltextProcessingInstr(deltext)»;

        «ELSE»
            if(«createDeltextProcessingInstr
             (deltext)»){
            «IF deltext.target ! = null»
                holdIn(«deltext.target.name.
                  toUpperCase»,
                  «deltext.deltextTimeAdvance»);
            «ENDIF»
            }
        «ENDIF»
        }
    «incrementDeltext»
'''

def String deltextTimeAdvance(Deltext deltext){
    if(deltext.res ! = null){
        return compile(deltext.res.setSig.timeAdv).
         toString.trim
    } else if (deltext.cont ! = null){
         return "sigma-e"
    } else{
        return "sta.get("+deltext.target.name.
          toUpperCase+")"
    }
}

def compile(TimeAdv ta)'''
    «IF ta.inf ! =null»INFINITY
    «ELSE»
        «IF ta.^var ! = null»«ta.^var.name»
        «ELSE»«ta.tav»
        «ENDIF»
    «ENDIF»
'''
```

Step 11 unpacks the confluent function defined as the `Confluent` element (Listing 5.11). The description is self-explanatory.

LISTING 5.11 CODE-GEN FOR THE DELTCON FUNCTION

```
//11.  override deltcon(){
       @Override
       public void deltcon(double e, message x){
       «FOR atBeh: e.atBeh»
             «IF atBeh.confluent ! = null»
             //unpack AtomicBeh.confluent
             «atBeh.confluent.compile(engineType)»
             «ENDIF»
       «ENDFOR»
       }

def compile(Confluent deltcon, String engineType)'''
      «IF deltcon.con.equals("ignore-input")»
             deltint();«ENDIF»
      «IF deltcon.con.equals("input-only")»
             deltext(e,x);«ENDIF»
      «IF deltcon.con.equals("input-first")»
             deltext(e,x);
             deltint();«ENDIF»
      «IF deltcon.con.equals("input-later")»
             deltint();
             deltext(e,x);«ENDIF»
'''
```

Step 12 unpacks the output function defined as the `Outfn` element (Listing 5.12). It has three elements: a state, a set of output messages, and a code segment. It maps a specific state with a set of nonempty output messages and executes the code specified as String.

LISTING 5.12 CODE-GEN FOR THE OUT FUNCTION STRUCTURE

```
//12.  override out(){
       @Override
       public message out(){
             message m = new message();
             «FOR atBeh: e.atBeh»
                   «IF atBeh.outfn ! = null»
                   //unpack AtomicBeh.outfn
                   «atBeh.outfn.compile(engineType)»
                   «ENDIF»
              «ENDFOR»
              return m;
       }
```

Each element of the `Outfn` type is compiled as per the function specified below. It checks first for the current state of the atomic model. It then creates a processing function of the form `processOutStateS()`. The procedure is similar to the one described for `Deltext` element processing. Finally, it then semantically anchors it with the DEVS message structures by creating contents within a DEVS message (Listing 5.13).

LISTING 5.13 CODE-GEN FOR THE OUT FUNCTION

```
def compile(Outfn outFn, String engineType)'''
      «getElsePrefixOutfn»
      if(phaseIs(«outFn.state.name.toUpperCase»)){
      «FOR out: outFn.out»
            «out.ref.fullyQualifiedName.toString»
            «out.name.toFirstLower» = make«out.name.
              toFirstUpper»();
            «createMakeOutProcessingFunction(out,
                "make", "", null)»
      «ENDFOR»
      «createOutfnProcessingInstr(outFn)»;
      «FOR out: outFn.out»
            m.add(makeContent("out«out.name.
              toFirstUpper»",
            «out.name»));
      «ENDFOR»
      }
      «incrementOutfn»
'''
```

Step 13 deals with accessing the library logging functions of the underlying DEVS engine. It provides a placeholder to semantically anchor the information the end modeler would like to get logged as allowed by the DEVS engine. In DEVSVJAVA, since we are using the simulation viewer, the logging is available as tooltips on atomic components. Consequently, the code (Listing 5.14) generates the function that helps create the dynamic tooltip for each atomic model. Other than the basic information about an atomic model such as model name and sigma, we add the atomic variables defined by the user.

LISTING 5.14 TOOLTIP FUNCTION IN DEVSJAVA FOR VIEWABLEATOMIC

```
//13. override logging function
      @Override
      public String getTooltipText(){
            StringBuilder sb = new StringBuilder();
            sb.append(super.getTooltipText());
```

```
          «FOR f: e.variables»
                sb.append("\n«f.name.toFirstUpper»"+
                  «f.name»);
          «ENDFOR»
          return sb.toString();
     }
```

Step 14, the last step, unpacks all the processing functions that were created while unpacking the Deltint, Deltext, and Outfn elements. The functions were stored in a collection as the atomic model was being unpacked. They are serialized in this step (Listing 5.15).

LISTING 5.15 CODE-GEN TO WRITE ALL THE PROCESSING FUNCTIONS

```
def getProcessingFunctions()'''
      «FOR s: fns»
            «s»
      «ENDFOR»
'''
```

5.2.2.1 Atomic XML Generation

Similar to the EntityName.xml, an AtomicName.xml is generated as shown in Listing 5.16.

LISTING 5.16 CODE-GEN FROM ATOMIC XML

```
def compileXml(Atomic e, String engineType)'''
<atomic name = "«e.name»" package = "

      «IF e.eContainer ! = null »
      «engineType».«e.eContainer.fullyQualifiedName»
        «ENDIF»">
      <extends type = "

      «IF e.superType ! = null» «e.superType.name.
        toFirstUpper»
      «ELSE»«e.superBaseType(engineType)»
      «ENDIF»"/>

      <imports>
           «FOR f: e.getImports(engineType)»
           <import>«f»</import>
           «ENDFOR»
      </imports>
```

```
    <vars>
         «FOR f: e.variables»
         <var name = "«f.name»" type = "«f.type.
           compileXml»"/>
         «ENDFOR»
    </vars>
    <states>
         «FOR f: e.stas»
         <state>«f.name»</state>;
         «ENDFOR»
    </states>
    <inports>
         «FOR f: e.msgs»
         «IF f.type.equals("input")»
           <inport>in«f.
           name.toFirstUpper»</inport>«ENDIF»
         «ENDFOR»
    </inports>
    <outports>
         «FOR f: e.msgs»
         «IF f.type.equals("output")»
         <outport>out«f.name.toFirstUpper»</
           outport>«ENDIF»
         «ENDFOR»
    </outports>
    <initial state = "«e.init.state.name»">
         «IF e.init.code ! = null»
         <code>
                «e.init.code.str»
         </code>
         «ENDIF»
    </initial>
    <time-advance>
         «FOR f: e.stas»
         <ta state = "«f.name»" sigma = "
         «IF f.timeAdv.inf ! = null»INFINITY«ELSE»
               «IF f.timeAdv.^var ! = null»
               «f.timeAdv.^var.name»«ELSE»«f.
                 timeAdv.tav»
               «ENDIF»
         «ENDIF»"/>
         «ENDFOR»
    </time-advance>
    <internal-transition-func>
         «FOR atBeh: e.atBeh»
         «IF atBeh.deltint ! = null»
         <deltint>
               «atBeh.deltint.compileXml»
               «IF atBeh.deltint.code ! = null»
               <code>
                     «atBeh.deltint.code.str»
               </code>
```

```
                «ENDIF»
          </deltint>
                «ENDIF»
          «ENDFOR»
      </internal-transition-func>
      <external-transition-func>
            «FOR atBeh: e.atBeh»
                  «IF atBeh.deltext ! = null»
            <deltext>
                  «atBeh.deltext.compileXml»
                  «IF atBeh.deltext.code ! = null»
                  <code>
                       «atBeh.deltext.code.str»
                  </code>
                  «ENDIF»
            </deltext>
                  «ENDIF»
            «ENDFOR»
      </external-transition-func>
      <out-func>
            «FOR atBeh: e.atBeh»
                  «IF atBeh.outfn ! = null»
            <lamda>
                  «atBeh.outfn.compileXml»
                  «IFatBeh.outfn.code ! = null»
                  <code>
                        «atBeh.outfn.code.str»
                  </code>
                  «ENDIF»
            </lamda>
                  «ENDIF»
            «ENDFOR»
      </out-func>
      <confluent-func
      «FOR atBeh: e.atBeh»«IF atBeh.confluent ! = null»
        type = "«atBeh.confluent.compileXml»"«ENDIF»
      «ENDFOR»/>
</atomic>
'''
```

This completes the semantic anchoring of atomic DEVSML to a specific DEVS engine such as DEVSJAVA. Next, we will look at the code generation features for a coupled model.

5.2.3 Coupled

Generation of a DEVS coupled model is similar to that of the atomic model. It is a five-step process:

1. Create class
2. Create components

3. Create inports/outports
4. Create test inputs
5. Create couplings

The Coupled element is unpacked as shown in Listing 5.17.

LISTING 5.17 CODE-GEN FROM COUPLED DEVS

```
def compile (Coupled e, String engineType)'''

      //1. Create class
      «IFe.eContainer ! = null »
           package «e.eContainer.fullyQualifiedName»;
      «ENDIF»
      import java.util.*;
      import GenCol.*;
      import simView.*;
      import genDevs.modeling.*;

      public class «e.name» extends«IF e.superType !
        = null»
      «e.superType.fullyQualifiedName»«ELSE»
        ViewableDigraph«ENDIF»{

             public «e.name»(){
                   this("«e.fullyQualifiedName»");
             }

             public «e.name»(String name){
                    super(name);

                    //2. Create components
                    «FOR f: e.components»
                         «f.compile(engineType)»
                    «ENDFOR»

                    //3. Create inports/outports
                    «FOR f: e.msgs»
                           «IF f.type.equals("input")»
                           addInport("in«f.name.
                             toFirstUpper»");«ENDIF»
                           «IF f.type.equals("output")»
                           addOutport("out«f.name.
                             toFirstUpper»");«ENDIF»
                    «ENDFOR»

                    //4. Create test inputs
                    «FOR f: e.msgs»
                          «IF f.type.equals('input')»
                          addTestInput("in«f.name.
                            toFirstUpper»",
```

```
                    new «f.ref.fullyQualifiedName.
                      toString»());
                    «ENDIF»

                    «IF f.type.equals('input')»
                    addTestInput("in«f.name.
                      toFirstUpper»",
                    new «f.ref.fullyQualifiedName.
                      toString»(),1);
                    «ENDIF»
                «ENDFOR»

                //5. Create Couplings
                «FOR f: e.couplings»
                    «f.compile»
                «ENDFOR»
            }
        }
    '''
```

Step 2 unpacks a component, which may be an atomic or a coupled component as per DEVS semantics. The Xtext framework is seamlessly integrated with the EMF workbench. As a result, searching by the `name = ID` construct and the cross references to the actual object (using imports) are readily available. Since in DEVSML we use the Xtext references, we gain access to the actual atomic or coupled model that may be specified in some other file. We are then able to semantically anchor it to the DEVS coupled specification by instantiating the appropriate component. Finally, we add it to the current coupled model (Listing 5.18).

LISTING 5.18 CODE-GEN FOR RETRIEVING A SUBCOMPONENT

```
def compile(Component e, String engineType)'''
    «e.eCrossReferences.get(0).fullyQualifiedName.
       toString» «e.name»
    = new «e.eCrossReferences.get(0).fullyQualifiedName»
    ("«e.fullyQualifiedName.toString.toFirstUpper»");
    add(«e.name»);
```

The unpacking of inports, outports, and test inputs is similar to the `Atomic` element and is omitted for obvious reasons. The last step, Step 5, is the unpacking of the `Coupling` element. It has three parts: (1) external input coupling (EIC element), (2) external output coupling (EOC element), and (3) the internal coupling (IC element). Each coupling has a source component, a source port, a destination component, and a destination port. Again, a heavy use of the `name = ID` construct of Xtext is leveraged here (Listing 5.19).

LISTING 5.19 CODE-GEN FOR COUPLING SPECIFICATIONS

```
def compile(Coupling e, String engineType)'''
      «IF e.eic ! = null»
      addCoupling(this,
      "in«e.eic.msgtype.name.toFirstUpper»",
        «e.eic.dest.name»,
      "in«e.eic.destMsgType.name.toFirstUpper»");
      «ENDIF»

      «IF e.eoc ! = null»
      addCoupling(«e.eoc.src.name»,
        "out«e.eoc.msgtype.name.toFirstUpper»",this,
        "out«e.eoc.destMsgType.name.toFirstUpper»");
      «ENDIF»

      «IF e.ic ! = null»
      addCoupling(«e.ic.src.name»,
        "out«e.ic.msgtype.name.toFirstUpper»",
         «e.ic.dest.name»,
         "in«e.ic.destMsgType.name.toFirstUpper»");
      «ENDIF»
'''
```

5.2.4 Execution of DEVS Models

Finally, once the entities, atomics, and coupled models are generated, next is the simulation viewer or coordinator that executes the root-coupled model. As we are semantically linked to the DEVSVJAVA engine, we add the DEVSJAVA engine as a dependency for our generated code and create a dummy wrapper class that invokes the simulation viewer SimView. The generated codebase also contains a Main.java file that invokes the SimView DEVSJAVA viewer (Listing 5.20).

LISTING 5.20 MAINCLASS FILE TO INVOKE DEVSJAVA SIMVIEWER

```
def compile(Model m, String engineType)'''
      package run;

      import simView.SimView;

      public class Main {

            public static void main(String… args){
                  new SimView();
            }

      }
'''
```

This section provided details about the semantic anchoring of various DEVSML types to a DEVS simulation engine such as DEVSJAVA. The next section will describe an example in detail that shows the usage of the DEVSML DSL and its integration with an execution framework.

5.3 ILLUSTRATION

5.3.1 Installation and Setup

Check the website http://duniptechnologies.com/book/sos for setting up the fds editing and execution environment.

5.3.2 EFP Model

To illustrate the capability of the DEVSML, we use the classic EFP hierarchical model. The model is first described in Chapter 4. EFP contains two models: coupled EF and atomic Proc. The coupled EF internally contains two models: atomic Genr and atomic Transd. The generator Genr sends the entity Job at a periodic rate. The processor Proc receives the generated Job and gets busy processing it. On completion, it reports it to the transducer Transd, which keeps a count of generated and processed jobs. The Transd and Genr are part of the Experimental frame EF. The Transd has an observation time after which it reports the throughput.

5.3.2.1 Entities

There are four entities in the system: (1) Job, (2) Start, (3) Stop, and (4) Result. Job is exchanged between Genr, Proc, and Transd. Start is received by Genr to begin generating jobs. Stop is received by Genr to stop generation. Result is sent by Transd once the observation period is complete. The entities exchanged between the components are shown in Figure 5.1. The entities are in the package ent. The outline view shows the package containment. Figure 5.2 shows the generated Java code for the entity Result.

5.3.2.2 Generator (Genr) Model

Figure 5.3 shows the Genr DEVSML model in the Eclipse workbench. The first column shows the gpt.fds file and the package *gpAtomics* containing the atomic Genr

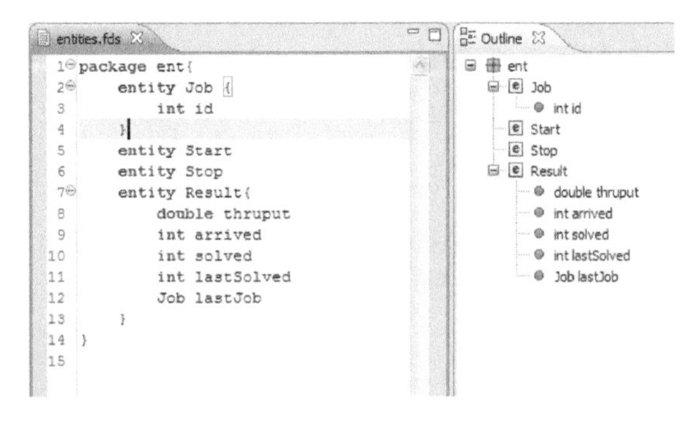

FIGURE 5.1 DEVSML entity.

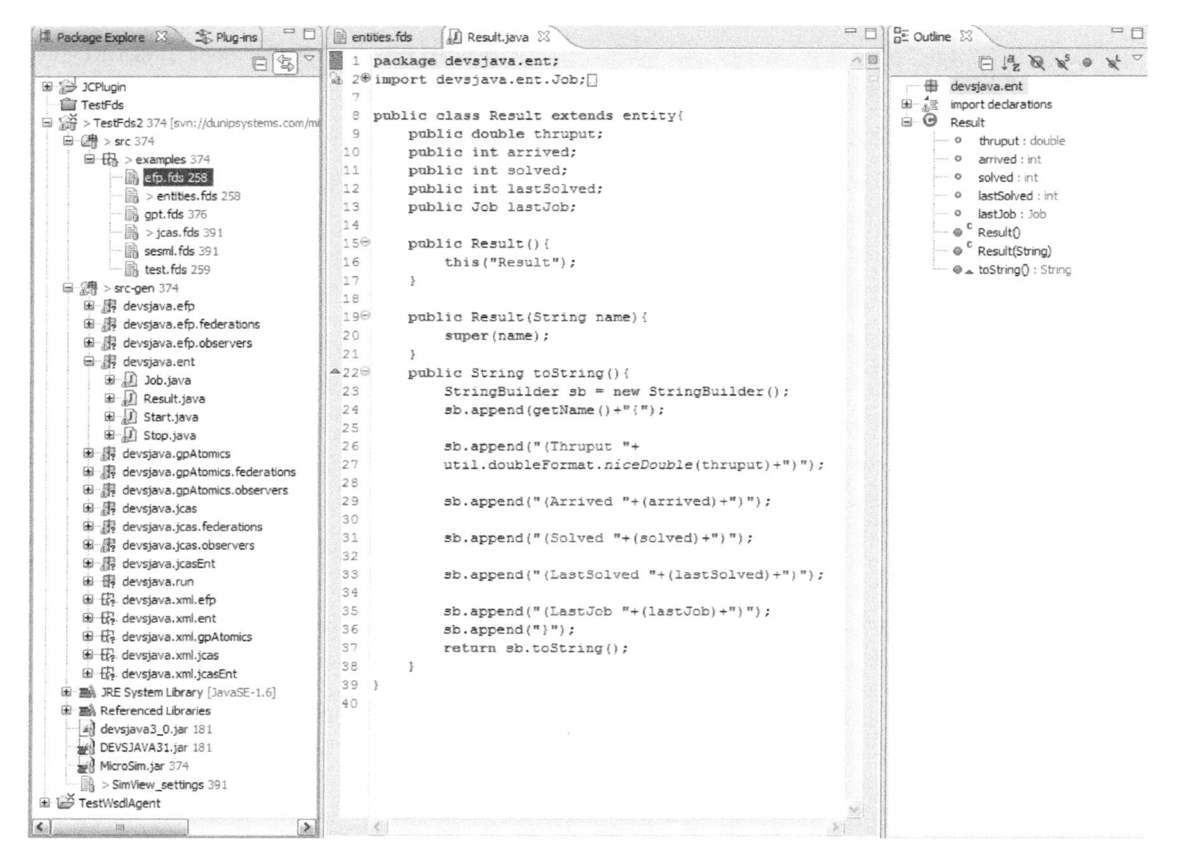

FIGURE 5.2 Generated Result entity.

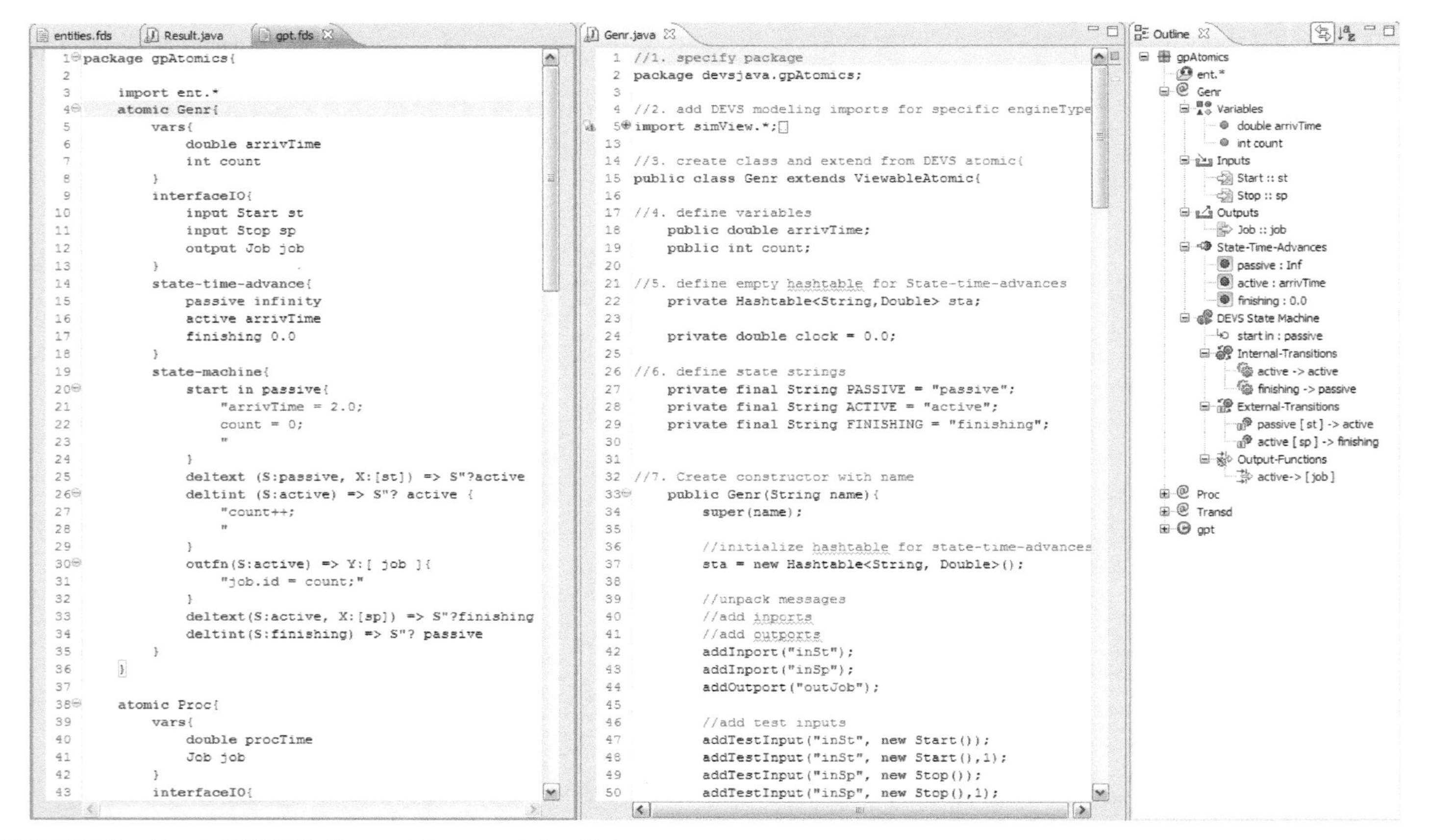

FIGURE 5.3 Atomic DEVSML Genr and the generated artifacts.

DEVSML specification. Notice the keyword highlighting based on the fds EBNF grammar and state-machine specification statements starting with keywords deltint, deltext, outfn, and so on. The second column shows the auto-generated Genr.java file that is synchronized with the Genr.fds file. Every Save operation results in code sync from `.fds` PIM to Java PSM. Notice the generated Java code in the second column. It synthesizes the inports, the outports, and the test-inputs. Further, the state-time-advances are stored in a hashtable and the time-advance for a state is retrieved from it at runtime. The last column on the right shows the hierarchical outline structure of the Genr.fds file with rich icons and other supplemental information. The generated files are held in the src-gen folder that is compiled using the DEVSJAVA libraries on the project classspath.

The complete generated Java code for Genr.java is shown in Listing 5.21.

LISTING 5.21 AUTO-GENERATED CODE FOR GENR

```
//1.  specify package
package devsjava.gpAtomics;

//2.  add DEVS modeling imports for specific engineType
import simView.*;
import devsjava.ent.Job;
import devsjava.ent.Stop;
import java.util.*;
import GenCol.*;
import java.awt.*;
import devsjava.ent.Start;
import genDevs.modeling.*;

//3.  create class and extend from DEVS atomic{
public class Genr extends ViewableAtomic {

        //4.  define variables
        public double arrivTime;
        public int count;

       //5.  define empty hashtable for State-time-advances
         private Hashtable<String, Double> sta;

       private double clock = 0.0;

       //6.  define state strings
       private final String PASSIVE = "passive";
       private final String ACTIVE = "active";
       private final String FINISHING = "finishing";

       //7. Create constructor with name
       public Genr(String name) {
            super(name);
```

```
        //initialize hashtable for state-time-advances
          sta = new Hashtable<String, Double>();

        //unpack messages
        //add inports
        //add outports
        addInport("inSt");
        addInport("inSp");
        addOutport("outJob");

        //add test inputs
        addTestInput("inSt", new Start());
        addTestInput("inSt", new Start(), 1);
        addTestInput("inSp", new Stop());
        addTestInput("inSp", new Stop(), 1);
    }

    //8. override initialize()
    @Override
    publicvoid initialize() {
        //unpack initialization code
        arrivTime = 2.0;
        count = 0;
        ;

        //unpack state-time-advances
        sta.put(PASSIVE, INFINITY);
        sta.put(ACTIVE, arrivTime);
        sta.put(FINISHING, 0.0);

        //hold in initial state
        holdIn(PASSIVE, sta.get(PASSIVE));
        clock = 0.0;
        showState();
    }

    //9. override deltint()
    @Override
    public void deltint() {
        clock+ = sigma;
        //unpack AtomicBeh.deltint
        if (phaseIs(ACTIVE)) {
            if (processDeltintActiveActive()) {
                holdIn(ACTIVE, sta.get(ACTIVE));
            } else {
                passivate();
        }
    }
```

```
            //unpack AtomicBeh.deltint
            else if (phaseIs(FINISHING)) {
                 if (processDeltintFinishingPassive()) {
               holdIn(PASSIVE, sta.get(PASSIVE));
            } else {
                   passivate();
            }
      } else
         passivate();
      showState();
   }

   //10. override deltext()
   @Override
   public void deltext(double e, message x) {
          //advance elapsed time
          Continue(e);
          clock + = e;
          //unpack AtomicBeh.deltext
          if (phaseIs(PASSIVE) && somethingOnPort
            (x,"inSt")) {
                 java.util.List<Start> stList = new
                   ArrayList<Start>();
                 for (int i = 0; i < x.getLength(); i++) {
                        if (somethingOnPort(x, "inSt")) {
                               Start st = (Start)
                                 getEntityOnPort(x, "inSt");
                               stList.add(st);
                               showInput(st);
                        }
                 }
                 processDeltextPassiveSt(stList);
                 holdIn(ACTIVE, sta.get(ACTIVE));
   }
   //unpack AtomicBeh.deltext
   elseif (phaseIs(ACTIVE) && somethingOnPort(x, "inSp")) {
                 java.util.List<Stop> spList = new
                   ArrayList<Stop>();
                 for (int i = 0; i < x.getLength(); i++) {
                      if (somethingOnPort(x, "inSp")) {
                               Stop sp = (Stop)
                                 getEntityOnPort(x, "inSp");
                               spList.add(sp);
                               showInput(sp);
                      }
                 }
                 processDeltextActiveSp(spList);
                 holdIn(FINISHING, sta.get(FINISHING));
```

```
        }
        showState();
}

//11. override deltcon(){
@Override
public void deltcon(double e, message x) {
}

//12. override out(){
@Override
public message out() {
        message m = new message();
        //unpack AtomicBeh.outfn
        if (phaseIs(ACTIVE)) {
                Job job = makeJob();
                processOutActiveJob(job);
                m.add(makeContent("outJob", job));
                showOutput(job);
        }
        return m;
}

//13. override logging function
@Override
public String getTooltipText() {
       StringBuilder sb = newStringBuilder();
       sb.append(super.getTooltipText());

       sb.append("\nArrivTime" + arrivTime);
       sb.append("\nCount" + count);
       return sb.toString();
}

public void showOutput(entity en) {
        System.out.println((clock + sigma) + "\t\t" +
          getName() +
        "\t\tOUTPUT:"+ en.getName());
}

public void showInput(entity en) {
        System.out.println(clock + "\t\t" + getName() +
          "\t\tINPUT:"+ en.getName());
}
public void showState() {
        System.out.println(clock + "\t\t" + getName() +
          "\t\tSTATE:"
          + getFormattedPhase() + ",SIGMA:" +
```

```
            getFormattedSigma());
  }

  //14.  write processing function

  private boolean processDeltintActiveActive() {
        boolean returnVal = true;
        count++;
        return returnVal;
  }

  private boolean processDeltintFinishingPassive() {
        boolean returnVal = true;
        return returnVal;
  }

  private boolean processDeltextPassiveSt
    (java.util.List<Start> st) {
        boolean returnVal = true;
        return returnVal;
  }

  private boolean processDeltextActiveSp
    (java.util.List<Stop> sp) {
        boolean returnVal = true;
        return returnVal;
  }

  private Job makeJob() {
        Job job = newJob();
        return job;
  }

  private boolean processOutActiveJob(Job job) {
        boolean returnVal = true;
        job.id = count;
        return returnVal;
  }

  //15. End class
  }
```

5.3.2.3 Processor (Proc) Model

The DEVSML Proc model with an outline view is shown in Figure 5.4.

The generated Java code is shown in Listing 5.22.

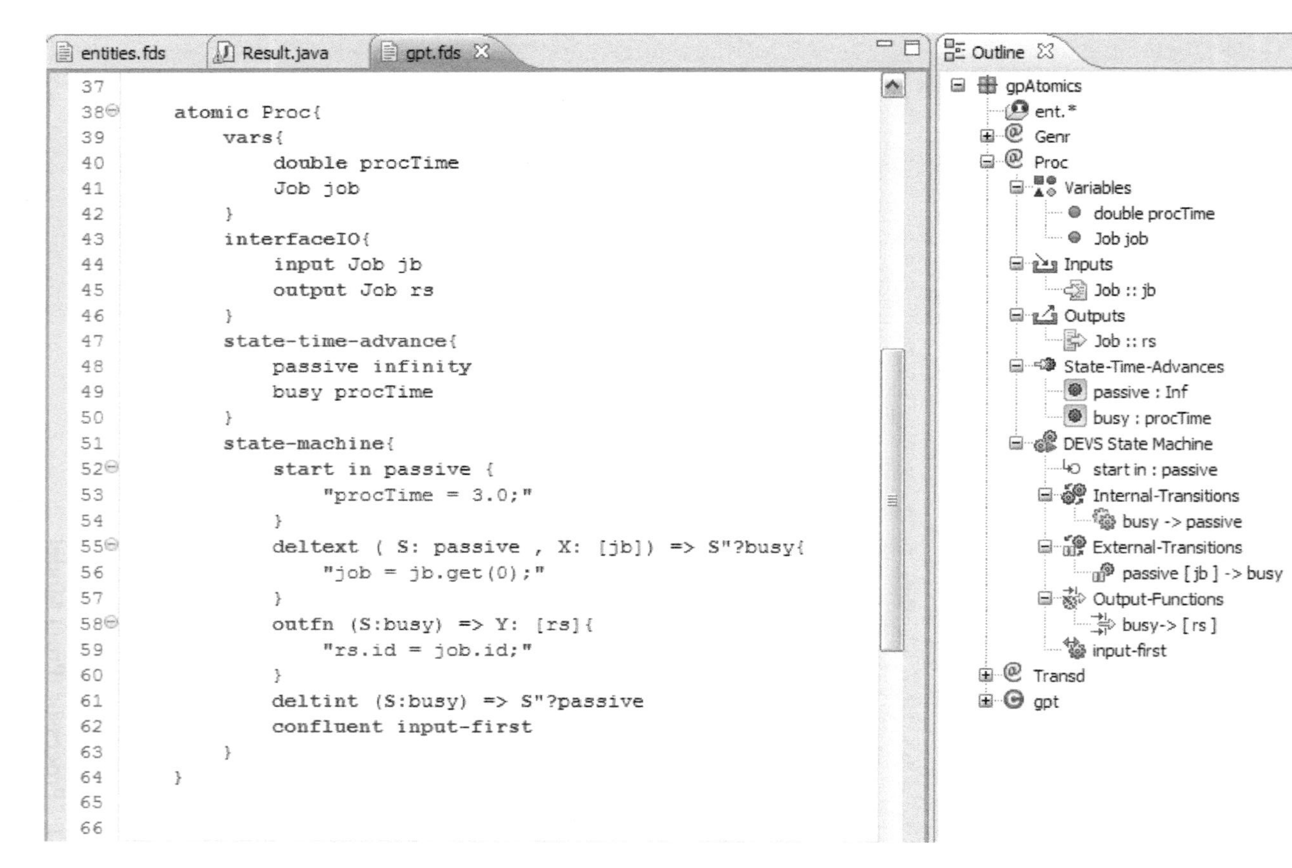

FIGURE 5.4 Proc model.

LISTING 5.22 AUTO-GENERATED CODE FOR PROC

```
//1. specify package
package devsjava.gpAtomics;

//2. add DEVS modeling imports for specific engineType
import simView.*;
import devsjava.ent.Job;
import java.util.*;
import GenCol.*;
import java.awt.*;
import genDevs.modeling.*;

//3. create class and extend from DEVS atomic{
public class Proc extends ViewableAtomic {

     //4. define variables
     public double procTime;
     public Job job;

     //5. define empty hashtable for State-time-advances
     private Hashtable<String, Double> sta;

     private double clock = 0.0;

     //6. define state strings
     private final String PASSIVE = "passive";
     private final String BUSY = "busy";

     //7. Create constructor with name
     public Proc(String name) {
       super(name);

       //initialize hashtable for state-time-advances
       sta = new Hashtable<String, Double>();

       //unpack messages
       //add inports
       //add outports
       addInport("inJb");
       addOutport("outRs");

       //add test inputs
       addTestInput("inJb", new Job());
       addTestInput("inJb", new Job(), 1);
     }
```

```
//8. override initialize()
@Override
public void initialize() {
    //unpack initialization code
    procTime = 3.0;
    ;

    //unpack state-time-advances
    sta.put(PASSIVE, INFINITY);
    sta.put(BUSY, procTime);

    //hold in initial state
    holdIn(PASSIVE, sta.get(PASSIVE));
    clock = 0.0;
    showState();
}

//9. override deltint()
@Override
public void deltint() {
  clock + = sigma;
  //unpack AtomicBeh.deltint
  if (phaseIs(BUSY)) {
          if (processDeltintBusyPassive()) {
                  holdIn(PASSIVE, sta.get(PASSIVE));
          } else {
                passivate();
          }
  } else
        passivate();
  showState();
}

//10. override deltext()
@Override
publicvoid deltext(double e, message x) {
  //advance elapsed time
  Continue(e);
  clock + = e;
  //unpack AtomicBeh.deltext
  if (phaseIs(PASSIVE) && somethingOnPort(x, "inJb")) {
          java.util.List<Job> jbList = new
            ArrayList<Job>();
          for (int i = 0; i < x.getLength(); i++) {
                  if (somethingOnPort(x, "inJb")) {
                  Job jb = (job) getEntityOnPort(x,
                    "inJb");
```

```
                    jbList.add(jb);
                    showInput(jb);
                    }
            }
            processDeltextPassiveSt(stList);
            holdIn(ACTIVE, sta.get(ACTIVE));
    }
    showState();
}

    //11. override deltcon(){
    @Override
    public void deltcon(double e, message x) {
        //unpack AtomicBeh.confluent
        deltext(e, x);
        deltint();
    }

    //12. override out(){
    @Override
    public message out() {
          message m = new message();
          //unpack AtomicBeh.outfn
          if (phaseIs(BUSY)) {
                Job rs = makeRs();
                processOutBusyRs(rs);
                m.add(makeContent("outRs", rs));
                showOutput(rs);
          }
          return m;
    }
    //13. override logging function
    @Override
    public String getTooltipText() {
        StringBuilder sb = new StringBuilder();
        sb.append(super.getTooltipText());

          sb.append("\nProcTime" + procTime);
          sb.append("\nJob" + job);
          return sb.toString();
    }

    public void showOutput(entity en) {
        System.out.println((clock + sigma) + "\t\t" +
          getName() + "\t\tOUTPUT:"
                    + en.getName());
    }
```

```
    public void showInput(entity en) {
        System.out.println(clock + "\t\t" + getName() +
          "\t\tINPUT:"
                    + en.getName());
    }

    public void showState() {
        System.out.println(clock + "\t\t" + getName() +
          "\t\tSTATE:"
        + getFormattedPhase() + ",SIGMA:" +
          getFormattedSigma());
    }

   //14.  write processing function

   private boolean processDeltintBusyPassive() {
       boolean returnVal = true;
       count++;
       return returnVal;
   }

   private boolean processDeltextPassiveJb(java.util.
     List<Job> jb) {
       boolean returnVal = true;
       job = jb.get(0);
       return returnVal;
   }

   private Job makeJob() {
         Job rs = new Job();
         return rs;
   }

   private boolean processOutBusyRs(Job rs) {
         boolean returnVal = true;
         rs.id = job.id;
         return returnVal;
   }

   //15. End class
}
```

5.3.2.4 Transducer (Transd) Model

The DEVSML model with an outline view is shown in Figure 5.5.

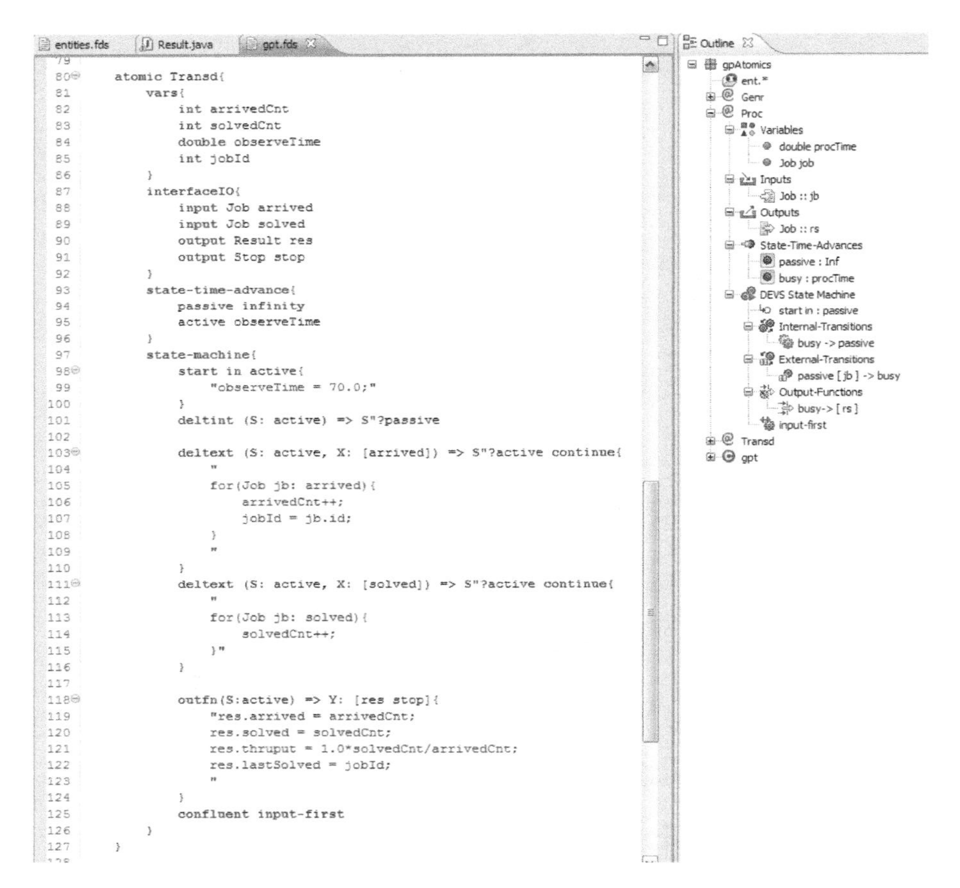

FIGURE 5.5 Transd model.

The generated Java code for Transd is shown in Listing 5.23.

LISTING 5.23 AUTO-GENERATED TRANSD MODEL

```
//1.  specify package
package devsjava.gpAtomics;

//2.  add DEVS modeling imports for specific engineType
import simView.*;
import devsjava.ent.Result;
import devsjava.ent.Job;
import devsjava.ent.Stop;
import java.util.*;
import GenCol.*;
import java.awt.*;
import genDevs.modeling.*;

//3.  create class and extend from DEVS atomic{
public class Transd extends ViewableAtomic {

     //4.  define variables
     public int arrivedCnt;
     public int solvedCnt;
     public double observeTime;
     public int jobId;

     //5.  define empty hashtable for State-time-advances
     private Hashtable<String, Double> sta;

     private double clock = 0.0;

     //6.  define state strings
     private final String PASSIVE = "passive";
     private final String ACTIVE = "active";

     //7.  Create constructor with name
     public Transd(String name) {
        super(name);

        //initialize hashtable for state-time-advances
        sta = new Hashtable<String, Double>();
        //unpack messages
        //add inports
        //add outports
        addInport("inArrived");
        addInport("inSolved");
        addOutport("outRes");
        addOutport("outStop");

        //add test inputs
```

```
        addTestInput("inArrived", new Job());
        addTestInput("inArrived", new Job(), 1);
        addTestInput("inSolved", new Job());
        addTestInput("inSolved", new Job(), 1);
    }

    //8. override initialize()
    @Override
    public void initialize() {
        //unpack initialization code
        observeTime = 70.0;
        ;

        //unpack state-time-advances
        sta.put(PASSIVE, INFINITY);
        sta.put(ACTIVE, observeTime);

        //hold in initial state
        holdIn(ACTIVE, sta.get(ACTIVE));
        clock = 0.0;
        showState();
    }

    //9. override deltint()
    @Override
    public void deltint() {
        clock+ = sigma;
        //unpack AtomicBeh.deltint
        if (phaseIs(ACTIVE)) {
             if (processDeltintActivePassive()) {
                    holdIn(PASSIVE, sta.get(PASSIVE));
             } else {
                    passivate();
             }
        } else
             passivate();
        showState();
    }

    //10. override deltext()
    @Override
    public void deltext(double e, message x) {
        //advance elapsed time
        Continue(e);
        clock+ = e;
        //unpack AtomicBeh.deltext
    if (phaseIs(ACTIVE) && somethingOnPort(x,
      "inArrived")) {
            java.util.List<Job> arrivedList = new
              ArrayList<Job>();
```

```
        for (int i = 0; i < x.getLength(); i++) {
                if (somethingOnPort(x, "inArrived")) {
                        Job arrived = (Job)
                          getEntityOnPort(x, "inArrived");
                        arrivedList.add(arrived);
                        showInput(arrived);
                }
        }
        processDeltextActiveArrived(arrivedList);
        holdIn(ACTIVE, sigma - e);
    }
    //unpack AtomicBeh.deltext
    else if (phaseIs(ACTIVE) && somethingOnPort(x,
      "inSolved")) {
        java.util.List<Job> solvedList =
          new ArrayList<Job>();
        for (int i = 0; i < x.getLength(); i++) {
                if (somethingOnPort(x, "inSolved")) {
                        Job solved = (Job)
                          getEntityOnPort(x, "inSolved");
                        solvedList.add(solved);
                        showInput(solved);
                }
        }
        processDeltextActiveSolved(solvedList);
        holdIn(ACTIVE, sigma - e);
    }
    showState();
}

    //11. override deltcon(){
    @Override
    public void deltcon(double e, message x) {
        //unpack AtomicBeh.confluent
        deltext(e, x);
        deltint();
    }

    //12. override out(){
    @Override
    public message out() {
        message m = new message();
        //unpack AtomicBeh.outfn
        if (phaseIs(ACTIVE)) {
              Result res = makeRes();
              Stop stop = makeStop();
              processOutActiveResStop(res, stop);
              m.add(makeContent("outRes", res));
              showOutput(res);
              m.add(makeContent("outStop", stop));
```

```
        showOutput(stop);
      }
      return m;
    }

    //13. override logging function
    @Override
    public String getTooltipText() {
      StringBuilder sb = new StringBuilder();
      sb.append(super.getTooltipText());

      sb.append("\nArrivedCnt" + arrivedCnt);
      sb.append("\nSolvedCnt" + solvedCnt);
      sb.append("\nObserveTime" + observeTime);
      sb.append("\nJobId" + jobId);
      return sb.toString();
    }

    public void showOutput(entity en) {
      System.out.println((clock + sigma) + "\t\t" +
        getName() + "\t\tOUTPUT:" + en.getName());
    }

    public void showInput(entity en) {
      System.out.println(clock + "\t\t" + getName() +
        "\t\tINPUT:" + en.getName());
    }

    public void showState() {
      System.out.println(clock + "\t\t" + getName() +
        "\t\ tSTATE:"
        + getFormattedPhase() + ",SIGMA:" +
        getFormattedSigma());

    }

    //14. write processing function
    private boolean processDeltintActivePassive() {
      boolean returnVal = true;
      return returnVal;
    }

    private boolean processDeltextActiveArrived(java.
      util.List<Job> arrived) {
      boolean returnVal = true;

      for (Job jb : arrived) {
          arrivedCnt++;
          jobId = jb.id;
      }
```

```
            return returnVal;
        }

        private boolean processDeltextActiveSolved(
          java.util.List<Job> solved) {
            boolean returnVal = true;

            for (Job jb : solved) {
                    solvedCnt++;
            }
            return returnVal;
        }

        private Result makeRes() {
            Result res = new Result();
            return res;
        }

        private Stop makeStop() {
            Stop stop = new Stop();
            return stop;
        }

        private boolean processOutActiveResStop(Result res,
          Stop stop) {
            boolean returnVal = true;
            res.arrived = arrivedCnt;
            res.solved = solvedCnt;
            res.thruput = 1.0 * solvedCnt/arrivedCnt;
            res.lastSolved = jobId;
            return returnVal;
        }

        //15. End class
}
```

5.3.2.5 EF and EFP Coupled Models

Figure 5.6 shows the two coupled models EF and EFP and is self-explanatory. The outline is shown in the last column.

The generated code for EF and EFP is shown in Listing 5.24 and 5.25, respectively.

5.3.2.6 Executable File

A handle to the DEVS simulation viewer (Listing 5.26) is also generated that the user can run as an application to access the DEVSJAVA library.

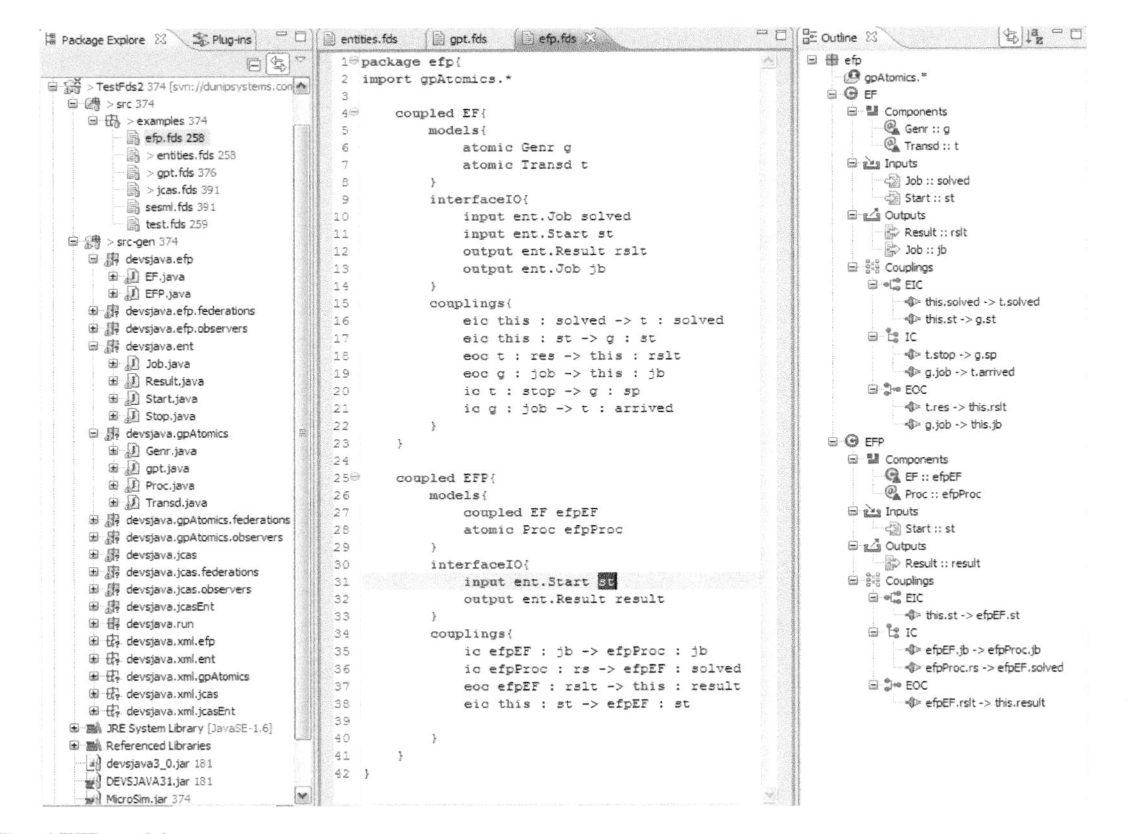

FIGURE 5.6 EF and EFP models.

LISTING 5.24 AUTO-GENERATED CODE FOR EF

```
package devsjava.efp;

import genDevs.simulation.coordinator;
import simView.*;
import java.util.*;
import GenCol.*;
import java.awt.*;
import devsjava.gpAtomics.Transd;
import genDevs.modeling.*;
import devsjava.gpAtomics.Genr;

public class EF extends ViewableDigraph {

      public static void main(String… args) {
            coordinator c = new coordinator(new EF());
            c.initialize();
            c.simulate(2000);
}

public EF() {
          this("efp.EF");
      }

      public EF(String name) {
            super(name);
            Genr g = newGenr("Efp.EF.g");
            add(g);
            Transd t = newTransd("Efp.EF.t");
            add(t);

            addInport("inSolved");
            addInport("inSt");
            addOutport("outRslt");
            addOutport("outJb");

            addTestInput("inSolved", new devsjava.ent.
              Job());
            addTestInput("inSolved", new devsjava.ent.
              Job(), 1);
            addTestInput("inSt", new devsjava.ent.
              Start());
            addTestInput("inSt", new devsjava.ent.
              Start(), 1);

            addCoupling(this, "inSolved", t, "inSolved");
            addCoupling(this, "inSt", g, "inSt");
            addCoupling(t, "outRes", this, "outRslt");
```

```
                addCoupling(g, "outJob", this, "outJb");
                addCoupling(t, "outStop", g, "inSp");
                addCoupling(g, "outJob", t, "inArrived");
        }
}
```

LISTING 5.25 AUTO-GENERATED CODE FOR EFP

```
package devsjava.efp;

import genDevs.simulation.coordinator;
import devsjava.efp.EF;
import simView.*;
import java.util.*;
import GenCol.*;
import java.awt.*;
import devsjava.gpAtomics.Proc;
import genDevs.modeling.*;

public class EFP extends ViewableDigraph {

    public static void main(String... args) {
        coordinator c = new coordinator(new EFP());
        c.initialize();
        c.simulate(2000);
    }

    public EFP() {
        this("efp.EFP");
    }

    public EFP(String name) {
        super(name);
        EF efpEF = new EF("Efp.EFP.efpEF");
        add(efpEF);
        Proc efpProc = new Proc("Efp.EFP.efpProc");
        add(efpProc);

        addInport("inSt");
        addOutport("outResult");
        if (observed) {

        addTestInput("inSt", new devsjava.ent.Start());
        addTestInput("inSt", new devsjava.ent.Start(), 1);

        addCoupling(efpEF, "outJb", efpProc, "inJb");
```

```
        addCoupling(efpProc, "outRs", efpEF, "inSolved");
        addCoupling(efpEF, "outRslt", this, "outResult");
        addCoupling(this, "inSt", efpEF, "inSt");
    }
}
```

LISTING 5.26 MAIN CLASS TO INVOKE DEVSJAVA SIMVIEWER

```
package devsjava.run;

import simView.SimView;

public class Main {

      public staticvoid main(String… args){
            new SimView();
      }

}
```

Figure 5.7 shows the fully functional model along with tooltip for Transd.

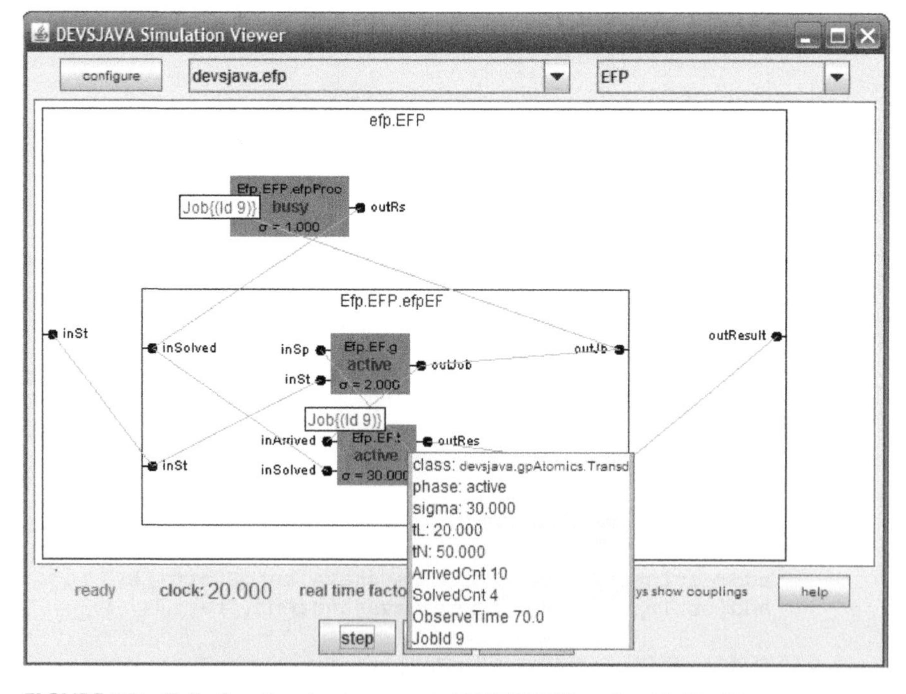

FIGURE 5.7 Fully functional auto-generated DEVSJAVA code with SimView.

5.4 NATURAL LANGUAGE DEVS: ANOTHER DEVS DSL

Now, having described the DEVSML platform-independent language that is semantically anchored to multiple DEVS engines, such as DEVSJAVA, MicroSim, and MicroNet, we would like to go a step further and design another DSL that can be transformed to DEVSML.

We designed another grammar based on a structured natural language and call it Natural Language DEVS (NLDEVS). The earlier version of NLDEVS was bounded by the XFDDEVS specification. Consequently, it had many inherent limitations. NLDEVS is mentioned here to establish the concrete nature of the DEVSML. NLDEVS can be directly mapped either into the platform-independent DEVSML or into the DEVS platform-specific implementation. The relationship between NLDEVS, DEVSML, and DEVS is summarized in Figure 5.8. Figure 5.9 shows transformations where NLDEVS is transformed to the DEVSML. A sample of atomic NLDEVS in Figure 5.9 shows the state-machine description of the Genr atomic component.

As can be seen, the keywords of NLDEVS are totally different from that of the DEVSML. The NLDEVS editor is replete with code completion and model checking such that the designed model is DEVS-correct. The auto-generated DEVSVML Genr.fds can be seen in the right column, and any validation checks at the DEVSML layer are made visible during the incremental save operations.

5.5 SUMMARY

Almost every DSL at some point is transformed to an executable code in a programming language, such as C, C++, Java, Python, etc. While coding in a programming language gives the developer full control of the underlying platform, it is no different from coding in an assembly language in older days. With computers, an integral part

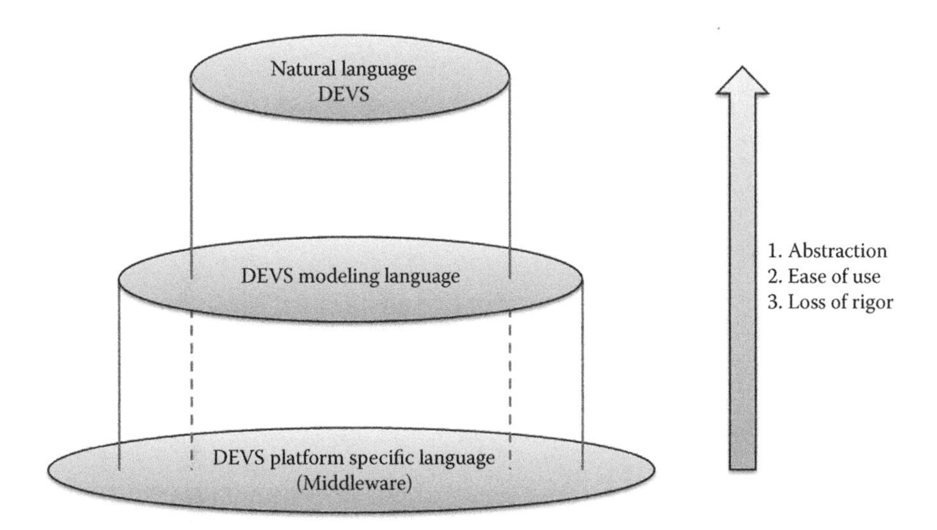

FIGURE 5.8 Moving up the abstraction and loss of rigor.

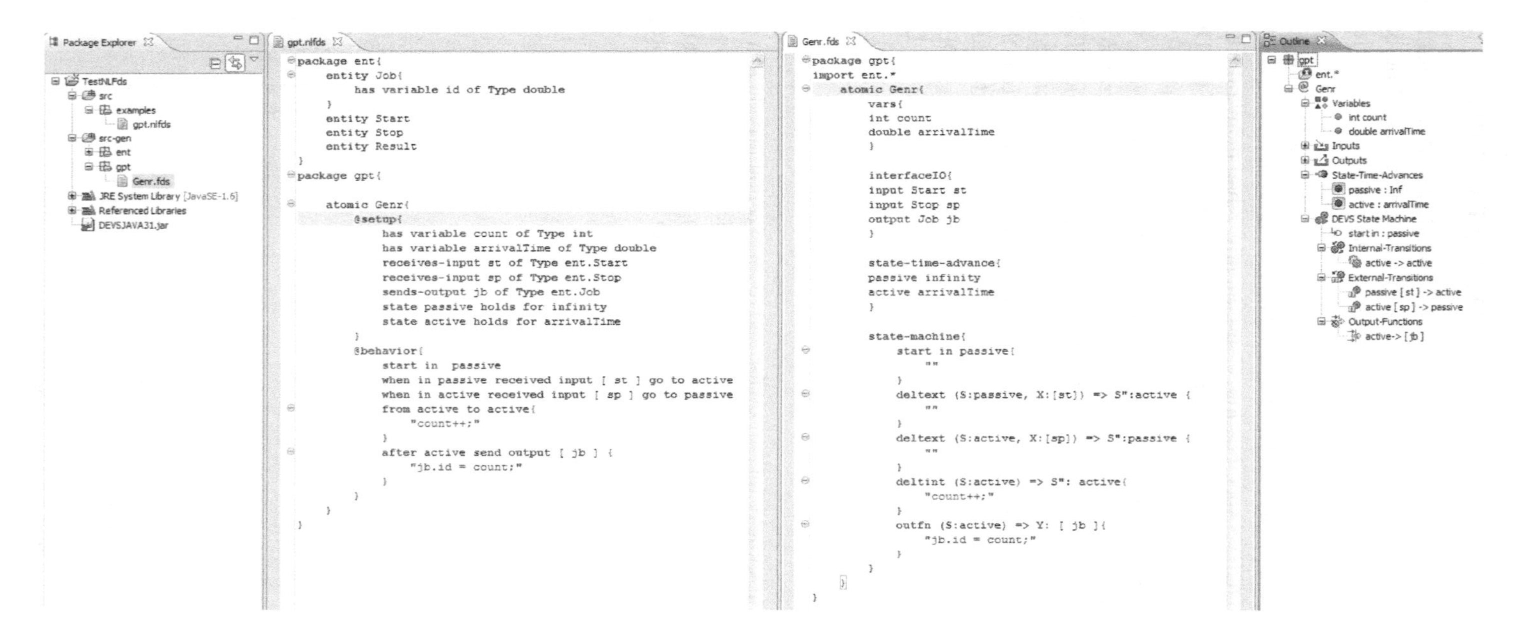

FIGURE 5.9 NLDEVS description of atomic Genr.

of almost every daily activity and every business enterprise, the code developers are more domain-focused than ever before. Lack of tools that help developers to focus on the domain of interest is proving to be an impediment in developing a semantically rich code. Agreed, one loses power and control when abstractions are developed. However, what is gained is the domain context and usability that is transparent to the programming language underneath.

The XML-based XFDDEVS was a significant development toward the development of the netcentric DEVS Unified Process. However, it had many shortcomings, such as no confluent function, no multiple inputs, no multiple outputs, no complex message types, and no state variables. With the latest DEVS Modeling Language, all the shortcomings of XFDDEVS have been removed. With an advanced code generation framework such as Xtext, DEVS executables can be generated using any general-purpose programming language. In the next chapter, we will see how the DEVS DSL introduced in this chapter helps define the core of the DEVS Unified Process through the DEVSML 2.0 stack.

Exercise 5.1

Execute the code-gen for the MicroSim engine for the EFP example.

Exercise 5.2

Execute the code-gen from the MicroNet engine for the EFP example.

Exercise 5.3

EF is a coupled model. Develop an atomic model called EFAtomic using DEVSML that has the same behavior as the EF coupled model and the same state-space. Verify the closure under coupling property. Simulate using the DEVS engine of your choice.

Exercise 5.4

Write the NLDEVS description of Proc and Transd. Run the code generators and execute the simulation. Verify the simulation results with the code generated from DEVSML.

Exercise 5.5

Read Chapter 18.

Exercise 5.6

Understand the automated test observer code generation mechanism using Xpand in SVN for the JCAS example in Chapter 18.

REFERENCES

Hong, K., & Kim, T. (2006). DEVSpecL-DEVS specification language for modeling, simulation and analysis of discrete event systems. *Information and Software Technology, 48*(4), 221–234.

Hwang, M. H., & Zeigler, B. P. (2007). Reachability graph of finite deterministic DEVS. *IEEE Transactions on Automation Science and Engineering, 6*(3), 468–478.

Mittal, S. (2007). *DEVS Unified Process for Integrated Development and Testing of Service Oriented Architectures*. PhD Dissertation. Tucson, AZ: University of Arizona.

Mittal, S., & Douglass, S. (2012). DEVSML 2.0: The language and the stack. *DEVS Symposium*. Orlando, FL.

Mittal, S., & Douglass, S. A. (2011). Net-centric ACT-R-based cognitive architecture with DEVS unified process. *Proceedings of the DEVS Symposium, Spring Simulation Multiconference—SpringSim'11*. Boston, MA.

Mittal, S., Zeigler, B., & Hwang, M. (2012, June 12). *XFDDEVS: XML-Based Finite Deterministic DEVS*. Retrieved from duniptechnologies.com: www.duniptechnologies.com/research/xfddevs. Last accessed November 2012.

Xtext. Retrieved from www.eclipse.org/Xtext. Last accessed November 2012.

6 DEVS Unified Process

6.1 OVERVIEW

The DEVS Unified Process (DUNIP) is based on *open systems* concept. An open system is a dynamical system that can exchange energy, material, and information with the outside world through its reconfigurable interfaces. An open system also possesses the capability to form complex hierarchical structures enabling them to compete and cooperate at the same time. In fact, the mechanism to reorganize in a hierarchical structure is one of the basic requirements to manage complexity. Open systems are also characterized by emerging behavior and evolving structure. These two facets are functions of an open system's permeability to outside influence, inherited guidelines, ability to self-govern, and the degrees of synergistic efforts as it interacts with other systems and with its environment. To have an executable adaptive system of systems (SoS), the framework must provide capabilities to model an open system. In addition, a process also needs to be defined that allows the development of an executable open system. Much of the open system development hinges on the variable structure capability within a component-based system. The ability to add or remove hierarchical components, change connections between components, and, lastly, modify the behavior of a component as it evolves per its surroundings are the desired characteristics of an open system modeling framework. While the first two capabilities are structural in nature and have been documented in DEVS literature, the third one is behavioral modification at runtime. This capability is the most difficult one to achieve. Using the latest advances in variable structure DEVS described in Chapters 7 and 23, runtime behavioral modification in DEVS could be achieved. The DEVS open systems approach underlying the DUNIP gives it strong formal foundation to develop modeling and simulation (M&S) complex systems software capable of designing emergent behaviors.

While designing an open system has its own challenges, there is an acute need for a new testing paradigm that could provide answers to several challenges described in a three-tier structure (Carstairs, 2005). The lowest level, containing the individual systems or programs, does not present a problem. The second tier, consisting of SoS in which interoperability is critical, has not been addressed in a systematic manner. The third tier, the enterprise level, where joint and coalition operations are conducted, is even more problematic. Although current test and evaluation (T&E) systems are approaching adequacy for tier-two challenges, they are not sufficiently well integrated with defined architectures focusing on interoperability to meet those of tier three. To address pragmatic (e.g. mission threads) testing at the second and third tiers, this chapter provides a collaborative distributed environment based on DUNIP, which can also be seen as federation of new and existing facilities or services from commercial, military, and not-for-profit organizations in a netcentric environment. In such an environment, M&S technologies can be exploited to support model-continuity and model-driven design developments, making T&E an integral part of the design and operations life cycle. The development of such a distributed testing environment would have to comply with Department of Defense (DoD) mandates requiring that the DoD Architecture Framework (DoDAF) (DoDAF Working Group, 2003) be adopted to express high-level system and operational requirements and architectures (Chairman of the Joint Chiefs of Staff Instruction, 2004). Unfortunately, DoDAF and DoD netcentric mandates pose significant challenges to T&E since DoDAF specifications must be evaluated to see if they meet requirements and objectives, yet they are not expressed in a form that is amenable to such evaluation (Leach, 2007; Mittal, 2006).

In an SoS, systems and/or subsystems often interact with each other because of interoperability and overall integration of the SoS. These interactions are achieved by efficient communication among the systems using either peer-to-peer communication or through a central coordinator in a given SoS. Since the systems within an SoS are operationally independent, interactions among systems are generally asynchronous in nature. A simple yet robust solution to handle such asynchronous interactions (specifically, receiving messages) is to throw an event at the receiving end to capture the messages from single or multiple systems. Such system interactions can be represented effectively as discrete event models. In discrete event modeling, events are generated at random time intervals as opposed to some predetermined time interval seen commonly in discrete time systems. More specifically, the state change of a discrete event system happens only upon arrival (or generation) of an event, not necessarily at equally spaced time intervals. To this end, discrete event model is a feasible approach in simulating the SoS framework and its interactions. Several discrete event simulation engines such as Omnet+, NS2, DEVS, and so on are available that can be used in simulating interaction in a heterogeneous mixture of independent systems. The advantage of DEVS is its effective mathematical representation and its support to distributed simulations using middleware such as DoD's high level architecture (HLA).

DEVS formalism has been in existence for over 40 years now. It has been applied to multiple domains and as described in Chapter 3, many of the formalisms can be reduced to DEVS formalism. DEVS is based on Systems theory with its hierarchy of system specifications and closure under coupling properties. The DUNIP is the consummation of how DEVS can be applied to SoS design and analysis in full systems engineering life cycle setup (Mittal, 2007). DUNIP is not a single concept but an integration of various concepts that have been developed over the years in DEVS research. These concepts have now evolved into an integrated process that facilitates systems M&S. Combining the systems theory, M&S framework, and model-continuity principles lead naturally to a life cycle development process, originally referred to as bifurcated model-continuity-based life cycle methodology (Zeigler et al., 2005), as shown in Figure 6.1. The process can be applied to development of systems using model-based design from scratch or as a process of reverse engineering in which requirements have already been developed in an informal manner.

The precursor to DUNIP has the following characteristics:

- *Behavior requirements at lower levels of system specification*: The hierarchy of system specification as laid out by Zeigler et al. (2000) offers

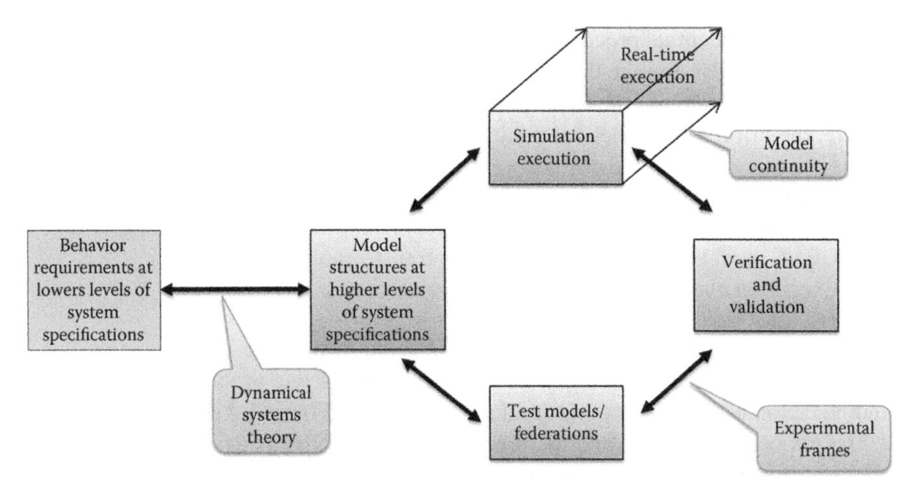

FIGURE 6.1 Bifurcated model-continuity-based system life cycle process.

well-characterized levels at which the requirements for system behavior can be stated. The process is essentially iterative and leads to increasingly rigorous formulation resulting from the formalization in subsequent phases.

- *Model structures at higher levels of system specification*: The formalized behavior requirements are then transformed to the chosen model implementations, for example, DEVS-based transformation in C++, Java, C#, and others.
- *Simulation execution*: The model base that may be stored in a model repository is fed to the simulation engine. It is important to state the fact that separating the model from the underlying simulator is necessary to allow independent development of each. Many legacy systems have both the model and the simulator tightly coupled to each other, which restricts their evolution. DEVS categorically separates the model from the simulator for the same simple reason.
- *Real-time execution*: The simulation can be made executable in real-time mode and in conjunction with model-continuity principles, the model itself becomes the deployed code.
- *Test models/federations:* Branching in the lower path of the bifurcation process, the formalized models give way to test models that can be developed at the atomic level or at the coupled level where they become federations. It also leads to the development of experiments and test cases required to test the system specifications. DEVS categorically aids the development of experimental frames at this step towards the development of the test suite.
- *Verification and validation* (V&V): The simulation provides the basis for correct implementation of the system specifications over a wide range of execution platforms and the test suite provides basis for testing such implementations in a suitable test infrastructure. Both these phases of systems engineering come together in the V&V phase.

The above conceptual process does not take into account the fundamental systems engineering process and the development life cycle, the prime objective of which is continuous satisfaction of the requirements across a system's life cycle. To define the DUNIP, the bifurcated model-continuity process is extended to include

- Requirement specifications from disparate systems
- Development of required domain-specific languages (DSLs) along with DEVS modeling language (DEVSML) transformations
- Execution over a transparent simulation netcentric infrastructure

DUNIP is a universal process and is applicable in multiple domains. However, the understated objective of DUNIP is to incorporate discrete event formalism as the binding factor at all phases of this development process. Figure 6.2 illustrates DUNIP.

The important concepts and the process within DUNIP, as described in Parts II and III of this book, are listed as follows:

1. *Requirements specification using DSLs*: How DSLs are used to specify system requirements and definitions. We will see how Unified Modeling Language (UML), DoDAF, Extended Backus Naur Form (EBNF) grammars, and other DSL generation languages like Groovy, Scala, and so on can be used to define requirements.
2. *Platform-independent modeling at lower levels of system specification using DEVS DSL*: Chapter 5 described a DEVS DSL that is based on finite deterministic DEVS. It will be presented in the context of DEVSML 2.0 stack in Section 6.2 ahead.

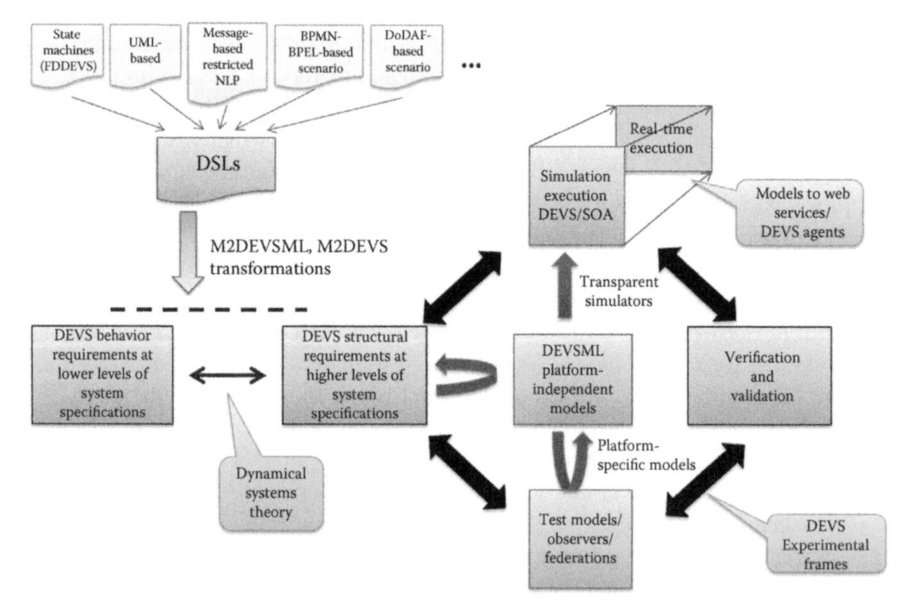

FIGURE 6.2 The DEVS Unified Process (DUNIP).

3. *Model structures at higher levels of system resolution using system entity structures (SESs)*: Chapter 10 discusses the role of SES at higher levels of systems specification and a model-based repository framework in which components stored in a repository can be used for systems development.
4. *Platform-specific modeling*: DEVS implementations on different platforms—Chapter 5 illustrated how DEVS platform-independent models (PIMs) can be implemented in a platform-specific language such as JAVA, C#, or C++.
5. *Automated test model generation using DEVS PIMs*: Chapter 18 discusses the automated generation of DEVS observers and test agents from DEVS PIMs through an example based on Joint Close Air Support. We will show how DEVS DSL plays a critical role in achieving this capability.
6. *Netcentric execution in a distributed setup*: Chapters 14, 15, and 16 will present the netcentric framework for DEVS execution. It will provide details on the DEVS simulation architecture, message serialization, and cross-platform execution of DEVS PIMs. It will describe a DEVS virtual machine that the user can either run locally or in a cluster to achieve parallel execution.
7. *Interfacing of models with real-time systems*: Chapter 16 is at the heart of model-continuity principle. We will show through our netcentric systems example how DEVS can act as a production system and can interface with live web services. It will also describe a test instrumentation system deployed on a service-oriented architecture (SOA) and integration with Event Driven Architectures (EDA).
8. *Verification and validation*: Chapter 20 emphasizes the experimental frame design in the context of a large system such as GENETSCOPE.

We will now describe the underlying platform transparent M&S layered stack called DEVSML 2.0 and contrast the DUNIP with model-driven architecture (MDA). Further, DUNIP is established as an agile process and how it maps to the Open Unified Technical Framework (OpenUTF). We will also enumerate the personnel required to execute DUNIP in a larger systems modeling endeavor.

6.2 DEVSML 2.0 STACK

The DEVSML 2.0 framework has two pieces: the stack and the language itself. We have described the language, a DEVS DSL, in Chapter 5. A DSL is a dedicated language for a specific problem domain and is not intended to solve problems outside it. For example, HTML for web pages, Verilog and VHDL for hardware description, and so on are DSLs for very specific domains. A DSL can be a textual or a graphical language or a hybrid one. A DSL builds abstractions so that the respective domain experts can specify their problem well suited to their domain understanding without paying much attention to the general-purpose computational programming languages, such as C, C++, Java, and so on, which have their own learning curve. The notion of domain-specific modeling arises from this concept, and the DSL designers are tasked with creating a domain-specific modeling language. If a DSL is also meant for simulation purposes, then one more task of mapping a specific DSL to a general-purpose computational language is also on the cards. In this section, we propose a DEVS DSL as a component of DEVSML 2.0 stack. The proposed stack also integrates the transparent modeling framework with the inclusion of DSLs and various transformations. We describe how platform-independent DSLs can be transformed in this framework and finally into the DEVS formalism.

Decoupling the model from the simulation platform has many benefits as it allows the modeler to construct models in a platform of his choice. The ability to execute DEVS models in multiple platforms (Windows, Linux, Mac, etc.) and languages (C++, Java, C#, Scheme, etc.) has already been achieved. Figure 6.3 shows the relationship between the DEVS modeling and the DEVS simulation layers through a "simulation relation." It also shows that the design of DEVS models was dependent on the language where the underlying assumption always has been "everything is an object." In the new framework, we migrate from this concept to "everything is a model." With this paradigm shift, we find ourselves in the world of concepts and abstractions that are very specific to the domain which is "to be modeled." This gives rise to the discipline of domain-specific modeling and the metamodels that help specify these domain models. It also forces us to develop various kinds of transformations that take these DSLs to the execution stage. Now, there are two choices. Either the DSL designer takes the DSL directly to the execution code, which involves no transformations but only code generation to a programming language, or he works with an existing framework that guarantees execution in formal systems theoretic way. If the DSL designer opts for the second option, choosing DEVS as a framework is recommended due to its rich history of model specification and simulator development. This ability provides a solution to scalability, integration, and interoperability, as described in Risco-Martín et al. (2009); Mittal (2007);

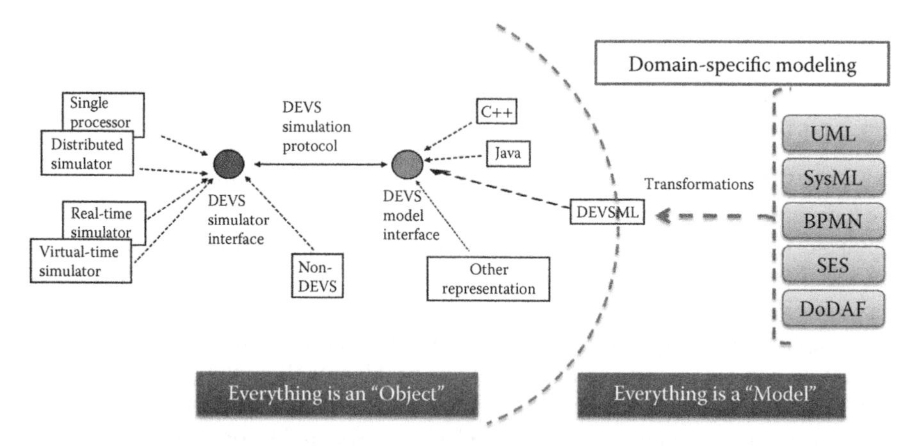

FIGURE 6.3 Separation of modeling and simulation layers with DEVS simulation protocol.

Mittal et al. (2007); Mittal and Zeigler (2009); Mittal et al. (2009); and Zeigler et al. (2008). Having a process to transform any DSL to DEVS components, especially to the DEVSML platform-independent specification, then have obvious advantages.

6.2.1 Stack

The latest version for the stack was proposed as a part of Air Force Research Laboratory's Large Scale Cognitive Modeling initiative by Mittal and Douglass (2011a,b; 2012). As a part of the proposed stack DEVSML 2.0, the DEVSML described in Chapter 5 is based on EBNF grammar and is supported by DEVS middleware Application Programming Interface (API). The middleware is based on DEVS M&S standards compliant (under evaluation) API and interfaces with a netcentric DEVS simulation platform such as an SOA that offers platform transparency. With the maturation of technologies like Xtext 2.3 (Xtext-Eclipse) and Xtend 2.0 (Eclipse-Xtend), we now have extended the earlier concept of XML-based DEVSML to a much broader scope wherein other DSLs can continue to be expressed in all their richness in a language-independent manner that is devoid of any DEVS and programming language constructs (Figure 6.4).

We need to make a clear distinction here that the DEVS modeling "language" is a DEVS modeling specification language that is anchored to DEVS simulation layer using the simulation relation in DEVS middleware API. Consequently, a DEVSML-specified model is DEVS executable. The idea of including other DSLs at the top

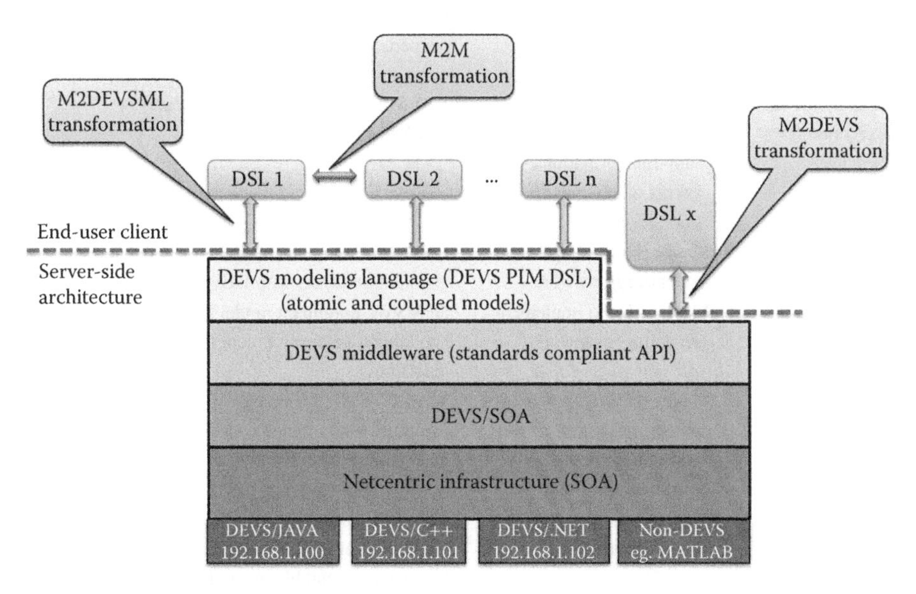

FIGURE 6.4 DEVSML 2.0 stack employing M2M and M2DEVS transformations for model and simulator transparency.

layer of the stack is a major addition in the DEVSML 2.0 stack. In addition, the stack in Figure 6.4 adds three transformations at the top layer:

1. Model-to-model (M2M)
2. Model-to-DEVSML (M2DEVSML)
3. Model-to-DEVS (M2DEVS)

The key idea being domain specialists (the end-user) need not delve in the DEVS world to reap the benefits of DEVS framework. The end-user as indicated in Figure 6.4 will develop models in their own DSL, and the DEVS expert along with the DSL designer will help develop the M2M and M2DEVSML transformations to give a DEVS backend to the DSL models. While M2DEVSML transformation delivers an intermediate DEVS DSL (the DEVSML DSL), the M2DEVS transformation directly anchors any DSL to platform-specific DEVS. On a reverse note, a DEVS expert is ideally suited to develop DSLs in other domains as developing transformations like M2DEVS and M2DEVSML need not be negotiated with the DSL expert. A DEVS expert with DEVSML skill set can perform a dual job of both the DSL and DEVSML experts. Table 6.1 summarizes the state of DEVSML as viewed from the perspectives of a DEVSML expert and other DSL expert.

TABLE 6.1
Comparing DSL Expert with DEVS Expert

Features	DSL Expert	DEVS Expert
Domain understanding	Develops domain modeling language with domain experts.	Develops M2DEVSML or M2DEVS with DSL expert.
Simulation/DSL execution	Has to ensure the computational mapping and transformations taking all abstractions to code level.	DEVS simulation is "given" as a part of DEVSML 2.0 framework.
Scalability	Maybe. Has to implement an entire framework.	DEVSML compliant model is scalable at simulation layers with DEVS/SOA.
Collaboration	Maybe. Has to implement an entire framework.	DEVSML compliant model is ready for collaborative development using netcentric platform.
Integration	Maybe. Has to implement an entire framework and component-based infrastructure.	The DEVSML 2.0 framework with DEVS system foundation provides model integration and composition features as a part of collaborative modeling effort.
Platform neutrality	Maybe. Has to implement an entire framework.	DEVSML compliant model is platform independent both in M&S layers. The same model is executable in any DEVS simulator in any language.

The addition of M2M, M2DEVSML, and M2DEVS transformations to the DEVSML stack adds true model and simulator transparency to a netcentric M&S SOA infrastructure. The transformations yield DEVS PIMs that can be developed, compared, and shared in a collaborative process within the domain. Working at the level of DEVS DSL allows the models to be shared among the broad DEVS community that brings additional benefits of model integration and composability. DEVSML 2.0 stack allows DSLs to interact with DEVS middleware through an API. This capability enables the development of simulations that combine and execute DEVS and non-DEVS models (Martín et al., 2009). This hybrid M&S capability facilitates interoperability. The scale is provided by the underlying SOA infrastructure that is largely made of virtualization technologies and utilizes platform-as-a-service (PaaS) capabilities provided by enterprise containers such as Glassfish 4.0 (Oracle), as described in Section III of this book.

6.3 MDA AND DUNIP

DUNIP is built on the paradigm of Model-Driven Architecture or MDA, as described by Object Modeling Group (OMG). However, the scope of DUNIP goes beyond the MDA objectives. Potential concerns with the current MDA's state of art include the following:

- MDA approach is underpinned by a variety of technical standards, some of which are yet to be specified (e.g., executable UML).
- Tools developed by many vendors are not interoperable.
- MDA approach is considered too idealistic, lacking the iterative nature of software engineering process.
- MDA practice requires skilled practitioners, and design requires engineering discipline not commonly available to code developers.

Further, MDA does not have any underlying systems theory, and groups like INCOSE (International Council on Systems Engineering) are working with the Object Management Group to adapt UML to systems engineering. Testing is included only as an extension of UML, known as executable UML, for which there is no current standard. Consequently, there is no testing framework that binds executable UML and simulation-based testing.

Despite these shortcomings, MDA has been adopted by the Joint Single Integrated Air Picture (SIAP) Systems Engineering Organization (JSSEO) and various recommendations have come forth to enhance the MDA process. JSSEO is applying the MDA approach toward the development of aerospace command and control (C2) capabilities, for which a SIAP is foundational (Krikeles & Merenyi, 2004). The data-driven nature of C2 SoS means that powerful MDA concepts adapt well to collaborative SoS challenges.

Current DoD enterprise level approaches for managing SoS interoperability, like the Net-Centric Operations Warfare Reference Model (NCOW/RM), DoDAF, and the Joint Technical Architecture (JTA), simply do not have the technical strength to deal with the extremely complex engineering challenges. We proposed enhanced DoDAF for DoDAF version 1.0 (available in Chapter 11) to provide DEVS-based model engineering. MDA as implemented by industry and adapted by JSSEO does have the requisite technical power, but requires innovative engineering practices. Realizing the importance of MDA concepts and the executable profile of UML, the basic objective of which is to simulate the model, JSSEO is indirectly looking at the M&S domain as applicable to SoS engineering (Dutchyshyn, 2005). Table 6.2 brings out the shortcomings of MDA in its current state and the capabilities provided by DEVS technology and, in turn, the DUNIP process.

Model-driven software development is based on the notion of DSLs for modeling various types of models. Once DSL has been identified, its metamodel is created that represents a particular modeling domain. Metamodels are defined in terms of meta-metamodel. In UML, this is the Meta-Object Facility. A meta-metamodel is created that would define both the UML2 metamodel and their selected DSL extensions. The whole objective is to find a common ground and a way to express the relationship between a metamodel and an implementation code. This kind of capability where a single meta-metamodel can be used to integrate two different DSLs toward a common model allowing specific constraints of each metamodel is very much needed in SOA domain as multiple tools and standards exist preventing such integration. To integrate two models with different DSLs, the models are first decomposed at the

TABLE 6.2
Comparison of MDA and DUNIP

Desired M&S Capability	MDA	DUNIP
Need for executable architectures using M&S	Yes, although not a standard yet	Yes, underlying DEVS theory.
Applicable to Global Information Grid (GIG) SOA	Not reported yet	Yes.
Interoperability and cross-platform M&S using GIG/SOA	—	Yes, DEVSML and SOA DEVS provides cross-platform M&S using simulation web services.
Automated test generation and deployment in distributed simulation	—	Yes, based on formal systems theory and test models autogeneration at various levels of system specifications.
Test artifact continuity and traceability through phases of system development	To some extent, model becomes the application itself	Yes.
Real-time observation and control of test environment	—	Dynamic model reconfiguration and runtime simulation control integral to DEVS M&S. Enhanced model view controller (MVC) framework is designed to provide this capability.

metamodel level, required information extracted and supplemented (on the basis of meta-metamodel), which results in an integrated model. This concept is also central to the Model-Integrated Computing (MIC) initiative, and DUNIP along with the DEVSML 2.0 stack incorporates DSLs as a product artifact. In DUNIP, such collaboration comes naturally due to the proposed DEVS atomic and coupled document type definitions and the EBNF grammar of DEVSML 2.0 that specify any DEVS model in any DSL implementations and vice versa through the M2DEVS and M2DEVSML transformations. The underlying DEVSML metamodel is used for validating any DEVS model. The current DEVSML implementation has successfully integrated three DSL implementations (DEVSJAVA-ACIMS, Microsim-DUNIP, and xDEVS-Spain) on common DEVSML metamodels.

6.4 AGILITY IN DUNIP

Agile software methodologies have taken quite a notice these recent years primarily due to factors such as volatile ever-changing requirements, dynamic technological landscape, high employee turnover, and most importantly, satisfying business needs. It is summarized as a strategic capability that can respond to change, is adaptive, balances flexibility and structure, draws about the creativity and innovation of the development team, and ultimately leads the organization through turbulence and uncertainty.

There is a fundamental shift in the approach of delivering the product by hardline requirement specifications supported by methodologies like the waterfall model, capability maturity model (CMM) and CMM integration, and the agile practices. While the former delineates defined repeatable processes so that the performance can be measured within very close tolerances, the agile methodologies are more geared toward employing the latest advancement in technologies to explore and to deliver the product as soon as possible. The key point of agile practices is the inclusion of software engineering life cycle in each iteration so that the features delivered are production ready at the end of each iteration. While the visions of most projects are clear, what remain fuzzy are the exact requirement specifications that the developers are faced with. With agile practices, a constant dialogue with the customer or the subject matter experts (SMEs), repeatable testing procedures, incremental development, and using the latest technology, the requested feature can be delivered in the next iteration without changing the entire project vision. Similarly, the DUNIP is based on agile methodology. Table 6.3 lists the similarities with each phase of agile development methodology.

Table 6.3 establishes that DUNIP has all the needed phases of being agile, and the model continuity enables any DEVS artifact to be real software. With DUNIP's SOA edge, we have any DEVS model that is available as a web service. M&S in today's world is more than just software. It is an enabling technology that has far reaching impacts on any nations' progress and advances the forefront of various technologies in many domains such as biology, chemistry, physics, space science, and so on. While there are customized M&S software for different problem sets and different domains, an agile methodology is another ace that when employed could incorporate the latest advancements in software engineering discipline and apply it to the M&S solution at hand.

TABLE 6.3
Summary of Agile Methodology Practices within DUNIP

Phase	Agile Methodology	DUNIP
Model	Identify the domain and business use-case requirements and specify in DSLs such as UML.	DUNIP begins by taking requirements in various DSLs and through M2DEVS, M2M, and M2DEVSML transformations and transform them into DEVS specifications. It also constructs various experimental frames from requirement specifications for later V&V operations.
Implementation	Transform your models into executable code with running unit tests.	From PIMs, the DUNIP engine generates code in platform-specific models such as Java, C++, C#, and so on. With strong DEVS theory underlying each atomic model, they can be mathematically verified. Unit testing for each transition or an event is inherent in DEVS.
Test	Identify defects, ensure quality, and verify requirements.	The development of test suite is done in parallel with that of the DEVS PIM. The test models verify the atomic model's operation at various levels of system specifications such input and output (I/O) pair, I/O function, and so on. The experimental frames are also enhanced at this stage.
Deployment	Plan the delivery and make it available to end-users.	With ready deployment capabilities per model-continuity principles to SOA infrastructure and zero transition times, the model is the actual software that is readily moved to the production servers.
Configuration management	Managed access to project artifacts.	DUNIP is very well positioned to reuse and contribute to model repository. PIMs are strong contenders for such tracking and version management. PSMs can very well be source versioned using tools like Subversion.
Project management	Manage people, project iterations, and budget.	These qualities are universal and due to the component nature of DEVS technology, the project plan can be partitioned into iterative cycle and milestones.
Environment	Ensure that the proper process, guidance, and tools are available for the team.	DEVS has been in existence for over 30 years, and there is a large community support in basic theory and toolsets.

The execution of DUNIP across a system's development life cycle can best be summarized in Figure 6.5 as a spiral development process, with DUNIP at its core as the integrative factor. With each spiral, new requirements (including experimental frames and validation requirements), new system components, and new test components are added. At the end of each spiral, there is a working system model that

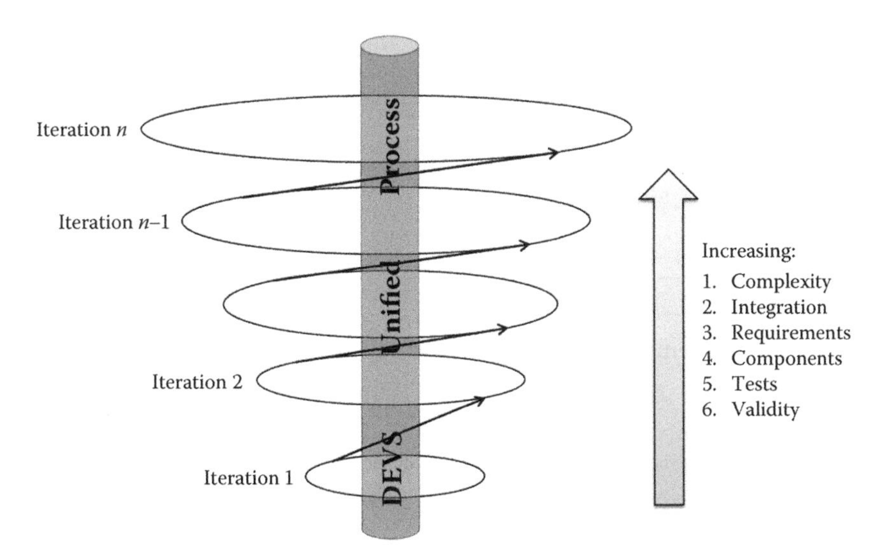

FIGURE 6.5 Spiral nature of DUNIP.

satisfies the requirements of the current iteration as well as of previous iterations, increasing the system complexity and requirements traceability.

6.5 ALIGNING WITH OPENUTF

OpenUTF by Steinman, Lammers, and Valinski (2008) is proposed as a cost-effective, open-source, high-performance, scalable, service-oriented, software development framework for implementing netcentric M&S and service-oriented application for the Chemical and Biological Defense (CBD) program and is described in CBD M&S Strategic Plan (Joint Program Executive Office [JPEO] Chemical and Biological Defense [CBD] Modeling and Simulation Strategic Plan). Some of the constituent elements with the OpenUTF are as follows:

- OpenUTF kernel
- Models
- Services
- Data
- Development tools
- Composability tools
- Visualization tools
- Analysis tools
- Web service interfaces
- LVC interoperability standards

The proposed UTF is a government-managed collection of systems and tools developed and maintained by individual organizations and various groups like the Naval Surface Warfare Center; Air Force Research Laboratory, Rome, NY; High Performance Computing Modernizing Program at SPAWAR; and Simulation

Standards Interoperability Organization study groups, which are seriously considering solutions with OpenUTF.

To participate in OpenUTF, a list of requirements has been proposed by Steinman, Lammers, and Valinski (2009).

Table 6.4 aligns these requirements with the DUNIP and how DUNIP is positioned to address each of these.

TABLE 6.4
OpenUTF Requirements as Addressed by DUNIP

Requirement	OpenUTF	DUNIP	Meets Requirement
1	Separation between engine and model	DEVS, since the 1970s, have been advocating the separation of the model and the simulator. There are separate M&S layers in the DEVSML 2.0 stack.	Yes
2	Optimized communications	A discrete event system abstracts all the network layer communication details behind an SOA interface.	Yes
3	Abstract time	DEVS has notion of real time, logical time, local time, and global time in its formalism since its inception.	Yes
4	Scheduling constructs	DUNIP provides a client to schedule the simulation runs on local or remote DEVS virtual machines.	To some extent
5	Time management	DEVS provides mechanisms dealing with abstract simulators defined in abstract time.	Yes
6	Encapsulated software components	The inclusion of DSLs along with M2M, M2DEVS, and M2DEVSML transformations provides model transparency and plug and play capabilities to various software components.	Yes
7	Hierarchical composition	DEVS models are hierarchical by nature.	Yes
8	Distributable composites	DEVSML 2.0 stack with the DEVS virtual machine can be deployed to any remote note in a cluster or in a netcentric environment.	Yes

TABLE 6.4 (*Continued*)
OpenUTF Requirements as Addressed by DUNIP

Requirement	OpenUTF	DUNIP	Meets Requirement
9	Abstract interfaces	DEVS is a component-based system with a clear-defined logical port value-based formalism. Any software components or a web service can be wrapped in DEVS formalism.	Yes
10	Interaction constructs	DEVS formally specifies the I/O hierarchy in its system of specification as I/O frame, I/O pair, and I/O function on a time-basis.	Yes
11	Publish/subscribe	This is a need-to-have feature. DUNIP has the capability to provide model-as-a-service and simulation-as-a-service leading to DEVS-as-a-service. These models can be deployed on high-performance nodes to act as publish/subscribe exchanges.	Yes
12	Data translation services	Data translation has been a part of model transformation, and specific technologies like Apache Camel can be deployed in conjunction with DEVS components to implement event-based transforms.	Yes
13	Execution of multiple applications	With M2DEVS and M2DEVSML transformations, disparate DSLs and ultimately domain-specific application can interoperate with any DEVS collaborative system.	Yes
14	Platform independence	DUNIP is geared to provide model interoperability and transparent simulation infrastructure.	Yes
15	Scalability	The DEVS/SOA component is well integrated with PaaS capability that allows DEVS virtual machines to get deployed in high-performance virtualized environment, on a need basis at runtime.	Yes

(*Continued*)

TABLE 6.4 (*Continued*)
OpenUTF Requirements as Addressed by DUNIP

Requirement	OpenUTF	DUNIP	Meets Requirement
16	Support Live, Virtual and Constructive (LVC) interoperability standards	DEVS has been shown to integrate well with HLA, TENA (test and training enabling architecture), and DIS (distributed interactive simulation) systems.	Yes
17	Web services	DUNIP is netcentric since conception.	Yes
18	Cognitive behavior representation	While this is not a component per se in DUNIP, the likelihood of integrating an intelligent system based on a cognitive DSL is a possibility.	Maybe
19	Stochastic modeling constructs	The design of experimental frames provides such capability. However, there is not dedicated component as such.	In future
20	Geospatial representations	A mobility component can be integrated with the DEVS system. It is a very specific requirement and DEVS would develop a wrapper around an existing service.	In future
21	Software utilities	Utilities like version control, document management, and so on are quite a norm today.	Yes
22	External modeling framework	Again, this goes back to DSLs and the transformations at metamodeling level.	Yes
23	Output data formats	DUNIP has a formal process to define entities in the domain ontology for every solution. Consequently, this is a very rudimentary requirement albeit a necessary one.	Yes
24	Test framework	DUNIP automatically generates a component test framework.	Yes
25	Community-wide participation	DEVS has been in existence for a good 40 years now.	Yes

6.6 PERSONNEL REQUIREMENT TO REALIZE DUNIP

Any M&S effort is a multidisciplinary effort that requires the knowledge of SMEs, software developers, architects, and the simulation experts. If a simulation tool is a component-off-the-shelf (COTS) element, then the only personnel required is the end-user or the end-modeler that uses this component for designing models, simulation experiments and conducting analysis. While we are not dealing with such COTS elements in this book, we are surely moving toward COTS components based on DUNIP, that is, DEVS-based plug and play components. The type of personnel we require in developing any DUNIP-based solution are as follows:

1. *Business/requirements analyst*: This person will engage in conversations with the end-user/end-modeler who will be using the simulation tool under development. This person will help define the experimental frame with the DEVS developer, the analysis domain, the measures of effectiveness, and the measure of performance. This person will produce a set of UML use-case descriptions along with SysML (system, operational, end-user, and acceptance) requirements set in collaboration with the systems engineer.
2. *SME*: This person will be responsible to validate the test suite and various experiments defined in collaboration with the business analyst. This person will also collaborate with the DEVS expert to develop abstractions and operational model using DEVSML.
3. *DEVS developer*: This person will be responsible for developing the operational model in collaboration with the SMEs and will help transition to large-scale-distributed simulation efforts if need be. This person will also integrate the inputs received from the requirements analyst and ensures that the experimental frame provides the latitude in answering customer questions. This person will collaborate with the data fusion developers to design the data flow model.
4. *Middleware specialist*: This person is responsible for developing the M2DEVS and M2DEVSML transformations with respect to the end-user requirements and will work closely with the SMEs and the DEVS developer. This role may be shared with the DEVS developer capabilities if this person has sufficient background in middleware and graph transformation technologies.
5. *Data/ontology developer*: This person will be tasked with developing the data structures and the domain ontologies for structured information exchange and documentation and will work closely with SMEs, system engineer, and the DEVS developer.
6. *Data fusion developers*: This is one of the most resource intensive areas of the entire effort where a lot of simulation data are analyzed and visualized. Such effort will require a team of software, database, 2D, 3D, multimedia, web interface, and data mining developers that deal with distributed data processing and presentation capabilities. The efforts in this area are cumulative efforts and with time, the team would develop a library of such tools that will be integrated using defined interfaces in SOA for distributed use. The personnel in this category will work on the domain ontology and will work with the system engineer and the DEVS developer.
7. *Systems engineer*: This person will be responsible for developing the integration diagrams, interface validation and coordination between the DEVS

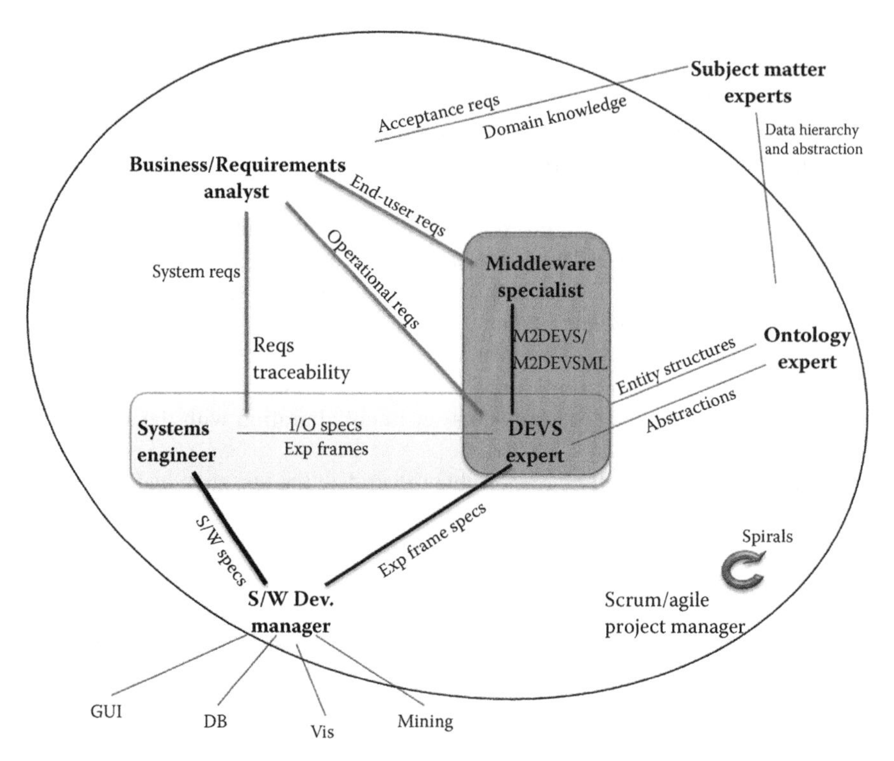

FIGURE 6.6 People and interactions involved in DUNIP project.

developer, middleware specialist, ontology developer, and the fusion developers. The primary job of this person is to ensure data sanity, componentization, and facilitate communication between the above personnel.

8. *Scrum/agile expert*: This person will ensure the iterations, delivery, and feedback on the customized solution and will keep all the above seven personnel aligned. This person is well versed with various modeling technologies, such as Eclipse modeling framework, Eclipse graphical modeling framework, enterprise computing, and distributed simulation and data fusion technologies.

The above roles and interactions have been summarized in Figure 6.6. The elliptical region shows the management scope of the project manager.

6.7 SUMMARY

With the capability presented of DEVSML 2.0 stack in this chapter, we are now in a position to achieve model transparency with M2DEVS and M2DEVSML transformations. It should be noted that the complexity of these transformations are on a case-by-case basis depending on the abstractions of a particular DSL. However, as a DEVS expert, the design of such transformations will reap far-reaching benefits to the execution, scalability, interoperability, integration, and collaborative capabilities for any DSL. Further with DEVS as common denominator in multiformalism hybrid modeling effort, DEVSML 2.0 framework provides many advantages. We highlighted the roles of a DSL expert and a DEVS expert in the context of a DSL and a simulation DSL.

We have discussed the advantages of employing an M&S-integrated framework such as DUNIP and its supporting DEVS/SOA infrastructure. As components comprising SoSs are designed and analyzed, their integration and communication is the most critical part that must be addressed by the employed SoS M&S framework. DUNIP, in analogy to the rational unified process based on UML, offers a process for integrated development and testing of systems that may rest on the SOA infrastructure. The proposed methodology encompasses the advantages of several interrelated concepts such as the systems theory, DEVSML and DEVS/SOA, M&S framework, and the model-continuity concepts, especially because it separates models from their underlying simulators, enables real-time execution and testing at multiple levels and over a wide range of execution platforms, uses open standards, supports collaborative development, and has the potential to provide additional SoS architectural views.

We also described how DUNIP satisfies most of the requirements of OpenUTF. Finally, we laid out what kind of personnel is required to realize the DUNIP for any problem domain. While the personnel required are classic of any software development effort today, the key people with DEVS and simulation expertise are required to work on concepts like experimental frame, simulation engine design, model V&V, and netcentric model integration. It is true that a DEVS expert is a software engineer first; in fact every simulation developer must be a software engineer first, however, only simulation theory skills are not sufficient for a successful M&S product delivery. The team engaged in DUNIP has a good mix of DEVS experts and seasoned software developers working in an agile environment. A simulation solution worked under the guidelines of DUNIP is a component-based, multiplatform, scalable, collaborative, and an agile solution that continuously meets the changing requirements of the problem domain in today's netcentric environment.

REFERENCES

Carstairs, D. (2005). Wanted: A new test approach for military net-centric operations. *ITEA, 26*(3), 7–9.

Chairman of the Joint Chiefs of Staff Instruction. (2004). *Joint Capabilities Integration and Development System*. JCS Instruction 3170.01H.

DoDAF Working Group. (2003). *DoD Architecture Framework, Ver. 1.0 Vol.III: Deskbook*. DoDAF Working Group, US Department of Defense.

Dutchyshyn, H. (2005). *Engineering System-of-Systems Capabilities: A SIAP Perspective*. Joint SIAP System Engineering Organization (JSSEO), USAF.

Eclipse-Xtend. (Eclipse.org) Retrieved June 5, 2012, from http://www.eclipse.org/xtend.

Joint Program Executive Office (JPEO) Chemical and Biological Defense (CBD) Modeling and Simulation Strategic Plan.

Krikeles, B., & Merenyi, R. (May 17–20, 2004). Use of MDA in the SIAP Program. *MDA Implementers' Workshop: Succeeding with Model Driven Systems*. Orlando, FL.

Leach, R. J. (October 2007). Issues using DoDAF to engineer fault-tolerant system of systems. *Crosstalk: The Journal of Defense Software Engineering*, 22–27. Last accessed: October 10, 2012, Available from: http://www.crosstalkonline.org/storage/issue-archives/2007/200710/200710-Leach.pdf

Risco-Martín, J. L., Moreno, A., Aranda, J., & Cruz, J. M. (2009). Interoperability between DEVS and non-DEVS models using DEVS/SOA. *Proceedings of the 2009 Spring Simulation Multiconference*. San Diego, CA.

Mittal, S. (2006). Extending DoDAF to allow DEVS-based modeling and simulation. *Journal of Defense Modeling and Simulation: Applications, Methodology, Technology, 3*(2).

Mittal, S. (2007). DEVS unified process for integrated development and testing on service oriented architectures. PhD Dissertation. Tucson, AZ: University of Arizona.

Mittal, S., & Douglass, S. (2012). DEVSML 2.0: The language and the stack. *DEVS Symposium*. Orlando, FL.

Mittal, S., & Douglass, S. A. (2011a). From domain specific languages to DEVS components: Application to cognitive M&S. *Proceedings of the Workshop on Model-Driven Approaches for Simulation Engineering—SpringSim'11*. Boston, MA.

Mittal, S., & Douglass, S. A. (2011b). Net-centric ACT-R-based cognitive architecture with DEVS unified process. *Proceedings of the DEVS Symposium, Spring Simulation Multiconference—SpringSim'11*. Boston, MA.

Mittal, S., Risco-Martín J. L., & Zeigler, B. (2007). DEVS-based web services for Net-centric T&E. *Summer Computer Simulation Conference*, San Diego, CA.

Mittal, S. M., Risco-Martín, J. L. & Zeigler, B. (2009). DEVS/SOA: A cross-platform framework for net-centric modeling and simulation in DEVS unified process. *Transactions of SCS International, 85*(7).

Mittal, S., Zeigler, B., & Risco-Martín J. L. (2009). Implementation of formal standard for interoperability in M&S/system of systems integration with DEVS/SOA. *International Command and Control Journal*, Special Issue: Modeling and Simulation in Support of Network-Centric Approaches and Capabilities; 3(1).

Oracle Wiki: Glassfish PaaS Functional Specification. (Oracle) Retrieved June 5, 2012, from https://wikis.oracle.com/display/GlassFish/GlassFish+PaaS+FSD.

Steinman, J., Lammers, C., & Valinski, M. (2008). A unified technical framework for net-centric system of system test and evaluation, training, modeling and simulation, and beyond. *Fall 2008 Simulation Interoperability Workshop*, Orlando, FL.

Steinman, J., Lammers, C., & Valinski, M. (2009). Composability requirements for the Open Unified Technical Framework. *Fall 2009 Simulation Interoperability Workshop, Orlando, FL.*

Xtext-Eclipse. (Eclipse.org) Retrieved June 5, 2012, from http://www.eclipse.org/xtext.

Zeigler, B. P., Fulton, D., Hammonds, P., & Nutaro, J. (2005). Framework for M&S-based system development and testing in a net-centric environment. *ITEA Journal of Test and Evaluation, 26*(3).

Zeigler, B. P., Mittal, S., & Hu, X. (2008). Towards a formal standard for Interoperability in M&S/System of Systems Integration with DEVS/SOA. *GMU-AFCEA Symposium on Critical Issues* in C4I, Fairfax, VA.

Zeigler, B. P., Praehofer, H., & Kim, T. G. (2000). *Theory of modeling and simulation*. New York: Academic Press.

Section II

Modeling and Simulation-Based Systems Engineering

7 Reconfigurable DEVS

7.1 OVERVIEW

System requirements change. Component behaviors change. System structures change. Life by definition changes every day, every moment. Growth or degeneration is the only evidence of life. If one is in the business of modeling and developing abstractions for an artificial system, one has to address this *change*. Change happens over time. This aspect makes it interesting. Modeling is an art that builds abstractions at the right level of study, and simulation is the mechanism that allows models to display behavior over time. A modeling and simulation framework should be agile to formally specify change and how the system transforms to a new configuration. This is the focus of this chapter.

According to Brown and Wallnau (1998: pp. 38), a component is "a nontrivial, nearly independent, and replaceable part of a system that fulfills a clear function in the context of a well-defined architecture. It conforms to and provides the physical realization of a set of interfaces." A component system is built by composition of various independent components and by establishing relationships among them. As each component has a high degree of autonomy and has well-defined interfaces, the variable structure of components can be achieved during runtime. For the component-based modeling and simulation, the variable structure provides several advantages:

- It provides a natural and effective way to model the complex systems that exhibit structure and behavior changes to adapt to different situations. Examples of these systems include distributed computing systems, reconfigurable computer architectures (Zeigler & Reynolds, 1985; Zeigler & Louri, 1993), fault tolerant computers (Chean & Fortes, 1990), and ecological systems (Uhrmacher, 1993).
- Structure changing and component upgrading is an essential part of these systems. Without the variable structure capability, it is very hard, if not impossible, to model and simulate them, let alone study the transition effect that the system incurs when new components are added in a real deployed system.
- From the design point of view, the variable structure provides the additional flexibility to design and analyze a system under development. For example, it allows one to design and simulate a system in which the components are added or removed incrementally and form dynamic relationships with existing components.
- It allows one to load only a subset of a system's components during simulation. This is very useful to simulate very large systems with a tremendous number of components, as only the active components need to be loaded dynamically to conduct the simulation. Otherwise, the entire system has to be loaded before the simulation begins.

In general, there are six forms of reconfiguration of component-based systems:

1. Addition of a component
2. Removal of a component
3. Addition of a connection between two or more components
4. Removal of a connection between two or more components
5. Migration of a component
6. Update of a component

The first two operations result in an update of the modeled system where there is a change in the number of components in the system. The next three result in a reconfiguration of the existing system. The last one results in the modification of the component itself, either its behavior or its interface structure. In Discrete Event Systems Specification (DEVS), these are collectively known as variable-structure modeling. As the variable structure changes a component-based simulation during runtime, boundary conditions and the limits to which a component affects other components need to be specified with the said operation.

The dynamic structure outcomes have been adequately dealt with in our earlier work (Hu et al., 2005), formally by Barros (1995, 1997, 1998), Uhrmacher (2001), Uhrmacher and Priami (2005), and Uhrmacher et al. (2007). In all these undertakings, the DEVS formalism was extended and mathematical proofs were given. Barros' DSDEVS used the concept of a network executive that knows the entire substructure and could initiate structural changes. On the other hand, the formal undertaking of dynamic structure by Uhrmacher (2001) and Uhrmacher et al. (2011) initiated the structural change from within the system components. While these efforts were more geared toward the theoretical aspect of variable-structure DEVS, our focus in this chapter is to approach the variable structure through the system design requirements. The DEVS unified process is a systems engineering process, and requirements play a vital role in any system design. We approach our solution through the *experimental frame* that is outside the system entirely. The experimental frame may contain Barros' network executive and associate it with the system requirements at both the modeling and the simulation layers.

DEVS systems have a continuous time base, but their execution is event based. A variable-structure discrete event system adds a temporal nature to the structure of the system itself. The structure of a system can be dynamic at three levels:

1. Component level: Entire substructures are removed or added in a live system.
2. Connection level: Interactions are reconfigured in a live system.
3. Interface level: Interface of the component itself is subject to reconfiguration.

The behavior of a system can be dynamic in four ways:

1. State space
2. Time advances of each state

3. Transition functions (e.g., δ_{int}, δ_{ext}, δ_{con}, λ)
4. Initial state

To better understand change and how modeling of change is handled in DEVS, let us understand where change occurs in a DEVS hierarchical system specification. More technically, such dynamism must be traceable to the DEVS levels of system specification as described in Chapter 2. Table 7.1 provides the mapping of how dynamism is introduced at various levels. Table 7.1 shows the manner in which change happens at different levels of specifications. It presents basic concepts on how a framework like DEVS is reconfigurable in a changing environment.

In the sections ahead in the chapter, we will discuss how changes at these different levels are addressed in the DEVS modeling and simulation layers, as derived from the basic DEVS elements of modeling and simulation. Since DEVS theory is implemented as software, we must leverage various software engineering frameworks that help implement the theory efficiently. One such framework that gained widespread acceptance with the enterprise computing is the model view controller (MVC) paradigm. The model in MVC paradigm is the actual software implemented that holds the business logic. In relation to the simulation software, the model in the MVC paradigm implies the model and the simulation together. We will separate the notion of model in MVC to a more relevant model simulator view controller (MSVC) paradigm, where it incorporates the simulator as a different component altogether.

TABLE 7.1
Introducing Dynamism at Various Levels of System Specifications

Level	Name	How Dynamism Is Introduced	Outcome
4	Coupled systems	1. System substructure 2. System couplings 3. Subsystem I/O interfaces 4. Subsystem active/dormant	1. Dynamic component structures 2. Dynamic interaction
3	I/O system	1. Addition/removal of states 2. Augmentation of transitions with constraints/guard conditions	1. Dynamic states 2. Dynamic transitions 3. Dynamic outputs
2	I/O function	1. Initial state 2. Addition/removal of initial state 3. Addition/removal of I/O pairs	Dynamic initial state
1	I/O behavior	1. Time scale between the I/O behavior 2. I/O mapping changing the behavior itself 3. Allowed behavior 4. Addition/removal of I/O pairs	Dynamic I/O behavior
0	I/O frame	1. Allowed values 2. I/O to port mapping	Dynamic interfaces

7.2 MSVC PARADIGM AND DEVS FRAMEWORK

Although a number of commercial and academic simulators are available for complex studies, few have the capability to tune the simulation while it is in execution. Due to tight coupling between the model and the simulation engine in such simulation applications, the capability to introduce changes in parameter values during execution is limited or nonexistent. This leads to monolithic applications that fail to integrate and interoperate. The work described here has the objective of developing a DEVS-based modeling and simulation environment with dynamic simulation control and visualization. The DEVS modeling and simulation framework separates model, experimental frame, and simulator. This modularity facilitates the development of a simulation framework supporting runtime simulation tuning. The motivation behind providing real-time intervention is to support a rapid feedback cycle that allows experimentation with system parameters and structures. This can result in an effective configuration that is difficult to achieve when turnaround requires hours or days. Furthermore, such instantaneous observation and control enables important transient situations to be recognized and considered.

7.2.1 Real-Time Control and Visualization Limitations of Existing Simulators

Some of the limitations of existing simulation packages are as follows:

- Everything has to be programmed before running the simulation.
- User interfaces are not easily customized.
- There is no support for changing parameters and component structures during a simulation.
- Simulation runtimes tend to be long (a few hours); more importantly, if a run ends in a crash, there is no way to intervene and readjust the system.
- There is little runtime visualization of the system behavior to aid understanding and to steer the simulation in a productive direction.
- Model and simulation calibration is a new concept largely unattended by the legacy and current simulators.
- Model-driven design and development is a new technology supported by only a handful of simulation frameworks.
- Distributed M&S and concepts like model repository are not supported in most of the frameworks.
- Treating an M&S T&E framework as an online system by itself is nonexistent and unaddressed by current simulators.
- Performance-oriented simulation frameworks are nonexistent. Most are bounded by initial model configuration.

We discuss the layered architecture underlying the simulation environment ahead. After describing this architecture, we discuss some proposed run control and visualization techniques intended to greatly improve user understanding of, as well as the ability to control, the complex structural and behavioral relationships characteristic of large system behaviors.

MSVC paradigm was designed as an extension of MVC. The separation of model and simulator has emphasized many advantages that come about with this idea—most important being the reuse of simulation software, especially in the context of distributed simulations. The other problems that are solved by this paradigm are as follows:

- Distributed simulation protocol changes can be encapsulated within the controller (input and time management policies) and viewer (output policies) objects.
- By separating the viewer and controller, it is straightforward to add displays, logging tools, and other output-processing devices to the simulator.
- Modeling, simulation, distribution or parallelization, and user interface issues can be addressed separately.

Nutaro and Hammonds (2004) demonstrated the MSVC paradigm with an application of middleware simulation, wherein the simulator was tuned to display the behavior of certain middleware by incorporating effects such as Run-Time Infrastructure (RTI) latency (with reference to the distributed simulation High Level Architecture (HLA) framework). In their methodology, the simulator is a thread derived from the controller thread that contains the platform (RTI latency) delay parameter. As the controller thread generates this event, it is communicated to the simulator as well as to the viewer using interthread communication. Although Nutaro and Hammonds did not consider model updating or model control, their work constitutes a part of our enhanced MSVC framework, where there is full capability in the controller to modify the model as well as the simulator.

Our work is implemented in DEVSJAVA and has a superthread that runs at the root coordinator level that monitors the experimental frame for any user-generated activity controls. There exists no viewer thread, as the viewer objects are created hierarchically as delegated classes of the model as well as of the simulator object. Any modification in their state is also reflected in the contained viewer object.

7.3 ENHANCED MSVC

Figure 7.1 provides the graphical representation of an enhanced MSVC paradigm. It has been represented with respect to the DEVS M&S framework components. Model and view take their usual functions and meanings. The control in MVC is explored in more detail and is mapped to the DEVS experimental frame. Internally, the experimental frame has a modular structure with a basic control component and controller A and B as derived components.

The basic control component translates the information contained in the parameter set coming from requirement specifications. It is specialized into two components: one dedicated to simulator middleware control and the other dedicated to model control. It also assigns different parameters to the appropriate controller. In (Nutaro & Hammonds, 2004), controller A provides tools to control the DEVS simulator—more appropriately, the middleware aspect of simulation. Controller B provides the toolset to control the model. Controller B provides functionality to vary

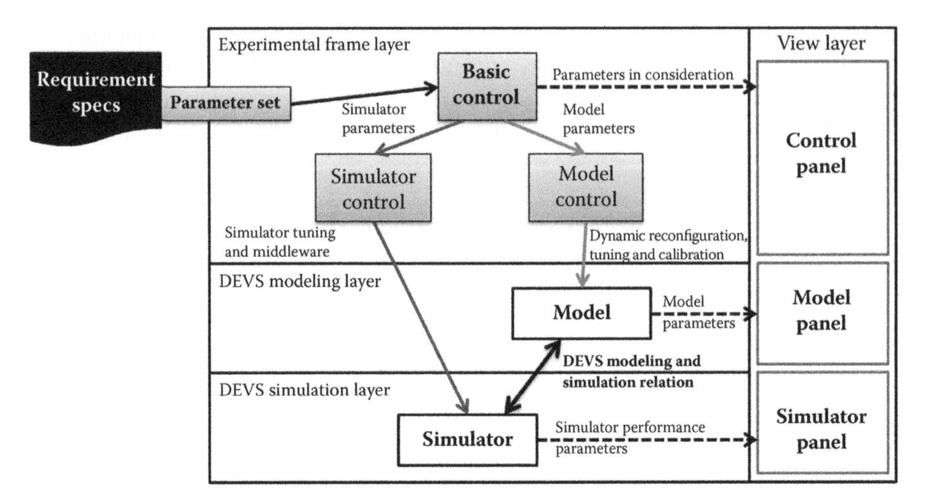

FIGURE 7.1 Enhanced MSVC architecture.

the number of components, in addition to the parameters in a component, both at the component and at the subsystem level. The parameter set for both the controllers is made available to the user (e.g., a sliding bar) in the controller frame in the view panel that enables the user to tune the active simulation toward optimum performance.

The enhanced MSVC has exhaustive control expressed in the experimental frame domain. The experimental frame component in the DEVS M&S framework is a key construct that enables the user to drive and maneuver the simulation in the right direction. The concept of experimental frame, that is, a mechanism by which an experimental scenario is designed for the model architecture, is further enhanced to enable the user to reconfigure and tune the simulation itself. Given that the user has the capability to control the simulation parameters, the issue of extraction and identification of those parameters is taken care by the basic control component that interfaces with the requirement specifications document. Consequently, the experimental frame now provides rich control equipment that the operational test designers can use to their advantage.

7.4 DYNAMIC MODEL AND SIMULATION RECONFIGURATION

7.4.1 Implementation of the Variable Structure in Extended MSVC

The variable structure essentially deals with modification of the component as well as of the number of components that specify the modeled system. Its power lies in its runtime implementation that gives us the capability to study the transition effects when the system is presented with a different number of components and interrelationships. This is entirely a modeling issue and is independent of how the system is simulated when presented with such changes. With the DEVS modeling approach, this is brought to fruition in its modeling layer. With the proposed MSVC approach, as is obvious, this is implemented in the modeling layer that is in control of the experimental frame controller layer. The modeling layer that holds the system model, its configuration, and the intercomponent relationships receives commands from the experimental frame on modifying the system. The users are in charge of the experimental frame. Consequently, if they wish to modify the system structure, they are given the toolbox to modify the model from the experimental frame. Of course, the toolbox is also designed by the modeling designer, who decides if the system is to be analyzed and the chosen component plays a significant role in system dynamics and performance. With the closure-under-coupling property inherent in the DEVS formalism, an entire subsystem or an individual component in the system can be added as a component in the model, in addition to its relationships with other existing components. This property aids in adding a complete system model as a component in a running simulation. With reference to Figure 7.1, the experimental frame view will contain the controls that the user can perform to modify the structure of the model.

7.4.2 Notion of System Steady State

Evolution is a discipline by which one can understand the growth of a system with respect to time. Modeling growth is a difficult concept, let alone simulating growth. Biological evolution is studied through looking into the past and seeing how different species have changed according to their environment. In computer systems, the Internet is one such system that has evolved over time and has resulted in the World Wide Web that now sustains heterogeneous components sustaining together. Evidently, no one could foresee during its conception days that it had the potential to become the Internet of today with over two billion hosts. To model growth, one has to have the capability to modify the structure of constituent components—its interfaces on how it changes when the component is placed in different environments. Biological organisms survive by a process of adaptation and transmitting this information to progeny with encoded information unlike the computer systems. The computer systems are characterized by rigid interfaces through which they communicate with the environment. Certainly, we are not focused toward modeling adaptation, though it can be done with the current DEVS suite, but trying to understand the response of the system when another component is introduced in the system is of prime importance. The response time of a system is defined as the time taken by the system to display any significant effect once the model has been modified. There are legacy systems, and the new technology is bringing new components that need to be backward compatible.

The steady state of any IT network system can be defined as the situation when the computer network is stable and there is constant throughput, there is network latency, and there are no overflowing buffers in routers. In essence, it boils down to the efficient utilization of bandwidth across all links such that there are no blockages. Total data transmitted from network components are received at the designated destinations, with allowable errors. Consequently, capacity planning is one study that results in quantifying the bandwidth to make the system stable with a specified number of components. Looking at it in an inverse perspective, finding the number of components that can be sustained by any particular deployed network is of equal interest. The question arises: How can we model a network system in which the system can simulate the growth of this network, arriving at a steady state and providing us with the result that the network can sustain a particular number of components? The current variable structure capability provides us with the needed functionality in which the experimental frame is given the control to arrive at a steady state. What this actually means is that once a small model of the network system is simulated and utilization is reported, the system continues to keep adding new (preordained) components, along with their relationships, to the existing system until the system reaches a specified network throughput. At what rate the new components are added is a tunable parameter, made available in the experimental frame. This whole exercise shows, given a certain system exhibiting a certain behavior, how the system would perform and evolve if let loose, or what the maximum number of components is that the system can be loaded with so that it maintains a steady state. To determine at what result-set the system would break, or if it has a survivable nature, is worth analyzing. The runtime capability gives us a window to monitor the effects the system incurs when it is modified by external effects like the rate of growth of the system.

7.5 DYNAMIC SIMULATION CONTROL

7.5.1 DEVS Simulation Engine

DEVS has been erected on a framework that exploits the hierarchical, modular construction underlying the high level of system specifications. The basic specification structure in all the associated DEVS-derived formalisms, for example, DTSS and DESS, is supported by a class of atomic models. An atomic model is an irreducible component in the DEVS framework that implements the behavior of a component. It executes the state-machine and interacts with other components using its defined inports and outports. Each such atomic class has its own simulator class. A network of these atomic models constitutes a coupled model that maintains the coupling relationships between the constituent atomic components. Each such coupled model class has its own simulator class, known as a *coordinator.* Assignment of coordinators and simulators follows the lines of the hierarchical model structure, with simulators handling the atomic-level components and coordinators handling the successive levels until the roots of the tree are reached. These simulators and coordinators form the DEVS simulation engine, and they exchange messages by adhering to the DEVS simulation protocol (see Figure 7.2). The message exchange is depicted in Figure 7.2. For more details about the simulation protocol, refer to Chapters 3 and 4.

Figure 7.3 shows the mapping of a hierarchical model to an abstract simulator associated with it. Atomic simulators are associated with the leaves of the hierarchical model. Coordinators are assigned to the coupled models at the inner nodes of the hierarchy. At the top of the hierarchy, there is a root coordinator that is in charge of initiating the simulation cycles.

Since the DEVS model is based on the DEVS formalism that is based on mathematical systems theory, the behavior expressed through DEVS can be translated to any other formalism, though there exists very few other theoretical M&S frameworks. With the separation of the model from the simulator, the advantage is that

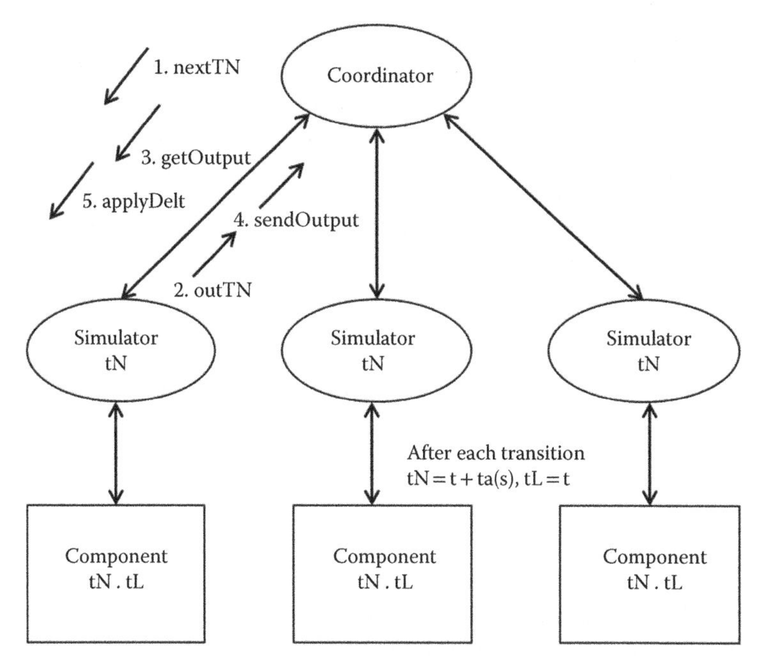

FIGURE 7.2 The DEVS simulation protocol.

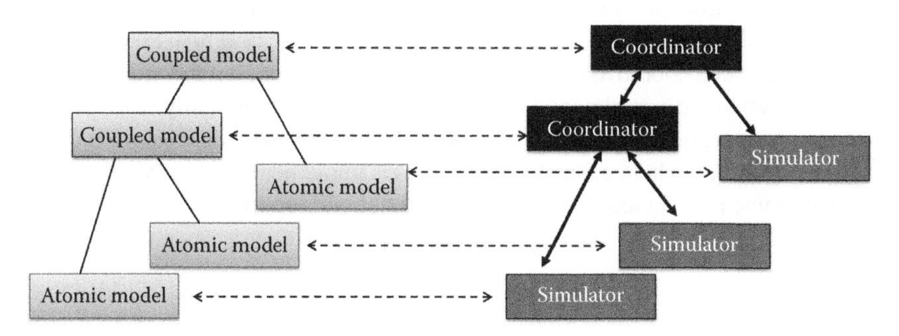

FIGURE 7.3 Assignment of model to a corresponding simulator.

it *supports formalism interoperability*. The next subsection throws light on how an experimental frame intervenes in the DEVS simulation protocol by causing interrupts, and how it implements dynamic simulation control.

7.5.2 Interrupt Handling

The controller frame is built on top of a root coordinator in DEVSJAVA, shown in Figure 7.3. We developed interfaces to enable the DEVS engine to take into account the change of experimental frame parameters during the simulation run. It generates interrupts, which are handled by the coordinator in DEVSJAVA. The event from the controller frame is handled by the root coordinator that holds the simulation at that instant, taking care of the simulation state. The event then is channeled through the hierarchical simulator network to the intended model. Once the model has been updated, the root coordinator resumes the simulation by reinitiating the DEVS simulation protocol. Consequently, the model is updated in between the running simulation with other events still being held in different component simulators. Only the intended model is updated, which then participates accordingly as before. How this event (parameter update inside a model) brings change or how the system responds to this change can be seen very well in different visualizers. More details can be found in the GENETSCOPE case study described in Chapter 20.

7.5.3 The Notion of Simulation Control Explored

Having laid out the framework to implement the dynamic simulation control, we now explore different methodologies in which the simulation can be controlled. Following are the three ways by which the simulation can be interjected and brought to successful execution.

7.5.3.1 Automated Control

In this methodology, we have stored procedures, basically a predefined event list stored as a file that is being read actively during the running simulation and generates events that send interrupts to the coordinator.

This does not require a controller frame that is used to provide real-time interrupts. The experimental frame takes the shape of this file in which different scenarios are preloaded along with simulation parameters. Certainly, execution of a scenario can be considered as one simulation run or a session, but the introduction of a parameter set in the experimental frame that is injected dynamically in the running simulation is of prime interest. This methodology is verily extended toward the setup shown in Figure 7.4, where a family of test cases is implemented as an XML file. The sequence of test is executed in a sequential manner and reported.

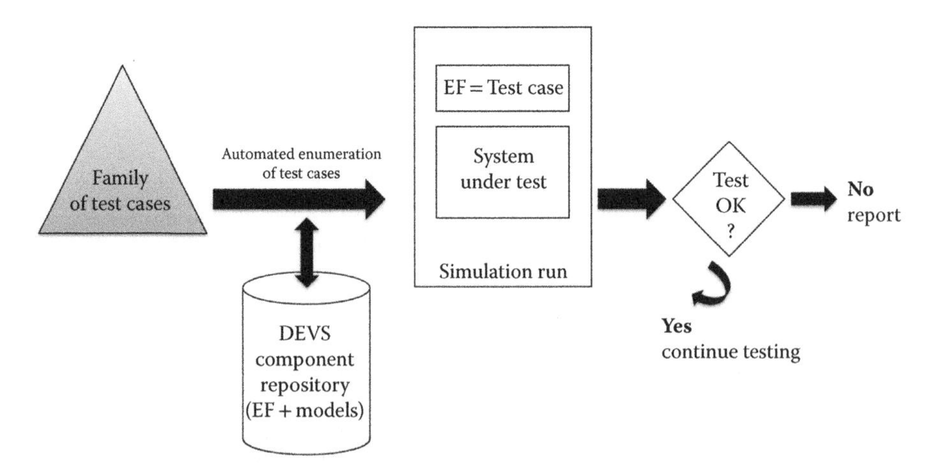

FIGURE 7.4 Automated simulation control with experimental frames stored in SES/Model-base.

7.5.3.2 Manual Reactive Control

In this methodology, the *experimental frame* is operated through a *controller frame* that is designed by the system test designers. This methodology provides us with a mechanism to manually interject in the running simulation to introduce modifications. It also provides us with the capability to steer the simulation if the simulation is moving toward a crash or if the user wants to see the temporal effects of any parameter update. The capability to steer and study the effects of any single parameter is a powerful capability and is almost nonexistent in current simulators, in both the academic and the commercial arenas. There is, however, some software available in the business analytics domain that provides this capability. We implemented this capability in one of our projects. Refer to the GENETSCOPE case study described in Chapter 20.

7.5.3.3 Hybrid Control

As the name suggests, this methodology takes the best of the above two approaches. This methodology has an automated scenario generation/modification capability as well as reactive control through the controller frame. The main purpose of the controller frame in this approach is to study the temporal effects and steer the simulation toward optimum performance.

7.5.4 Parameter Control

This subsection presents some ideas on the selection and categorization of parameters. Two classes of parameters that were identified for any system are the tunable parameter set and the result parameter set.

7.5.4.1 Tunable Parameter Set

This set is comprised of the parameters that are to be included in the experimental frame.

This set is termed "tunable" for obvious reasons, as the simulation analysis is conducted to study their effects on the system performance when their values are modified. The data type of such values is either *float* or *double*. These parameters are called tunable parameters because these parameters are implemented as a slider component in the controller frame with definite bounds. The user can control this slider to tune the system for optimum performance. In the network system terminology, link capacity, router buffer, and so on, can be classified as tunable parameters. With reference to the complex systems where finding significant parameters is an arduous task, this makes more sense, as we need to understand the impact of the identified significant parameter on the overall system performance. This identification is achieved with the help of subject matter experts.

7.5.4.2 Result Parameter Set

This set is comprised of the aggregated result values that provide the overall system performance estimates. They relate to the holistic values that are synthesized after processing various output streams from a simulation run. The aggregated parameters in a network system can be thought of as latency and network throughput. This parameter takes leverage from various top-level mission effectiveness criteria that require holistic assessment of the entire system and where such analyses cannot be attributed to a single component.

7.6 SYNOPSIS

This chapter has described various capabilities that the DEVS simulation framework provides through the use of the implemented experimental frames. It also shows that an experimental frame is the place where the user can initiate the model modification and can modulate the simulation according to need. From the basic capability of creating an experimental scenario for the modeled system, we have enhanced it by providing more features like simulation control and parameter tuning. We have also explored various ways simulation control could be performed and how parameters are categorized to find their way in the experimental frame. Together with the variable structure capability, the experimental frame becomes an all-encompassing user interface to a complex hierarchical system model under simulation. It gives the user more power to observe and visualize the simulation by isolation at the parameter level and the component level, as well as on the subsystem level.

REFERENCES

Barros, F. (1995). Dynamic structure discrete event system specifications. *Proceedings of the 1995 Winter Simulation Conference*. Arlington, VA.

Barros, F. (1997). Modeling formalisms for dynamic structure systems. *ACM Transactions on Modeling and Computer Simulation, 7*(4), 501–515.

Barros, F. (1998). Abstract simulators for the DSDE formalism. *Proceedings of the 1998 Winter Simulation Conference*. Washington, DC.

Brown, A. W., & Wallnau, K. C. (1998). The current state of CBSE. *IEEE Software, 15*(5), 37–46.

Chean, M., & Fortes, L. (1990). A taxonomy of reconfigurable techniques for fault-tolerant processor arrays. *IEEE Computer, 23*(1), 55–69.

Hu, X., Zeigler, B., & Mittal, S. (2005). Variable structure in DEVS component-based modeling and simulation. *Transactions of SCS, 81*(2), 91–102.

Nutaro, J., & Hammonds, P. (2004). Combining the model/view/control design pattern with the DEVS formalism to achieve rigor and resuability in distributed simulation. *Journal of Defense Modeling and Simulation*, 1(1), 19–28.

Uhrmacher, A. (2001). Dynamic structures in modeling and simulation—A reflective approach. *ACM Transactions on Modeling and Simulation, 11*(2), 206–232.

Uhrmacher, A. (1993). Variable structure models: Autonomy and control—Answers from two different modeling approaches. *Proceedings on AI, Simulation and Planning in High Autonomy Systems*. Tucson, AZ.

Uhrmacher, A., & Priami, C. (2005). Discrete event systems specification in systems biology—A discussion of stochastic Pi calculus and DEVS. *Winter Simulation Conference*. Orlando, FL.

Uhrmacher, A., Ewald, R., John, M., Maus, C., Jeschke, M., & Biermann, S. (2007). Combining micro and macro modeling in DEVS computational biology. *Winter Simulation Conference*. Washington, DC.

Uhrmacher, A., Himmelspach, J., & Ewald, R. (2011). Effective and efficient modeling and simulation with DEVS variants. In G. Wainer & P. Mostermann, *Discrete Event Modeling and Simulation: Theory and Applications* (pp. 139–176). Boca Raton, FL: CRC Press.

Zeigler, B., & Louri, A. (1993). A simulation environment for intelligent machine architectures. *Journal of Parallel and Distributed Computing, 18*, 77–88.

Zeigler, B., & Reynolds, R. (1985). Towards a theory of adaptive computer architectures. *Proceedings of the 5th International Conference on Distributed Computer Systems* (pp. 468–475). Denver, CO.

8 Real-Time DEVS and Virtual DEVS

8.1 INTRODUCTION

Within the real-world environment, hardware and software systems are subject to real-time constraints such as operational deadlines relative to an event or microsecond timing responses. Real-time systems must guarantee responses within strict time limits. Real-time software requires synchronous programming languages, real-time operating systems, and real-time networks to build real-time software applications. Real-time computations are said to fail whenever they are not completed before their deadline. Hence, real-time deadlines must be satisfied, regardless of the system load. Real-time simulation denotes the framework's ability to interact with the surrounding environment, such as software components or human operators. Some examples are the human-in-the-loop simulations such as flight simulation for training airline pilots.

Real-time discrete event simulation is a way to verify real-time systems where the simulation model interacts with the surrounding environment. Another example of model executing within a real-world environment is embedded systems employing Discrete Event Systems Specification (DEVS) models to build software systems that interact with their entourage. The RT-DEVS formalism provides a framework for development of software systems that fulfill strict real-time constraints.

8.2 RT-DEVS MODEL FORMAL SPECIFICATION

The RT-DEVS formalism is an extension of the DEVS formalism for real-time systems simulation. An atomic model in RT-DEVS formalism is defined as follows (Hong et al., 1997):

$$\text{RTAM} = < X, S, Y, \delta_{\text{int}}, \delta_{\text{ext}}, \lambda, \text{ta}, \psi, A >$$

where

X is the set of input values

S is the state space

Y is the set of output values

$\delta_{\text{ext}} : Q \times X^b \rightarrow S$ is the external transition function

$Q = \{(s, e) : s \in S, 0 \leq e \leq \text{ti}(s)\}$ is the total state set, where e is the time elapsed since last transition.

X^b is a set of bags over elements in X

$\delta_{int} : S \times R^+_{0,\,inf} \rightarrow S$ is the internal transition function

$\lambda : S \times R^+_{0,\,inf} \rightarrow Y$ is the output function where $R^+_{0,\,inf}$ is the non-negative real numbers with ∞ adjoined.

Note that $R^+_{0,\,inf}$ denotes the set of time in terms of ticks that a model stays in a certain state.

$$\text{ti}(s) : S \rightarrow R^+_{0,\,inf} \times R^+_{0,\,inf} \text{ is the time interval advance function.}$$

Note that a time advance, $\text{ti}(s)$, is given by an interval $\text{ti}(s)|_{min} \leq \text{ti}(s) \leq \text{ti}(s)|_{max}$.

where
$\psi: S \rightarrow A$ is an activity mapping function

$$A : \text{a set of activities } A = a \mid t(a) \in R^+_{0,\,inf} \text{ and } t(a) \leq \text{ta}|_{max} \} \cup \varnothing$$

The RT-DEVS extension replaces virtual (logical) time advance in the original DEVS formalism with real-time advance. A restriction of the specification is that a model cannot receive events when its current state executes an activity function.

An atomic RT-DEVS has activities, a mapping function, and an internal time advance function that are not defined in the classic DEVS formalism. The RT-DEVS formalism defines a set of activities associated with a state. These are executable functions whose completion is required by an atomic RT-DEVS model for transition from one state to another. Activities are modeled tasks in a real system and require actual execution time to be spent during simulation. Execution for each task is specified by the time advance interval function defined for the state. Sending, receiving, and modifying states are excluded from the set of activities. The execution time for an activity is nondeterministic. Hence, the time advance function is given by an interval $\text{ti}(s)|_{min} \leq \text{ti}(s) \leq \text{ti}(s)|_{max}$ that represents minimum and maximum real time defined for each state. The real-time DEVS simulator validates the specified time elapsed of the model against the real-time clock within the simulator. The real numbers in time advance specify computation time of an activity in terms of ticks. This specification enables the user to define the computation time adaptable to the given computing system.

A real-time DEVS coupled model connects real-time DEVS models together to build a coupled model. As in the traditional DEVS formalism, a set of component models makes up a new coupled model. The real-time DEVS coupled model is defined as follows:

$$\text{RTCM} = < D, \{M_i\}, \{I_i\}, \{Z_i\} >$$

where

D, I_b, and Z_b are same as original

M_i is a real-time DEVS model

A real-time DEVS coupled model is defined in the same way as in the original DEVS formalism with one unique exception. The *SELECT* function no longer appears in RT-DEVS, defined in DEVS formalism, to break ties among simultaneous events. The reason why there is no longer a *SELECT* function is that such simultaneous events cannot happen in a real-time simulation environment.

8.3 REAL-TIME DEVS SIMULATION

In real-time simulation, models must progress according to the real-world clock. Real-world processes have no connection with the virtual simulation time. Hence, the single method to synchronize the current time of a simulation model with the environment is to apply the real-world time as the simulation time. Models specified by the RT-DEVS formalism are not simulated along the virtual time base as other classic DEVS models. Instead, they are executed in real-world time. The execution of RT-DEVS models requires one simulator per model, as with simulators for classic DEVS. The RT-DEVS simulator handles two types of messages or events: internal and external. A model receives messages from other models and processes in a situated environment. The following pseudo code illustrates the simulator algorithm that processes these messages.

RT Simulator

when receive $(*, t)$

if $(\mathrm{ti}_N|_{\min} \le t \le \mathrm{ti}_N|_{\max})$ then

$y := \lambda(s);$

$s := \delta_{\mathrm{int}}(s);$

$t_L := t;$

$\mathrm{ti}_N := [t_L + \mathrm{ti}(s)|_{\min}, t_L + \mathrm{ti}(s)|_{\max}];$

$a := \psi(s);$

else

error;

end if;

when receive (x, t)

if $(\mathrm{ti}_L \le t \le t_N|_{\max})$ then

$e := t - t_L;$

$s := \delta_{\mathrm{ext}}(s, e, x);$

$t_L := t;$

$t_N := [t_L + \mathrm{ti}(s)|_{\min}, t_L + \mathrm{ti}(s)|_{\max}];$

$a := \psi(s);$

else

error;

end if;

The first part of the pseudo code specifies how the RT simulator reacts to an internal time event or an empty message coming from the higher level RT coordinator. The RT coordinator only launches an internal time event when the activity related to the current state of the RT-DEVS model has completed and no external message has been received. The response of the RT simulator to a time event is quite similar to the one defined in the classic DEVS formalism, except for the validation of the execution time, whether t lies between the specified time boundaries. On the other hand, the second part of the pseudo code denotes the behavior of the simulator to an external event or message. Likewise, the reaction of the RT simulator is quite similar to the classic DEVS simulator performance, except that the current activity is cancelled and replaced by the corresponding activity of the new state.

Note that the execution time of transition and output functions are not included in the execution time of activities, and that the execution time for each activity varies with every change in the atomic model state. Hence, time inconsistencies can arise during the simulation since RT-DEVS models require executing in real-time internal and external transition functions, output functions, and activities. In the worst-case scenario, timing errors are dragged or accumulated during the simulation. These errors can be removed or compensated during the scheduling of the simulation.

8.3.1 Real-Time DEVS Simulation Framework

A real-time DEVS coordinator program must consist of at least the execution of two concurrent control threads for monitoring the internal activities and the occurrence of external events. A commonly extended simulation framework for RT-DEVS models execution is the one defined by Cho and Kim (2001). They propose a simulation methodology that employs the event-driven scheduling policy to meet timing requirements of RT-DEVS models. This framework consists of prioritizing the execution order of simulation models to prevent unacceptable delays violating time requirements. An RT-DEVS model with a smaller t_N has higher priority, and events are processed with the highest priority under the scheduling policy. The model scheduler creates a thread for each simulation model after flattening all coupled models. The following pseudo code illustrates the algorithm applied for this simulation methodology.

RT Simulator

when receive $(*, t)$

  if $(ti_N|_{min} \leq t \leq ti_N|_{max})$ then

    $y := \lambda(s)$;

    send message (y, t) to associated ports;

    $s := \delta_{int}(s)$;

    $t_L := t$;

    $ti_N := [t_L + ti(s)|_{min}, t_L + ti(s)|_{max}]$;

    $P(t_N|_{max})$;

    $a := \psi(s)$;

  else

    error;

  end if;

when receive (x, t)
    if $(\mathrm{ti}_L \le t \le t_N|_{max})$ then
        $e := t - t_L$;
        $s := \delta_{ext}(s, e, x)$;
        $t_L := t$;
        $t_N := [t_L + \mathrm{ti}(s)|_{min}, t_L + \mathrm{ti}(s)|_{max}]$;
        $P(t_N|_{max})$;
        $a := \psi(s)$;
    else
        error;
    end if;

Note that the difference between this algorithm and the one described in the previous subsection is the inclusion of the function $P(t_N|_{max})$, which modifies the execution priority of all simulation models based on the current t_N and t_L values. The RT-DEVS model with the lower value of the difference between its maximum t_N and the time t_L is set to the highest priority for execution.

8.3.2 Real-Time Reactive DEVS Simulation

Real-time simulation involves completion in physical time for the execution of a simulated model. Such simulations deal with a reactive real-time system. The environment consists of live modules that may include humans, who have occasional interactions with the system. In simulation, a time selectivity problem appears in a reactive simulation environment where simulation models interact with real-world objects (see Figure 8.1).

The importance of an event in a reactive simulation usually depends on the time spent between the occurrence from an external source and the event consumption

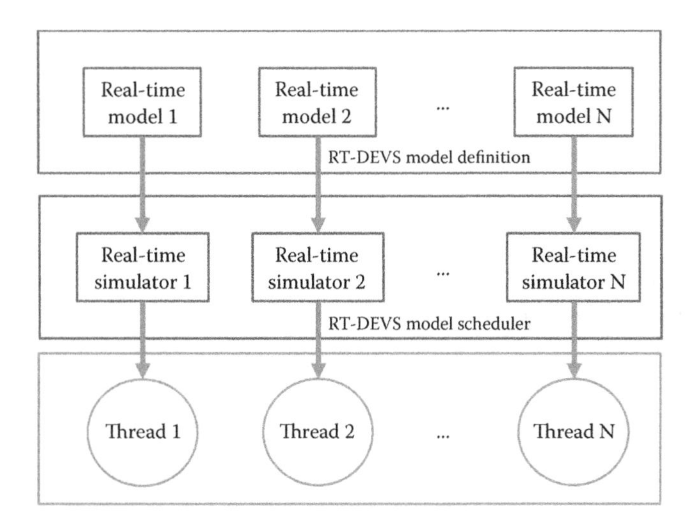

FIGURE 8.1 RT-DEVS simulation framework.

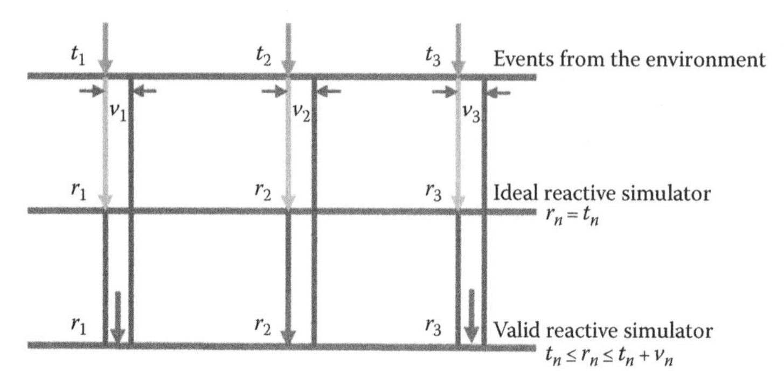

FIGURE 8.2 Time in event handling.

at the simulation model. Thus, time events should have an associated attribute that denotes a validity time interval. Figure 8.2 shows the timeline in event handling, where *t* represents the event occurrence time, *r* is the event consumption time, and *v* is the valid time interval.

According to Cho and Kim (1998), the time selectivity problem is solved to a certain extent by executing simulation models and the interactive environment concurrently. They define a driver model associated with each real-world object. These models act as interfaces among simulation models and the real world. Both driver models and simulation models are executed concurrently to respond to external events from the real world. Driver models monitor the events generated by the corresponding real-world object. Whenever an external event that is originated by a real-world object is triggered, a driver model converts it into an event comprehensible by the associated simulation model. Similarly, the driver model converts output events from simulation models to an input for the corresponding real-world object. A real-time driver model is defined as follows:

$$\text{RTDM} = < X, Y, T_{\text{ME}}, T_{\text{EM}} >$$

where

$X = X_{\text{M}} \cup X_{\text{E}}$ is an input event set
X_{M} is input events from model
X_{E} is input events from environment
$Y = Y_{\text{M}} \cup Y_{\text{E}}$ is an output event set
Y_{M} is output events from model
Y_{E} is output events from environment
T_{ME} is $X_{\text{M}} \rightarrow Y_{\text{E}}$ is an event translation function from a model to an environment
T_{EM} is $Y_{\text{E}} \rightarrow X_{\text{M}}$ is an event translation function from an environment to a model

Driver models have no internal or external transition functions or a time advance function. It translates events from the environment to the simulation and vice-versa. The simulation environment employs multithread to enable all simulation and driver models to execute concurrently (Figure 8.3).

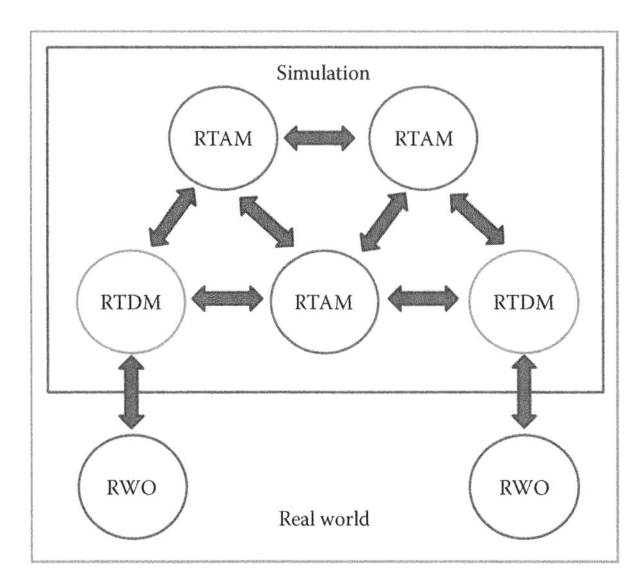

FIGURE 8.3 Real-time reactive simulation.

8.3.3 Real-Time Transformational DEVS Simulation

On the other hand, in real-time transformational DEVS simulations, models symbolize environmental and simulation modules, and the specific computing system. The virtual time advanced in the model is synchronized with the user-defined physical times. This logical time can be homogeneously increased or decreased without altering the required event ordering. Hence, transformational models can apply either RT-DEVS or the classic DEVS formalism. A modeler should select one or the other depending on their simulation goals. For example, if the modelers are more interested in validating the interactions among modeling components in relation to real time, they should apply the RT-DEVS extended formalism. On the contrary, if the simulation worth is based on performance analysis, DEVS models can be simulated over virtual time.

8.4 RT-DEVS IMPLEMENTATION

Now we describe with some Java snippets an example implementation of an RT-DEVS reactive simulation framework. The code in Listing 8.1 shows the interface for atomic RT-DEVS models where *phase* indicates the current state in words and *sigma* represents the value of the time advance function value. Note that the sigma value is a real number, not an interval. That is because we define a valid delay named as *phi*. Hence, the interval defined for each state is bounded by sigma-to-sigma plus phi. Moreover, activities are not explicitly defined; instead, the duration of an activity is assumed to be the current-state sigma value.

The code in Listing 8.2 illustrates the driver model interface that translates stimulus from the environment to the model and from the model to the environment. Note that we included two more functions that are triggered whenever a translation fails in order to handle this kind of potential error.

LISTING 8.1 ATOMIC RT-DEVS INTERFACE

```
public abstract class RTAtomic extends Atomic {

   protected double phi;

   public RTAtomic(String name) {
     super(name);
     super.passivate();
     this.phi = Constants.INFINITY;
   }

   public double getPhi() {
    return phi;
 }

   public void setPhi(double phi) {
     this.phi = phi;
   }
}
```

LISTING 8.2 DRIVER INTERFACE

```
public abstract class ComponentDriver extends Component {

   protected PortDriver portDriver;

   public ComponentDriver(String name,
                                    PortDriver portDriver) {
     super(name);
     this.portDriver = portDriver;
   }

   public PortDriver getPortDriver() {
     return portDriver;
   }

   public abstract void tME();

   public abstract void tEM();
```

```
  public abstract void failureME();

  public abstract void failureEM();
}
```

Listing 8.3 shows the code for an RT-DEVS simulator. As in the simulator for classic DEVS, we have the time for the next event (*tN*), the time of the last event (*tL*), and the atomic model. In addition, we also include the real-time world clock. Functions *deltint* and *deltext* are executed with time and external events, respectively. The validation of the internal timeout is extracted on to the main process function *run*.

LISTING 8.3 RT-DEVS SIMULATOR

```
public class RTSimulator extends DevsRTSimulator<RTAtomic>
                              implements Runnable{
  private static final Logger logger =
         Logger.getLogger(RTSimulator.class.getName());
  protected Thread myThread = null;

  public RTSimulator(RTAtomic model) {
    super(model);
    myThread = new Thread(this);
  }

  @Override
  public void initialize(double t) {
    tL = t;
  }

  public void deltint(double t) {
    model.lambda();
    propagateOutput();
    model.deltint();

    this.tL = System.currentTimeMillis()/1000.0;
    this.tN = t + model.ta();
  }

  public void deltext(double t, Event event) {
    double e = t - tL;
    model.setSigma(model.getSigma() - e);
    event.updatePortTo();
    model.deltext(e);

    tL = System.currentTimeMillis()/1000.0;
    tN = t + model.ta();
  }
```

```
  @Override
  public void run() {
    RTAtomic modelRT = (RTAtomic) model;
    startTime = System.currentTimeMillis()/1000.0;
    Event event = null;
    while (!stop) {
      double tW = super.tN - super.tL;
      if (tW < -modelRT.getPhi()) {
        logger.severe("Validity time failed for an
                       internal event for: "
                       + (-tW - modelRT.getPhi()));
      }
      try {
        event = queue.poll((long) (1000 * tW)
                           ,TimeUnit.MILLISECONDS);
      } catch (InterruptedException ex) {
        logger.log(Level.SEVERE, null, ex);
        super.stop();
      }
      if (event ! = null) {
        double e = System.currentTimeMillis()/1000.0
                   - startTime
                   - event.getTimeMillis()/1000.0;
        if (e > modelRT.getPhi()) {
          logger.severe("Validity time failed for an
                         internal event for: "
                         + (e - modelRT.getPhi()));
        }
        deltext(System.currentTimeMillis()/1000.0
                            - startTime, event);

      } else {
        deltint(System.currentTimeMillis()/1000.0
                        - startTime);
      }
    }
  }

  @Override
  public void start() {
    myThread.start();
  }

  @Override
  public double ta(double t) {
    return model.ta();
  }

  @Override
```

```
public void lambda(double t) {
  if (t = = tN) {
    model.lambda();
  }
}

@Override
public void deltfcn(double t) {
  throw new UnsupportedOperationException("Not
                                  supported yet.");
  }
}
```

The main function *run* loops over time, triggering internal time and external message events. First, it computes the time to be spent in the current activity, that is, the difference between the time for the next event and the last one. Next, it checks the validity of the current activity time interval. Moving ahead, the program is blocked waiting for a specific period to execute an internal time event and for an external message event. If a message is received, the current activity is interrupted, and it checks whether the gap between generation and processing time of the message exceeds the defined threshold *psi*. If there is no error, then the external transition function is executed; otherwise, an error is thrown. On the other hand, if the whole time of an activity is consumed, then the internal transition function is triggered.

The code in Listing 8.4 shows a part of the root program for the simulator that launches and initiates every simulator for each model in a new thread. Once started, it waits for the user-defined simulation time to be completed, and finally stops each simulator thread to end cleanly, terminating the simulation.

LISTING 8.4 RT-DEVS ROOT SIMULATOR

```
public class RTRoot implements Runnable {

  private static final Logger logger =
              Logger. getLogger(RTRoot.class.getName());
  protected HashMap<String, DevsRTSimulator> simulators
              = new HashMap<String, DevsRTSimulator>();
  protected Thread myThread;
  protected long timeIntervalMillis;

  public RTRoot(Coupled model) {
    Collection<Component> components =
                       model.getComponents().values();
    for (Component component : components) {
      if (component instanceof Coupled) {
```

```
            RTCoordinator simulator =
                    new RTCoordinator((Coupled) component);
            simulators.put(component.getName(), simulator);
        } else if (component instanceof RTAtomic) {
            RTSimulator simulator =
                     new RTSimulator((RTAtomic) component);
            simulators.put(component.getName(), simulator);
        }
    }
    LinkedList<Coupling> couplings = model.getIC();
    for (Coupling c : couplings) {
      simulators.get(c.getComponentFrom())
                    .addConnection(model.getComponent(c
                    .getComponentFrom()).getOutPort(c
                    .getPortFrom()), simulators.get(c
                    .getComponentTo()), model.getComponent(c
                    .getComponentTo()).getInPort(c
                    .getPortTo()));
    }
    myThread = new Thread(this);
  }

  public void initialize(double t) {
    for (DevsRTSimulator simulator : simulators.values()) {
      simulator.initialize(t);
    }
  }

  public void simulate(double timeInterval) {
    timeIntervalMillis = (long) (1000 * timeInterval);
    myThread.start();
  }

  @Override
  public void run() {
    for (DevsRTSimulator simulator : simulators.values()) {
      simulator.start();
    }
    try {
      Thread.sleep(timeIntervalMillis);
    } catch (InterruptedException ex) {
      logger.log(Level.SEVERE, null, ex);
    }
    for (DevsRTSimulator simulator : simulators.values()) {
      simulator.stop();
    }

   }

  }
```

8.5 RT-DEVS EXAMPLE

This section describes a simple human-in-the-loop reactive simulation example using the RT-DEVS simulation framework defined above. Figure 8.4 illustrates an example of a simulated vehicle being controlled and monitored by external elements to the simulation. A user controls the simulation with the help of a joystick, the way a controller commands, in order to control the simulated vehicle, and receives feedback through an external graphical interface that displays the current state of such vehicle. The RT-DEVS simulation components are a driver model (control), an atomic model (vehicle), and the real-world objects (joystick and external graphical interface). The driver model translates environment signals to the simulation and vice-versa. The atomic model simulates the dynamics of a vehicle that receives control commands and the driver model and sends its current state to the same driver model. The communication with the environment can be easily performed using the Java Native Interface (JNI).

Next, we show the driver model (Listing 8.5) that translates signals from the joystick to simulation commands and the simulation vehicle state to values accepted by the external interface. The driver sends a stop signal to the simulation vehicle if the failure function (environment to model) is triggered to avoid losing control of the simulation, for example, if the joystick is unplugged.

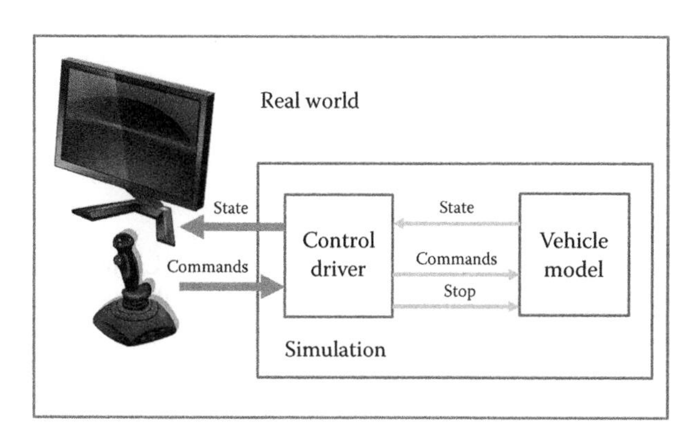

FIGURE 8.4 RT-DEVS example.

LISTING 8.5 CONTROL DRIVER

```
public class Control extends ComponentDriver {

  protected Port<State> iportState =
                    new Port<State>(State.class, "state");
  protected Port<Command> oportCommand =
               new Port<Command>(Command.class, "command");
  protected Port<Boolean> oportStop =
               new Port<Boolean>(Boolean.class, "stop");
```

```
    public Control(String name) {
      super(name,
            new ServerPort("control", 4050, "127.0.0.1"));
      this.addInPort(iportState);
      this.addOutPort(oportStop);
      this.addOutPort(oportCommand);
    }
  @Override
  public void tME() {
    portDriver.write(iportState.getSingleValue());
  }

  @Override
  public void tEM() {
    oportCommand.addValue((Command) portDriver.read());
  }
  @Override
  public void failureEM() {
    oportStop.addValue(Boolean.TRUE);
  }

  @Override
  public void failureME() {
    }
  }
```

To finish, we show the RT-DEVS-based vehicle model (Listing 8.6) that updates vehicle's state (integrates) based on the last control commands every time the internal transition function is triggered. The procedure updates the value of sigma (the duration of the current activity). Every time it receives a message, the external transition function is launched. If it receives a stop signal, it sets the control values correspondingly; otherwise, it updates the control values according to the inputs. In both the cases, it resumes the value of sigma. Finally, whenever the output function is executed, it sets the output port the value of its current state only if the new state is different from the last one sent. In this manner, signaling the environment with useless information is avoided.

LISTING 8.6 VEHICLE MODEL

```
public class Vehicle extends RTAtomic {

  protected Port<Command> iportCommand =
            new Port<Command>(Command.class, "command");
  protected Port<Boolean> iportStop =
            new Port<Boolean>(Boolean.class, "stop");
```

```
  protected Port<State> oportState =
            new Port<State>(State.class, "state");
  IntegratorRK4 integrator;
  double velocity;
  double rudder;
  double h;

  public Vehicle(String name, double h, double phi,
            double x, double y, double psi) {
    super(name);
    this.setSigma(h);
    this.setPhi(phi);
    this.addInPort(iportCommand);
    this.addOutPort(oportState);
    this.addInPort(iportStop);
    this.velocity = 0;
    this.rudder = 0;
    double[] s0 = {x, y, psi, 0, 0};
    this.integrator = new IntegratorRK4();
    this.h = h;
    this.integrator.setInitialState(s0, 0, h/1000);

  }

  @Override
  public void deltint() {
    double[] ctl = {this.rudder, this.velocity};
    this.integrator.update(ctl);
    this.setSigma(h);
  }

  @Override
  public void deltext(double t) {

    if (!iportStop.isEmpty()) {
      this.velocity = 0;
      this.rudder = 0;
    } else {
      Command command = iportCommand.getSingleValue();
      this.velocity = command.getStarboardSpeed();
      this.rudder = command.getStarboardAngle();
    }
}

  @Override
  public void lambda() {
    oportState.addValue(new State());
  }
}
```

REFERENCES

Cho, S. M., & Kim, T. G. (1998). Real time DEVS simulation: Concurrent, time-selective execution of combined RT-DEVS model and interactive environment. *Proceedings of the Summer Computer Simulation Conference* (pp. 410–415), Reno, Nevada (USA).

Cho, S. M., & Kim, T. G. (2001). Real time simulation framework for RT-DEVS models. *Transactions of the Society for Computer Simulation International*, *18*(4), 203–215.

Hong, J. S., Song, H. S., Kim, T. G., & Park, K. H. (1997). RT-DEVS executive: A seamless realtime software development framework. *Discrete Event Dynamic Systems*, *7*, 355–375.

9 Model-Driven Engineering and Its Application in Modeling and Simulation

9.1 INTRODUCTION

Model-driven engineering (MDE) is a system development approach that uses models to support various stages of the development life cycle (Atkinson & Kühne, 2003). MDE relies on model transformations to enable the automated development of a system. It produces well-structured and maintainable systems because of its focus on formally defined models, metamodels, and meta-metamodels.

A model is an abstract representation of a system, which is specified in a modeling language (Buede, 2009). Different models are used for different purposes during the system development life cycle. Each model represents a different view of the system. Model transformations enable the reuse of information that was earlier described in another model. To increase the effectiveness of model transformations, domain-specific languages (DSLs) are preferred in MDE applications. DSL is a language designed for specifying a system in a specific domain.

MDE has been successfully applied in software engineering, and it has been advocated as a cost- and effort-saving development approach for software projects (Selic, 2003). In software engineering, the application of MDE principles is generally called model-driven development (MDD) (a.k.a. model-driven software development). MDD principles are usually described informally and applied by different standards such as model-driven architecture (MDA) (Object Management Group [OMG], 2003), model-integrated computing (MIC) (ISIS, 1997), Microsoft domain-specific languages (Microsoft, 2010), and Eclipse Modeling Framework (EMF) (Eclipse, 2009).

MDA and metamodeling in an MDE context were introduced to modeling and simulation (M&S) in 2002 (Tolk, 2002; Vangheluwe & de Lara, 2002). Since then, many research studies appeared in simulation literature that applied an MDE approach and utilized existing MDD tools. In many applications, metamodeling was used to specify a DSL and model transformation was used for automatic code generation from the models that were specified in that DSL. Although these specific applications have shown the applicability of the MDE principles in simulation projects, the research community has recently started to move toward a general framework for MDD of simulation models such as MDD4MS (model-driven development for modeling and simulation) and DEVSML 2.0 (discrete-event system specification modeling language). These frameworks fill the gaps between different stages of the M&S life cycle and provide model continuity in simulation projects through formal model transformations. MDD4MS introduces three different metamodels to represent the simulation model implementation at different abstraction levels. The DEVSML 2.0 stack brings various domain-specific models along with their metamodels, and it provides specified transformations that semantically anchor in formal DEVS theory (Mittal & Douglass, 2012). The DEVSML 2.0 stack shows that an underlying DEVS metamodel provides model reusability, integration, and interoperability between different platforms.

This chapter integrates the MDD4MS framework with DEVSML 2.0 metamodel, and it uses the defined M2DEVSML (model-to-DEVSML) transformations for bringing particular applications to the DEVS M&S arena.

Section 9.2 provides a general overview about MDE, and it highlights various acronyms such as MIC, MDD, MDA, and many others. Section 9.3 presents an overview on DSLs. MDE applications in the M&S field are examined in Section 9.4, and the MDD4MS framework is described in Section 9.5. Lastly, MDD4MS and DEVSML 2.0 are integrated in Section 9.6.

9.2 MDE, FLAVORS, AND TECHNIQUES

In MDE, models are the primary artifacts of the system development process and source models are transformed into destination models at different stages to (semi) automatically generate the final system. Although the idea seems simple, the concept and its technical implementation are quite hard and there are mismatches between various terms used in the M&S discipline and software engineering. This section starts with the definitions of the key terms which are adapted from Çetinkaya and Verbraeck (2011).

A model is an abstract representation of a system defined in a modeling language, where a modeling language is a means of expressing systems in a precise way by using diagrams, rules, symbols, signs, letters, numerals, and so on. A modeling language consists of abstract syntax, concrete syntax, and semantics (Atkinson & Kühne, 2003). The abstract syntax describes the vocabulary of the concepts provided by the modeling language and how they can be connected to create models, that is, the grammar of the language. The abstract syntax consists of concepts, relationships, and well-formedness rules, where well-formedness rules state how the concepts may be combined. The concrete syntax presents a model either in a diagram form (visual syntax) or in a structured textual form (textual syntax). The semantics of a modeling language is the additional information to explain what the abstract syntax actually means. Semantics explain the abstraction that is made during the modeling process.

The relation between a modeling language and a model expressed in that language is called the <<*conformsTo*>> relation, such as "*the model M* <<*conformsTo*>> *the modeling language L*". Figure 9.1 illustrates the relationship between a model and a modeling language with a simple example. A modeling language for defining simple state diagrams is presented with its syntax and semantics. Then a sample model is shown, which conforms to this modeling language.

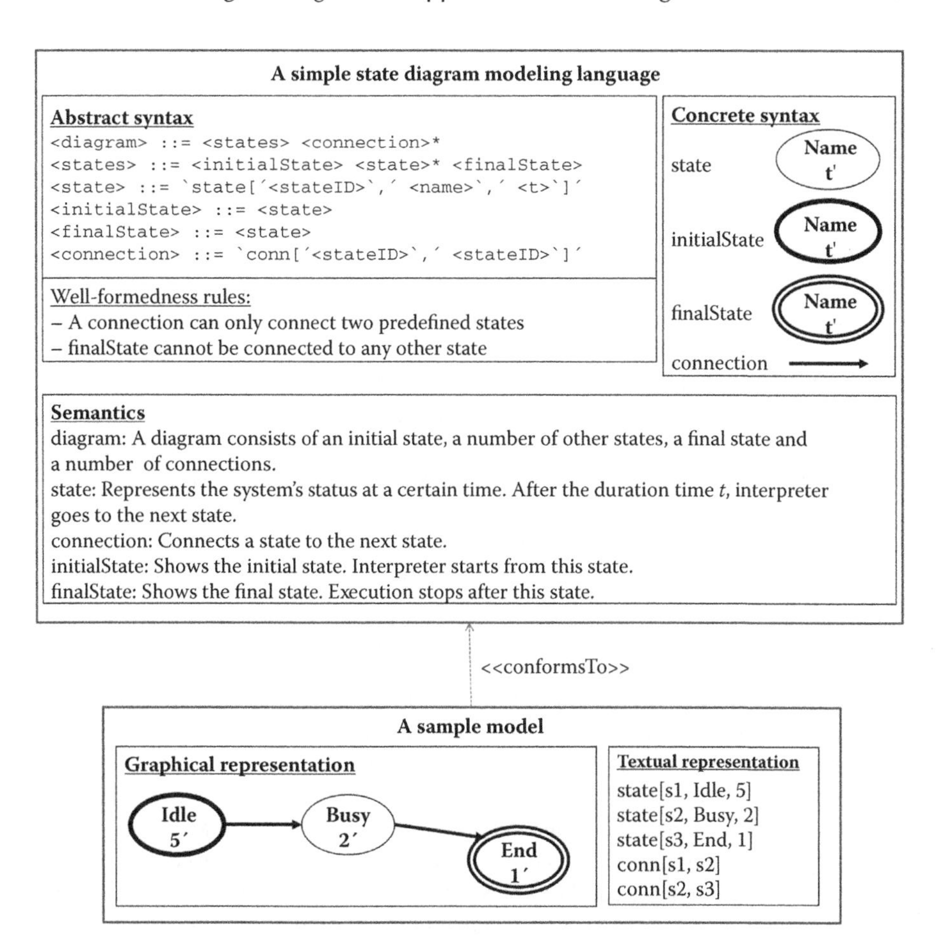

FIGURE 9.1 Relationship between a model and a modeling language.

Different models are used for different purposes during the system development life cycle. Each model represents a different view of the system. As a commonly used example, MDA (OMG, 2003) introduces three types of viewpoints: (1) The computation-independent viewpoint focuses on the environment of the system and the requirements for the system. (2) The platform-independent viewpoint focuses on the structure and operation of a system while hiding the details necessary for a particular platform. (3) The platform-specific viewpoint focuses on the use of a specific platform by a system. Based on these viewpoints, three types of models are used in MDA: (1) A computation-independent model (CIM) is a representation of a system that shows a conceptual domain-specific model without any implementation details. (2) A platform-independent model (PIM) is a representation of a system that exhibits a specified degree of platform independence so as to be suitable for use with a number of different platforms. (3) A platform-specific model (PSM) is a representation of a system that combines the specifications in the PIM with the details that specify how that system uses a particular type of platform.

The MDE approach requires that the models and modeling languages are well defined. Metamodeling is the most commonly used method to describe modeling languages in a formal way. The following section explains metamodeling.

9.2.1 Metamodeling

Metamodeling is the process of complete and precise specification of a metamodel. A metamodel is a definition of a modeling language in the form of a model. In other words, a metamodel is itself a model and has to be defined in a modeling language. The higher level modeling language that is used to describe a modeling language is called a metamodeling language. The metamodeling language can also be defined with a metamodel. The metamodel of the metamodeling language is called a meta-metamodel. To avoid an infinite stack in the number of metamodeling levels, meta-metamodels are often designed to be self-reflexive, and therefore, the modeling language of the meta-metamodel is the meta-metamodel itself. Most approaches implementing MDE define a three-level metamodeling stack for model, metamodel, and meta-metamodel. Figure 9.2 shows the metamodeling pattern in MDD.

The metamodeling pattern can be applied several times, and a hierarchy of abstraction levels can be built, where models at level n are specified using the language defined as a metamodel at level $n + 1$. However, the general application of metamodeling is performed with four levels, which was first described in the Unified Modeling Language (UML) specification by OMG (1999). Table 9.1 shows the four-level metamodeling stack in practice. The Meta-Object Facility (MOF) is the de-facto metamodeling language in OMG specifications for the M_3 level.

In a metamodeling approach, the *<<conformsTo>>* relation between the model and the modeling language can be expressed via an *<<instanceOf>>* relation, such as "*the model is an <<instanceOf>> the metamodel*". While the *<<conformsTo>>*

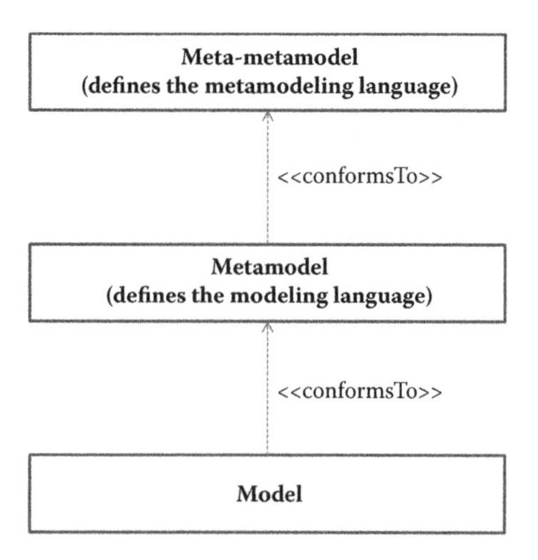

FIGURE 9.2 Metamodeling pattern in MDD.

TABLE 9.1
Four-Level Metamodeling Stack in Practice

Levels	Description	Example
M_3 level: meta-metamodel	Defines the language for specifying metamodels.	**MetaClass** and **MetaAttribute** in the MOF specification
M_2 level: metamodel	Defines the language for specifying models. An instance of the meta-metamodel.	**Class** and **Attribute** in the UML specification
M_1 level: model	Defines the model without user data. An instance of the metamodel.	**Car** class and **Name** attribute in a UML model
M_0 level: user data	Defines a specific user model with user data. An instance of the model.	**Car.3** Instance with **Name** = **"abc"** in a specific user model

relation only guarantees a valid model, the <<*instanceOf*>> relation requires that every element of a model must be an instance of some element in the metamodel.

Once the models are constructed conforming to their specific metamodels, model transformations can be performed to generate new models based on the existing information throughout the MDE process. The following section explains model transformations in MDE.

9.2.2 Model Transformations

Instead of creating the models from scratch during the different development life cycle stages and activities, model transformations enable the reuse of information that was once modeled. A model transformation is the process of converting one or more source models into one or more target models according to a set of transformation rules. The rules are defined with a model transformation language. A transformation rule consists of two parts: a left-hand side that accesses the source model and a right-hand side that generates the target model. The source model conforms to the source metamodel, and the target model conforms to the target metamodel. During the transformation, the source model remains unchanged. Figure 9.3 shows a model transformation pattern, which is adapted from the MDA guide (OMG, 2003).

According to the type and complexity of the process, the model transformation can be named a model-to-model (M2M) transformation, a model-to-text (M2T) transformation, model merging, model linking, model synthesis, or model mapping. As the name implies, an M2M transformation converts a source model into a target model, which are the instances of the same or different metamodels (Völter et al., 2006). However, an M2T transformation converts a source model into a text file. M2T transformation is generally used for final code generation or supportive document creation.

If the source and target metamodels are identical, the transformation is called endogenous; otherwise it is called exogenous (Mens & Van Gorp, 2006). If the level

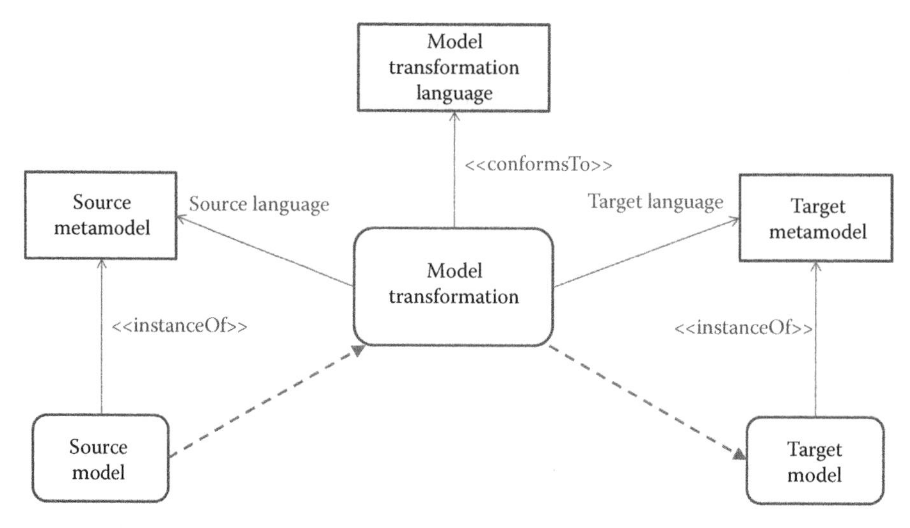

FIGURE 9.3 Model transformation pattern in MDD, based on OMG (2003).

of abstraction does not change, the transformation is called a horizontal transformation. If the level of abstraction does change, the transformation is called a vertical transformation. These concepts relate very well to the system homomorphism concepts described in Chapter 2.

Although there are different solutions for transforming models to text, this is not the case for transforming models to models (Czarnecki & Helsen, 2003). Since M2M transformation is a relatively young area, very limited and often ad hoc transformation capabilities are offered by the existing metamodeling environments. Popular M2M transformation languages are QVT (Query/View/Transformation), ATL (ATLAS Transformation Language), GReAT (Graph Rewriting and Transformation Language), and Xtend.

The MOF 2.0 QVT specification is a set of model transformation languages defined by OMG (2011). The QVT specification has a hybrid declarative/imperative nature, and it defines three related transformation languages: Relations, Operational Mappings, and Core. The Relations metamodel and language supports complex object pattern matching and object template creation with a high-level user-friendly approach. Operational Mappings can be used to implement one or more Relations from a Relations specification when it is difficult to provide a purely declarative specification of how a Relation is to be populated. The Core metamodel and language is defined using minimal extensions to MOF and Object Constraint Language (OCL) on the lower level.

ATL is one of the most popular M2M transformation languages (Jouault et al., 2008). Once the target metamodel and source metamodel are available, an ATL transformation file can produce a target model from a source model. During the transformation, an attribute of a target model instance receives a return value of an OCL expression that is based on a source model instance. The ATL integrated development environment (IDE), which is developed on top of the Eclipse platform, provides a number of tools such as a text editor with syntax highlighting, a debugger, and an interpreter that aim to ease the development of ATL transformations.

There are many other model transformation languages. Besides, projects like Kermeta, Fujaba, and GReAT (these projects apply an MDD approach) provide M2M transformation methods. Detailed explanation about the model transformation languages and methods can be found in Czarnecki and Helsen (2003) and Dehayni et al. (2009).

In an MDD process, different intermediate models are needed for expressing different views of the system. Hence, there is usually a chain of several M2M transformations between these models and a final model-to-code transformation.

9.2.3 Various Flavors

The "model-based (MB)" and "model-driven (MD)" terms and initials have been used in a variety of system and software related acronyms, such as MBD, MDSD, MDD, MDA, MBSE, MDE, and many others. Although there is a consensus that these approaches suggest the systematic use of models as the primary means of a process and facilitate the use of DSLs, there is not a common understanding of the terminology. In this section, the definitions of the frequently used acronyms and the objectives of those approaches are presented. A comparative analysis between the system development approaches is presented at the end of the section.

1. MBE
   Model-based engineering (MBE) focuses on creating higher level models for the final implementation of a system (Bapty & Sztipanovits, 1997). The main goal in MBE is to support the system development process during the design, integration, validation, verification, testing, documentation, and maintenance stages. Sometimes, a specific activity can be labeled as model-based testing, model-based design (MBD), model-based integration, model-based analysis, and so on. MBE originated in the 1980s in parallel with the evolution of the computer-aided design and MBD techniques. MBD provides a model-based visual method for designing complex systems, originally for designing embedded and distributed systems. The work by Zeigler et al. (2000) in the area of systems entity structure/model base was also introduced in the same era.
2. MBSE
   In systems engineering, the application of MBE principles is called model-based systems engineering (MBSE) (Oliver et al., 2009). MBSE provides the required insight in the analysis and design phases; it enhances better communications between the different participants of the project; and it enables effective management of the system complexity. INCOSE (International Council on Systems Engineering) has identified the institutionalized use of MBSE tools and techniques as an integral part of its vision (Estefan, 2008).
3. MDE
   As stated in the Section 9.1, MDE is a system development approach that uses models as the primary artifacts of the development process

(Schmidt, 2006). Hence, it can be seen as a subcategory of MBE. However, MDE introduces model transformations between different abstraction levels. In MDE, source models are transformed into destination models at different stages to (semi)automatically generate the final system. The main goal in MDE is increasing productivity through automated model transformations.

4. MDD or MDSD
   The application of the MDE principles in software engineering is called model-driven development (MDD) or model-driven software development (MDSD) (Völter et al., 2006). The modern era of MDD started in the early 1990s and now offers a notable range of methods and tools. Different specifications such as MDA (OMG, 2003), MIC (ISIS, 1997), Eclipse Modeling Project (Eclipse, 2008), and Microsoft Software Factories (Microsoft, 2005) are some of the conceptual applications of MDD principles.
5. MDA
   Model-driven architecture (MDA) is a software design and development approach that provides a set of guidelines for specifying and structuring models (OMG, 2003). MDA relies on the MOF to integrate the modeling steps and provide the model transformations. Although DSLs play a significant role in all MDD approaches, MDA prescribes the use of metamodels and meta-metamodels for specifying the modeling languages without any necessity to be domain specific.
6. MIC
   Model-integrated computing (MIC) refines the MDD approaches and provides an open integration framework to support formal analysis tools, verification techniques, and model transformations in the development process (ISIS, 1997). MIC allows the synthesis of application programs from models by using customized model-integrated program synthesis (MIPS) environments. The metalevel of MIC provides metamodeling languages, metamodels, metamodeling environments, and metagenerators for creating domain-specific tool chains on the MIPS level. The fully integrated metaprogrammable MIC tool suite provides open-source tools such as Generic Modeling Environment (GME).
7. Software Factories
   Microsoft Software Factories is a unified software development approach, which tries to synthesize ideas from DSLs, MDD, software product lines, and development by component assembly (Microsoft, 2005). When compared to MDA, Software Factories are less concerned with portability and platform independence and more concerned with productivity.
8. Eclipse Modeling Project
   The Eclipse Modeling Project focuses on the evolution and promotion of the MDD technologies within the Eclipse community (Eclipse, 2008). It provides a unified set of modeling frameworks and open source tools. The following projects are all included in Eclipse Modeling Project: EMF, generative modeling technologies, M2M transformation, M2T transformation, and model development tools.

9. Executable architectures
   Executable models refer to those higher level models, where a full model-to-implementation transformation can be applied and an executable final implementation of a system can be generated. The "executable model" term is generally used when MDD is performed to generate final compilable source code. "Executable architecture" is also a term used for MDD in the military domain (Garcia & Tolk, 2010).

As a result, MB and MD approaches place models in the core of an entire system development process and try to increase the system specification quality. MB approaches offer various support methods based on models during the development process, while MD approaches suggest better and faster ways of developing systems through automated model transformations between models, which are specified with well-defined modeling languages. A comparative analysis is presented in Table 9.2.

As a final remark, due to the fact that different MDD tools apply the same principles, an MDD expert can easily think about combining different approaches and tools to perform an MDD process. For example, the MDA concepts such as PIM and PSM metamodels can be defined with the MIC tool suite and GME in particular.

9.2.4 MDD Tools and Techniques

A modeling tool provides ways, generally a graphical user interface or a text editor, to develop models based on a modeling language. In most cases, the modeling tools are extended with extra features, such as model verification, syntax highlighting, and model interpretation.

A metamodeling environment (a.k.a. a domain-specific modeling [DSM] environment in MDD literature) is used to develop a modeling tool by defining its modeling language (as a metamodel) in a predefined metamodeling language. A metamodeling environment provides a graphical or textual editor to define metamodels, and it automatically generates a modeling tool from the specified metamodel. The resulting tool may either work within the metamodeling environment or less commonly be produced as a separate stand-alone program. Well-known metamodeling languages are MOF, UML, EMF Ecore, MetaGME, and Kermeta, which will be briefly discussed below.

The MOF specification is the foundation of OMG's industry standard environment, where models can be created, integrated, and transformed into different formats. MDA relies on the MOF to integrate the modeling steps and provide the model transformations (OMG, 2003, 2006). The earlier version of the MOF meta-metamodel is a part of the ISO/IEC 19502:2005 standard (ISO/IEC, 2005). The MOF meta-metamodel is referred as the MOF model, and MOF specification uses numbered layers instead of using the meta- prefix. Although the classical MOF specification is based on four layers, the later versions allow more or less meta-levels than this, depending on how MOF is deployed. However, the minimum number of layers is two. The MOF model is self-describing; that is, it is formally defined using its own metamodeling constructs. In 2006, OMG introduced the Essential Meta-Object Facility (EMOF), which is a subset of MOF with simplified classes. The primary goal of EMOF is to allow defining metamodels by using simpler concepts while supporting extensions.

TABLE 9.2
Various MB and MD Approaches

	System/Software Development Approaches								
Features	**MBE**	**MBSE**	**MDE**	**MDD/ MDSD**	**MDA**	**MIC**	**Software Factories**	**Eclipse Modeling Project**	**Executable Architecture**
Use of DSLs	Y	Y	Y	Y	Y	Y	Y	Y	Y
DSL representation with metamodeling	—	—	Y	Y	Y	Y	Y	Y	Y
Guidance for model transformations	—	—	Y	Y	Y	Y	Y	Y	—
Support for component reusability	—	—	—	—	—	Y	Y	—	—
Code generation mechanism	—	—	—	—	Y	Y	Y	Y	Y
Tool support for overall process	—	—	—	—	—	Y	Y	Y	—
Applicable to all domains	Y	Systems engineering	Y	Software engineering	Software engineering	Software engineering	Software engineering	Software engineering	Military applications

The Eclipse Ecore meta-metamodel is the main part of the Eclipse EMF project (Eclipse, 2009). The EMF project is a modeling framework and code generation facility for building tools and other applications based on a structured data model. Ecore is based on the EMOF specification, but renames the metamodeling constructs like EClass, EAttribute, EReference, or EDataType. A metamodel defined with Ecore needs to have a root object, and a tree structure represents the whole model.

MetaGME is the top level meta-metamodel of the MIC technology (ISIS, 1997). MIC defines a technology for the specification and use of DSLs. It refines and facilitates MDD approaches and provides an open integration framework to support formal analysis tools, verification techniques, and model transformations in the development process.

Kermeta is a metamodeling language, which allows describing both the structure and the behavior of models (IRISA, 2011). Kermeta is built as an extension to EMOF, and it provides an action language for specifying the behavior of models. The Kermeta development environment provides a comprehensive tool support for metamodeling activities, such as an interpreter, a debugger, a text editor with syntax highlighting and code autocompletion, a graphical editor, and various import/export transformations. Kermeta is fully integrated with Eclipse, and it is available under the open-source Eclipse public license.

The MDD principles are implemented by different metamodeling environments such as Microsoft's DSL tools for Software Factories, MetaEdit+, AndroMDA, GME, XMF-Mosaic, and the Eclipse generative modeling technologies project. For a successful system development and efficient metamodeling process, it is important to select the most convenient tools and languages. Detailed discussion about various metamodeling environments can be found in Achilleos et al. (2007) and Amyot et al. (2006).

9.3 DOMAIN-SPECIFIC LANGUAGES

In MDE, a DSL is a language designed for a particular problem domain. The term "domain-specific" is generally used for expressing different application domains such as logistics, health care, airports, and container terminals. The abstractions defined in a DSL map directly to the problem domain concepts that it chooses to represent. Consequently, it is at the same semantic level as the problem domain and may select a syntax that facilitates end-user to better use the computational representation of the problem domain. These abstractions enable the modeler to focus on the core aspects of the subject, ignoring the unnecessary details. A DSL is meant to hide the complexities of the computational domain and highlight the complexities of the domain it is designed for. Ghosh (2010) describes four qualities of a good abstraction:

1. *Minimalism*: Publish only those behaviors that the client wants and have been promised and negotiated
2. *Distillation*: Keep the implementation free of nonessential details

3. *Extensibility*: Design abstractions and primitives that can be extended and lead to better DSL evolution
4. *Composability*: Abstraction should be able to assimilate lower level abstractions leading to higher order abstractions

A DSL is specified using a metamodel, and the metamodeling process is called DSM. From the implementation point of view, DSLs are often classified as internal and external DSLs:

- *Internal DSLs*: An internal (sometimes called embedded) DSL is an idiomatic way of writing code in general-purpose programming languages, such as Ruby, Lisp, Scala, Groovy, Clojure, and many other functional languages. No special-purpose parser is necessary for internal DSLs. Instead, they are parsed just like any other code written in the language. Internal DSLs are easier to create because they do not require a special purpose parser. The execution platform for such a DSL is also given as no computational mapping needs to be done. The constraints of the underlying language, however, limit the options for expressing domain concepts.
- *External DSLs*: An external DSL is a custom language with its own infrastructure for lexical analysis, parsing techniques, interpretation, compilation, and code generation. Further, an external DSL can be textual, graphical, and hybrid.

Now, having given the general-purpose definition of a DSL, we inherently assume that a DSL is an executable piece of code that is semantically anchored to a computational programming language. This assumption is based on the premise that the DSL and its execution platform are considered together when we talk about DSM. We would like to position some concepts to separate these concerns, in line with other chapters of this book. As has been stated, the model and the simulator are two different entities (Zeigler et al., 2000) in the DEVS-based M&S discipline. With such a distinction, we are in a position to formally select the execution platform of a DSL. In Chapter 5, the DEVSML was described, which is a DSL for DEVS models. The DEVS DSL is semantically anchored with the DEVS simulation platform that executes a DEVS model in a DEVS simulator using the DEVS simulation relation.

Hence, while DEVSML is a DSL from the point of an MDD expert, it is not domain specific from the point of a simulation expert since it can be used for any kind of application domain. We have described various aspects of the benefits of such distinction in the DEVSML 2.0 stack in Chapter 6. The forthcoming section presents various other model-driven approaches for the M&S discipline.

9.4 MODEL-DRIVEN APPROACHES IN M&S

MDE is increasingly gaining the attention of both industry and research communities in the M&S field. The metamodeling term has been used in simulation for many years in a different context. Metamodeling referred to constructing simplified models for simulation models that are quite complex (Barton, 1998), and therefore slow to execute. In this context, a metamodel is known as a surrogate model. Surrogate models mimic the complex behavior of the underlying simulation model. Metamodeling in the MDE context has been introduced to simulation in the last decade, and research on MDE has emerged in the last 5 years (Garcia & Tolk, 2010; Ighoroje et al., 2012; Mittal & Douglass, 2011; Mooney & Sarjoughian, 2009).

Bakshi et al. (2001) present a model-based extensible framework to facilitate embedded system design and simulation called MILAN (Model-based Integrated simuLAtioN framework). MILAN is a collaborative project between the University of Southern California and Vanderbilt University. MILAN provides a formal paradigm for the specification of structural and behavioral aspects of embedded systems, an integrated model-based approach, and a unified software environment for system design and simulation.

Vangheluwe and de Lara (2002) introduced metamodeling and model transformation research into DEVS-based M&S in 2002. De Lara and Vangheluwe (2002) present a tool for metamodeling and model transformations for simulation, namely AToM3. They introduce the combined use of multiformalism modeling and metamodeling to facilitate computer-assisted modeling of large systems.

Iba et al. (2004) propose a simulation model development process and present an example for agent-based social simulations. The proposed process consists of three major phases: conceptual modeling, simulation design, and verification. In the conceptual modeling phase, the modeler analyzes the target world and describes the conceptual model. In the simulation design phase, the modeler designs and implements the simulation model, which is executable on the provided simulation platform. The modeler translates the conceptual models into simulation models according to the suggested framework. In the verification phase, the modeler runs the simulation and inspects whether the simulation program is coded right. If necessary, the modeler returns to the first or second phase and modifies the models.

Guiffard et al. (2006) provide a study that aims at applying a model-driven approach to the M&S in military domain. This work has been carried out in the context of a larger research program (High Performance Distributed Simulation and Models Reuse) sponsored by the Delegation Generale pour l'Armement of the French Ministry of Defense. The paper presents a prototype implementation as well. The prototype demonstrates that the automated transformation from a source model to executable source code is possible. The authors state that the amount of work needed for writing correct and complete set of transformation rules is extremely large.

Lei et al. (2007) present a metamodel-based method to represent reusable simulation models, where they define the simulation model representation methods as the methods of expressing simulation model specifications. They classify the existing methods into three groups: representation by programming language, representation by simulation language, and representation by generic modeling language.

Duarte and de Lara (2009) show the application of MDD to agent-based simulation. They present the design of a DSL for M&S of multiagent systems. The language uses four different diagram types to define agents' types, their behavior, their sensors and actuators, and the initial configuration. A customized modeling tool developed in Eclipse and a code generator for a Java-based agent simulation language are presented as well.

Topçu et al. (2008) propose the Federation Architecture Metamodel (FAMM) for describing the architecture of a High Level Architecture (HLA) compliant federation. FAMM supports the behavioral description of federates based on live sequence charts, and it is defined with metaGME. FAMM formalizes the standard HLA object model and federate interface specification.

Monperrus et al. (2008) report on an industrial experiment at Thales to use MDA for system engineering. The experiment consisted of setting up a model-driven simulation environment for a maritime surveillance system.

D'Ambrogio et al. (2010) introduce a model-driven approach for the development of DEVS simulations. The approach enables the UML specification of DEVS models and automates the generation of DEVS simulations that make use of the DEVS/service-oriented architecture (SOA) implementation. An example application for a basic queuing system is also presented.

Levytskyy et al. (2003) present two DEVS metamodels that are used to automatically generate a tool that allows the graphical definition of DEVS models. The tool is capable of generating a representation suitable for the simulation by an external DEVS interpreter.

Garro et al. (2011) propose an MDA-based process for agent-based modeling and simulation (MDA4ABMS) that uses the agent-modeling framework of the Eclipse agent modeling platform project. The Acore metamodel of the project is similar to EMF Ecore and defined in Ecore, but it provides higher level support for complex agents. MDA4ABMS process allows (automatically) producing platform-specific simulation models starting from a platform-independent simulation model obtained on the basis of a CIM. Then, the source code can be automatically obtained with significant reduction of programming and implementation efforts.

Although the aforementioned efforts show the applicability of the MDD approach in the simulation field, the research community has recently started to move toward a comprehensive theoretical framework. Çetinkaya et al. (2011) propose a general MDD framework for modeling and simulation (MDD4MS). The MDD4MS framework defines three metamodels for conceptual modeling, simulation model specification, and simulation model implementation. In order to obtain the continuity throughout the M&S life cycle, model transformations are suggested as well. The MDD4MS Research Project consists of three parts:

1. MDD4MS framework
2. MDD4MS case studies
3. MDD4MS prototype

The MDD4MS framework defines the concepts and is explained in the following section. The MDD4MS case studies are the different applications of the framework. The MDD4MS prototype provides an Eclipse-based prototype implementation, which is illustrated with a case study for business process modeling and simulation in Chapter 22.

9.5 MDD4MS FRAMEWORK

The MDD4MS framework (Çetinkaya et al., 2011) addresses the steps in a simulation study to apply an MDD approach from conceptual design through final implementation. The framework includes:

- A model-driven simulation model development life cycle
- Model and metamodel definitions for different stages
- Model transformations between the models, which are the instances of specific metamodels
- Domain-specific solutions with DSLs, model templates, and component libraries
- A tool architecture for the overall process

The first four aspects will be described in this section while the tool architecture will be presented in Chapter 22.

9.5.1 Model-Driven Simulation Model Development Life Cycle

The MDD4MS framework suggests a life cycle with five stages and presents an integrated solution from problem definition to final evaluation. Figure 9.4 shows the MDD4MS life cycle.

9.5.1.1 Problem Definition

If a problem owner identifies a need for a simulation study, he/she defines the purpose of the simulation study and his/her requirements. This essentially consists of setting the boundaries of the problem and choosing the value system (key performance indicators) according to which the performance of the system will be assessed.

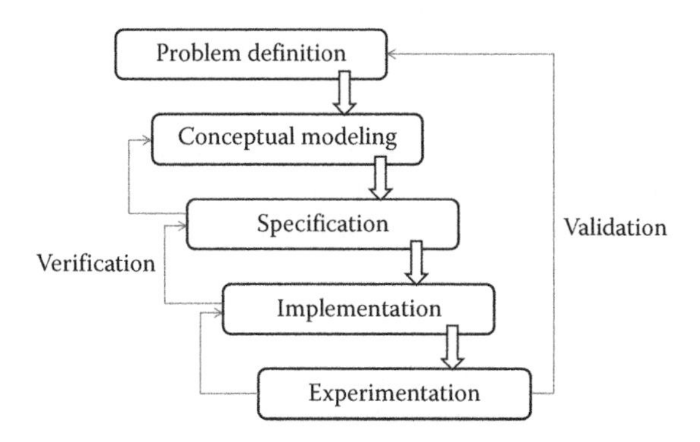

FIGURE 9.4 MDD4MS life cycle.

9.5.1.2 Conceptual Modeling

When the problem owner defines the requirements and starts the simulation study, the conceptual modeler makes a high-level abstraction of the real system or the future system according to a given worldview and prepares a conceptual model (CM) for the future executable simulation model. A CM serves as a bridge between problem owner and simulation modeler. A CM can be mapped to a CIM in MDA. The conceptual modeler prepares the CM using his/her knowledge, design patterns and CM metamodel. Conceptual modeling is a difficult stage in the simulation project, and it directly affects the quality of the project. Proper development of the CM is critical for expressing the purpose of the simulation study. The CM should be defined in a well-defined conceptual modeling language. It can be described using either textual or graphical presentations, or both.

A range of methods have been used for representing CMs, such as UML, SysML (Systems Modeling Language), Business Process Modeling and Notation (BPMN), process flow diagrams, event graphs, and simulation activity diagrams. However, there is no commonly accepted conceptual modeling language for simulation (Robinson, 2006). Some of these conceptual modeling frameworks are presented in Part IV of this book as example case studies.

9.5.1.3 Specification

After the problem owner and the conceptual modeler agree on a CM, the simulation modeler transforms it into a platform independent simulation model (PISM) according to a formal specification language for a certain formalism such as DEVS, Petri Nets, state charts, partial differential equations, or finite state automata. At this stage, the simulation modeler defines the system functionality without taking into account any specific platform on which the model could later be implemented. Besides, he/she collects the required data, defines the experimental frame, and prepares the scenario templates. A PISM can be described using either textual or graphical presentations, or both. A PISM can be mapped to a PIM in MDA. This is the stage of formalizing the CM so that mathematical analyses can be conducted and computational representation can be achieved. Formal rigor is added at this stage, and any lack of information leads back to the previous step where requirements are refined. A PISM is expected to be mathematically complete, however to be able to simulate it on a specific simulator, it should be transformed to a PSSM.

9.5.1.4 Implementation

After the conceptual modeler and the simulation modeler agree on a PISM, the simulation programmer develops the platform specific simulation model (PSSM). A PSSM is an implementation model of a PSIM for a specific platform, and it is expected to be computer executable. A PSSM can be mapped to a PSM in MDA. Although a PSSM can be directly written in a programming language, it is suggested to define it in a higher level representation of a programming language first and then, a full compilable and executable source code can be automatically generated from it. In this way, the PSSM will have two different representations, which are programmatically the same. However, to make a clear distinction, the executable source code will be termed as PSSM code, when it is needed to make a separation. Otherwise,

PSSM represents both the higher level implementation model and the executable simulation model. This stage involves generation of an executable model through the transformation process from a modeling formalism to a specific implementation language.

9.5.1.5 Experimentation

Once the executable PSSM is ready, the simulation expert runs the PSSM with the collected data according to some scenarios and analyzes the experimentation results. Validation experimentations can be performed to validate if the input/output behavior of the simulation model matches to the purpose of the simulation study.

9.5.1.6 Comparison to MDA

The MDD4MS framework introduces new acronyms instead of using the MDA definitions of CIM, PIM, and PSM. The terms are refined for the M&S domain. A PSSM is accepted to be a computer-executable model. A CM is not executable in any way. A PISM is executable mathematically but not on a specific computer platform. Table 9.3 presents a summary for the output of the MDD4MS life cycle stages.

In its current version, MDD4MS does not introduce models for the problem definition and experimentation stages. It focuses on the conceptual modeling and model

TABLE 9.3
Models of the MDD4MS Framework

Stage	Output	Output Format	MDA Model	MDD4MS Model
Problem definition	Generally a written document that expresses the system requirements	Expressed in a natural language or in first-order propositional logic.	—	—
Conceptual modeling	A visual and/or textual CM	Defined by a well-defined conceptual modeling language such as a DSL.	CIM	CM
Specification	A visual and/or textual simulation specification	Defined by a specification language for a certain formalism to semantically anchor a DSL for formal rigor.	PIM	PISM
Implementation	Compiled or executable simulation model source code	Written in a programming language or automatically generated using various transformations.	PSM	PSSM
Experimentation	Output of the simulator	Depends on the problem definition and simulator implementation. The output is augmented by various data analysis mechanisms.	—	—

development like MDA, but models for requirements engineering and output analysis can be introduced.

9.5.2 Metamodeling in MDD4MS

Building up the CM, PISM, and PSSM manually from scratch can be a time-consuming and error-prone task. MDD brings great advantages for defining formal modeling languages via metamodels. The MDD4MS framework introduces three metamodels to support model transformations.

1. CM is an instance of simulation conceptual modeling metamodel (CM metamodel)
2. PISM is an instance of simulation model specification metamodel (PISM metamodel)
3. PSSM is an instance of simulation model implementation metamodel (PSSM metamodel)

All the metamodels are instances of a higher level meta-metamodel, which needs to be self-reflexive so that automated model transformations can be supported from CM to PISM and from PISM to PSSM. Specifications described in the PISMs are adapted to specific platforms by means of PSSMs from which the code is automatically generated. Figure 9.5 shows the metamodeling stacks in the MDD4MS. There are three metamodeling stacks that are illustrated vertically in the picture for the three middle stages of the MDD4MS life cycle.

9.5.3 Model transformations

Model transformations lay on an orthogonal axis to metamodeling as shown in Figure 9.6. After defining the metamodels, model transformations between the

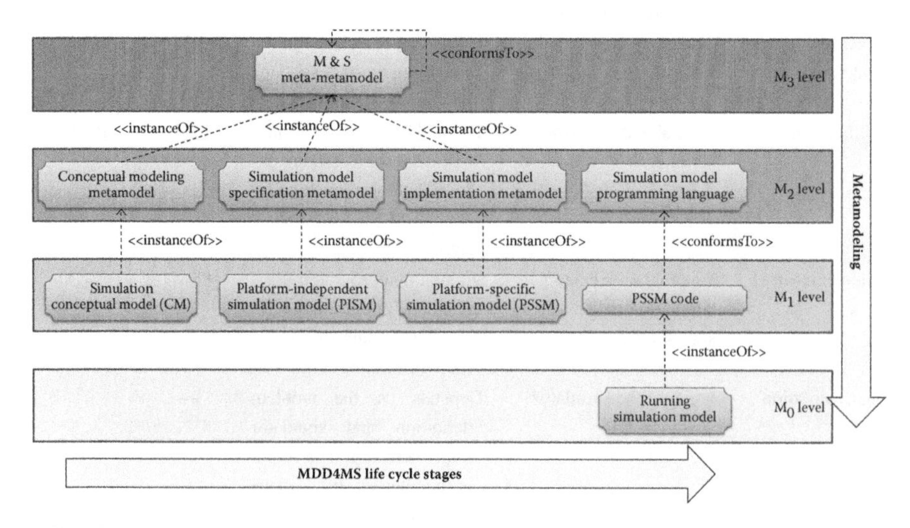

FIGURE 9.5 Metamodeling in MDD4MS.

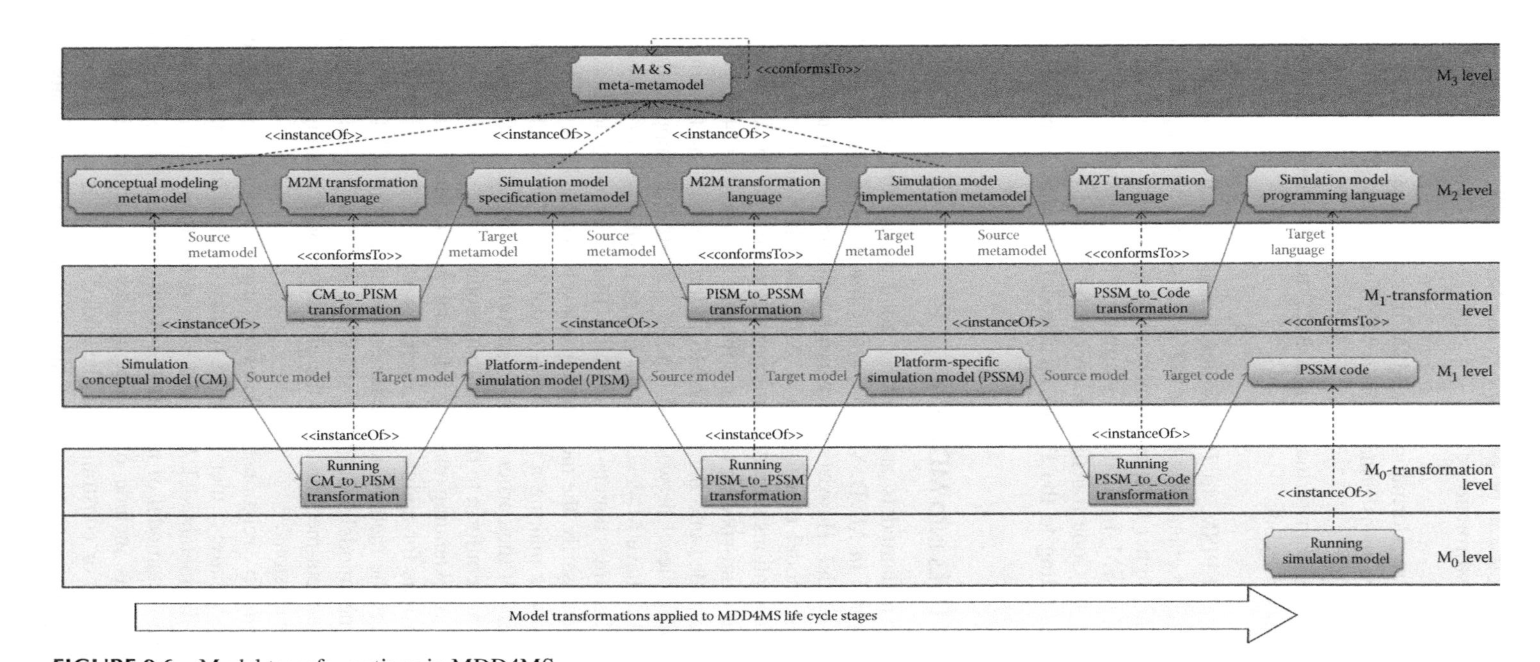

FIGURE 9.6 Model transformations in MDD4MS.

models of the life cycle can be pe formed. The MDD4MS framework suggests the following model transformations:

- CM_to_PISM transformation is a M2M transformation from CM to PISM.
- PISM_to_PSSM transformation is a M2M transformation from PISM to PSSM.
- PSSM_to_Code transformation is a M2T transformation from PSSM to executable source code.

CM_to_PISM and PISM_to_PSSM transformations can be defined in terms of a mapping between metamodels by an M2M transformation language. The PSSM represents both the higher level implementation model and the executable simulation model. Thus, PSSM_to_Code transformation is suggested to obtain the source code. The PSSM_to_Code transformation can be defined by an M2T transformation language according to the PSSM metamodel and the selected programming language.

9.5.4 Adding DSLs into MDD4MS

Defining precise metamodels and writing good model transformation rules are challenging activities in MDD. Various model transformations can be defined for the same source model. However, there is no guidance to define a good solution or choose the best model transformation. MDD4MS introduces two ways to support model transformations with domain-specific constructs. These are either adding DSLs or adding domain-specific template libraries into the MDD4MS.

A DSL can be added into the MDD4MS in three ways. The first one is adding a new metamodeling level between M_1 and M_2. In order to support this, M_2-level metamodels need to have M_3-level metamodeling capabilities. In this case, the M_2-level metamodel becomes a meta-metamodel so that a new metamodel can be generated from that. Necessarily, the metamodeling levels need to be renumbered.

The second way is making a new metamodel for M_2 level. This solution sacrifices the reusability of the earlier work. Predeveloped models with the old M_2-level metamodels will not conform to the new metamodel.

Another way is extending existing M_2-level metamodels via domain-specific metamodel extensions for adding new domain-specific metamodels for any stage. Figure 9.7 presents the metamodel extension mechanism in MDD4MS. Every element of the new metamodel should extend an element from the old metamodel while conforming to the meta-metamodel of the old metamodel. Adding new elements is not allowed for extensions.

The <<*extends*>> relationship between two metamodels expresses that metamodel-B <<*extends*>> metamodel-A if each element in metamodel-B extends an element from the metamodel-A. In this way, metamodel-B is also an <<*instanceOf*>> the meta-metamodel of metamodel-A (cf. Section 9.2.1). Domain-specific metamodel extensions include old concepts and new domain-specific concepts. However, restrictions or constraints can be added to old or new concepts, such as writing rules that restrict the modeler to using only new DSM elements.

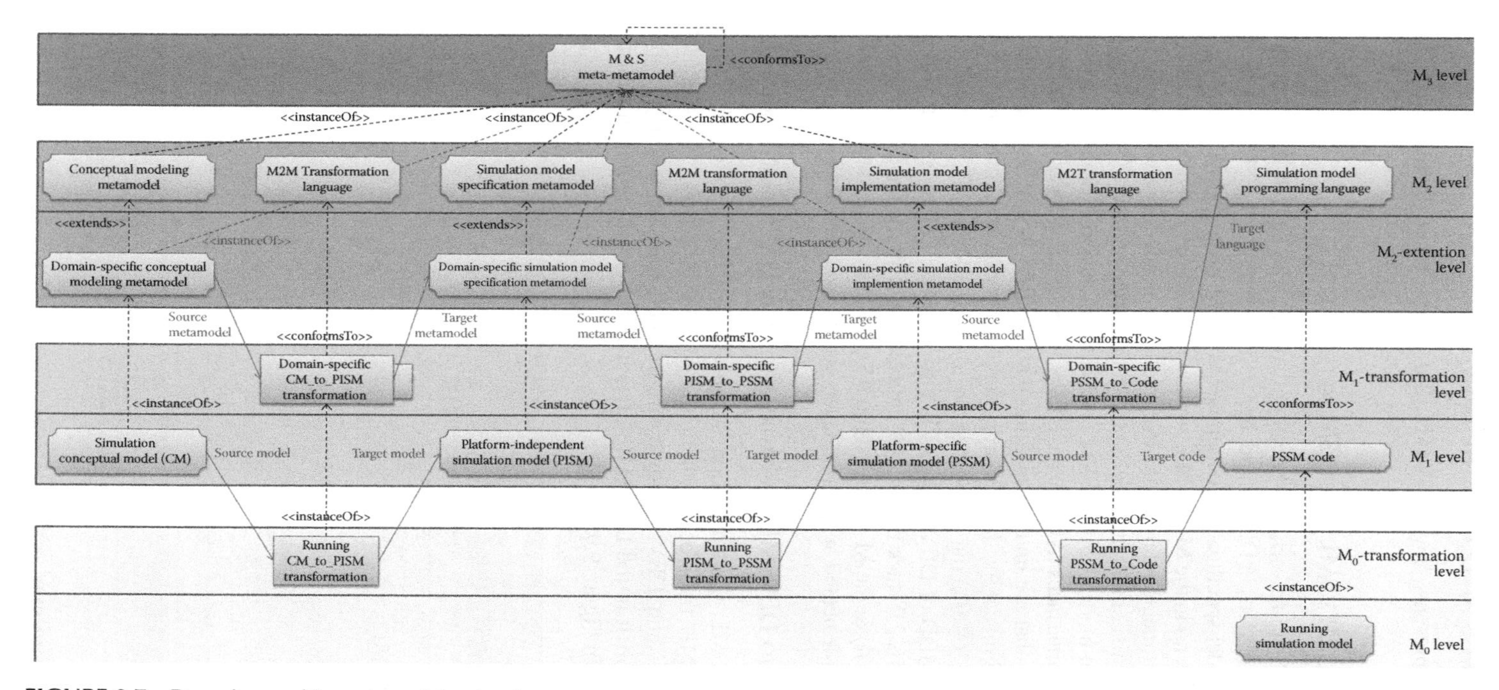

FIGURE 9.7 Domain-specific metamodel extensions.

9.6 MDD4MS AND DEVSML 2.0

This section presents how DEVSML 2.0 is positioned in the larger MDD4MS framework.

9.6.1 Using DEVSML as a PISM metamodel

The MDD4MS framework can be supported with the DEVSML 2.0 metamodel for particular applications in DEVS modeling and simulation. DEVSML can be used as the simulation model specification metamodel (a PISM metamodel that needs to be platform independent) in the MDD4MS, as shown in Figure 9.8. Then, a simulation model implementation metamodel (PSSM metamodel) is needed for the definition of DEVS/SOA middleware compliant models. One of the prime reasons for choosing the DEVS/SOA compliant models at the PSSM level is the inclusion of standardized application programming interfaces (APIs) at the DEVS M&S layers. DEVS/SOA is based on standardized APIs that are under evaluation with the DEVS Standardization Committee, and any DEVS model that is standards-compliant is executable on a netcentric DEVS/SOA platform. Overlaying MDD4MS on top of the DEVSML 2.0 stack opens new ways for the MDD of DEVS models in a formal way. Figure 9.8 can also be viewed as the DEVSML 2.0 stack rotated 90° anticlockwise. While the DEVSML 2.0 stack is defined primarily at the M_2 level, Figure 9.8 provides actual detail how the DEVSML 2.0 stack is implemented at different layers within the MDD4MS and MDE at large.

9.6.2 CM to DEVSML transformation for DEVS-based simulation

The M2DEVSML transformation suggested in Chapter 6 is incorporated as a CM-to-DEVSML transformation in the MDD4MS context. Further, DEVSML-to-Middleware and DEVS_Middleware-to-Code transformations are introduced, which are implicit in the DEVSML 2.0 stack. The resulting model transformations are presented in Figure 9.9.

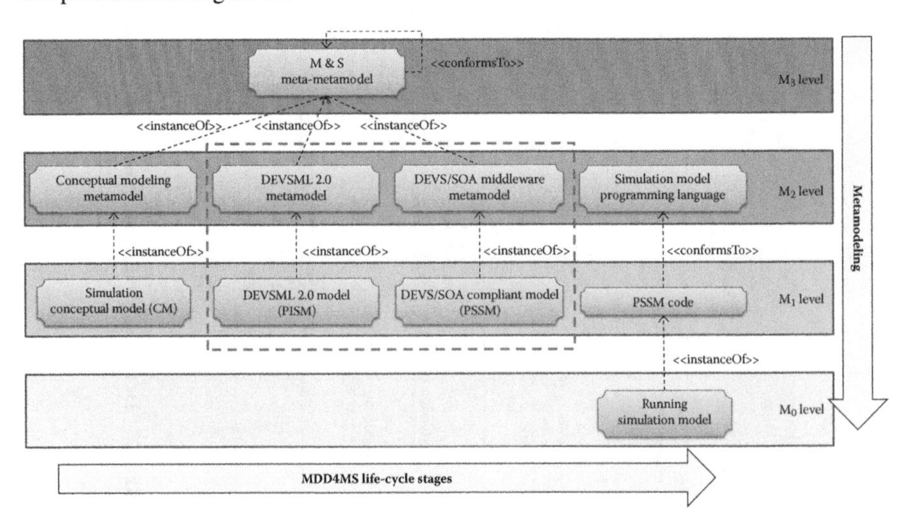

FIGURE 9.8 DEVSML as PISM metamodel.

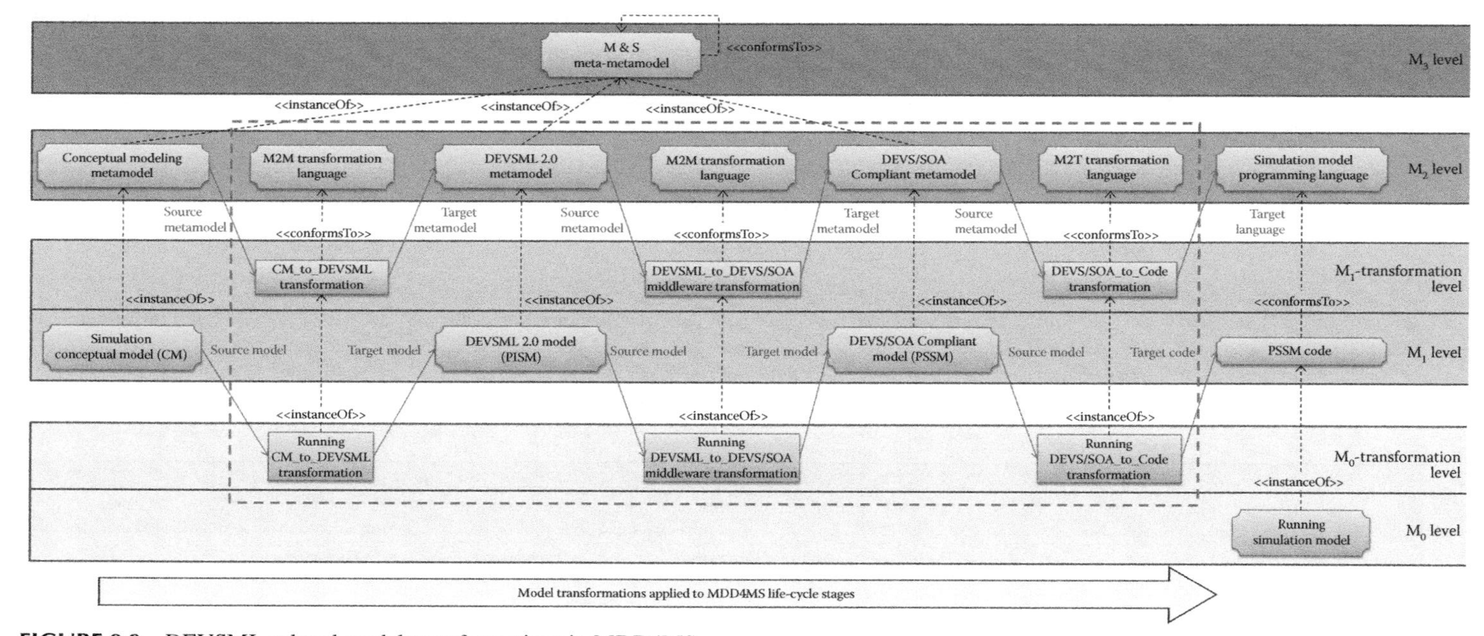

FIGURE 9.9 DEVSML-related model transformations in MDD4MS.

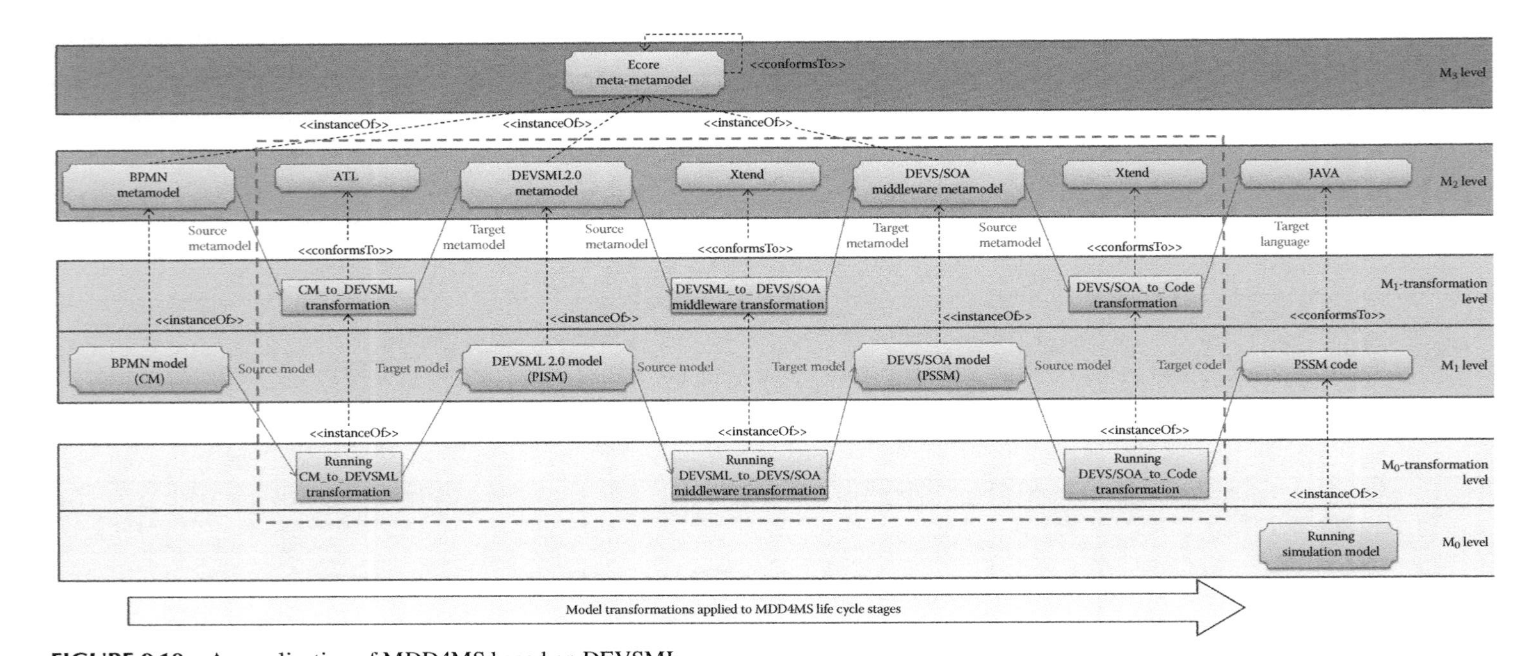

FIGURE 9.10 An application of MDD4MS based on DEVSML.

An example application for MDD of DEVS models from BPMN models and their execution on a JAVA-based simulation platform is illustrated in Figure 9.10. The DEVS/SOA PSSM compliant model is chosen as DEVSJAVA. An implementation of the example is presented in Chapter 22.

9.7 SUMMARY

The MDE approach brings great advantages to M&S. It provides ways to formally define simulation models and modeling languages. Since models are defined formally and free of implementation details, the conceptual modelers and domain experts can understand the models more easily and they can play a direct role in the simulation model development process. The simulation model implementation becomes more efficient and reliable since a large part of the code can be generated automatically. Besides, the availability of the existing MDD tools and techniques is a significant advantage.

This chapter presented a general overview of MDE and DSLs. MDE applications in the M&S field were discussed, and the MDD4MS framework was described as a general MDD framework for M&S studies. Due to the fact that both the MDD4MS framework and the DEVSML 2.0 stack rely on formal underpinnings, integration of these frameworks is suggested based on the DEVSML 2.0 metamodel and M2DEVSML transformations. Integrating the DEVSML 2.0 metamodel in the MDD4MS framework provides a way for introducing different applications to the DEVS M&S arena.

REFERENCES

Achilleos, A., Georgalas, N., & Yang, K. (2007). An open source domain-specific tools framework to support model driven development of OSS. In D. H. Akehurst, R. Vogel, & R. F. Paige (Eds.), Proceedings of the 3rd European Conference on Model Driven Architecture—Foundations and Applications (ECMDA-FA) (Vol. 4530, pp. 1–16). Springer-Verlag Berlin Heidelberg.

Amyot, D., Farah, H., & Roy, J.-F. (2006). Evaluation of development tools for domain-specific modeling languages. In R. Gotzhein & R. Reed (Eds.), Proceedings of the 5th Workshop on System Analysis and Modeling: Language Profiles (Vol. 4320, pp. 183–197). Springer-Verlag Berlin Heidelberg.

Atkinson, C., & Kühne, T. (2003). Model-driven development: A metamodeling foundation. *IEEE Software*, *20*(5), 36–41.

Bakshi, A., Prasanna, V. K., & Ledeczi, A. (2001). MILAN: a model based integrated simulation framework for design of embedded systems. *ACM SIGPLAN Notices*, *36*(8), 82–87.

Bapty, T. A., & Sztipanovits, J. (1997). Model-based engineering of large-scale real-time systems. Proceedings of the International Workshop on Engineering of Computer-Based Systems (pp. 467–474). IEEE Computer Society Press, Washington, DC, USA.

Barton, R. R. (1998). Simulation metamodels. In D. J. Meideros, E. F. Watson, J. S. Carson, & M. S. Manivannan (Eds.), Proceedings of the 30th Winter Simulation Conference (pp. 167–176). Los Alamitos, CA: IEEE Computer Society Press.

Buede, D. M. (2009). *The Engineering Design of Systems: Models and Methods*. New York, NY: Wiley-Blackwell.

Çetinkaya, D., & Verbraeck, A. (2011). Metamodeling and model transformations in modeling and simulation. *Proceedings of the Winter Simulation Conference; AZ, USA* (pp. 3043–3053).

Çetinkaya, D., Verbraeck, A., & Seck, M. D. (2011). MDD4MS: A model driven development framework for modeling and simulation. Proceedings of the Summer Computer Simulation Conference (SCSC); The Hague, Netherlands (pp. 113–121).

Czarnecki, K., & Helsen, S. (2003). Classification of model transformation approaches. Proceedings of the 2nd OOPSLA Workshop on Generative Techniques in the context of Model Driven Architecture; CA, USA (pp. 1–17).

Dehayni, M., Barbar, K., Awada, A., & Smaili, M. (2009). Some model transformation approaches: A qualitative critical review. *Journal of Applied Sciences Research, 5*(11), 1957–1965.

De Lara, J., and Vangheluwe, H. (2002). AToM3: a tool for multi-formalism and meta-modelling. In R.-D. Kutsche & H. Weber (Eds.), Proceedings of the Fundamental Approaches to Software Engineering (Vol. 2306, pp. 174–188). Springer-Verlag Berlin Heidelberg.

Duarte, J. N., & de Lara, J. (2009). ODiM: A model-driven approach to agent-based simulation. In J. Otamendi, A. Bargiela, J. L. Montes, & L. M. D. Pedrera (Eds.), Proceedings of the 23rd European Conference on Modelling and Simulation; Madrid, Spain (pp. 68:1–68:8).

D'Ambrogio, A., Gianni, D., Risco-Martín, J. L., & Pieroni, A. (2010). A MDA-based approach for the development of DEVS/SOA simulations. Proceedings of the Spring Simulation Multiconference; Orlando, FL, USA (pp. 142:1–142:8). Society for Computer Simulation International San Diego, CA, USA.

Eclipse. (2008). *Generic Eclipse Modeling System (GEMS)*. Retrieved April 24, 2012, from http://www.eclipse.org/gmt/gems.

Eclipse. (2009). *Eclipse Modeling Framework (EMF) Project*. The Eclipse Foundation. Retrieved April 24, 2012, from http://www.eclipse.org/modeling/emf/.

Estefan, J. A. (2008). Survey of model-based systems engineering (MBSE) methodologies Rev. B. Technical Report, INCOSE MBSE Focus Group.

Garcia, J., & Tolk, A. (2010). Adding executable context to executable architectures: Shifting towards a knowledge-based validation paradigm for system-of-systems architectures. Proceedings of the Summer Simulation Multiconference; Ottawa, Canada (pp. 593–600). Society for Computer Simulation International San Diego, CA, USA.

Garro, A., Parisi, F., & Russo, W. (2011). A model-driven architecture approach for agent-based modeling and simulation. In J. Kacprzyk, N. Pina, & J. Filipe (Eds.), Proceedings of the International Conference on Simulation and Modeling Methodologies, Technologies and Applications (SIMULTECH); Noordwijkerhout, The Netherlands (pp. 74–83). SciTePress-Science and Technology Publications.

Ghosh, D. (2010). *DSLs in action*. Manning Publications, Stamford, CT, USA.

Guiffard, E., Kadi, D., Mochet, J.-P., & Mauget, R. (2006). CAPSULE: Application of the MDA methodology to the simulation domain. Proceedings of the European Simulation Interoperability Workshop (EURO SIW); Stockholm, Sweden (pp. 181–190). SISO-Simulation Interoperability Standards Organization, Bedford, MA, USA.

Iba, T., Matsuzawa, Y., & Aoyama, N. (2004). From conceptual models to simulation models: Model driven development of agent-based simulations. Proceedings of the 9th Workshop on Economics and Heterogeneous Interacting Agents; Kyoto, Japan.

Ighoroje, U. B., Maiga, O., & Traoré, M. K. (2012). The DEVS-driven modeling language: Syntax and semantics definition by meta-modeling and graph transformation. Proceedings of the Theory of Modeling & Simulation: DEVS Integrative M&S Symposium; Orlando, FL, USA (TMS-DEVS). Society for Computer Simulation International San Diego, CA, USA.

IRISA. (2011). *Kermeta workbench*. Retrieved April 7, 2012, from http://www.kermeta.org/.

ISIS. (1997). *Model Integrated Computing (MIC)*. Retrieved May 10, 2012, from http://www.isis.vanderbilt.edu/research/MIC.

ISO/IEC. (2005). *ISO/IEC 19502:2005, International Standard: Information Technology—Meta Object Facility (MOF)*. Retrieved May 10, 2012, from http://www.iso.org/iso/home/store/catalogue_tc/catalogue_detail.htm?csnumber = 32621

Jouault, F., Allilaire, F., Bézivin, J., & Kurtev, I. (2008). ATL: A model transformation tool. *Science of Computer Programming*, *72*(1–2), 31–39.

Lei, Y., Song, L., Wang, W., & Jiang, C. (2007). A metamodel-based representation method for reusable simulation model. In S. G. Henderson, B. Biller, M.-H. Hsieh, J. Shortle, J. D. Tew, & R. R. Barton (Eds.), Proceedings of the 39th Winter Simulation Conference (pp. 851–858). Piscataway, NJ: IEEE.

Levytskyy, A., Kerckhoffs, E. J. H., Posse, E., & Vangheluwe, H. (2003). Creating DEVS components with the metamodelling tool ATOM3. In A. Verbraeck & V. Hlupic (Eds.), Proceedings of the 15th European Simulation Symposium; Delft, The Netherlands (pp. 97-103). SCS-Europe BVBA: Society for Computer Simulation International San Diego, CA, USA.

Mens, T., & Van Gorp, P. (2006). A taxonomy of model transformation. *Electronic Notes in Theoretical Computer Science*, *152*, 125–142.

Microsoft. (2005). *Software Factories*. Retrieved May 13, 2012, from http://msdn.microsoft.com/en-us/library/bb977473.

Microsoft. (2010). *Visual Studio visualization and modeling SDK (Domain-Specific Languages)*. Retrieved May 12, 2012, from http://msdn.microsoft.com/en-us/library/bb126259.

Mittal, S., & Douglass, S. (2011). From domain specific languages to DEVS components: application to cognitive M&S. Proceedings of the Theory of Modeling & Simulation: DEVS Integrative M&S Symposium (TMS-DEVS); Boston, MA, USA (pp. 256–265). Society for Computer Simulation International San Diego, CA, USA.

Mittal, S., & Douglass, S. (2012). DEVSML 2.0: The language and the stack. Proceedings of the Theory of Modeling & Simulation: DEVS Integrative M&S Symposium (TMS-DEVS); Orlando, FL, USA. SCS Society for Computer Simulation International San Diego, CA, USA.

Monperrus, M., Jaozafy, F., Marchalot, G., Champeau, J., Hoeltzener, B., & Jézéquel, J.-M. (2008). Model-driven simulation of a maritime surveillance system. Proceedings of the 4th European Conference on MDA: Foundations and Applications (pp. 361–368). Springer-Verlag Berlin, Heidelberg.

Mooney, J., & Sarjoughian, H. S. (2009). A framework for executable UML models. Proceedings of the Spring Simulation Multiconference (pp. 160:1–160:8). San Diego, CA: Society for Computer Simulation International.

Oliver, D. W., Andary, J. F., & Frisch, H. (2009). Model-based systems engineering. In A. P. Sage & W. B. Rouse (Eds.), *Handbook of Systems Engineering and Management* (2nd ed., pp. 1361–1400). John Wiley & Sons, Inc, Hoboken, NJ, USA.

OMG. (1999). UML Specification Version 1.3. Retrieved October 21, 2012, from http://www.omg.org/spec/UML/1.3/.

OMG. (2003). *Model Driven Architecture (MDA) Guide Version 1.0.1*. Retrieved October 21, 2012, from http://www.omg.org/mda/specs.htm.

OMG. (2006). *Meta Object Facility (MOF) Core Specification, Version 2.0*. Retrieved October 21, 2012, from http://www.omg.org/spec/MOF/2.0/.

OMG. (2011). *Meta Object Facility (MOF) 2.0 Query/View/Transformation (QVT)*. Retrieved October 21, 2012, from http://www.omg.org/spec/QVT/1.1/.

Robinson, S. (2006). Conceptual modeling for simulation: Issues and research requirements. In L. F. Perrone, F. P. Wieland, J. Liu, B. G. Lawson, D. M. Nicol, & R. M. Fujimoto (Eds.), Proceedings of the Winter Simulation Conference (pp. 792–800). Monterey, CA: IEEE.

Schmidt, D. C. (2006). Model-driven engineering. *IEEE Computer*, *39*(2), 25–31.

Selic, B. (2003). The pragmatics of model-driven development. *IEEE Software*, *20*(5), 19–25.

Tolk, A. (2002). Avoiding another green elephant—A proposal for the next generation HLA based on the Model Driven Architecture. Proceedings of the Fall 2002 Simulation Interoperability Workshop; Orlando, FL, USA (pp. 02F-SIW-004: 1–12). SISO-Simulation Interoperability Standards Organization, FL, USA.

Topçu, O., Adak, M., & Oğuztüzün, H. (2008). A metamodel for federation architectures. *ACM Transactions on Modeling and Computer Simulation*, *18*(3), 10:1–10:29.

Vangheluwe, H., & de Lara, J. (2002). Meta-models are models too. In E. Yucesan, C. H. Chen, J. L. Snowdon, & J. M. Charnes (Eds.), Proceedings of the 34th Winter Simulation Conference (pp. 597–605). Piscataway, NJ: IEEE.

Völter, M., Stahl, T., Bettin, J., Haase, A., & Helsen, S. (2006). *Model-driven software development: Technology, engineering, management*. John Wiley & Sons, West Sussex, England.

Zeigler, B. P., Praehofer, H., & Kim, T. G. (2000). *Theory of modelling and simulation: Integrating discrete event and continuous complex dynamic systems* (2nd ed.). Academic Press, San Diego, CA, USA.

10 System Entity Structures and Contingency-Based Systems

10.1 INTRODUCTION

System entity structures (SESs) compose a formal ontology framework that captures system aspects and their properties (Zeigler, 1984; Zeigler & Zhang, 1989). In the early 1990s, researchers working at the intersection of artificial intelligence and modeling and simulation began to design and implement environments that automated the process of design space exploration (Rozenblit et al., 1990; Kim et al., 1990; Zeigler et al., 1991; Zeigler & Chi, 1993) and solved engineering problems (Rozenblit & Huang, 1991). SES ontologies were originally used to represent system configuration alternatives in these environments. These SESs were in no way specifying the internal behavior of the entities that were modeled through some other mechanism (e.g., Discrete Event Systems Specification [DEVS] modeling). The SES was primarily used to specify the relations between these entities (Rozenblit & Zeigler, 1993). In addition to capturing aspects, entities, taxonomic relationships, variable values, and structural/configuration alternatives, these SESs included information about how entities in the SES could be realized in the DEVS formalism and composed into an executable model (Zeigler et al. 2000). To systematically explore design spaces in these environments: (1) rule-based search processes were used to derive all valid pruned entity structures (PES) captured by the SES; (2) information based on entities and aspects in each PES was used to compose an executable model using DEVS components stored in a model repository; (3) each composed model was simulated; and (4) simulation results were analyzed in order to identify the most desirable design alternative. The scope of SES described a complete solution set that contained all the permutations and combinations available for determining the actual design in a formal manner. This actual design is a result of the intersection of the requirements set and the optimal solution from the solutions set. The solutions set is determined by the pruning process on the SES and the optimal solution was determined by the simulation of each of the designs in the solutions set.

Pruning is a process applied to an SES that results in the reduction of design choices. It is an iterative procedure, which, on application of system design requirements, increasingly reduces the specializations and refinement of aspects (especially, multi-aspects). A pruned SES is called a PES and conforms to the SES axioms. Consequently, during the iterative process, the successive prunes represent a design space as well. The end result of the pruning process results in a PES that has no choices left. A final PES is strictly an entity structure that can be computationally realized as a component model in an M&S framework, such as DEVS. Figure 10.1 illustrates this process. Figure 10.2 illustrates the concept.

In this chapter, we will look at the formal specifications of SES. We will describe an SES editor to work our way through as well as a constraint specification language (CSL) which, when applied to SES, results in a system that is contingency-driven. We will integrate CSL with DEVS level of system specification to add further rigor to the design process.

10.1.1 SES Properties and Axioms

SES semantics is constructed of the following elements:

- *Entity* (a physical entity or a concept)
- *Aspects* (decomposition: *is made up of*): Denoted by |
- *Specializations* (can be of type: *is a type of*): Denoted by ||
- *Multi-aspect* (decomposition into similar type: *is made up of many such*): Denoted by |||
- *Variables* (each entity has variables that have a range and value): Denoted by ~

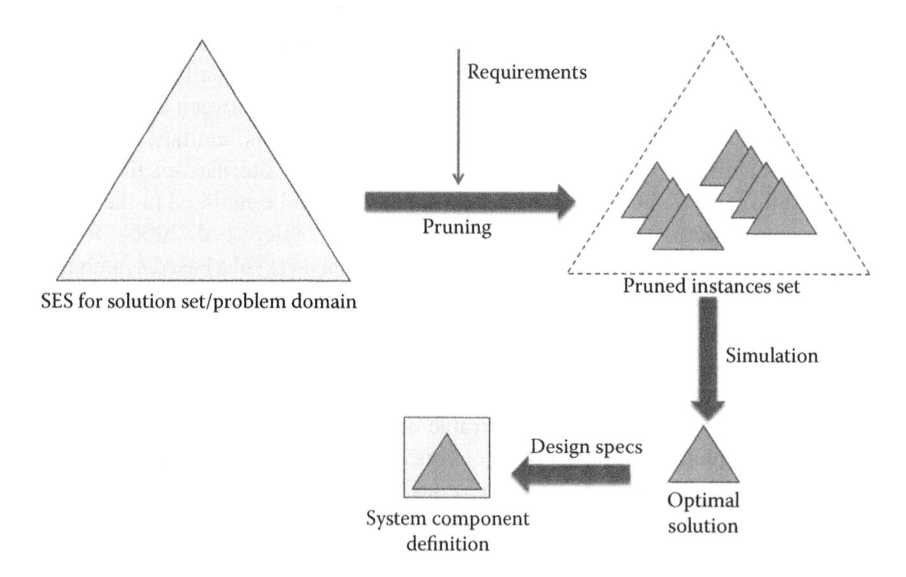

FIGURE 10.1 Early SES application.

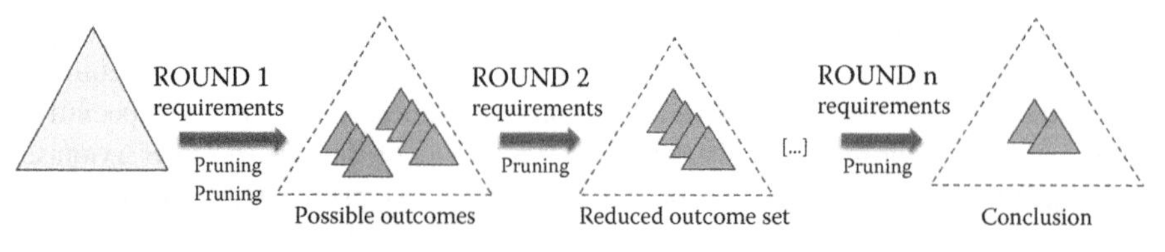

FIGURE 10.2 PES generation through an iterative process.

In addition to the above elements, an SES follows these axioms:

- *Uniformity*: Any two nodes with the same labels have identical, attached variable types and isomorphic subtrees.
- *Strict hierarchy*: No label appears more than once down any path of the tree.
- *Alternating mode*: Each node has a mode that is either entity, aspect, or specialization; if the mode of the node is entity, then the modes of its successors are aspect or specialization; if the mode of a node is aspect or specialization, then the modes of its children are entity. The mode of the root is entity.
- *Valid brothers*: No two brothers have the same label.
- *Attached variables*: No two variable types attached to the same item have the same name.
- *Inheritance*: the parent and any child of a specialization combine their individual variables, aspects, and specializations when pruning is activated.

Exercise 10.1

Read Chapters 4 and 5 of Zeigler & Hammonds (2007).

10.1.2 An Illustration

To illustrate the above concepts and axioms, let us revisit the experimental-frame processor (EFP) model discussed throughout the book. Figure 10.3 shows the elaborated version of EFP in SES notation. The root node, EFP, is decomposed through *efpDec* aspect into experimental frame (EF) and ProcBank entities. EF is decomposed through *efDec* aspect into Genr and Transd. Genr has two variables, *period* and *jobId*. Genr is specialized through *genSpec* into FixedInterval and RandomInterval, where RandomInterval has a variable *seed*. In this example, we have two types of job generators. FixedInterval generates at a fixed interval and inherits the variable *period* from Genr. RandomInterval generates a job at a random interval and has an additional variable *seed* that initializes the random interval sequence, thereby assigning a value to the variable *period* that is also inherited from Genr. Transd has variables such as *obsPeriod*, *lastJobSolved*, *throughput* and *arrived*. ProcBank is decomposed through multi-aspect *procBankMultiDec* into Proc. ProcBank also has a variable, *numProc*, expressing cardinality that describes how many Proc entities exist in a ProcBank. Proc has a variable *procTime*.

The EFP described in Figure 10.3 shows the entire design space of the EFP system. It details the various components: EF, ProcBank, Genr, Transd, FixedInterval, RandomInterval, and Proc. However, it does not inform us if all these components are used together in each of the EFP implementations or only a subset of them is used. For example, Genr is specialized into two types of generators. As a result, in any implementation, only one of them could be used, that is, either FixedInterval or RandomInterval. Another design choice is how many Proc entities are there in a ProcBank? As ProcBank is a container that holds homogeneous entities Proc, the number of Proc entities is bounded by the value

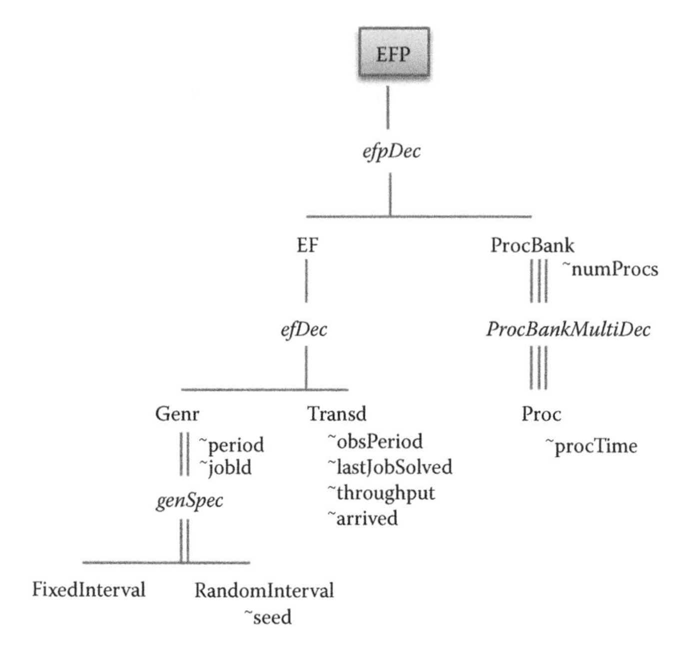

FIGURE 10.3 Elaborated EFP hierarchical system.

of variable *numProcs*. In order to obtain a realizable design, the design choices are reduced through the process of pruning, resulting in various PES. Figure 10.2 illustrated the concept.

The design choices are enumerated below:

A. Choice of Generator (FixedIntervalorRandomInterval)
B. Number of processors (Proc) in a processor bank (ProcBank): minimum 1 and maximum 10

When requirement A is applied to EFP SES, it generates two PESs, shown in Figures 10.4 and 10.5. The aspects and specialization names are removed for brevity purposes. However, when there is more than one aspect or specialization from an entity, then the usage of aspect and specialization labels become mandatory, in fact, useful.

When only requirement B is applied, it results in PES III, shown in Figure 10.6. As one can see from this figure, even though the multi-aspect has been decoded to satisfy the designed requirement B, we still do not have an exact number of processors that implements a particular design. Requirement B only provides information about the minimum and maximum values of the processors that can be implemented. This raises another question: Are the requirements A and B related in any manner that would help in creating a realizable system? This leads us to the subject of SES constraints and requirement–refinement issues.

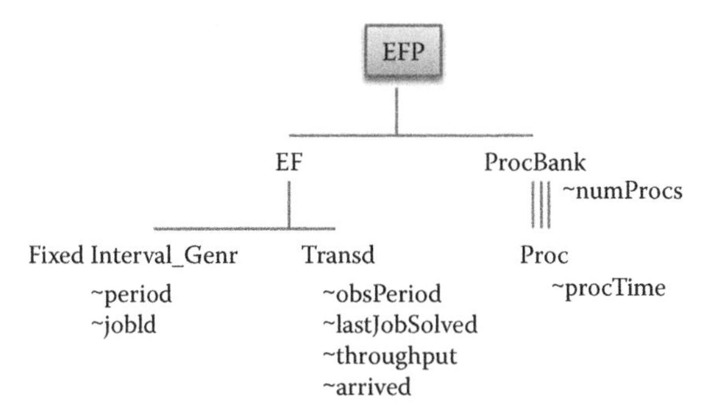

FIGURE 10.4 PES I for EFP-SES on application of requirement A.

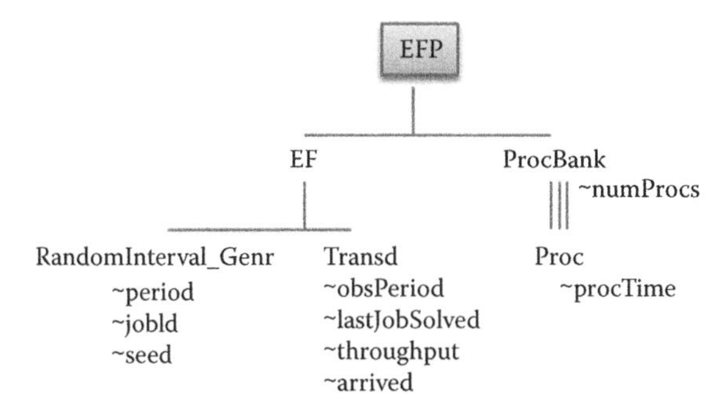

FIGURE 10.5 PES II for EFP-SES on application of requirement A.

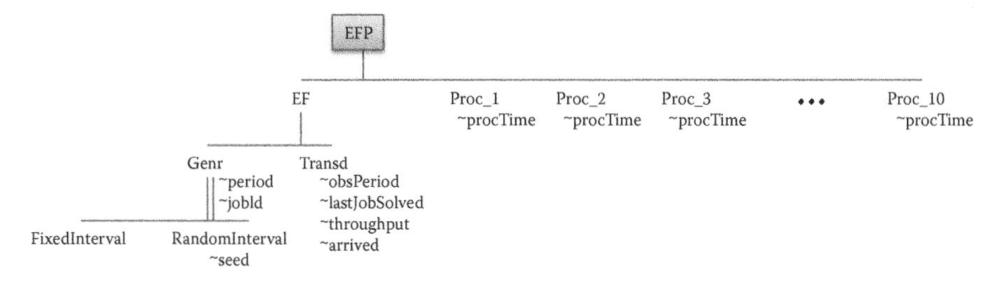

FIGURE 10.6 PES III for EFP-SES on application of requirement B.

As is clear from the above discussion, we are not able to obtain a realizable entity system with requirements A and B alone, and we add two more requirements to the list:

C. If FixedInterval generator, then number of processors is 2.
D. If RandomInterval generator, then number of processors is 4.

The application of requirement C, in conjunction with A and B, gives PES IV (Figure 10.7) and the application of requirement D, in conjunction with A and B, gives PES V (Figure 10.8). Both PES IV and V are component-entity structures (CESs) that are realizable in the component-based DEVS M&S framework.

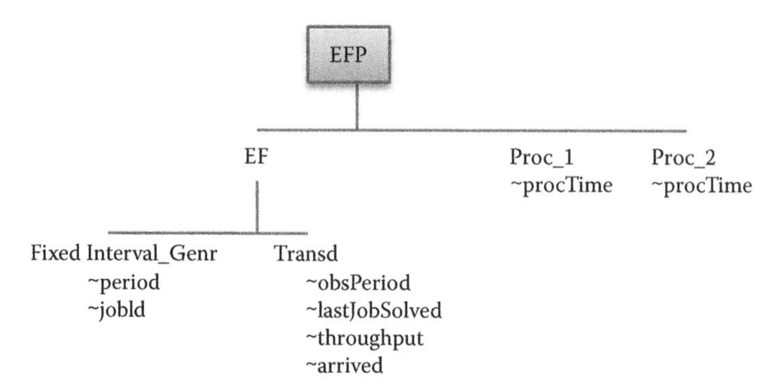

FIGURE 10.7 PES IV for EFP-SES on application of requirements A, B, and C, resulting in CES I.

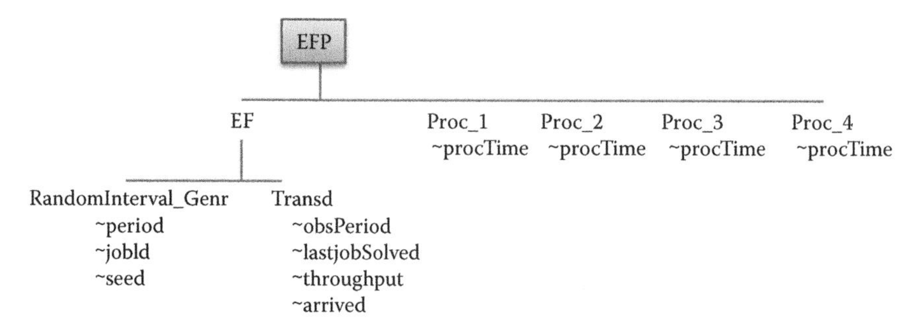

FIGURE 10.8 PES V for EFP-SES on application of requirements A, B, and D, resulting in CES II.

The process outlined above unearthed gaps in our requirements specification even though the example is quite limited. The same process can be applied to a larger component-based system and it can be projected that any PES generation process that does not result in a CES requires more information—either as new requirements or bounds in the variable range specification. Had the *numProc* variable been of fixed value, say 2, then the requirements C and D are not needed at all. PES III, IV, and V will never be available and PES I and PES II will be classified as CES I and II, respectively. The analysis is summarized in Table 10.1.

Now, given a design space expressed as EFP-SES, Table 10.1 gives us a window into two possible options—CES I and CES II—that are realizable component models. It also gives us information about designing an experimental frame. Assuming the components FixedInterval_Genr, RandomInterval_Genr, Proc, and Transd are already in the model-repository and that their behaviors have been specified as atomic models, the EFP-SES gives us various options as to how the models could be put together as a coupled model to achieve a desired objective. The specification of requirements guides the design process. Rows E3 and E4 define an initial starting point towards two experimental frames that yield scenarios to experiment with CES I and CES II.

Let us now understand the SES formal structure and its computational realization.

TABLE 10.1
Requirements Matrix Resulting in PES and CES

		Requirements				
S.No.	**Requirements**	**A**	**B**	**C**	**D**	**Outcome**
E1	A	X				PES I, PES II
E2	B		X			PES III
E3	C	X	X	X		PES IV, **CES I**
E4	D	X	X		X	PES V, **CES II**

10.2 FORMAL REPRESENTATION OF SES

Similar to DEVS formalism, SES is also based on set theory. A formal SES (Zeigler & Hammonds, 2007) is defined by the structure:

SES = < *Entities, Aspects, MultiAspects, Specializations, rootEntity, entityHasAspect, entityHasMultiAspect, entityHasSpecialization, aspectHasEntity, multiAspectHasEntity, multiAspectHasVariable, specializationHasEntity, entityHasVariable, variableHasRange* >

where

- Entities, Aspects, multiAspects, Specializations, Variables are finite mutually disjoint sets
- rootEntity $\in$ Entities

and the following relations hold:

entityHasAspect $\subseteq$ *Entities* $\times$ *Aspects*
entityHasMultiAspect $\subseteq$ *Entities* $\times$ *multiAspects*
entityHasSpecialization $\subseteq$ *Entities* $\times$ *Specializations*
aspectHasEntity $\subseteq$ *Aspects* $\times$ *Entities*
multiAspectHasEntity $\subseteq$ *multiAspects* $\times$ *Entities* (is a functional relation)
multiAspectHasVariable $\subseteq$ *multiAspects* $\times$ *Variables*
specializationHasEntity $\subseteq$ *Specializations* $\times$ *Entities*
entityHasVariable $\subseteq$ *Entities* $\times$ *Variables*
variableHasRange $\subseteq$ *Variables* $\times$ *RangeSpec*

As an example, the EFP SES shown in Figure 10.3 is formally expressed as

EFP_SES = <
Entities = {*EFP, EF, ProcBank, Proc, Genr, Transd, FixedInterval, RandomInterval*}

Aspects = {*efpDec, efDec*}
MultiAspects = {*procBankMultiDec*}
Specializations = {*genSpec*}
rootEntity = {*EFP*}
entityHasAspect = {(*EFP, efpDec*), (*ef, efDec*)}
entityHasMultiAspect = {(*ProcBank, procBankMultiDec*)}
entityHasSpecialization = {(*Genr, genSpec*)}
aspectHasEntity = {(*efpDec, EF*), (*efpDec, ProcBank*), (*efDec, Genr*), (*efDec, Transd*)}
multiAspectHasEntity = {(*procBankMultiDec, Proc*)}
multiAspectHasVariable = {(*procBankMultiDec, numPages*)}
specializationHasEntity = {(*genSpec, FixedInterval*), (*genSpec, RandomInterval*)}
entityHasVariable = {(*Genr, period*), (*Genr, jobId*), (*RandomInterval, seed*), (*Transd, obsPeriod*), (*Transd, lastJobSolved*), (*Transd, throughput*), (*Transd,* arrived), (*Proc, procTime*)}

variableHasRange = {

$$period \in R_{0+}^{\infty},$$

$$jobId \in N,$$

$$obsPeriod = 70,$$

$$lastJobSolved \in N,$$

$$througput \in R_{0+}^{\infty},$$

$$arrived \in N$$

$$procTime \in R_{0+}^{\infty},$$

$$numProcs = [1,10],$$

$$seed \in N$$

}

The algorithm to generate a directed graph in Figure 10.3 is given by Zeigler and Hammonds (2007) to computationally represent the formal SES representation.

Similarly, the PES is defined by the structure:

PES = < *OriginalSES, Entities, Aspects, MultiAspects, Specializations, rootEntity, entityHasAspect, entityHasMultiAspect, entityHasSpecialization, aspectHasEntity, multiAspectHasEntity, multiAspectHasVariable, specializationHasEntity*>

with the same constraints as in the SES, except that

$OriginalSES \subseteq SES^{+}$

$multiAspectHasEntity \subseteq multiAspects \times PES^{+}$

where

SES^{+} = set of all system entity structures, and
PES^{+} = set of all prunable entity structures

To have an in-depth theoretical understanding of SES and PES set theoretical analysis, refer to Zeigler and Hammonds (2007).

Exercise 10.2

Read Chapter 6 of Zeigler and Hammonds (2007).

Exercise 10.3

Read Chapters 7 and 8 of Zeigler and Hammonds (2007).

Exercise 10.4

Formally represent PES (I–III).

Exercise 10.5

Formally represent CES I and II.

10.3 SES EDITOR

Similar to DEVS Modeling Language in Chapter 5, we have used Xtext as the tooling framework to develop the SES Modeling Language (SESML). Xtext requires an Extended Backus Naur Form (EBNF) grammar. The SES EBNF grammar is shown below.

```
AbstractElement:
      Import | SesEntity | SesElement | DataEntity | SesComm ;

SesEntity:
      '@entity' name = ID ('{'
            pairs + = Variable*
            elementRefs + = SesElementRef*
            msgs + = Msg*
            (atomic = Atomic)?
            '}')? ;

SesElement:
      Aspect | Spec | MultiAspect ;
```

```
Aspect:
      '@aspect' name = ID 'is-made-of' entityRefs +=
         SesEntityRef*  ;

Spec:
      '@spec' name = ID 'can-be' entityRefs += SesEntityRef* ;

MultiAspect:
      '@multi' name = ID 'is-made-of-many' singleEntity =
        SesEntityRef
      'and atleast ' count = INT ('and atmost ' count =
        INT)?;

SesEntityRef:
      sesEntity = [SesEntity | QualifiedName] ;

SesElementRef:
      AspectRef | SpecRef | MultiAspectRef ;

AspectRef:
      '@aspect-ref' ref = [Aspect|QualifiedName] ;

SpecRef:
      '@spec-ref' ref = [Spec | QualifiedName] ;

MultiAspectRef:
      '@multi-ref' ref = [MultiAspect | QualifiedName] ;

Variable:
      type = VarType name = ID ('has' range = Range)? ;

VarType:
      simple = ('int'|'double'|'String'|'boolean') ;

Range:
      'min' min = FLOAT 'max' max = FLOAT ;
```

As an example, the EFP system in Figure 10.3 is realized as per the EBNF notation described above. The reusability of aspects, specializations, and multiAspects is inherent in the developed framework. Figure 10.9 shows the syntax highlighting (in the left) implemented in the editor and the hierarchical outline (in the right) that results from the application of algorithm described in Zeigler & Hammonds (2007).

Exercise 10.6

Install the SES plug-in and explore the various features of the Xtext editor.

Exercise 10.7

Implement CES I and CES II in the SES editor.

CES I and CES II are the entity structures that can be realized as a DEVS system. The structure has been defined with the EFP SES and consequently, with CES I and II.

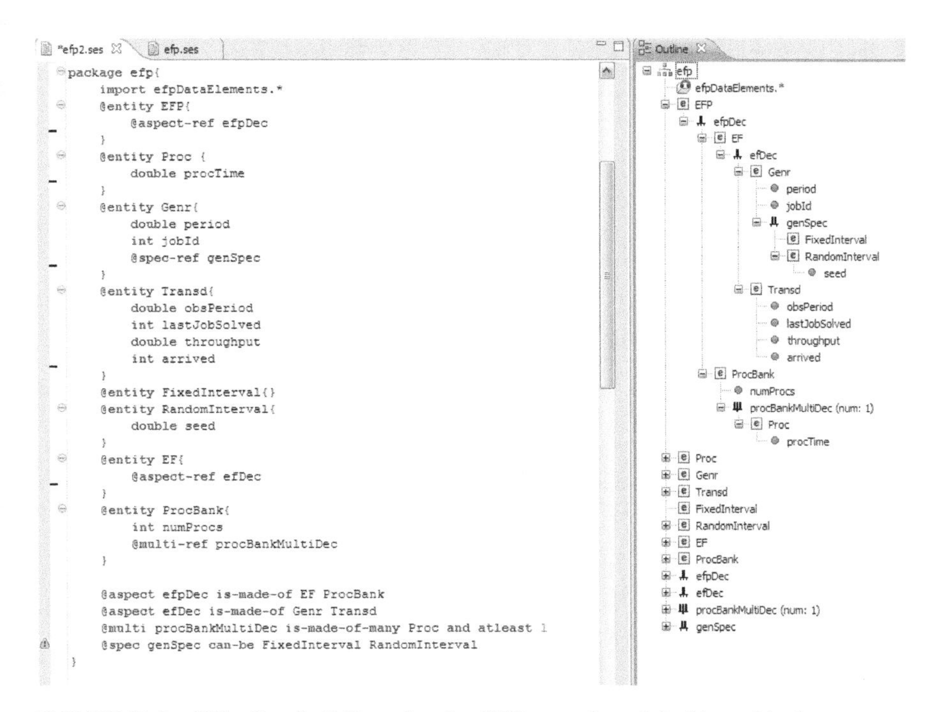

FIGURE 10.9 SES editor in Eclipse showing EFP example and the hierarchical structure.

The next step is to supply the behavior and communication links between the entities. The behavior is expressed as atomic models and the entities that have atomic behavior are: FixedInternval_Genr, RandomInterval_Genr, Proc, and Transd. The coupled entities, as described in CES I and II, are EF and EFP. The SES formalism utilizes the DEVSML models of Proc, Genr, and Transd, described in Chapter 5 as placeholders.

In order to implement the data communication and DEVS behavior, SESML was augmented with various DEVSML constructs, such as data entities, I/O interface, and DEVSML behavior specifications. This augmentation now makes SESML a logical extension of DEVSML. As an example, the Proc entity is shown in Figure 10.10. It shows four data entities: Start, Stop, Job, and Result. The Proc entity is augmented with input and output interfaces that use these dataentities. The EBNF grammar is added with the following constructs:

```
DataVariable:
      type = DataVarType name = ID ;

DataVarType:
      simple = ('int' | 'double' | 'String' | 'boolean') |
      complex = [DataEntity | QualifiedName] ;

DataEntity:
      '@data-entity' name = ID('{'
            pairs += DataVariable*
      '}')? ;
```

FIGURE 10.10 SESML as an extension of DEVSML incorporating data entities and atomic behavior.

Next, the coupled entities EF and EFP must have an I/O interface along with the means to specify the couplings. SESML is augmented with the following EBNF constructs:

Msg:

```
type = ('input'|'output') ref = [DataEntity |
  QualifiedName] name = ID ;
```

SesComm:

```
ic = IC | eoc = EOC | eic = EIC ;
```

EIC:

```
'@comm-dn' src = SesEntityRef 'sends' msg = [Msg]
  'in-to'
dest = [SesEntity|QualifiedName] 'at' msgDest = [Msg] ;
```

```
EOC:
'@comm-up' src = SesEntityRef 'sends' msg = [Msg] 'out-to'
      dest = [SesEntity|QualifiedName] 'at' msgDest = [Msg] ;

IC:
      '@comm-ic' src = SesEntityRef 'sends' msg = [Msg] 'to'
      dest = [SesEntity|QualifiedName] 'at' msgDest = [Msg] ;
```

As an example, the EF and EFP entities with augmented I/O interfaces are shown in Figure 10.11. The data entities have been refactored into a separate package, efp-DataElements, which is now imported. All the other entities have been augmented with I/O interfaces as well. Another thing to note is the specialization entities, namely, FixedInterval and RandomInterval, do not have any I/O interface. As shown in CES I and CES II, the Genr is replaced by either of the two: the pruning process

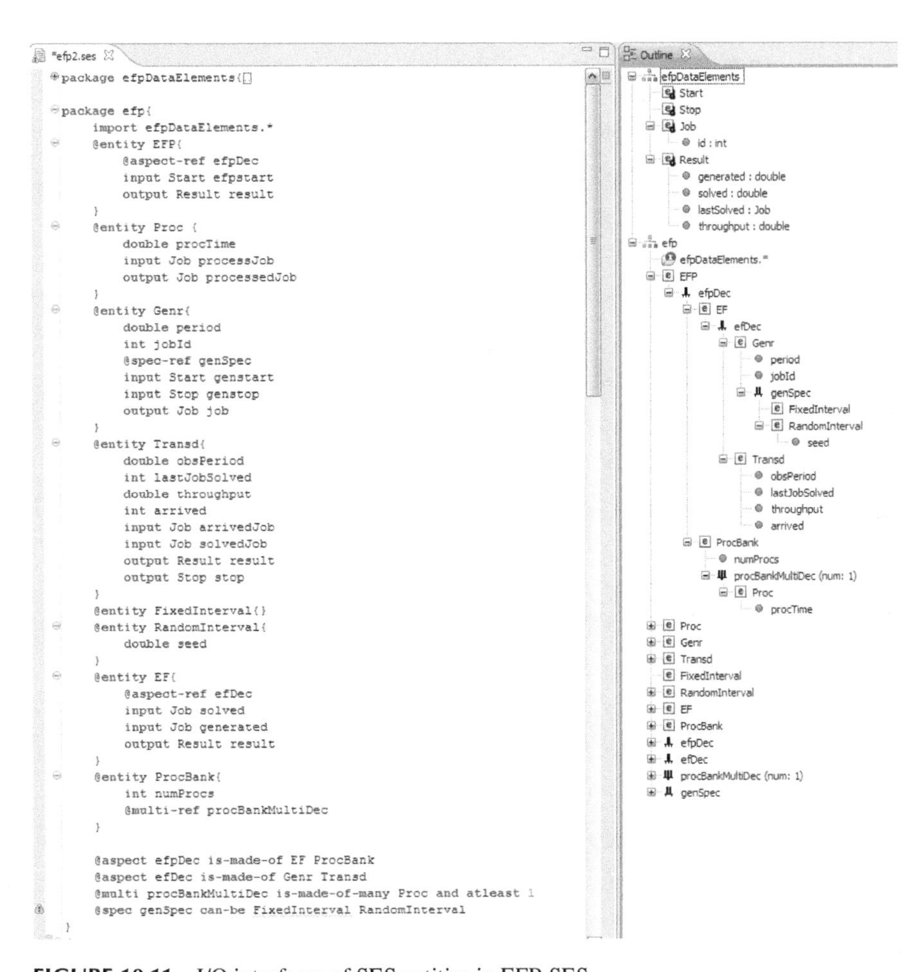

FIGURE 10.11 I/O interfaces of SES entities in EFP-SES.

replaces the Genr entity with either FixedInterval or RandomInterval and when such a replacement occurs, all the properties defined for Genr are then inherited by the specialized entity. Consequently, the interface gets substituted for each of the specialized entities as well.

The coupling specifications are only meaningful in a CES that is devoid of any specializations and multi-aspects, as they represent design choices. Figure 10.12 shows the CES I description in SESML. It shows Genr replaced by FixedInterval_Genr and ProcBank replaced by Proc_1 and Proc_2. The atomic DEVSML behavior description in Proc_1 and Proc_2 is omitted in Figure 10.11 and Figure 10.12 for brevity. The coupling specification is shown in Figure 10.13. The external input couplings (EIC) are expressed with the keyword `'@comm-dn'`, the external output couplings (EOC) with `'@comm-up'`, and the internal couplings with `'@comm-ic'`.

Figure 10.14 shows the realized DEVSJAVA model, ready for simulation.

Exercise 10.8

Work out the CES II in SESML and generate the DEVSJAVA code.

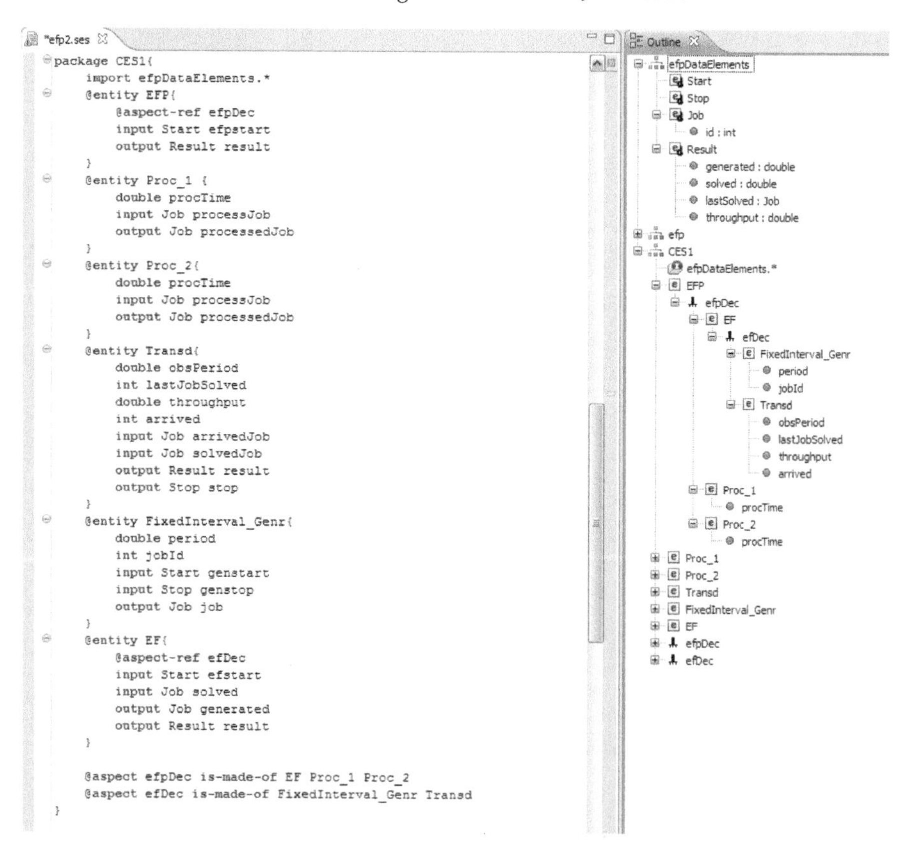

FIGURE 10.12 SESML representation of Component Entity Structure (CES) I.

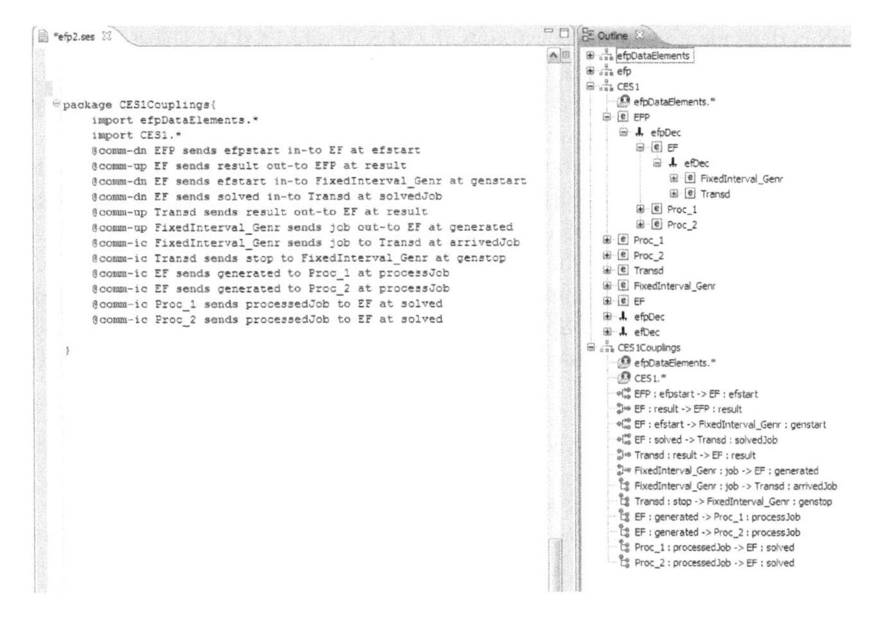

FIGURE 10.13 Coupling specification integrated with dataentities and SES grammar.

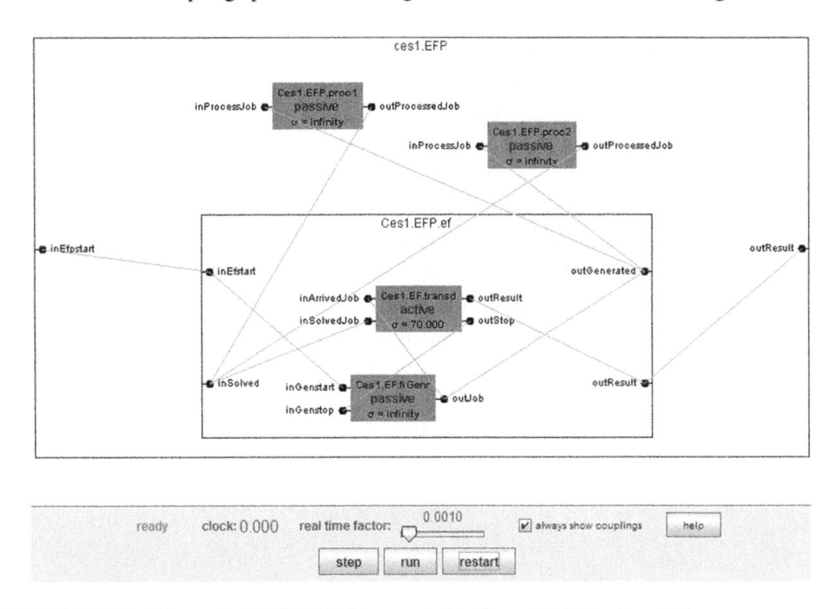

FIGURE 10.14 Generated DEVSJAVA executable from SESML description.

10.4 CONSTRAINT-BASED PRUNING

An SES is a collection of entities linked through various aspects, multi-aspects, and specializations. In its entirety, it captures the design space of the problem domain and holds promise that one of the PESs may be the design solution. The outcome as a PES is dependent upon the constraints the SES is subjected to. These constraints are generally expressed as relations between various entities, spread across the SES, which limit or allow the selection of other entities in a generated PES. We have seen, in Table 10.1, how requirements A and B did produce PES I and II, but were unable to lead to a CES. Additional requirements (C and D) were needed that aided the existing requirements (A and B) towards CES I and II. In the relation-based pruning, the requirements are expressed in terms of entity relations using logical handlers—and, or, and not—between entities and various aspects, specializations, and multi-aspects. As an example, Table 10.1 is casted as relation-based constraint specification in Table 10.2. Specialization *genSpec* is related with multi-aspect *procBankMultiDec* for identification of value of the variable *numProcs*. Table 10.2 reads as follows:

```
C1: if Select FixedInterval from genSpec
     then numProcs from procBankMultiDec equals 2
C2: if Select RandomInterval from genSpec
     then numProcs from procBankMultiDec equals 4
```

Such rules, expressed as first order logic (FOL), can be easily represented computationally and result from refining the requirements A, B, C, and D. However, it still does not guarantee that the resulting PES will be a CES. These refinements are only an exercise on how to link requirements to a more meaningful and concise computational representation that can be realized in a programming language, and processed in an automated manner. Much work by Zeigler and Hammonds (1989, 2007) has been in the area of entity selection through such relation-based constraints and the reader is encouraged to refer to Zeigler and Hammonds (2007) for additional examples and mathematical consideration.

To extend the rule-based constructs, Douglass and Mittal (2012) proposed cognitive domain ontology (CDO) formalism—a logical extension of SES with a constraint specification language containing additional operators in alignment with various SES axioms. These operators in the constraint specification language are implemented in Lisp. CDO is implemented in Lisp. Table 10.3 enumerates these operators.

For detailed implementation, refer to Douglass and Mittal (2012). The list presented in Table 10.3 is not exhaustive and many other complex operators can be conceived based on the domain and on the foundation of FOL and basic operators.

TABLE 10.2
Relations Between genSpec and procBankMultiDec in EFP-SES

			Multi-aspect	Variable
Constraints	**Specialization**	**Entities**	***procBankMultiDec***	***numProcs***
C1	*genSpec*	FixedInterval	X	2
C2	*genSpec*	RandomInterval	X	4

TABLE 10.3
Operators for a Constraint Specification Language in CDO

Basic Operator	Meaning	Example
`and`	Conjunction	`(and p q)`
`or`	Disjunction	`(or p q)`
`not`	Negation	`(not q)`
`= =>`	Implication	`(= => p q)`
`< = =>`	Biconditional	`(< = => p q)`
`false`	Logical falsity	`(and (not p) q)`
`true`	Logical truth	`(or p (not q))`
`e@`	Entity located	`(e@ EF genSpec)`
`v@`	Variable attached to CDO entity	`(v@ Genr period)`
`equale`	Entities are equal	`(equale genSpec Fixed Interval)`
`equalev`	Variable has a value	`(equalv jobID 0)`
`let`	Variable/value binding	`(let ((p its-raining)` `(q ground-gets-wet)))` `(= => p q)`
Complex Operators		
`assert!`	Set value of a constraint variable	`(assert!` `(equalv (v@ (weight) kg) 100))`
`andv`	Conjunction with variables	`(andv (equale` `(e@ aspect sport) golf)` `(equale (e@ aspect size)` `small))`
`orv`	Disjunction with variables	`(orv` `(equale (e@ aspect size)` `small)` `(equale (e@ aspect size)` `large))`
`notv`	Negation with variables	`(notv (equale (e@ Genr` `period))`
`ifv`	Implication with variables	`(ifv (equale (e@ genSpec)` `FixedInterval)` `(let equale(v@` `ProcBank numProcs))` `(= =>equalv numProcs 2)`
`fail`	Failure with backtracking	`(ifv` `(andv` `(equale (e@ ensemble) soloist)` `(equale (e@ style) symphonic)` `(assert! (fail))`
`a-member-of`	Non-deterministic selection from a list	`(assert! v` `(a-member-of)(a s d f)))`
`either`	Non-deterministic selection from arguments	`(either small medium large)`

(Continued)

TABLE 10.3 (*Continued*)
Operators for a Constraint Specification Language in CDO

Basic Operator	Meaning	Example

`an-integer-above`
`an-integer-between`
`an-integer-below`
	Define integer ranges	`(equalv (an-integer-above 10))`
		`(assert!`
		`(equalv (v@(weight)kg)`
		`(an-integer-between 75 105)))`

`a-real-above`
`a-real-between`
`a-real-below`
	Define real ranges	`(equalv pi`
		`(a-real-between 3.0 4.0))`
		`(ifv (notv (equalv pi 3.141592))`
		`(fail))`

>v, > = v, <v, < = v	Comparison functions that accept constraint variables	`(ifv (< = v v 10) (fail))`

These operators are declarative. While the list has its own advantages, there are better options available, such as XML (Zeigler & Hammonds, 2007), and DSL authoring tools, such as Xtext, that allow better readability and syntactical development.

Exercise 10.9

Implement the basic and complex operators in SESML Xtext environment.

There is no end to modifications to the list of complex operators. The usage is manageable in small SES examples. However, in a large SES, the management of such operators becomes cumbersome. These operators help define constraints that link entities with each other, with aspects, specializations, multiaspects, and variables spread all over the SES. What is an efficient way to manage the resulting constraint network in a PES?

Figure 10.15 shows the SESML in perspective with DEVSML. It shows how SES extends the DEVS level of system specifications at Level 4. SES development is a top-down phenomenon while DEVS is bottom-up, with I/O frame at its lowest level. Constraints narrow down the design space as they are successively applied. The inverse cone reflects the impact of constraints.

An SES is built from domain knowledge from subject matter experts. At this level, the SES is a conceptual representation. To use SES for a particular problem, constraints are added so that the design space is more manageable and relevant to the context and system dynamics. Scenarios are conceptualized with the help of domain constraints. The next downward level comprises the pragmatics, a design space consisting of PES that may or may not be transformed to CES. At this point, the context is very clearly defined and experimental frames can be designed. Semantics that relate to information exchange between the entities at various levels of hierarchy are added at the next level. More constraints are added, which help transform a PES to a CES. At this point, a DEVS syntactic model is generated. SES facilitates the top-down design by the specification of constraints at pragmatic, semantic, and syntactic levels, leading to a system component structure. Such a demarcation of concerns is also in congruence with the Levels of Conceptual Interoperability Model (LCIM) in Tolk and Muguira (2003), shown in Table 10.4. The last column in Table 10.4 describes the multi-level constraints in relation to LCIM.

Lastly, the constraint specifications are related to DEVS levels of systems specification to incorporate them at the behavior level (Table 10.5). The discussion so far has used constraints at the system structure level (Level 4) and analyzed how

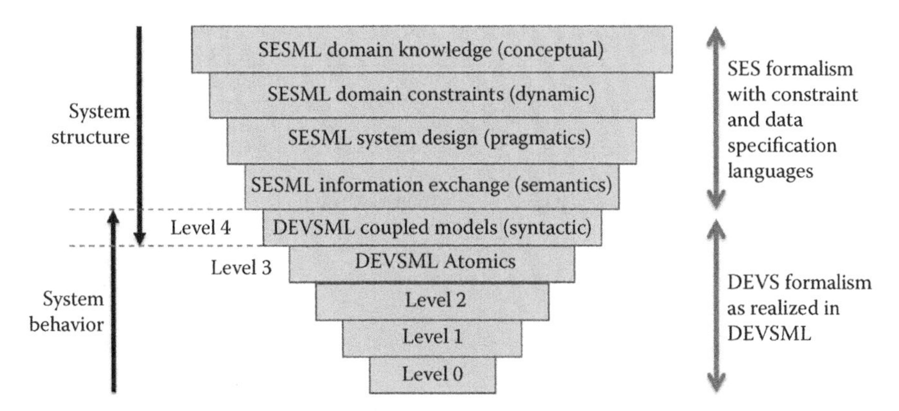

FIGURE 10.15 Layered architecture with DEVSML and SESML.

TABLE 10.4
Role of Constraints at Various Levels of Conceptual Interoperability

Level of Conceptual Interoperability	Characteristic	Key Condition	Constraint Specifications
Conceptual	The assumptions and constraints underlying the meaningful abstraction of reality are aligned	Requires the conceptual methods be documented based on engineering methods, enabling their interpretation by other engineers	Constraints specify hard truths, physics, and other base level considerations that are uniform over all contexts
Dynamic	Participants are able to comprehend changes in system state, assumptions, and constraints that each is making over time and able to use to their advantage	Requires common understanding of system dynamics	Context is added so that additional constraints enable the system be applicable to the context

(Continued)

TABLE 10.4 (*Continued*)
Role of Constraints at Various Levels of Conceptual Interoperability

Level of Conceptual Interoperability	Characteristic	Key Condition	Constraint Specifications
Pragmatic	Participants are aware of the methods and procedures that each is employing	Requires the use of data or the context of application	Pruning is performed and constraints are further added to refine PES and data exchange. Pragmatic frames and experimental frames are designed at this constraint level.
Semantic	The meaning of data is shared	Requires common information exchange model	Constraints are added to remove any ambiguities in PES towards the construction of CES. Data and coupling are the core focus.
Syntactic	Introduces common structure to exchange information	Requires a common data format or standard be used	Data and variable constraints refine the semantics of data exchange.
Technical	Data can be exchanged between participants	Requires that a communication protocol exists	Constraints on data format used for a particular communication.
Stand-alone	No interoperability		

TABLE 10.5
Constraint Specification Language as Applicable to DEVS Levels of Systems

Level	Name	System Specification	Constraint Specifications
4	Coupled System	System built by several component systems that are coupled together	Component selection and coupling selection.
3	I/O System	System with state and state transitions to generate behavior	Constraints that allow or limit transitions from a particular source state. Constraints that hide or enable state space based on environment contingencies or a specific source state.

TABLE 10.5 (*Continued*)
Constraint Specification Language as Applicable to DEVS Levels of Systems

Level	Name	System Specification	Constraint Specifications
2	I/O Function	Collection of input/output pairs partitioned according to the initial state the system is in when the input is applied	Constraints may force a specific state as an initial state, always or otherwise.
1	I/O Behavior	Collection of input/output pairs constituting the allowed behavior of the system from an external black box view	Constraints on mapping functions that map input trajectories to output trajectories, leading to fine or coarse I/O behavior. Quantization may be a result of such constraint.
0	I/O frame	Input and output variables and ports together with allowed values	Constraints to specify domain and range of variables.

variables may be constrained with respect to entity selection or any other variable value (Level 0). Integrating a constraint specification language with other DEVS levels of system specification (Level 1, 2, and 3) will make a constraint-specific language applicable to all levels of system design.

10.5 PRAGMATICS INTO ONTOLOGIES

Rather than being used to capture system alternatives to be explored through DEVS-based modeling and simulation, SESs are currently being used to capture formal structural and relational information about domains. The SES has advanced toward specifying entire ontologies rather than just an engineering problem (Lee & Zeigler, 2010; Zeigler & Hammonds, 2007). This current use exploits similarities between SES and general ontologies. Current research and modeling and simulation activities utilizing SES demonstrate that extraordinarily diverse domains can be formally captured and related to each other through formal structures, such as domain ontologies, pragmatic frames, and overlapping PES. The SES theory has been updated to incorporate various data engineering concepts that facilitate transformations and semantic mapping. Table 10.6 shows various operations that are currently available in the XML implementation of SES (Zeigler & Hammonds, 2007). Concepts like data harmonization and pragmatic frames have been introduced, which enable the comparison/fusion of two or more ontologies and facilitate the development of domain interoperability solutions. Data harmonization is the process by which a degree of commonness between two or more ontologies is studied. The pragmatic frame is the concept by which the data requirements of the receiving entity are studied. Integrating these two concepts together helps formally specify an SES framework that can encompass an entire set of pragmatic frames within a harmonized SES.

TABLE 10.6
SES and PES Operations

Operation	Summary	Key Applications
SES merging	More than two SES can be merged as components to form a larger SES	Synthesize large schemas. Integrate knowledge information from multiple subject matter experts to create a domain SES
SES substructuring	Substructures can be extracted as stand-alone SES	Decompose large schema into manageable parts
SES path labeling	All items have unique path identifiers	Provide context and access for elements in large schemas. No need for scope management
SES comparing	Set theory-based test can be done to test pairs of SES for inclusion, equality, and so on.	Harmonization of legacy formats and checking of profiles.
SES restructuring	Change in structure from one form or the other	Improving representation efficiency, reducing size, and so on. Reverse engineering applied to harmonization
SES pruning	Assign values to variables, select from specializations, and choose aspects. Allows partial pruning	Generates PES and CES to support DEVS based M&S
SES validation	Check that SES satisfies all axioms	Validation checking supports error discovery. Assures consistency of larger SES composed from multiple SES.
PES merging	PES can be merged as components to form larger PES	Synthesize larger PES
PES substructuring	Substructures can be extracted as stand-alone PES	Decompose larger PES to manageable parts
PES path labeling and value retrieval	Extract minimal information needed to disambiguate multiple occurrences and access values	Provide context and access to values in large schemas
PES differencing	Identify differences and their locations in a pair of PES	Provide basis for change-based information exchange
PES validation/ conformance	Validation applies to completely pruned structure; conformance accepts partial pruning stages	Validation checking supports error discovery and location; testing of standards, and collaborations
PES transformation to DEVS	Synthesize DEVS simulation models from PES	Synthesize executable architectures from SES

It must be noted here that a pragmatic frame is a pruned instance of a harmonized SES and is a form of "requirement." Figure 10.16 illustrates this. The main contributions of contemporary SES research are the increased expressive power of SES and the increased scalability that enables the specification of entire ontologies.

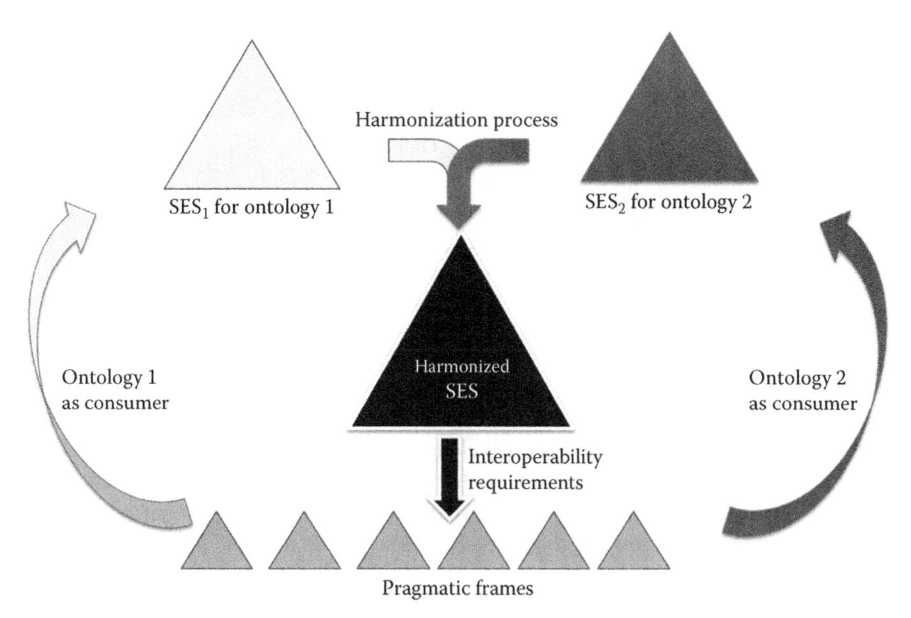

FIGURE 10.16 Formal SES harmonization process to deal with ontology and data engineering.

10.6 KNOWLEDGE-BASED CONTINGENCY-DRIVEN GENERATIVE SYSTEMS

In order to express a rich knowledge set that includes environment, contingencies, resources, possible actions, and much more, we need a framework like SES that allows us to represent knowledge in many facets or dimensions. For a multidimensional and multiresolutional knowledge representation, the SES knowledge framework allows constructions of this kind of representation. Ontology, in technical terms, is a graph of nodes, and information is presented in the relations that exist between these nodes. Of course, it is a great step, as the knowledge can now be presented in associative terms, more like a semantic network. It is now more amenable to data engineering efforts, but is essentially flat and not suitable for piecewise construction or layered methodologies for better manageability. The SES formal knowledge representation mechanism with its set of axioms and rules helps develop an ontology that can be constructed and deconstructed in a piecewise manner through SES aspects and specializations. The latest work in SES ontology domain is an evidence of such efforts (Lee & Zeigler, 2010). Figure 10.17 illustrates this.

We have shown earlier how a system can have its configuration in multiple aspects and specializations. Such aspects and specializations can be added or removed incrementally and intuitively without changing other facets of the system and still understandable by the common modeler. In other words, the modeler is not overwhelmed by the influx of new knowledge as it builds upon the existing ones. This is important because, in large systems, a large knowledge set often results in "information paralysis" at the modeler end. Such aspects and specializations give ontology

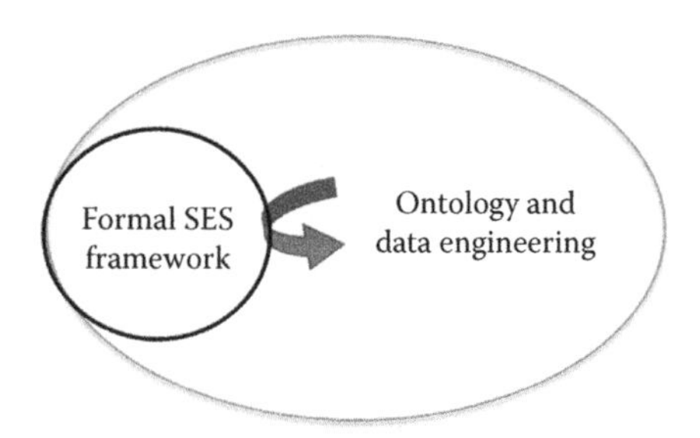

FIGURE 10.17 SES as an ontology and data engineering tool.

a multi-resolutional capability and can be called upon at real-time execution of the system. Also, note here that adding such elements is achieved through piecewise isolation and it is the defined rules that create relationships between the different SES elements at run-time, thereby managing complexity. It also implies that, while the general structure of the proposed ontology remains intact, it is the defined rules that dictate the association and affordances of the entire system at run-time. These rules then become dynamic and dictate how the knowledge entities interact. The realization of these rules by DEVS formalism in an SES-modeled system is much easier, manageable, and formally verifiable at run-time.

Another advantage of this piecewise construction is partitioning of the expert knowledge in the domain of interest. It now becomes much more feasible to integrate the expert knowledge of subject matter experts on aspects of such ontology. Once the structure of these aspects is laid out, it is easier to define and modify rules that link different aspects of the ontology. In a semantic network, the knowledge is stored in associations and, similarly, in SES ontology, the knowledge is stored in these SES rules that link different entities and are subject to negotiations and common understanding of the domain itself. It also leads to developing the entities with a defined set of interfaces, which will ultimately lead to the development of system components for further reuse and integration in larger systems. Let us now put our focus towards knowledge-based intelligent systems using the advanced SES capabilities.

> Intelligence-based systems should not be confused with the often narrowly used term intelligence system, which refer to a variety of Artificial Intelligence (AI) methods, such as neural networks, evolutionary algorithms, expert systems, diagnostic systems, symbolic AI, and other related topical areas. These systems are limited to AI applications, and intelligent systems engineering describes the engineering of such intelligent systems, not the use of intelligence to support systems engineering.

This quote is from Tolk et al. (2011: pp. 2) and we are in agreement with the view that we are interested in merging the state-of-the-art intelligence, as provided by AI methods to support systems engineering and, eventually, the system of systems engineering.

Tolk et al. also identified the components to capture the intelligence in-system within their book as an iterative cycle starting with the ontological specification of models, data, and knowledge; agents that act upon this data and knowledge;

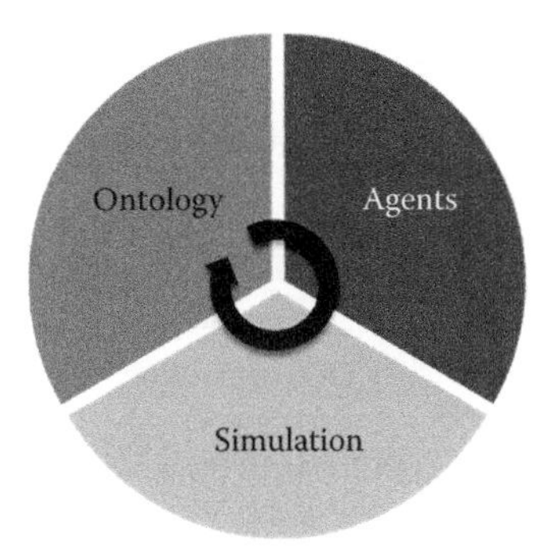

FIGURE 10.18 Intelligence-based systems engineering concept.

and lastly, the simulation that gives a dynamic executable environment to the agent. The cycle repeats, with the simulation results feeding back into the ontology (Figure 10.18).

Another recent book edited by Yilmaz and Oren (2009) is dedicated to agents-based systems engineering that supports the notion of an integrated use of modeling and simulation in agent-based systems and systems engineering.

To build on this foundation, the Knowledge-based Contingency-driven Generative Systems (KCGS) theoretical framework is based on three major areas, with the associated Levels of Constraint Specifications (based on SES formalism) at the center:

I. Ontology and data representation
II. Knowledge engineering and parallel distributed computational search mechanisms
III. DEVS Unified Process

I deals with knowledge representation and how data interoperability is achieved between different ontologies using SES foundational frameworks and how constraints are applicable to the ontology. In its current state, basic programmatic pruning mechanisms are used. II deals with the entire knowledge engineering and data-mining aspect of executing the pruning process that transforms data into information. This computational process has to align itself with the AI-based search mechanisms and real-time execution capabilities that will lead to formal PES. Finally, III takes the formal PES, constraints transformations and, using the DEVS M&S technology, provides the requirement traceability, platform-independent M&S, verification and validation and various other capabilities, such as SOA execution, and system component descriptions in the DEVS Unified Process (DUNIP).

In summary, we provide a formal framework to specify both the domain and agent's behavior ontology (lingua-franca) that is interoperable with other ontologies *at-large* (in italics) for system of systems integration and development. Such formal representation should lend itself to computational mechanisms for efficient real-time artificial behavior that originates from these ontologies. This formal ontological framework lends itself to formal modeling and simulation frameworks, such as DUNIP, for larger systems of systems and human-in-loop solutions.

TABLE 10.7
Mapping Requirements for M&S Framework for the Artificial Intelligent System with KCGS Components

S.No.	Framework Requirements	Technologies and Foundation	KCGS Component
1.	Based on general systems theory	DEVS systems theory	DEVS Unified Process
2.	Facilitate model-based development and engineering	DEVS M&S Framework, SES theory	SES theory
3.	Scalable and component-based	DEVS M&S framework	DEVSML 2.0, DEVS Unified Process
4.	Manage hierarchy and abstractions	DEVS systems theory	DEVS Unified Process
5.	Interoperable across implementation platforms	DEVS M&S framework	DEVSML 2.0
6.	Formal specification	DEVS M&S framework	DEVS Unified Process, SES theory
7.	Domain and platform neutral	DEVS systems theory, SES theory	Ontology and knowledge representation
8.	Agile and persistent	DEVS systems theory, SES theory	DEVS Unified Process, SES theory
9.	Interface with AI knowledge engineering methodologies	SES pruning	SES theory, data mining, Constraint Satisfaction Problem (CSP)

The capabilities defined above allow us to specify any kind of domain models and take the executing real models to live netcentric systems. A framework for modeling and simulation of the artificial intelligent situated agents in a dynamic environment must have these basic capabilities. Table 10.7 lists some of the requirements of such a framework and Figure 10.19 shows the KCGS framework that addresses these requirements. One such application of KCGS is implemented in Douglass and Mittal (2012), where behavior ontology is specified for a cognitive rational agent. Figure 10.19 also shows how different disciplines interact together and interface with the formal SES theoretical framework.

Ultimately, what is provided is an ontological framework, such as SES, that lends itself seamlessly to the simulation-based component modeling framework. Figure 10.20 presents the meta-meta-model of an autonomous system (Mittal, 2010). It formally captures real-world facets like environment and resources and agent-based facets like goals and behavior. Constraints play a dual role in an autonomous system's ontology. There will be two types of constraints. Type I constraints are physics-based (hard truths) and Type II constraints are situation-based. While Type I constraints are hard constraints, the Type II constraints are soft constraints that are dynamic. The Type II constraints are the ones that are responsible for

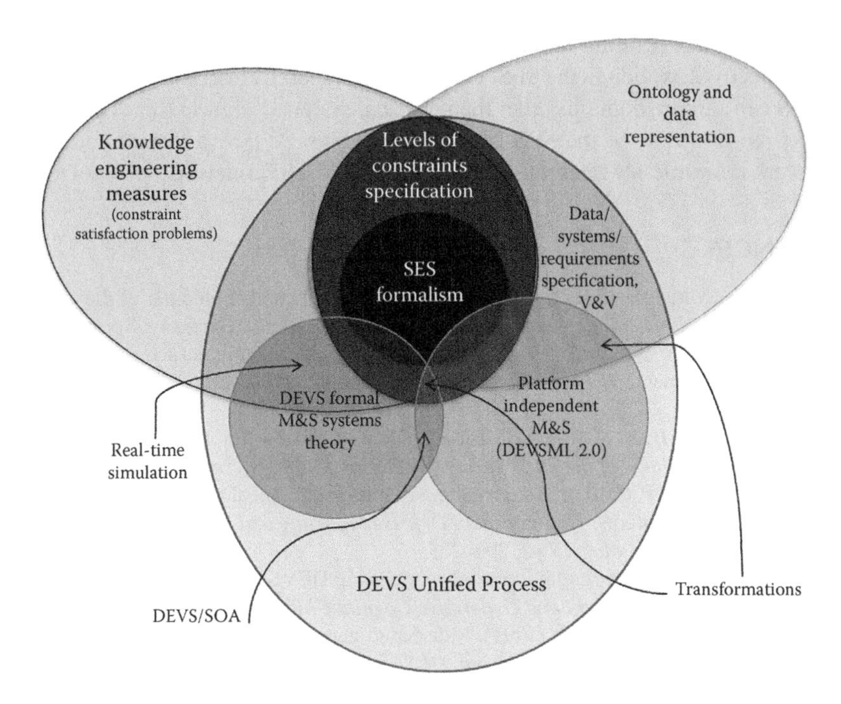

FIGURE 10.19 Knowledge-based contingency-driven generative systems framework.

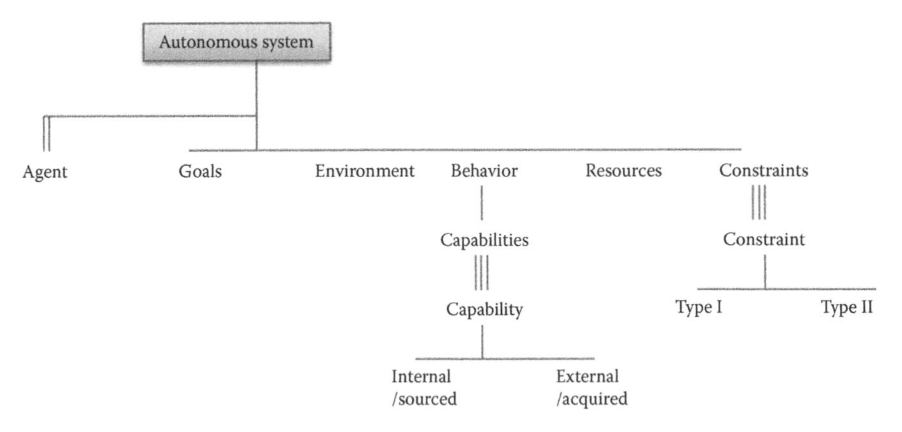

FIGURE 10.20 Meta-meta-model for an autonomous system.

contingency-based behavior and situated behavior. The pruning process will work on these Type II constraints to generate a behavioral SES that is "situated."

10.7 SUMMARY

Earlier SES research efforts demonstrated that SES, constraint-sensitive search processes, and conventional simulation can be used to capture, search through, and evaluate system configuration spaces. These efforts demonstrated how a rule-based search or the SES pruning process can derive system configurations that meet the design objectives explicit in the SES. Current SES research efforts have demonstrated that SES can capture domains other than physical system design spaces. This chapter provided an overview of the SES landscape and showed how a formal ontological framework is seamlessly integrated with a complex M&S framework such as DUNIP.

REFERENCES

Douglass, S., & Mittal, S. (2012). A framework for modeling and simulation of the artificial. In *Ontology, Epistemology and Telelogy (Eds. Andreas Tolk).* Germany: Spring Verlag.

Kim, T., Lee, C., Christensen, E., & Zeigler, B. (1990). System entity structuring and model base management. *IEEE Transactions on Systems, Man, and Cybernetics, 20*(5), 1013–1024.

Lee, H., & Zeigler, B. P. (2010). SES-Based ontology process for high level information fusion. *Proceedings of the 2010 Spring Simulation Multiconference.*

Lee, H., & Zeigler, B. P. (2010). System entity structure ontological data fusion process integrated with c2 systems. *The Journal of Defense Modeling and Simulation: Applications, Methodology, Technology, 7*(4), 206–225.

Mittal, S. (2010). Net-centric cognitive architecture using DEVS unified process. *Researching and Developing Persistent and Generative Cognitive Models Workshop.* Scottsdale, AZ.

Rozenblit, J. W., & Huang, Y. M. (1991). Rule-based generation of model structures in multifaceted modeling and system design. *ORSA Journal on Computing, 3*(4), 330–344.

Rozenblit, J. W., & Zeigler, B. P. (1993). Representing and constructing system specifications using the system entity structure concepts. *Proceedings of the 1993 Winter Simulation Conference*, (pp. 604–611).

Rozenblit, J. W., Hu, J., Kim, T. G., & Zeigler, B. P. (1990). Knowledge-based design and simulation environment (KBDSE): Foundational concepts and implementation. *Journal of the Operations Research Society, 41*(6), 475–489.

Tolk, A., & Muguira, J. (2003). The levels of conceptual interoperability model (LCIM). *Proceedings of the Fall Simulation Interoperability Workshop.*

Tolk, A., Adams, K., & Keating, C. (2011). Towards intelligence-based systems engineering and system of systems engineering. In A. Tolk, & L. Jain, *Intelligence-Based Systems Engineering* (pp. 1–22). Berlin, Hiedelberg: Springer-Verlag.

Yilmaz, L., & Oren, T. (2009). *Agent-directed simulation and systems engineering.*New York, NY: Wiley.

Zeigler, B. (1984). *Multifaceted modelling and discrete event simulation.* London, UK: Academic Press.

Zeigler, B. P., & Chi, S. D. (1993). Model-based architecture concepts for autonomous systems design and simulation. In *An Introduction to Intelligent and Autonomous Control* (pp. 57–78). Kluwer Academic Publishers.

Zeigler, B. P., Kim, T. G., & Praehofer, H. (2000). *Theory of modeling and simulation.* New York, NY: Academic Press.

Zeigler, B. P., Luh, C. J., & Kim, T. G. (1991). Model base management for multifaceted systems. *ACM Transactions on Modeling and Computer Simulation, 1*(3), 195–218.

Zeigler, B., & Hammonds, P. (2007). *Modeling* and *Simulation-Based Data Engineering: Introducing Pragmatics into Ontologies for Net-centric Information Exchange.* Academic Press.

Zeigler, B., & Zhang, G. (1989). The system entity structure: knowledge representation for simulation modeling and design. In L. Widman, K. Loparo, & N. Nielseen, *Artificial Intelligence, Simulation and Modeling* (pp. 47–73). Hoboken, NJ: John Wiley.

11 Department of Defense Architecture Framework *Version 1.0*

11.1 INTRODUCTION

The Department of Defense (DoD) mandate requires that the DoD Architecture Framework (DoDAF) be adopted to express high-level system and operational requirements and architectures (DoD, 2003). DoDAF is the basis for the integrated architectures mandated in DOD Instruction 5000.2 (2003) and provides broad levels of specification related to operational, system, and technical views (TVs). Integrated architectures are the foundation for interoperability in the Joint Capabilities Integration and Development System (JCIDS) prescribed in CJCSI 3170.01D (2004). DoDAF and other DoD mandates pose significant challenges to the DoD system and operational architecture development and testing communities since DoDAF specifications must be evaluated to see if they meet requirements and objectives, yet they are not expressed in a form that is amenable to such evaluation. However, DoDAF-compliant system and operational architectures do have the necessary information to construct high-fidelity simulations. Such simulations become, in effect, the executable architectures referred to in the DoDAF document. DoDAF is mandated for large procurement projects in the Command and Control domain, but its use in relation to modeling and simulation (M&S) is not explicitly mentioned in the documentation (Atkinson, 2004; Zeigler & Mittal, 2005). Operational views (OVs) capture the requirements of the architecture being evaluated, and system views (SVs) provide its technical attributes. Together, these views form the basis for the semi-automated construction of the needed simulation models.

DoDAF is a framework prescribing high-level design artifacts, but leaves open the form in which the views are expressed. A large number of representational languages are candidates for such expression. For example, the Unified Modeling Language (UML) and Colored Petri Nets (CPNs) are widely employed in software development and in systems engineering. Each popular representation has strengths that support specific kinds of objectives and cater to its user community needs. By going to a higher level of abstraction, DoDAF seeks to overcome the plethora of stove-piped design models that have emerged. Integration of such legacy models is necessary for two reasons. One is that, as systems, families of systems, and systems of systems become more broad and heterogeneous in their capabilities, the problems of integrating design models developed in languages with different syntax and semantics have become a serious bottleneck to progress. The second is that another recent DoD mandate also intended to break down this stove-piped culture requires the adoption of the Service Oriented Architecture (SOA) paradigm as supported in the development of Network

Centric Enterprise Services (CIO, DoD 2006; 2007). However, anecdotal evidence has suggested that a major revision of the DoDAF to support net-centricity has been widely considered to be needed. DoDAF 1.5 was proposed in 2007 to address those needs with a focus on SOA and net-centricity. Indeed, under DoD's direction, several contractors have begun to design and implement the NCES to support this strategy on the Global Information Grid (GIG). The result is that system development and testing must align with this mandate—requiring that all systems interoperate in a net-centric environment—a goal that can best be reached by having the design languages be subsumed within a more abstract framework that can offer common concepts to relate to. However, as stated before, DoDAF does not provide a formal algorithmically enabled process to support such integration at higher resolutions. Lacking such processes, DoDAF 1.0 is inapplicable to the SOA domain and GIG in particular.

The latest version of DoDAF, Version 2.0, was proposed in 2009. We are describing DoDAF 1.0 in this chapter to build the foundation of how complex systems of systems are specified in such a framework. Not only do we highlight the shortcomings in DoDAF 1.0, but we also present how the usage of System Entity Structure (SES) and DEVS-based M&S are much more capable to formally specify complex hierarchical structures and behavior. This chapter is based on our earlier work (Mittal, 2006).

11.1.1 DoDAF 1.0 Specifications

DoDAF, Version 1.0 (2003), defines a common approach for DoD architecture description development, presentation, and integration. The framework enables architecture descriptions to be compared and related across organizational boundaries, including joint and multinational boundaries. DoDAF is an architecture description and it does not define a process to obtain or build the description. The Deskbook provides one method for development of IT architectures that meet DoDAF requirements, focusing on gathering information and building models required to conduct design and evaluation of architecture. The DoDAF (2003) defines three elements for any architecture description. These are discussed below.

11.1.1.1 Operational Views

The OV is a description of the tasks and activities, operational elements, and information exchanges required to accomplish DoD missions. DoD missions include both warfighting missions and business processes. The OV contains graphical and textual products that comprise an identification of the Operational Nodes* and elements, assigned tasks, and activities, and information flows required between nodes. It defines the types of information exchanged, the frequency of exchange, which tasks and activities are supported by the information exchanges, and the nature of information exchanges.

* Operational node: A node specified in OV that performs one or more operations. A functional entity that communicates with another functional entity to implement a collective functionality or a capability.

11.1.1.2 System Views

The SV is a set of graphical and textual products that describe systems and interconnections providing for, or supporting, DoD functions. DoD functions include both warfighting and business functions. The SV associates systems resources to the OV. These systems resources support the operational activities and facilitate the exchange of information among operational nodes. Within this view, how the functionalities specified in OV will be met is elaborated.

11.1.1.3 Technical Views

The TV is the minimal set of rules governing the arrangement, interaction, and interdependence of system parts or elements, whose purpose is to ensure that a conformant system satisfies a specified set of requirements. Within this view, the delivery of systems and functionalities is ensured along with their migration strategies toward future standards.

These views provide three different perspectives for looking at the architecture. The emphasis of DoDAF 1.0 lies in establishing the relationship between these three elements ensuring entity relationships and supporting analysis (Figure 11.1). The DoDAF approach is essentially data-centric rather than product-centric. Table 11.1 enumerates various documents in each of the DoDAF 1.0 views.

11.1.2 Motivation for DoDAF-to-DEVS Mapping

The DoDAF 1.0 suffers from the following shortcomings:

1. Although there is mention of executable architectures in DoDAF, there is no methodology recommended by DoDAF that would facilitate the development of executable DoDAF models.
2. It has completely overlooked the model-driven development approach. Consequently, there is no formal M&S theory that DoDAF mandates.
3. DoDAF fails to address performance issues at the OV level.

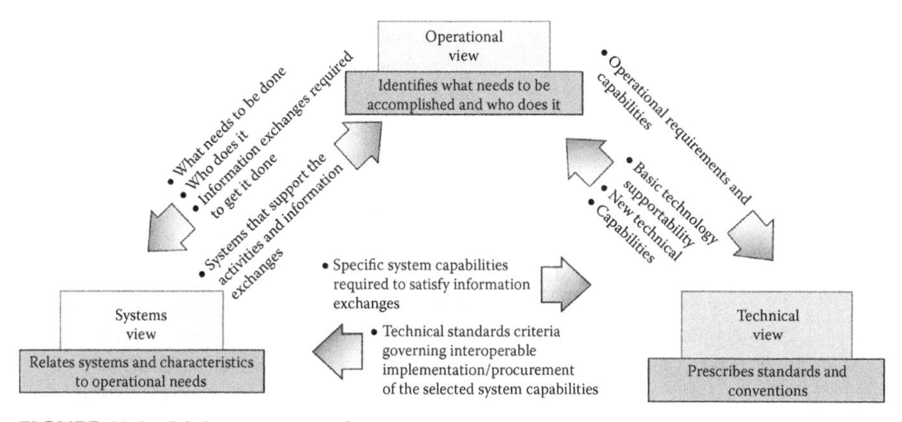

FIGURE 11.1 Linkages among views.

TABLE 11.1
Various DoDAF 1.0 Product Documents

Product	Name	Description
All View (AV)		
AV-1	Overview and Summary Information	Scope, purpose, users, environments, any analytical findings
AV-2	Integrated Dictionary	All terms used in DoDAF products
Operational View (OV)		
OV-1	High-Level Operational Concept Graphic	High-level graphical and textual description of operational concept
OV-2	Operational Node Connectivity Description	Operational nodes, activities performed at each node, and connectivities and information flow between nodes
OV-3	Operational Information Exchange Matrix	Information exchanged between nodes and the relevant attributes of that exchange
OV-4	Organizational Relationships Chart	Command, control, coordination, and other relationships among organizations
OV-5	Operational Activity Model	Activities, relationships among activities, inputs, and outputs
OV-6a	Operational Rules Model	Rules that constrain a particular activity
OV-6b	Operational State Transition Description	Operational activity sequence and timing that identifies responses of an activity to events
OV-6c	Operational Event-Trace Description	Operational activity sequence and timing that traces the actions in a scenario or critical sequence of events
OV-7	Logical Data Model	Data requirements and structural rules of the operational view
Systems View (SV)		
SV-1	Systems Interface Description	Depicts system nodes and systems resident at these nodes represented in OV-2. Also identifies the interface between systems and system nodes.
SV-2	Systems Communications Description	Information about communication systems, communication links, and communication networks. It automates the need lines presented in OV-2.
SV-3	System–System Matrix	Interface characteristics of different systems specified in a matrix form
SV-4	Systems Functionality Description	Systems functional hierarchies and system functions, and the systems data that flows between them
SV-5	Operational Activity to Systems Traceability Matrix	Maps the OV-5 with SV-5 in a matrix
SV-6	Systems Data Exchange Matrix	Specifies the characteristics of the data exchanged between systems
SV-7	Systems Performance Parameters Matrix	Specifies the performance parameters of each system, interface or system function, and the expected or required performance parameters at specified times in the future

TABLE 11.1 (*Continued*)
Various DoDAF 1.0 Product Documents

Product	Name	Description
SV-8	Systems Evolution Description	Captures evolution plans that describe how the system or the architecture in which the system is specified
SV-9	Systems Technology Forecast	Defines the underlying current and expected supporting technologies that have been targeted using standard forecasting methods
SV-10a	Systems Rules Model	Describe the rules under which the architecture or its systems behavior under specified constraints
SV-10b	Systems State Transition Description	State transition diagrams describing a system responses to various events by changing its state
SV-10c	System Event-Trace Description	Provides time-ordered examinations of the system data elements exchanged between systems, system functions, or humans in a particular scenario
SV-11	Physical Schema	Defines the structure of the various kinds of system data that are utilized by the systems in the architecture
Technical View (TV)		
TV-1	Technical Standards Profile	Extraction of Standards that applies to the given architecture
TV-2	Technical Standards Forecast	Description of emerging standards that are expected to apply to a given architecture in a specified timeframe

4. DoDAF fails to include measures of effectiveness (MoEs) that can be evaluated at the OV stage. If at all any performance measures are considered, they are at the SV level. System parameters and performance are at a totally different resolution from MoEs.
5. There is no mechanism to perform verification and validation (V&V) at the OV stage.
6. It fails to address M&S as a potent evaluation and acquisition tool.

This study provides a mapping of DoDAF architectures into a computational environment such as DEVS that incorporates dynamical systems theory and an M&S framework. The methodology will support complex information systems specification and evaluation using advanced simulation capabilities. We will see in the forthcoming sections that the proposed mapping will require augmentation of current DoDAF with more information set that is far from any duplication of the available DoDAF products. We will demonstrate how this information is added and harnessed from the available DoDAF products toward development of an extended

DoDAF-integrated architecture that is executable. This kind of augmentation has been attempted earlier (Lee et al., 2005) using CORE® of the Vitech Co. as a tool to develop the executable architecture. Lee et al. have developed architectural templates that elicit information for both the OVs and SVs that contained additional information compared with the usual DoDAF products. In another effort, a new model called Rosen–Parenti model adds another layer of abstraction to the existing DoDAF, augmenting the model with various user-oriented perspectives (Rosen et al., 2011). Rosen et al. further led on to develop the executable architecture with their proposed model and how V&V is applicable in their domain. Their model unearthed the shortcoming of DoDAF that it fails to address the performance issue at the OV level, which their model addresses in one of their perspectives.

The DoDAF is a mission-system reference architecture where the goal is to provide an architecture that can accomplish mission capabilities. It is a domain-specific architecture that is actually an instance of the parent reference architecture with specified applicable domain rules (Shelton et al., 2004). In one of our earlier works (Mittal et al., 2006), we proposed a rule-based meta-model structure that is applicable to such a reference architecture with M&S as a part of the design cycle. Even if the DoDAF architecture is same for two missions, it is the domain-specific rules that will eventually decide if the developed architecture is feasible and there are no rule incompatibilities. Levis and Wagenhals (2000) acknowledge that M&S can provide an integrated solution in evaluation of the designed architectures, but there is no explicit guidance on how to achieve this. Unfortunately, recent attempts to relate M&S to architecture frameworks such as C4ISR (a precursor to DoDAF) and model-driven architecture (Atkinson, 2004; Tolk & Muguira, 2004; Tolk & Solick, 2003) have established the need for including M&S in it, but have not provided a rigorous methodology for doing so. However, there have been efforts, such as NATO Active Layered Theatre Ballistic Missile Defense (ALTBMD) studies, that resulted in alignment of experimentation phases (Adshead et al., 2001). Despite all the information specified in the constructed OV, SV, and TV, Adshead et al.'s efforts required the generation of a new view, coping with which systems and connections were simulated in which systems, and based on which constraints, and so on. The ALTBMD study supports the view that DoDAF in its current state requires the addition of some added/pruned information set that is applicable to the M&S area.

In an attempt to evaluate the completeness of DoDAF document products in providing sufficient information set for an executable architecture, Zinn (2004) integrated the information contained in OV-5 and OV-6a to come up with an intermediate document that fed an agent-based simulation software called SEAS. Agent technology used by Zinn mitigates the solution as the independent agents (a plane, tank, etc.) can now behave independently and can modify behavior based on their decision rules. This essentially leads to an executable-state machine of a component whose behavior is adaptive to the changing environment. However, the problem arises in the absence of interface specifications to port data from these agent architectures into simulation software. In Zinn's thesis, DoDAF contained enough information to build an executable set, but it needed synthesis of intermediate documents. Zinn also acknowledged that the development of an executable architecture must address the following:

1. The 3rd Order Analysis as mentioned in DoDAF is a critical step in acquisition strategy, but there is no methodology to perform this analysis.
2. Legacy models are too monolithic to be disassembled and recomposed to model individual C4ISR effects (Gonzales et al., 2001).
3. CPNs (Wagenhals et al., 2000) provide a means to model and simulate the above-mentioned activity. However, they fall short in modeling an adaptive environment (in both structure and behavior), where the rules of engagement (ROEs) are constantly changing as the model learns and evolves. Another area where the CPNs lack as a technology itself is the inability to specify timing between states. Consequently, temporal effects cannot be considered in any executable architecture that bases its performance evaluation based on CPNs.

11.2 OVERVIEW OF THE ROLE OF DEVS-BASED TECHNOLOGY

In our attempt to augment the current DoDAF, our focus shall remain on adding minimal information that would enable DoDAF to become the executable architecture. There are potential advantages of making DoDAF a DEVS-compliant system.

The Air Force Chief Architect's Office (AFCAO) website lists three key impact areas where use of the architecture can provide real benefits:

1. Operations enhancement
   1.1 Requirement coherence and prioritization
   1.2 Better utilization of fewer personnel
   1.3 Deliberate exploitation of innovation
2. Programming and planning
   2.1 IT investment decisions (support for POM inputs)
   2.2 MIL-worth analysis (M&S executable architectures)
   2.3 AOA evaluation (trade study)
3. Acquisition support
   3.1 Enhanced warfighter/user capabilities ID
   3.2 Execution roadmaps
   3.3 Source selection
   3.4 Technology application/transition
   3.5 Test support (MOE/MOP)
   3.6 Interoperability and integration assurance

Even though it has been realized that M&S is necessary in performing evaluation and developing acquisition strategy, there is more opportunity for current simulation technology to help. Table 11.2 summarizes limitations of current M&S methodologies to support DoDAF and shows where the DEVS-based technology can contribute.*

* Although not all of the areas taken from the list above are considered in Table 11.1, the ones omitted are at a higher level of abstraction and will benefit indirectly from the application of M&S in general.

TABLE 11.2
Comparison of Current Technologies in Development with DEVS on Addressing M&S Issues

AFCAO Reference	Desired M&S Capability	Current Working Tools (Agent Based or CPNs)	Solutions Provided by DEVS Technology
1.1	Requirement Coherence and Prioritization	No formal methodology exists in defining architectures wherein the data model can be put directly to use for simulation modeling.	The present work aims to accomplish this, by injecting requirements quite early in the design stage of DoDAF architectures, specifically in OV phase.
2.2	MIL-Worth Analysis (M&S Executable Architectures)	Work is ongoing in this area. Due to the limitations of the technologies being used, the desired execution is not possible.	DEVS provide the capability to: 1. Reconfigure simulation on-the-fly. 2. Control simulation on-the-fly. 3. Provide dynamic variable-structure component modeling. 4. Separate model from the act of simulation itself, which can be executed on single or multiple platforms using DEVS/HLA. 5. Simulation architecture is layered to accomplish the technology migration or run different technological scenarios. 6. With DEVS Unified Process, automated test generation is integral to this methodology.
3.1	Enhanced Warfighter/User Capabilities	1. Deterministic CPNs. 2. Stochastic SEAS but too rigid to reconfigure on the fly. 3. Agent-based methodology again falls short in variable structure simulation model.	
3.2	Execution Roadmaps	No capabilities to control the ongoing simulation to steer it in the right direction	
3.3	Source Selection		
3.4	Technology Application/ Transition	No dynamic reconfiguration of model and simulation reported. The simulation architecture itself has to be layered enough to accomplish technology transition.	
3.5	Test Support	Agent-based technology, Zinn's work is essentially in this direction. CPNs are not capable of automated test generation.	
3.6	Interoperability and Integration Assurance	Limitations of the methodology itself. No mechanisms reported so far.	

Another way to look at it is through the pyramid in Figure 11.2, which provides the overview of this chapter—that is, incorporating DEVS M&S as an integral part of design and evaluation cycle based on requirement specifications at the top of the pyramid. The execution roadmap is as follows:

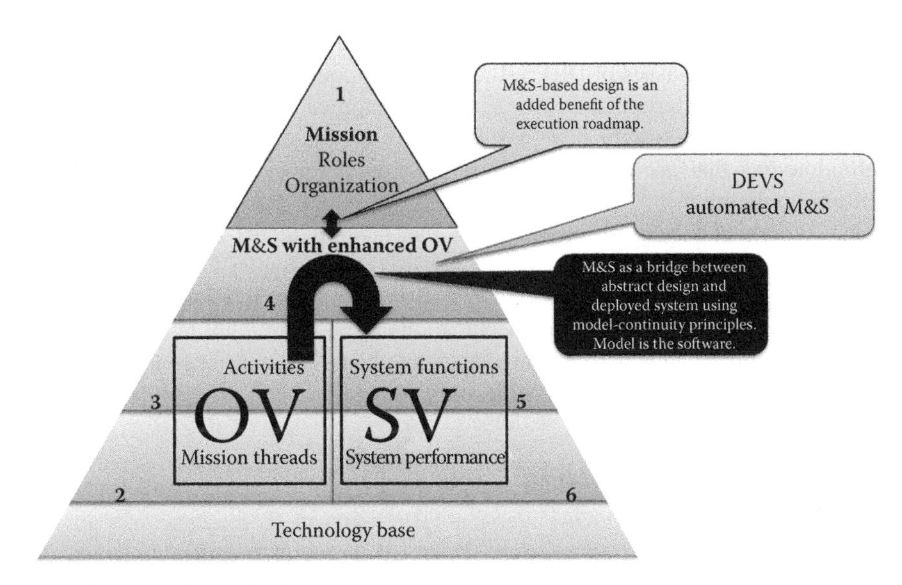

FIGURE 11.2 DoDAF/DEVS execution roadmap.

1. Define mission capabilities
2. Identify mission threads
3. Decompose into activities and information exchange needs
4. Perform M&S-based design evaluation
   a. Identification of scenarios (experimental frames)
   b. Identification of interfaces for key interface profile (KIP)
   c. Simulation based on key performance parameters (KPP)
5. Identify system using model continuity
6. Evaluate performance and do calibration based on simulation results of KPP in step 4.

We seek to employ the DoDAF-to-DEVS mapping to unify multiple model representations by expressing their high-level features within DoDAF and their detailed features as sub-classes of DEVS specifications. DEVS has been shown to be a universal embedding formalism, in the sense of being able to express any sub-class of discrete event systems, such as Petri nets, cellular automata, and generalized Markov chains (Zeigler et al., 2000). DEVS has also been employed to express a wide variety of more restricted formalisms, such as state machines, workflow systems, fuzzy logics, and others (Sarjoughian & Cellier, 2001). Moreover, DEVS environments have a long history of development and are now seeing ever-increasing use in the simulation-based design of commercial and military systems (Zeigler, 2003). Providing a DoDAF front-end to a back-end DEVS environment will appeal to military information system designers facing the DoDAF and NCES mandates. Such designers will be able to retain their skills with representations familiar to them, while complying with DoDAF abstractions. At the same time, they can see the results of their specifications evaluated via simulation-based execution of the model architecture. Moreover, since all mappings are into sub-classes of DEVS, the resulting models can be coupled together and therefore can interoperate at the systems dynamics level. Thus, this approach to the synthesis of system design formalisms leverages design and execution methodologies that are already used, or mandated for use, in commercial and military applications.

As a result of recent advances, DEVS can support model continuity through a simulation-based development and testing life cycle (Zeigler et al., 2005). This means that the mapping of high-level DoDAF specifications into lower-level DEVS formalizations would enable such specifications to be thoroughly tested in virtual simulation environments before being easily and consistently transitions to operate in a real environment for further testing and fielding.

11.3 FILLING GAPS IN DoDAF 1.0

We address four gaps in DoDAF Version 1.0 relating to (1) message flow among activities, (2) transition from OV-5 to OV-6, (3) temporal specifications, and (4) accountability for failure of activity execution.

11.3.1 Message Flow among Activities

AV-1 deals with the overall functionality of the system. AV-2 deals with the data dictionary and terms used in the DoDAF specification of the system. This leads to the first functional document OV-1, which gives way to OV-5 that describes the intended functionalities in fair amount of detail. It lays out the functionalities in terms of capabilities and activities in a hierarchical and a sequential manner. A capability is defined as a functionality comprising of many activities. These activities are linked by messages that are apparently flowing between the defined activities. What flows between these activities is not exactly defined in DoDAF. The abstraction level of information flow is not discussed in DoDAF. They can be top-level operational information exchanges (OIEs) or specific data messages. The first gap occurs at this point. DoDAF does not define the interactions between the activities. This is all the more ambiguous since activities are not entities that can be physically realized. However, these activities do exchange messages. There is no entity structure developed by this stage of functional development. An entity is defined as a component that can be physically realized. There is no mention of any entity taking responsibility for any activity. Were an activity to be mapped to the corresponding UML notation, it would map to use-case diagrams. However, unlike the use-case diagram, OV-5 does not take into account the entities (actors in UML) that take responsibility for that activity. Since there is no apparent entity structure, the interfaces are essentially absent. Structure in DoDAF does not appear until OV-2 and OV-7. The activities in OV-5 seem to exchange messages or events for that matter, but there is no mechanism specified to send or receive events. In other words, there is no specification of the activity interface. If activities are considered as potential components of the OVs, there is ample reason to consider them as components of a particular capability and provide interfaces to enable and define the message communication between activities. The advantages of considering an activity as a component are manifold:

1. Activity is grounded in the design at the DoDAF specification level.
2. It can be back-referenced to any particular capability or it may serve to more than one capability.
3. Activities with defined interfaces provide a way to develop a test suite to test the capability definition of the system.
4. Activities can be allotted to the defined entities such that a real-world entity or a group of entities can be held responsible for its execution.
5. It brings specificity to the component design by ensuring the interfaces defined in activities be mapped on one-to-one basis in the target component entity held responsible for this activity.
6. It provides a structure to the functional aspect of DoDAF that can feed the entity structure of the system, which can be then aggregated toward the SVs.
7. It paves way for DoDAF-DEVS mapping and how testing can be applied to the design process (as described in the previous section) at an operational stage.
8. It allows the designer and planner to define the needed entities in the OV phase of DoDAF specification of the system.
9. It provides the framework to incorporate modeling and simulation at two different levels of resolution to conduct feasibility studies:
   a. *At the capability level*
      At this level, the system can be modeled based on the functionality of the system. Rules and doctrines can be accounted for at this level of model. This provides a means to test the compatibility of existing rules and doctrines when testing the feasibility of any activity. There is no means to test and validate the compatibility and interoperability of various rules and doctrines that constrain any particular capability in DoDAF. The need for such consideration has been recognized by Dickerson and Soules (2002) in their proposed CV-6 document (Capability Evolution Document).
   b. *At the entity level*
      At this level, the system can be modeled based on the entity structure as developed from the capability model. New supporting entities that can support the existing capabilities or that are needed by specific capabilities such as fault-tolerance, scalability, and the like, can be introduced at this level.

Coming back to the discussion of current OV-5 in DoDAF, the activity diagrams are then detailed further in OV-6. The OV-6 consists of three parts. Our prime interest is in OV-6b (statechart diagrams) and OV-6c (timing sequencing and event-trace diagrams) as OV-6a deals with the rules and doctrines, basically a document to describe the constraints on different activities mentioned in OV-5. OV-6b can be mapped to the UML statechart diagrams, and OV-6c can be mapped to the UML timing sequence diagrams.

11.3.2 Transition from OV-5 to OV-6

The second gap comes during the transition from OV-5 to OV-6, specifically OV-6c event-trace diagrams. The OV process adds further details to activity diagrams in describing the sub-activities for OV-5 activities. It associates the current activities with the known operational nodes (which can be seen in OV-1). However, the nodes in consideration here may be as big as an organization itself and abstraction level is fairly high. Even if the node is the lowest level entity (an unlikely case), the complete behavioral life cycle of this entity is overlooked and only the activities in consideration are assigned to the operational node with a presumption that the node will execute this activity. The life cycle with respect to the attended activities is expressed in OV-6b. However, there is no hierarchy of these nodes present that could account for the hierarchical activities under question. Consider a typical activity diagram in OV-5 and imagine that in order to execute this activity, four different nodes are performing in tandem through a sequence of sub-activities and passing events to one another. The current setup is depicted in the OV-6c timing-sequence diagram describing the execution of an OV-5 activity. Similarly, the OV-6b diagram depicts the sequencing of these sub-activities. The problem of developing the life cycle of an operational node is easier if all the sub-activities are happening at one node, but that is usually not the case since multiple nodes are performing and synchronizing to execute the parent activity. The fragmented nature of activities to compose and define a parent activity or a capability makes the construction of OV-6b diagram for an operational node more difficult. The other drawback of this methodology is the possible occurrence of inconsistency between OV-6b and OV-6c, as the statechart in OV-6b is bounded by the activities called for in OV-5 and explored in a fragmented manner in OV-6c in specific event-trace diagrams. There is a possible loss of information here and DoDAF provides no means to ensure consistency. Although the development of OV-5 and OV-6 is an iterative process, it does not ensure foolproof life cycle of an operational node. This problem does not occur in the UML architecture, as the approach there is to start with use-case diagrams, and then move to the class diagrams that lead on to the activity diagrams, timing sequencing diagrams, and statechart diagrams of individual classes, in this specific order. This ensures consistency, as there are defined classes before making the timing sequencing diagrams or statecharts. DoDAF has a reverse approach to this design problem, where it groups and aggregates activities and defines classes (in OV-6) and then leads on to the OV-2; OV-7 and the data exchanges are simultaneously defined in OV-3. Developing statechart diagrams without developing the class entity structure is error-prone. The solution to this problem is to treat activity as a component and develop the activity–entity structure before OV-6b and after OV-5.

(It can be argued that the OVs are concerned with the functional description of the system that does not require any component structure definition. Furthermore, it is in the SVs that the structure is made available and interfaces are defined. The counter argument to this approach is that there is no mechanism provided in DoDAF to test the OVs development and conduct early feasibility studies.)

In order to fill the second gap and maintaining the order of OV-5 and OV-6 iterative development process, we suggest the following:

1. Considering the activities as components and placing them in a structure so that they are better defined and managed
2. Employing a methodology to shift from OV-6b and OV-6c to the DEVS behavioral model for operational nodes and to incorporate the activity components as constituent parts with defined interfaces
3. Incorporating the doctrines and ROEs specified in OV-6a to be implemented into the DEVS behavioral model of an activity component (see example in Section 10)

Section 11.3.3 describes a methodology to develop the DEVS state systems from OV-6c descriptions that are essentially time-sequencing diagrams. This, incorporated with the OV-6b statechart, will provide much-needed consistency between the two documents. Ultimately, it will result in a DEVS Model Repository (DMR) of Operational Nodes for modeling, simulation, testing, and control. The integrated solution to the above two gaps results in the introduction of two new OV documents:

1. OV-8: Activity Components Document
   This document lists the activities as components with defined port interfaces and significant parameters set for performance evaluation of the activity.
2. OV-9: Activity Interface Specifications
   This document describes the interface specifications between activities and entities. It holds information about the mapping of the sub-component inside an operational node that is responsible to execute any particular activity.

Details of development of these two documents can be seen in the next section. Figure 11.3 presents the OV setup in DoDAF and its extended version.

11.3.3 Temporal Information

One crucial piece missing in this existing tool set is the timing information that is absent in OV-5. Our proposal attempts to bridge this gap by giving adequate consideration to the interface structure and involving other views, namely, OV-6b and OV-6c, in conjunction with OV-5 and OV-6a. The communication information is not available until SV-2 and SV-6 are constructed, which is derived from OV-3 and is constructed pretty late in the OV phase itself. Introducing timing well before in the

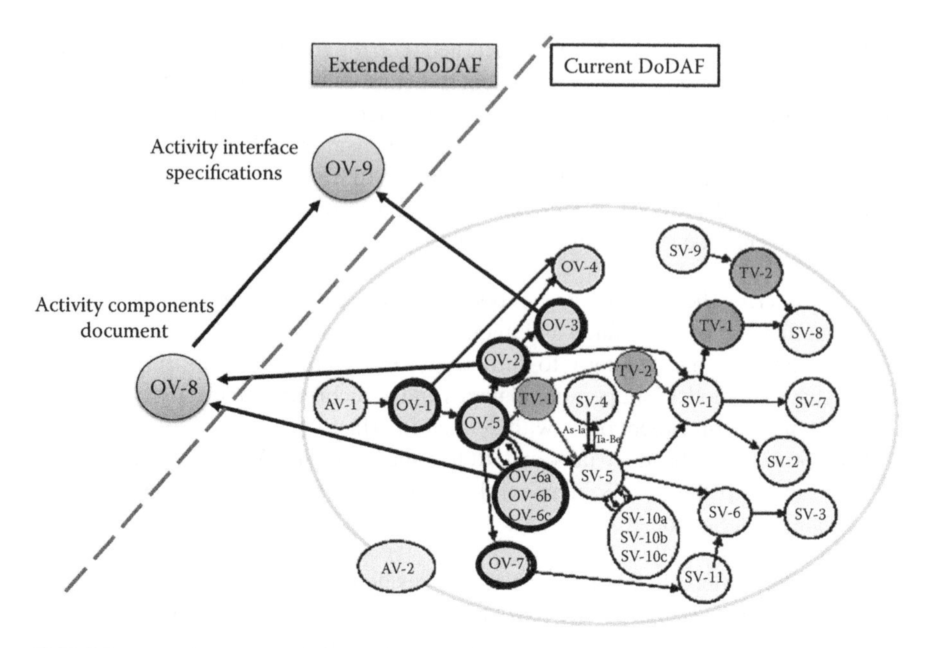

FIGURE 11.3 State of operational view documents in DoDAF.

design of OVs equips the acquisition strategy with information about those current systems that could meet these operational delay requirements.

A similar architecture called the Air Operations Center (AOC) Weapon System Block 10.1 architecture along the lines of DoDAF specifications being developed by MITRE for the Air Force. It is considered an integrated architecture based on the DoDAF 1.0 standard. It contains the required essential products along with data elements that are consistent across the views. Zinn (2004) developed a procedure wherein information from different AOC DoDAF views were handled manually and brought to a level where the doctrines can be utilized to impact the simulation model in question. This information is presented in if-else constructs in the form of a pseudo code that can be fed into any XML parser for further processing. More details about the procedure can be seen in his thesis.

Our proposal for executable DoDAF 1.0 can be seen as extending Zinn's procedures by giving more structure to the compiled information that includes temporal information in activities as an important part of the information exchange. Zinn used the IDEF process methodology to depict various OV and noted the inherent inadequacy of the IDEF3 process (Integrated Definition Methods: IDEF3 Process Description Capture Method, 2012), when being fed to agent-based simulation software, for example, SEAS, which has no rules defined in case of OR split that dictate which path to take to resume the activity. We believe that the problem stems from the fact that the underlying architecture does not consider timing an important concept. This problem would not have arisen in the first place and the situation would have resulted in time-out (as can be easily expressed in DEVS) in case of OR split. This is but one example of limitations of the modeling methodology that does not lead to simulatable models.

11.3.4 Accountability for Failure of Activity Execution

The final gap that we found was the lack of any accountability for failure of activity execution by operational nodes over time. There is an absence of accountability because there are no means provided in DoDAF to test the design principles in OVs. Since M&S is not systematically considered in the current DoDAF specification, there is no means to test the feasibility of the system. Furthermore, there is no support for modeling technologies like model-driven architectures and model-continuity principles.

Our effort is toward providing accountability to the design process by introducing M&S at the correct development stage, where it is possible to experiment and modify the operational architecture in question. This is very much needed, as it is not reaffirming to presume the capabilities of operational nodes (through COTS specifications) and then move on to the SVs without ensuring the functionality of the system in OVs. The reasons for choosing the DEVS formalism as a means to M&S are its expressive power and modularity support. The concept of elapsed time is one of the key aspects of the DEVS formalism, and it provides a means to evaluate the component behavior in a finite period. It enables the component to attend and respond to any external events in time intervals prescribed in its DEVS specifications. Incorporation of DEVS in DoDAF will make the design process more tractable and controllable.

11.4 FROM OV-6 UML DIAGRAMS TO DEVS COMPONENT BEHAVIOR SPECIFICATIONS

Figure 11.4 describes the development of DEVS description model from a simple time-sequencing thread in a time-sequencing diagram. It must be indicated here that OV diagrams are essentially drawn using UML, so we are thereby developing a methodology to transform UML diagrams to DEVS specifications. Developing DEVS models from UML specification has been undertaken separately in one of our case studies as well.

A simple timing-sequence diagram is considered to illustrate the DEVS activity component development process and how it fits into the DEVS description of an operational node. Consider that a hierarchical activity is being addressed by three operational nodes and they are exchanging events between sub-activities in order to perform this activity. In the first diagram in Figure 11.4 (leftmost), we can see them interacting with each other. The center part of the figure consists of the thread for one operational node and is enlarged for better analysis. The sequencing diagram is represented in the UML notation and this node has a lifeline during the course of which it receives events and sends output messages or events to other nodes.

In mapping to the DEVS formalism, we need to have information about the internal transitions (when no events are received) from one activity to another activity and the external transitions (when an event is received at this node sent by other node). The timeline of the node consists of sequence of activities that the node will undergo in the event of external transition or internal transition. The complete timeline is available in OV-6b, so there is more reason to maintain consistency and similar input and output trajectories of sequential activities. Different markings on the thread

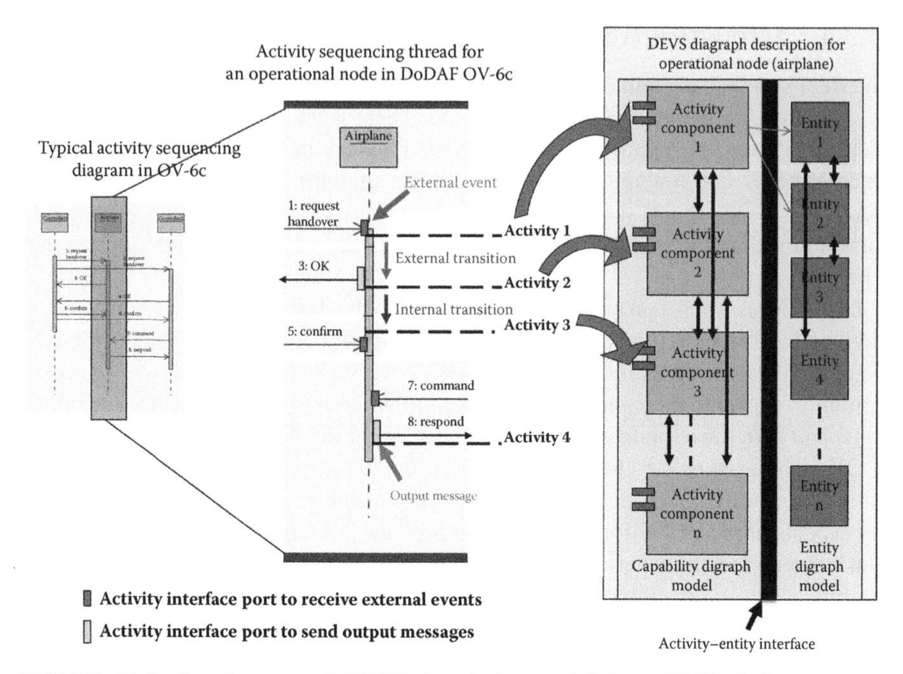

FIGURE 11.4 Development of DEVS description model from UML timing-sequence thread.

are self-explanatory. Activity 1 receives an external event and undergoes Activity 2 after generating an output message. Activity 2 undergoes internal transition toward Activity 3 in the absence of any external event. This particular thread displays only a subset of activities performed by this node. Since DEVS employs the port-based component structure system, we identify the input and output ports and assign them to specific activity components at this particular developmental stage. This results in the introduction of a new OV document (OV-8) that contains the mapping of ports and activity components. Finally, these activities, if not present in OV-6b, are then introduced in OV-6b for a comprehensive set of activities performed by this operational node. Another by-product of this stage is the mapping of activity components with entity components that constitute an operational node. This is specified in a new OV document called OV-9. This contains information about the activity ports, activity components, entity components, and entity ports. Introduction of these new OV documents modifies the overall DoDAF OV specification structure, which is illustrated in Figure 11.3.

11.4.1 DoDAF-to-DEVS Elements

The power of UML cannot be ignored. UML has matured to great levels and has become an integral part of any model-based system design. Even the most complex and encompassing DoDAF lends itself to UML in description of its various artifacts. However, the bridge to develop an executable code from imprecise UML constructs is under research. Executable UML with the aid of the action semantic approach is one effort that brings to light this important gap of "model to code" directly. DEVS, with its advanced model-continuity process, provides this capability readily. Further, the process of developing executable architectures from object-oriented designs is in place, but it has not being explored rigorously for testing software architectures.

As capable as the DEVS M&S framework is, it is still not in the mainstream industrial software system design and planning. The ideal progress path now is the development of a mechanism to employ DEVS M&S with a UML-based developmental methodology. Table 11.3 depicts a mapping of UML with the DEVS elements that provides just the same. Representation of DoDAF into corresponding UML has also been presented by Telelogic (Kobryn & Sibbald, 2004). Since UML is essentially an object-oriented methodology, work has been performed in the area of transforming the UML model to CPN executable architectures (Wagenhals et al., 2002).

We have introduced two new OVs, OV-8 and OV-9. These add features to enable M&S of the system under design. We have also demonstrated how these new documents will be created from the existing OVs. We aim to provide a structure to the OV process by shifting the perspective from describing functionality as an activity to an activity component with definite interfaces to other activity components as well as identified entities within an operational node. To what extent an operational node is decomposable is a subject requiring further research and capability requirements. We have developed a testing process for defined capabilities (which were defined during the conceptual design process in OV-5) and ways in which various rules and doctrines (in OV-6a) can be evaluated for interoperability with different capabilities. By purview of the information contained in OV-9, we have introduced the model repository as an important aspect of a DoDAF system specification that enhances DoDAF by making way for the M&S activity. Figure 11.5 shows the SES snapshot of the enhanced DoDAF with a focus toward the OVs. Table 11.3 provides the mapping of various DoDAF OV products into DEVS modeling constructs. UML is chosen as the preferred way of DoDAF representation. First, the UML element is mapped with the DoDAF product document and then the same UML element is mapped to the DEVS elements. Their representation included SV products as well. In Table 11.3, we have also incorporated the two new OV products: OV-8 and OV-9. UML elements are transformed to DEVS elements. The last column links the DEVS elements by categorizing them into model repository and semi-automated test-suite elements.

In the next section, we will illustrate by an example taken from Zinn's thesis how OV-8 and OV-9 are created from activity descriptions.

11.5 DoDAF-BASED ACTIVITY SCENARIO

11.5.1 Example: Implementation of an Activity Component

Consider an activity as mentioned in Zinn's work described in the IDEF0 format (Figure 11.6). This activity is governed by the doctrines specified in OV-6a, IDEF3 format. Figure 11.6 is a sample OV-5 diagram for a select contractor, and Figure 11.7

TABLE 11.3
Mapping of DoDAF with UML and DEVS M&S Elements

DoDAF Elements				
	Name	**Description**	**UML Elements**	**DEVS Elements (Generated Using XML)**
Operational View	OV-1	Top-level operational view	• Use-case diagrams	• Activity component identification • Top-level entity structure
	OV-5	Operational activity model	• Use-case • Activity-sequencing diagrams • Data-flow diagrams	• Activity component updating • Hierarchical organization of activities • Input–output pairs • Port identification
Operational View	OV-6	Operational Timing and Sequencing Diagrams	• Timing-sequencing diagrams • State-machine diagrams	• DEVS atomic model creation (initialize function, internal and external, transition functions, time advance and output functions) for activity components • Entity identification • Activity–entity component mapping
	OV-2	Operational Node Connectivity	• Composite structure diagrams	• Coupling information • Hierarchical component organization
	OV-8	Activity Component Description	• Composite structure diagrams • Statecharts • Significant activity parameters	• Activity component update • Activity port identification and refinement • Experimental frame design
	OV-3	Operational information matrix		• Input–output transaction pairs • Message formats • Activity interface and coupling information
	OV-9	Activity Interface Specifications	• Statecharts • Composite structure diagrams	• Activity–entity interface • Entity structure refinement • Activity–entity port mapping and refinement
	OV-7	Logical Data Model	• Packages (only for xUML) • Class diagrams	• Entity identification • Hierarchical structure

	OV-4	Organizational Relationship Chart	• Class diagrams	• Entity identification • Hierarchical entity structure
System View	SV-4	System Functional Description	• Use-case description • Activity sequencing diagrams	• Hierarchical functional components organization
	SV-5	System Functional Traceability Matrix (Based on OV-5)		• Coupling info refinement
	SV-10	System State Description and Event Trace (Based on OV-6)	• Sequence diagrams • Statecharts	• DEVS atomic model transition functions refinement
	SV-6	System Data-Exchange Matrix		• Input–output pair refinement
	SV-1	System Interface Description (Based on OV-2)	• Composite structure diagram	• Port assignment refinement • Entity refinement
	SV-2	System Communication Description	• Deployment diagrams	• Coupling info refinement (hierarchical management)
	SV-7	System Performance Parameters Matrix		• Experimental frame
	SV-3	System–System Matrix		• Hierarchical model organization • Entity refinement
	SV-11	Physical Schema	• Class diagrams	• Hierarchical model organization
Technical View	TV-1	Current Standards	• Timing response	• Basic DEVS model for COTS component
	TV-2	Future Standards		• Improved DEVS model for desired functionality

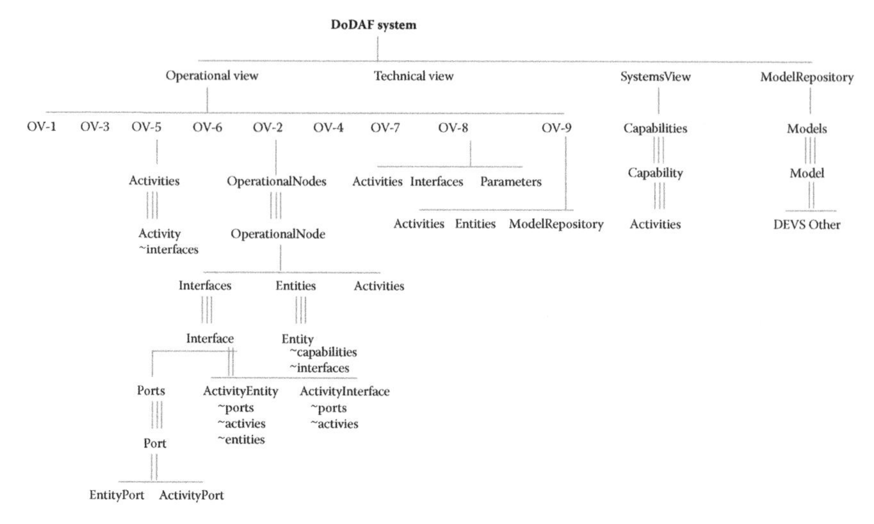

FIGURE 11.5 SES for extended DoDAF OV.

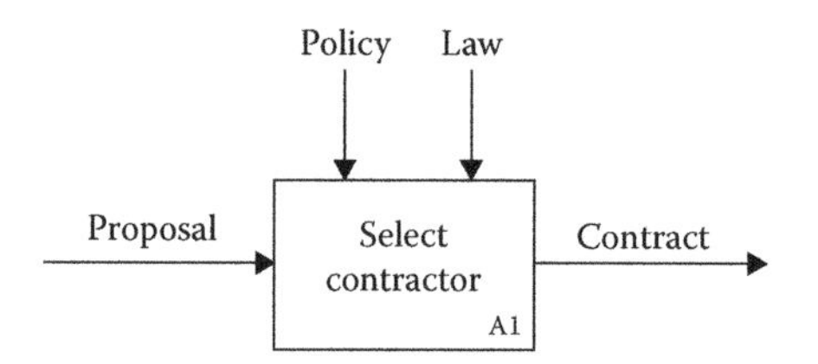

FIGURE 11.6 OV-5 diagram for select contractor in the IDEF0 notation.

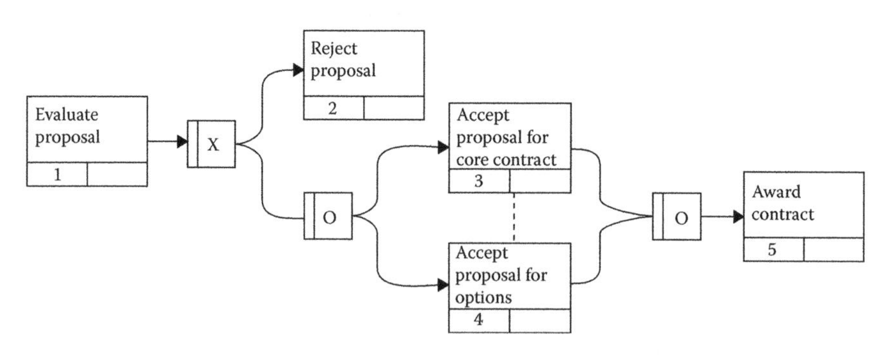

FIGURE 11.7 OV-6a diagram for a select contractor in the IDEF3 notation.

is the OV-6a description in the IDEF3 format, where X represents an XOR split and O represents an OR split. These are the critical decision-making points that impact the outcome of the activity based on the previous step. It is at this point, timing needs to be specified so that timeouts can occur without leading to any ambiguity. Zinn acknowledged this problem in the process.

Activity 1: Select Contractor

Description: The process used by the company to select the contractor for a new project

Inputs:	Proposal: contains the cost, schedule, and technical information as proposed by the contractor	
Outputs:	Contract	–the awarded contract
Controls:		
	Policy: Company contracting policy	
	Law: Federal, State and Local regulations	

Pseudocode for Activity 1

Evaluate Proposal

IF (cost > budget) THEN

Reject Proposal

ELSE

(Accept Proposal for Core Contract) OR

(Accept Proposal for Options) OR

((Accept Proposal for Core Contract) AND (Accept Proposal for Options))

FIGURE 11.8 Pseudo code as per Zinn's interpretation and the integration procedure.

The information from Figures 11.6 and 11.7 is compiled manually to generate the pseudo code in the format shown in Figure 11.8. This manual process amounts to the integration of OV-5 and OV-6a into a single document.

The graphical representation in Figure 11.6 is represented textually through the Popkin System Architect, as shown in Figure 11.8. Consequently, Figures 11.7 and 11.8 give us the comprehensive information about the activity, its purpose, its input–output information through ICOM* lines, and the pseudo code for operational rules (as defined in OV-6a). Figures 11.6 through 11.8 describe a general step approach to arrive at this pseudo code, which is then utilized by agent-based modeling software (e.g., SEAS) via Tactical Programming Language (TPL). Once the pseudo code has been made available, any software developer who is versed with TPL or any other language can interpret it. This process is then followed for a sample case study described in Section 11.5.2. Zinn brought forward the information expressed in the graphical format in OV-5 diagrams and OV-6a doctrines in the form of pseudo codes that are realizable into the software code. We utilize his efforts and demonstrate how this information can be used to feed the integrated DEVS methodology and development of OV-8 and OV-9.

11.5.2 Activity Taken from Zinn as an Example

Having described the modeling formalism used by Zinn, we move to a bit more complicated example composed of multiple activities. For illustration purpose, let us take one of the activities from a complex example described by Zinn, for ex. Activity 6: TCT-Determine Target Significance/Urgency. There are about 11 sub-activities that are being evaluated in Activity 6. Figure 11.9 provides the activity model report as generated by the Popkin System Architect.

This activity report is nothing but the interface descriptions for an activity in the OV-5 diagram. It tells us which other activities Activity 6 receives input from, and

* In IDEF0 diagrams, Inputs, Controls, Outputs, and Mechanisms are collectively referred to as ICOM arrows.

```
6 Operational Activity: TCT-Determine target significance/urgency
    (Track)
[Within OV-5 Diagram 'TCT-Level 1']
Glossary Text: Utilizing track data and other target information, C2 Warriors
    determine if the target/target set is threatening and/or fleeting, and esti-
    mate target availability, i.e., how long the target will remain susceptible
    to attack.
From 2005 C2 Constellation 3.2.5.2 and CAOC-4.5.2.7
ICOMline: Air Track (J3.2)
Output: going to TCT-Validate target/targetset(Target) as input
Glossary Text:
ICOMline: Current Intelligence - Dynamic Assessment/Target Status
Input: coming from <offpage>
Glossary Text:
ICOMline: Current intelligence - Target Classification
Input: coming from TCT-Define target/targetset (Fix) as output
Glossary Text:
ICOMline: Current intelligence - Target identification
Input: coming from <offpage>
Glossary Text:
ICOMline: line: Doctrine, Policy, LOAC, SROE, ROE
Control: coming from <offpage>
Glossary Text:
ICOMline: line: Dynamic Target Nomination
Output: going to <offpage>
Glossary Text:
ICOMline: line: Dynamic Targeting Execution Direction and Guideance
Control: coming from <offpage>
Glossary Text:
ICOMline: JMSNSTAT
Input: coming from <offpage>
Glossary Text:
ICOMline: Land (Ground) Point/Track (J3.5)
Output: going to TCT-Validate target/targetset (Target) as input
Glossary Text:
ICOMline: Reattack Recommendation
Output: going to TCT-Nominate engagement option (Target) as input
Glossary Text:
ICOMline: TRKREP
Output: going to TCT-Validate target/targetset (Target) as input
Glossary Text:
ICOMline: Track Management (J7.0)
Input: coming from <offpage>
```

FIGURE 11.9 Activity report model for Activity 6 generated through Popkin System Architect.

which activities Activity 6 sends outputs to. It also provides us the information about the control interfaces that are needed to execute the doctrines and rules. Figure 11.10 depicts the IDEF3 model that implements the OV-6a doctrines and rules for Activity 6.

The pseudo code for Activity 6 is provided in Figure 11.11, which is compiled manually from the information contained in OV-6a. For a complete description of Activity 6, refer to Zinn's work. Briefly, the context of Activity 6 in the TCT architecture is immediately after a target (or target set) is found and fixed. The upper half of Figure 11.11 shows an XOR junction that indicates only one path be taken. The resulting target update is then put through four simultaneous analyses indicated by the AND junction.

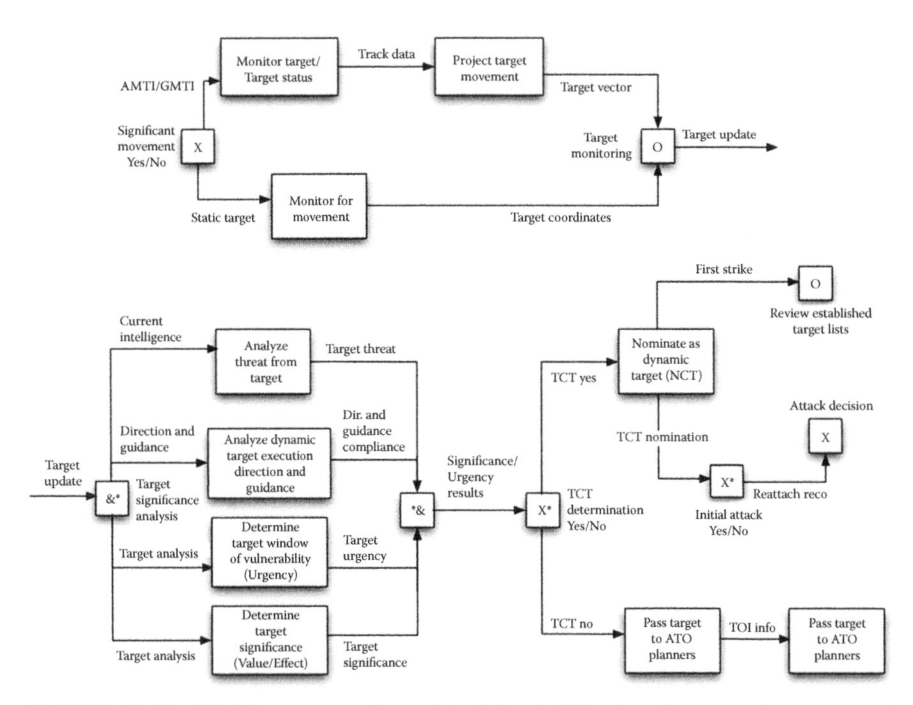

FIGURE 11.10 IDEF3 representation of Activity 6 ("Conduct Dynamic Assessment of Target" TCT 2005 Architecture, 2003: OV-6a).

IF Significant Movement of target
 Then Monitor Target/Target Status
 Project Target Movement
 Target Vector = ... ?
Else Monitor for Movement

Analyze Threat from Target (is the target closing Friendlies or Fleeing?)
Analyze Dynamic Targeting Ex Direction and Guidance (does this agree with the commander's requirements?)
Determine target window of vulnerability (urgency)
Determine target significance — partly based on above findings

IF it is determined to be a TCT based on the above info
 Then IF this is the first strike attempt on this target
  Then Goto Activity 7 (Validate Target/Target set)
  Else Goto Activity 8 (Nominate engagement option)
Else Pass target to ATO Planners

Monitor Target of Interest for Status Change

FIGURE 11.11 Pseudo code for Activity 6—based on IDEF3 diagram in Figure 11.10.

This results (after integrated processing) into "Is the target time critical?" If it passes this TCT test, it is again presented with a decision point, "Is the initial attack on the target?" The answer to this question results in two different modes of action, indicated by the XOR junction. Zinn acknowledges the fact that even though there is certain sequencing present, precise information about the rules defined are left to the imagination.

Section 11.5.1 demonstrated how the information in Figures 11.9 and 11.11 is transformed into a DEVS component-modeling framework. It also shows how OV-8 and OV-9 are populated. However, it must be realized that an operational node has not been defined with respect to the current example. Consequently, we will assume an entity structure that will illustrate the concept.

11.5.2.1 DEVS Interpretation of Activity 6

Based on the available information, let us assume that dynamic target assessment happens at a particular node. Assume that Activity 6 and its sub-activities are all happening at TCT. Let us call this Operational Node 1, with an ID O1. This will comprise our OV-2 diagram containing only one operational node executing all the 11 activities (Zinn, 2004). Again, a simple example with only one operational node has been considered to demonstrate the construction of new OV documents, namely, OV-8 and OV-9. Table 11.4 assigns identification numbers to various activities.

TABLE 11.4
Activity-ID Mapping for OV-8 and OV-9

S. No.	Activity	Sub-Activity	Internal Activity	ID
1.	Activity 6	Dynamic Target Assessment		A6
2.		Monitor Target/ Target Status		A6.1
3.		Monitor for Movement		A6.2
4.		Project Target Movement		A6.3
5.		Analyze Threat from Target		A6.4
6.		Analyze Dynamic Target Execution/Direction and Guidance		A6.5
7.		Determine Target Window of Vulnerability (Urgency)		A6.6
8.		Determine Target Significance (Value/Effect)		A6.7
9.		Nominate as Dynamic Target (NCT)		A6.8
10.		Pass Target to ATO Parameters		A6.9
11.		Pass Target to ATO Planners		A6.10
12.			Significant Movement Yes/No	A6.11
13.			Target Monitoring	A6.12
14.			Target Significance Analysis	A6.13
15.			Synthesize Results	A6.14
16.			TCT Determination Yes/No	A6.15
17.			Initial Attack Yes/No	A6.16
18.			Review Established Target Lists	A6.17
19.			Attack Decision	A6.18

Based on the IDEF3 diagram (graphical information for OV-6) in Figure 11.10 and our constructed OV-2 in the previous paragraph, we can construct our OV-8 document that lists activities and their logical interface information. We need such port information to be able to create components. Such a logical-port construction has been attempted by Telelogic where the focus was to create an SV executable model. Developing and specifying activity port interfaces at this level is a logical step toward an SV interface design as tractability is ensured. The OV-8 document below does not address the performance issue at the OV level in this current example. Inclusion of parameters for performance tracing and evaluation is shown by Mittal et al. (2006). A sample OV-8 document is shown in Table 11.5.

Based on the information provided in Figure 11.9, we have constructed and identified the interfaces that are being used by different activities to communicate. However, we have not considered the information contained in Figure 11.10 that describes how Activity 6 communicates with other activities. We did not explore the connectivity between other destination activities just to keep the example in the needed perspective. However, the procedure is essentially the same with more rows being added to the Table 11.5. To give a glimpse on how these interconnected activities (as components) will perform in tandem, notice the inports and outports of Activity 6 in Figure 11.12. The other activities shown in Figure 11.12 do not have any resemblance to the actual example in Zinn's work. They are just meant for illustration. To understand how Activity 6 works internally based on the different activities in Table 11.5, please look at Figure 11.10.

The coupling relations shown in Figure 11.13 are generated in an automated manner from the data presented in Table 11.5. Columns 2, 3, 4, and 5 provide sufficient information to generate the following lines of code (Listing 11.1) with simple string manipulations. Consequently, automated generation of the DEVS model is realizable. Hence, the OV-8 document provides sufficient information to

LISTING 11.1 CODE-GEN FROM TABLE 11.5

```
ViewableAtomic a61 = new ViewableAtomic("A6.1");
add(a61);
ViewableAtomic a62 = new ViewableAtomic("A6.2");
add(a62);
.....
ViewableAtomic a611 = new ViewableAtomic("A6.11");
add(a611);
....

a611.addOutport("outSigMovY");
a61.addInport("inSigMovY");
addCoupling(a611,"outSigMovY",a61,"inSigMovY");
a611.addOutport("outSigMovN");
a62.addInport("inSigMovN");
addCoupling(a611,"outSigMovN",a62, "inSigMovN");
.....
```

TABLE 11.5
Sample OV-8 Document

S. No.	Activity ID Component	Connection ID	Source Activity	Input Interface Name (Logical Port)	Message Description/OIEs	Container Op Node	Source Document/ Diagram
1.	A6					O1	
2.	A6.1	CA6.1	A6.11	inSigMovY	AMT/GMTI	O1	Figure 11.12/OV-6b,c
3.	A6.2	CA6.2	A6.11	inSigMovN	StaticTarget	O1	Figure 11.12/OV-6b,c
4.	A6.3	CA6.3	A6.1	inTrkData	TrackData	O1	Figure 11.12/OV-6b,c
5.	A6.4	CA6.4	A6.13	inCurrInte	Current Intelligence	O1	Figure 11.12/OV-6b,c
6.	A6.5	CA6.5	A6.13	inDirGuid	Direction and Guidance	O1	Figure 11.12/OV-6b,c
7.	A6.6	CA6.6	A6.13	inTarAnaly	Target Analysis	O1	Figure 11.12/OV-6b,c
8.	A6.7	CA6.7	A6.13	inTarAnaly	Target Analysis	O1	Figure 11.12/OV-6b,c
9.	A6.8	CA6.8	A6.14	inTctYes	TCT Yes	O1	Figure 11.12/OV-6b,c
10.	A6.9	CA6.9	A6.14	inTctNo	TCT No	O1	Figure 11.12/OV-6b,c
11.	A6.10	CA6.10	A6.9	inToiInfo	TOI Info	O1	Figure 11.12/OV-6b,c
12.	A6.11	CA6.11		inIsSigMov	Significant Movement	O1	Figure 11.12/OV-6b,c
13.	A6.12	CA6.121	A6.2	inTargCoord	Target Coordinates	O1	Figure 11.12/OV-6b,c
		CA6.122	A6.3	inTargVec	Target Vector	O1	Figure 11.12/OV-6b,c
14.	A6.13	CA6.13	A6.12	inTarUpdate	Target Update	O1	Figure 11.12/OV-6b,c
15.	A6.14	CA6.141	A6.4	inTarThreat	Target Threat	O1	Figure 11.12/OV-6b,c
		CA6.142	A6.5	inDGCompl	Direction Guidance Compliance	O1	Figure 11.12/OV-6b,c
		CA6.143	A6.6	inTarUrg	Target Urgency	O1	Figure 11.12/OV-6b,c
		CA6.144	A6.7	inTarSig	Target Significance	O1	Figure 11.12/OV-6b,c
16.	A6.15	CA6.15	A6.14	inSigUrgRes	Significance/Urgency Results	O1	Figure 11.12/OV-6b,c
17.	A6.16	CA6.16	A6.8	inTctNom	TCT Nomination	O1	Figure 11.12/OV-6b,c
18.	A6.17	CA6.17	A6.16	inFirstStr	First Strike	O1	Figure 11.12/OV-6b,c
19.	A6.18	CA6.18	A6.16	inReAtkRec	Reattack Recommendation	O1	Figure 11.12/OV-6b,c

develop a skeleton DEVS model that can make its entry into the model repository. Let us name the model for Activity 6 as MA6. The inner models are identified in the same predictable manner as MA6.1, MA6.2, …, MA6.18.

The next task in line is the inclusion of the pseudo code that contains the doctrines and rules from OV-6a (described in Figure 11.11). Consider the four initial lines from Figure 11.11 (Listing 11.2).

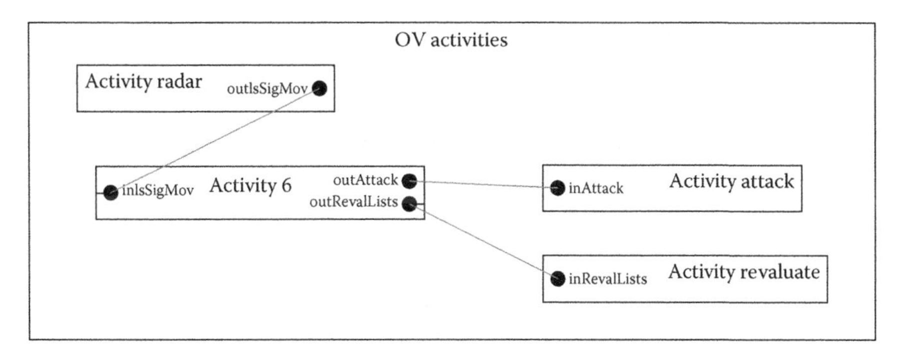

FIGURE 11.12 DEVS interrelationships of Activity 6 with other activities.

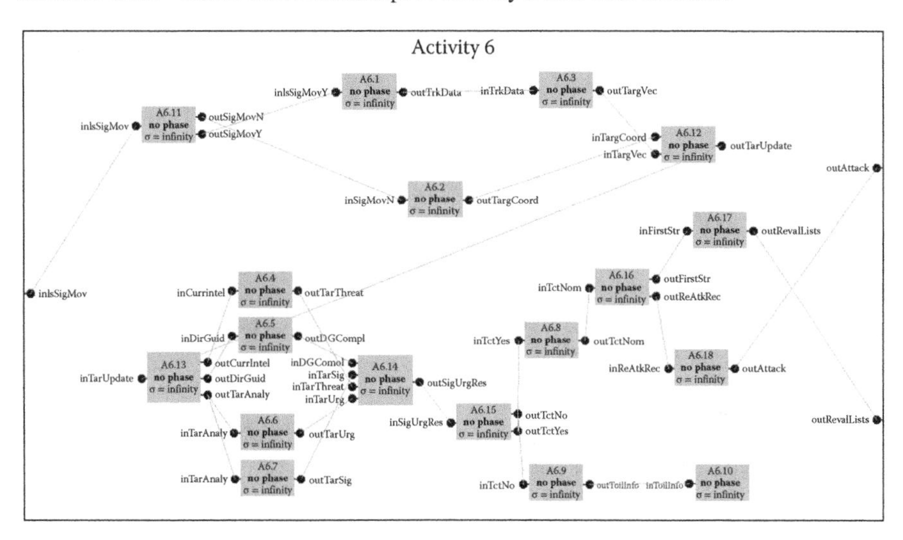

FIGURE 11.13 DEVS description of Activity 6 in relation to Table 11.5's Activity A6 components.

LISTING 11.2 SAMPLE RULE SNIPPET FROM FIGURE 11.11

```
IF Significant Movement of target
Then Monitor Target/Target Status
  Project Target Movement
  Target Vector =….. ?
Else Monitor for Movement
```

This particular doctrine is to be implemented at A6.11 (refer Table 11.5). This has far-reaching advantages. By assigning doctrines and rules to specific activity components, we are ensuring that each rule is formally implemented and is synchronized with the other rules that are in operation at that instant of time. In a sense, which rules are compatible and which can cause deadlocks can be determined by execution of the above Activity 6 DEVS model. The sample lines in Listing 11.3 are implemented in the *deltext()* function of component A6.11. The *deltint()* function defines the natural course of the activity.

LISTING 11.3 DEVS IMPLEMENTATION OF RULE SNIPPET

```
public void deltext(double e, messagex){
....
        for(int i = 0; i<x.length; i++){
        if(messageOnPort("inIsSigMov"){
        MessageTypeA msg = (MessageTypeA)x.getValOnPort(i,
          "inIsSigMov");
             If(msg.equals("yes"))
                  holdIn(0, "yesSigMov");
             else
                  if(msg.equals("no"))
                       holdIn(0, "noSigMov"));
             }
          .....
}

public message out(){
....
      if(phaseIs("yesSigMov")){
           m.add(makeContent("outSigMovY", new
             entity("start")));

      if(phaseIs("noSigMov"))
           m.add(makeContent("outSigMovN", new
             entity("start")));
......
}
```

Similarly, all other activities will receive inputs from other source activities in their *deltext()* functions that will contain the logic for implementation of doctrines. For convenience purposes, the execution time of these doctrines is considered zero. Notice the *holdIn()* function in the code above. However, this is an important place where we can tune and implement the realistic time in issuing commands by human commanders, for example, in a situation where the system is waiting for a command from an authority figure and decision has to arrive until a time-out occurs. In addition, consider that the activity component is executing a certain process with respect to its *deltint()* function and is in a certain phase waiting for any external input from other activities. In the situation of not receiving this input within allowable time-window, time-outs can very effectively guide the simulation to its completion and prevent the wait-to-infinity problem.

The OR split problem pointed out by Zinn in the IDEF3 methodology has no effect in the DEVS methodology. This problem is resolved by making the &, X, and O constructs in the IDEF3 methodology as internal-activity components (Table 11.5). Once they are componentized, time-outs can be implemented very easily that will completely eradicate this problem. These components are very well documented in DEVS *SimpArc* package Version 3.0. This solution also puts the focus back on the system-logic implementation and test if the communication delays are significant enough that time-outs are occurring frequently.

Finally, the last task is the description of the OV-9 document. This document contains information about the activities happening inside an operational node and how the sub-activities are mapped on to the components inside the operational node. For simplicity, we are working on the assumption that there is only one operational node, O1, in the example. As there is no information present on what its inner components are in Zinn's work, we will assume that there are, let us say, seven inner components that make up this node. Four of these seven components are associated with Activity 6, and the other three components are associated with some other activities, not considered for illustration purposes. The defined components are essentially COTS components with a defined behavior. They can even come from an SV document, SV-4. Consequently, each of them has their models for simulation purposes specified in the DEVS formalism. These models are essentially open-source models available to public through a common repository and are standardized. Table 11.6 depicts the information assumed for the construction of OV-9. The inner components depicted in Table 11.6 are only for illustration purposes.

Having Table 11.6 as available resources for OV-9, we have enough information to construct the activity–entity mapping in Table 11.7. We identify and define port interfaces that need to be added to the entity component models so that they can be coupled to the activity components. Once the OV-9 document is in place, the added interface information is used to update the models defined during the construction of these two documents. We saw in the construction of the OV-8 document that the resulting model is a stand-alone model that is capable to execute the simulation in the capability mode, testing the OV-5 and OV-6 description of the system. A sample OV-9 document looks as the one shown in Table 11.7.

The OV-9 document aids in bringing the systems perspective to the design and how the system's components initiate the designated activities. Assignment of an activity to an appropriate component entity is a job of an experienced designer, as per the definition of designer in the DoDAF document. This document ensures accountability that there is at least one component entity that is responsible for the execution of that particular activity. Notice that all the activity components addressed in the example have been assigned at least one operational node inner component entity. After the creation of the OV-9 document, the interface information, in the last column, is used to update the corresponding activity and the entity models in the model repository that were created during the construction

TABLE 11.6
Inner Components within Operational Nodes and Their Mapping with Standardized DEVS Models

S. No.	Operational Node	Inner Component Entities	Component Name	Associated Models Added to Repository	Hierarchical Parent/ Container	DEVS Model Type
1.	O1	OCE1	TCT	ME1	—	Digraph
2.		OCE1.1	Radar Tracking System	ME1.1	ME1	Atomic
3.		OCE1.2	Significance Analyzer	ME1.2	ME1	Atomic
4.		OCE1.3	Urgency Analyzer	ME1.3	ME1	Atomic
5.		OCE1.4	Vigilance Controller	ME1.4	ME1	Atomic
6.		OCE1.5	Attack Evaluator	ME1.5	ME1	Digraph
7.		OCE1.6	Attack Initiator	ME1.6	ME1.5	Atomic
8.		OCE1.7	Attack Terminator	ME1.6	ME1.5	Atomic

of OV-8. This is again an automated task with a simple string manipulation as described earlier, during the construction of OV-8 models.

Hence, during the creation of OV-8 and OV-9, we have populated the model repository with activity models (MA6.1–MA6.18) and operational node's inner component models (ME1, ME1.1–ME1.6) have created an interface between these two aspects of the DoDAF design.

11.5.3 Synopsis

Looking at Figure 11.4 from an activity component perspective, we have our defined inputs and outputs, and eventually the activity ports. In the example above, we have defined the interfaces of an activity that could be subjected to component coupling and testing. The coupling information can be integrated using the OV-3 document, as described in Table 11.6. The timing information is added using the OV-6b and OV-6c diagrams as we have defined components, the effects of which have been highlighted in Section 11.3. This information, along with the pseudo code provided by Zinn, is integrated to develop the DEVS model of the activity in question. The pseudo code is very well directed to the activity that is best responsible to execute those rules. At this point, the whole purpose of creating OV-8, the rule-activity mapping, is realized.

TABLE 11.7
OV-9 Description Document Mapping the Entity Component Inside Operational Node O1 with the Activity Components Defined in OV-8 with Port Interfaces

S. No.	Operational Node	Inner Component Entities	Component Name	Activity Component	Activity Component Name	Interface Description
1.	O1	OCE1	TCT			
		OCE1.1	Radar Tracking System	A6.1	Monitor Target/Target Status	monTarE
				A6.2	Monitor for Movement	monTarMovE
				A6.3	Project Target Movement	proTarMovE
				A6.11	Significant Movement Yes/No	sigMovYesNoE
				A6.12	Target Monitoring	tarMonE
				A6.10	Monitor Target of Interest for Status change	monTarInterE
		OCE1.2	Significance Analyzer	A6.13	Target Significance Analysis	tarSigAnalyE
				A6.4	Analyze threat from Target	analyThrTarE
				A6.5	Analyze Dynamic Target Execution Direction and Guidance	analyEDGE
				A6.7	Determine Target Significance	detTarSigE
				A6.14	Synthesize Results	syncE
		OCE1.3	Urgency Analyzer	A6.6	Determine Target Window of Vulnerability	detWinVulE
		OCE1.4	Vigilance Controller	A6.15	TCT Determination Yes/No	tctDetYesNoE
				A6.8	Nominate as Dynamic Target	nomDynTarE
				A6.9	Pass Target to ATO Planners	passTarAtoE
				A6.16	Initial Attack Yes/No	initAtckYesNoE
				A6.18	Attack Decision	atckDecE
				A6.17	Review Established Target Lists	revEstTarListsE
		OCE1.5	Attack Evaluator	A6.16	Initial Attack Yes/No	initAtckYesNoE
		OCE1.6	Attack Initiator			
		OCE1.7	Attack Terminator			

The OV-9 document deals with the mapping of the activity components with the entity components. Since Zinn did not define internal components for any operational node, we assumed certain inner components and mapped the activities to these components. Having ensured accountability for each of the activities, other areas that OV-9 contribute to are system design, reuse, and composability. We have with us a document that contains information of the functionalities any particular component can perform or participate in a collective functionality. Consider the situation when two or more inner components from systems perspective are thrown together to observe if the system is capable of performing something. This allows us to experiment with different systems that are claiming to exhibit certain functionality. It allows us to test interoperability.

Hence, the resulting integrated information from OV-3, OV-2, and OV-6 is converted to the information in documents OV-8 and OV-9, with the addition of logical ports, dedicated to the M&S area that are focused toward OVs. The manual/automated design of DEVS models, based on Table 11.3 interpretation, and semi-automated model-test-suite development stems from the DEVS Activity Model description documents, namely, OV-8 and OV-9.

11.6 DoDAF-DEVS DEVELOPMENT PROCESS

The process of synthesizing and augmenting the information contained in the DoDAF 1.0 documents can best be summarized as an integrated methodology executed in the following sequential manner:

1. Develop architecture requirements and define DoDAF All View AV-1 and OV-1 (a conceptual OV) showing the key capabilities.
2. Define the hierarchical capability functional description document OV-5 and provide more details in OV-6b,c leading to components identification.
3. Develop OV-8 and OV-9 documents that are dedicated to M&S.
4. Gather component and interface definition information and develop SV-4 and SV-5 documents that deals with identification of systems (COTS) that could provide the required capabilities. SV-4 deals with new proposed system identifications. SV-5 deals with systems mapping to operational activities. Their identification is continually refined, as development to deployment time is extended over long durations.
5. Specify the components, interface, nodes, and connectivity information from OV-8 and OV-9 documents into XML.
6. Put these XML DEVS component models in Web Model Repository.
7. From architecture requirements, develop an OV-6a ROE document description based on underlying meta-model and translate them into a meaningful code using NLP methods.
8. Gather the generalized-behavior DEVS model from the Web repository and apply the domain-specific rules/constraints specified in the previous step and develop run-time models ready for DEVS distributed simulation, automatically.
9. Gather performance results (and tune models if need be) and transform the code to actual system components using model continuity principles.

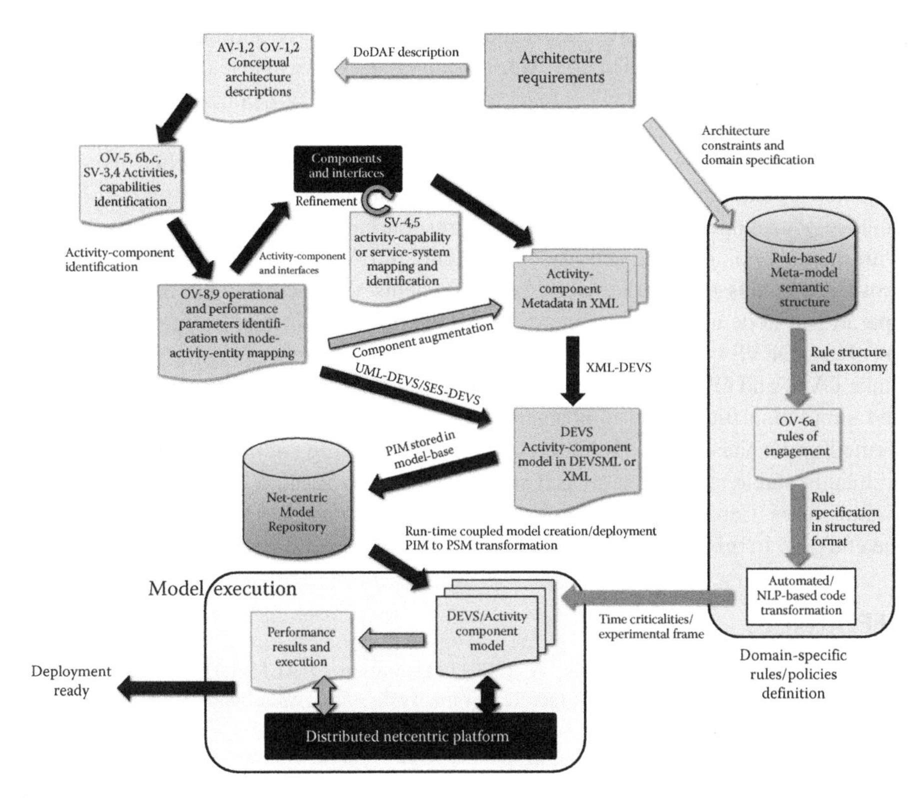

FIGURE 11.14 DoDAF executable architecture development process.

Figure 11.14 shows the integrated methodology. Hence, the resulting integrated information from OV-3, OV-2, and OV-6 is converted to the information in documents OV-8 and OV-9, with the addition of logical ports, dedicated to the M&S area, that are focused toward OVs.

11.7 SUMMARY

This chapter provides an overview of DoDAF Version 1.0 as a framework to model large, complex system architectures leading to systems and system or systems description. Although the DoDAF 1.0 specification provides an extensive methodology for system architectural development, it is deficient in several related dimensions—absence of integrated modeling and simulation support, especially for model continuity throughout the development process, and lack of associated testing support. To overcome these deficiencies, we describe an approach to support specification of DoDAF architectures within a development environment based on DEVS-based modeling and simulation. The result is an enhanced system life-cycle development process that includes model-continuity-based development and testing in an integral manner.

We introduce two new OVs, OV-8 and OV-9, to address the additional information that is needed to make the DoDAF M&S compatible. We have also demonstrated the process to create OV-8 and OV-9 from the existing OVs. OV-8 contains the information about the activity component structure, about significant parameters related to that activity, and how different activities are interfaced with each other using the specified logical interfaces. OV-9 contains information about the constituent components inside an operational node and its corresponding DEVS model structure along with their mapping to the activity components in OV-8. Together, OV-8 and OV-9 provide a means to correlate activity components with accountable entities in an operational node using logical interfaces. It is after the transformation of OV-8 and OV-9 into DEVS models that rules assigned to specific activity or entity components make OV-8 and OV-9 serve their complete purpose. Automation using XML, AST, and simulation tuning are important concepts that can be well executed and performed under the current DEVS technology. Composing simulations that are hierarchically stable and realizable is a step forward in evaluation of multi-resolutional architectures. Capability to objectify parameters and visualize them with respect to the end goal in mind is critical for success.

REFERENCES

Adshead, S., Kreitmair, T., & Tolk, A. (2001). Definition of ALTBMD Architectures by Applying the C4ISR Architecture Framework. *Fall Simulation Interoperability Workshop*. Orlando, FL.

AFCAO. Retrieved from AFCAO. https://cao.hanscom.af.mil/af-cio.htm.

Atkinson, K. (2004). Modeling and Simulation Foundation for Capabilities Based Planning. *Simulation Interoperability Workshop*. Arlington, VA.

Chairman, JCS. (2004). *Joint Capabilities Integration and Development*. Chairman JCS Instruction 3170.01D.

CIO, D. (2006). *Implementing the Net-Centric Data Strategy*. Washington, DC: DoD CIO.

CIO, D. (2007). *Department of Defense Net-Centric Services Strategy*. Washington, DC: DoD CIO.

DoD. (2003). *Operation of the Defense Acquisition System*. DoD Instruction 5000.2.

DoDAF Working Group. (2003). *DoD Architecture Framework, Ver. 1.0, Vol. III: Deskbook*. DoDAF Working Group.

Dickerson, C., Soules, S. (2002). Using Architecture Analysis for Mission Capability Acquisition, Command and Control Research and Technology Symposium, available at: http://www.dodccrp.org/Activities/Syposia/2002CCRTS/Proceedings/Tracks/pdf/123.pdf

DoDAF Working Group. (2004). *DoD Architecture Framework, Ver. 1.0, Vol. 1: Definitions and Guidelines*. DoDAF Working Group.

Gonzales, D., Moore, L., Pernin, C., Matonick, D., & Dreyer, P. (2001). *Assessing the Value of Information Superiority for Ground Forces—Proof of Concept. RAND Corporation*. Santa Monica, CA.

Integrated Definition Methods: IDEF3 Process Description Capture Method (June 10, 2012). Retrieved from www.idef.com.

Kobryn, C., & Sibbald, C. (2004). *Modeling DoDAF Compliant Architectures: A Telelogic Approach for Complying with DoD Architecture Framework*. Telelogic White Paper.

Lee, J., Choi, M., Jang, J., Park, Y., & Ko, B. (2005). The Integrated Executable Architecture Model Development by Congruent Process, Methods and Tools. *The Future of C2, 10th International Command and Control Research Technology Symposium*. Virginia Beach, VA.

Levis, A., & Wagenhals, L. (2000). C4ISR architectures: I. Developing a process for C4ISR architecture design. *Journal of Systems Engineering, 3*(4), 225–247.

Mittal, S. (2006). Extending DoDAF to allow DEVS-based modeling and simulation. *Journal of Defense Modeling and Simulation: Applications, Methodology, Technology, 3*(2), 95–123.

Mittal, S., Gupta, A., Mitra, A., & Zeigler, B. (2006). Strengthening OV-6a Semantics with Rule-Based Meta-models in DEVS/DoDAF-Based Life-cycle Architectures Development. *IEEE Information Reuse and Integration: Special Session on DoDAF.* Hawaii.

Mittal, S., Mak, E., & Nutaro, J. (2006). DEVS-based dynamic model reconfiguration and simulation control in the enhanced DoDAF design process. *Journal of Defense Modeling and Simulation, 3*(4), 239–267.

Rosen, J., Parenti, J., & Hamilton, J. (2011). The domain of interoperability in the US Department of Defense. In J. A. Hamilton (Ed.), *Modeling Command and Control Interoperability: Cutting the Guardian Knot.* Auburn, AL. Available at: www.eng.auburn.edu/users/hamilton/security/spawar/.

Sarjoughian, H., & Cellier, F. (2001). *Discrete Event Modeling and Simulation Technologies: A Tapestry of Systems and AI-Based Theories and Methodologies.* New York: Springer-Verlag.

Shelton, G., Case, R., & DiPalma, L. N. (May 2004). Advanced software technologies for protecting America. *CrossTalk: Journal of Defense Software Engineering,* 10–15.

Tolk, A., & Muguira, J. (2004). *M&S With Model-Driven Architecture. I/ITSEC.*

Tolk, A., & Solick, S. (2003). Using the C4ISR Architecture Framework as a Tool to Facilitate V&V for Simulation Systems Within the Military Application Domain. *Simulation Interoperability Workshop.*

Wagenhals, L., Haider, S., & Levis, A. (2002). Synthesizing Executable Models of Object Oriented Architectures. *Workshop on Formal Methods Applied to Defense Systems.* Adelaide, Australia.

Wagenhals, L., Shin, I., Kim, D., & Levis, A. (2000). C4ISR architectures: II. A structured analysis approach for architecture design. *Journal of Systems Engineering, 3*(4), 248–287.

Zeigler, B. P. (2003). *DEVS Today: Recent Advances in Discrete Event-Based Information Technology. MASCOTS.*

Zeigler, B. P., & Mittal, S. (2005). Enhancing DoDAF with DEVS-Based System Life Cycle Process. *IEEE International Conference on Systems, Man and Cybernetics.* Hawaii.

Zeigler, B. P., Fulton, D., Hammonds, P., & Nutaro, J. (2005). Framework for M&S-based system development and testing in a net-centric environment. *ITEA Journal of Test and Evaluation, 26*(3), 21–34.

Zeigler, B. P., Praehofer, H., & Kim, T. G. (2000). *Theory of Modeling and Simulation.* New York: Academic Press.

Zinn, A. (2004). *The Use of Integrated Architectures to Support Agent Based Simulation: An Initial Investigation,* MS Thesis. Wright-Patterson AFB, OH: Air Force Institute of Technology.

12 Modeling and Simulation-Based Testing and DoDAF Compliance

12.1 INTRODUCTION

Software testing is not a new area. Many texts have been written in this area and several methodologies have been developed. However, the idea of testing software architecture (SA) is comparatively new and requires a more rigorous effort. Testers must not only have good development skills but also be knowledgeable in formal language, graph theory, and algorithms (Whittaker, 2000). Software testing is usually approached in four phases:

1. Modeling the software's environment
2. Selecting test scenarios
3. Running and evaluating test scenarios
4. Measuring the testing process

This serves as a partition of the entire process of testing, similar to the STEP model given by Eickelmann & Richardson (1996) and Torkar (2005). There is a plethora of books on software testing since the first text by Myers (1978) that address tough testing issues, but the area of software architecture testing has not resulted in a mature methodology that is stable. Research is continuing in the current area. From code-level testing, the testing area has grown to include model-based testing, Unified Modeling Language (UML) as a means to support the modeling, to development of the Software Architecture Analysis Method (SAAM) framework (Kazman et al., 1994). However, the transition has not been smooth and appears as two separate classes of methodologies. The former is focused toward code level testing and coverage analysis, while frameworks like SAAM are focused toward the entire evaluation and effectiveness of any particular SA. This chapter summarizes the various efforts that have been put in the recent years in these two disparate classes, and argues that DEVS M&S provides an integral framework that helps align these two fields in coherence. Of the four-part process mentioned above, selecting test scenarios appears to be the most time-consuming, rigorous, and well attended in the literature. Test execution is assumed to be simpler until DEVS M&S provides a mathematical framework to conduct test-model execution in a formalized manner.

This chapter is structured in a manner to build the testing process from the ground up, that is, from code-level testing to the SA evaluation techniques and then introducing the DEVS M&S framework as a bridge between the two methodologies in the last section. The chapter discusses the classical testing methodologies, model-based testing process, architecture-based testing and evaluation, and DEVS Unified Process (DUNIP) in the described order. Finally, it provides a strategy to employ testing of architecture frameworks such as the Department of Defense Architecture Framework (DoDAF) and what a certification means for a DoDAF-compliant architecture.

12.2 BACKGROUND AND AVAILABLE TESTING METHODOLOGIES

The basic process that has withstood the pragmatics with respect to the testing of software systems (in a classical sense) consists of five steps:

1. Test creation
2. Test execution
3. Results collection
4. Results evaluation
5. Test quality analysis

Much of the literature is focused toward step 1, as we will see in Sections 12.2.1 through 12.2.4. Section 12.2.5 discusses the said approaches in contrast with DUNIP and how it can be aligned with the architectural frameworks, for which there is no mandated testing methodology.

12.2.1 Classical Testing Methods

Our focus for this chapter is oriented toward the first step, as it is the area that contains the most contributions in software testing. The remaining four steps can be developed on a case-by-case basis. This is by far the most exhaustive step of the testing process, as it deals with narrowing the domain space of the infinite input set. Two aspects contained therein—test data and test case—have not been dealt with in the literature and have been separate categories until recently (Torkar, 2005). We will see in later sections that this categorization is quite necessary when we are dealing with systems that are defined using high-level graphical specifications, especially UML. Most of the classical techniques that deal with test data generation are as follows:

- Myers (1978) defined *equivalence class partitioning* as a technique that partitions the input domain of a program into a finite number of classes (ets); it then identifies a minimal set of well-selected test cases to represent these classes. There are two types of input equivalence classes: valid and invalid. Regarding boundary value analysis, NIST (Adrian et al., 1981) defines it as a selection technique in which test data are chosen to lie along boundaries of the input domain (or output range) classes, data structures, and procedure parameters. Here the boundaries defined by NIST are analogous to partitions defined by Myers. Random data generation is another technique in which the test data are generated in a random manner (Duran & Ntfasos, 1999; Hamlet, 1994). Generated test cases can be automated to some extent (Clasessen & Hughes, 2000; Jamoussi, 1997).

- Another category of test data generation belongs to *path-oriented, constraint-oriented, and goal-oriented domains*. The path-oriented technique is defined by Beydeda and Gruhn (2003) as a data generation technique that makes use of control flow information to determine paths to cover and then generates test data for these paths. Constraint-based techniques are based on path-oriented techniques where the criteria are filtered by algebraic constraints to describe test cases to find particular type of faults (Beydeda & Gruhn, 2003; DeMillo & Offutt, 1991; Ramamoorthy et al., 1997). The goal-oriented technique is focused on finding the particular program input and the corresponding sequence on which the selected statement is executed. It is divided into two broad subcategories: chaining approach and assertion-based approach. The chaining approach uses program dependency analysis to identify statements that affect the execution of the selected statement (Diaz et al., 2003; Ferguson & Korel, 1996). An assertion-based technique that identifies the test case on which the assertion is violated is described by Korel (1990) and Koreland Al-Yami (1996).

Figure 12.1 summarizes the various test data generation methodologies (Torkar, 2005).

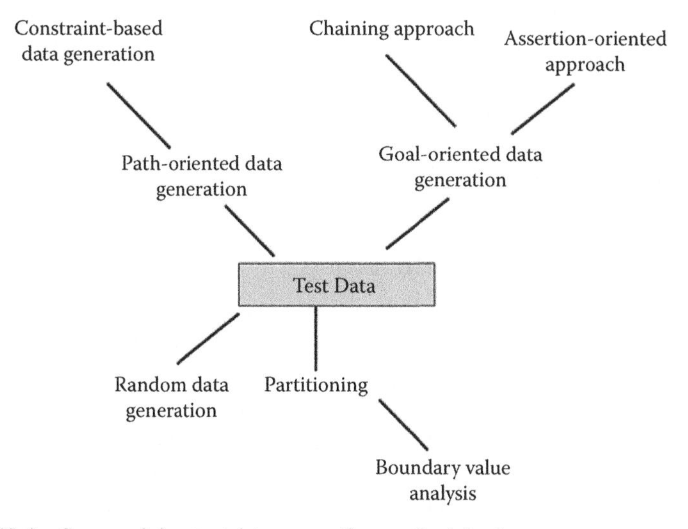

FIGURE 12.1 Summarizing test data generation methodologies.

The next step in the development of a test suite is the definition of test cases. Based on the technique used, the literature is classified into the following categories (Juristo et al., 2004) when generation of test cases is considered:

1. *Random*: Test cases are generated at random, and the process stops when there is enough, or a given number is reached, or a user-defined objective has been reached.
2. *Functional*: Same as the partitioning methodology described above.
3. *Control-flow*: Similar to path-oriented coverage described above. Test cases are generated until all the program sentences are executed at least once. However, a full execution is not recommended, as it is cost-prohibitive.
4. *Data-flow*: Test cases are generated to cover definitions of each variable for at least one use of the variable. Many variations of this particular process exist that limit the number of variables, and the number of paths traversed by this variable are considered.
5. *Mutation*: Test cases are generated based on the mutation operators defined for the programming language in question. Depending on the resources available, either all of the mutants are used or only a subset of them (after selective prioritization) is used.
6. *Regression*: Selection of test cases from an already-existing test suite is made through selection criteria or an all-inclusive methodology. Additions may be suggested that would contribute to the test suite itself.

Two broad categories cover the classical methodology section that involves automated procedures: specification-based and statistical approaches (Torkar, 2005). Specification-based test-case generation and selection technique can use a formal language (Avritzer & Weyuker, 1995; Offut, 1999) or a natural language (Lutsky, 2000) to automatically or semi-automatically generate test cases. The statistical-based techniques consist of mutation analysis (Baudry et al., 2002) and genetic algorithms (Jones et al., 1996; Lin & Yeh, 2001; Michael et al., 1997).

The next step that comes in line after generation of test data and test cases in automated or semi-automated manner is their selection. Prioritization of such test cases is discussed by Rothermel et al. (2001).

12.2.2 Model-Based Testing Techniques

Model-based testing is a variant of testing that relies on explicit behavior models that encode the intended behavior of the system and possibly the behavior of its environment (Utting et al., 2006). Pairs of inputs and outputs of the models of implementation are interpreted as test cases for this implementation: the output of the model is the expected output of the system under test (SUT). This testing methodology must take into account the involved abstractions and the design issues that deal with lumping different aspects, as these cannot be tested individually using the developed model.

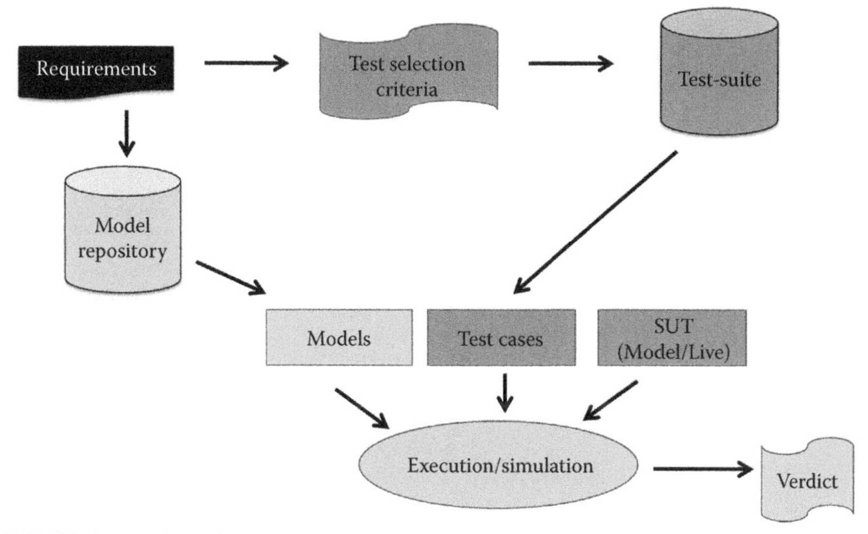

FIGURE 12.2 Graphical process extended further from Utting et al. (2006).

The process for model-based testing technique (Figure 12.2) is as follows (Utting et al., 2006):

1. A model of the SUT is built on existing requirements' specifications with desired abstraction levels.
2. Test selection criteria are defined with an objective to detect severe and likely faults at an acceptable cost. These criteria informally describe the guidelines for a test suite.
3. Test selection criteria are then translated into test-case specifications. It is an activity where a textual document is turned operational. Automatic test-case generators fall into this step of execution.
4. A test suite is generated that is built upon the underlying model and test-case specifications.
5. Test cases from the generated test suite are run on the SUT after suitable prioritization and a selection mechanism. Each run results in a verdict of "passed" or "failed" or "inconclusive."

Figure 12.3 provides a summary of contributions to the model-based testing domain. For more details, refer to Utting et al. (2006).

12.2.3 Automated Test-Case Generation Using UML Constructs

The performance and acceptance of any software system depends on the validation by the customer that is in part supported by the quality of the test suite that conducts tests on it. Consequently, it also depends on the quality of the test cases used during the validation process. In this particular methodology, the test cases are automatically generated that are created with respect to the software requirement

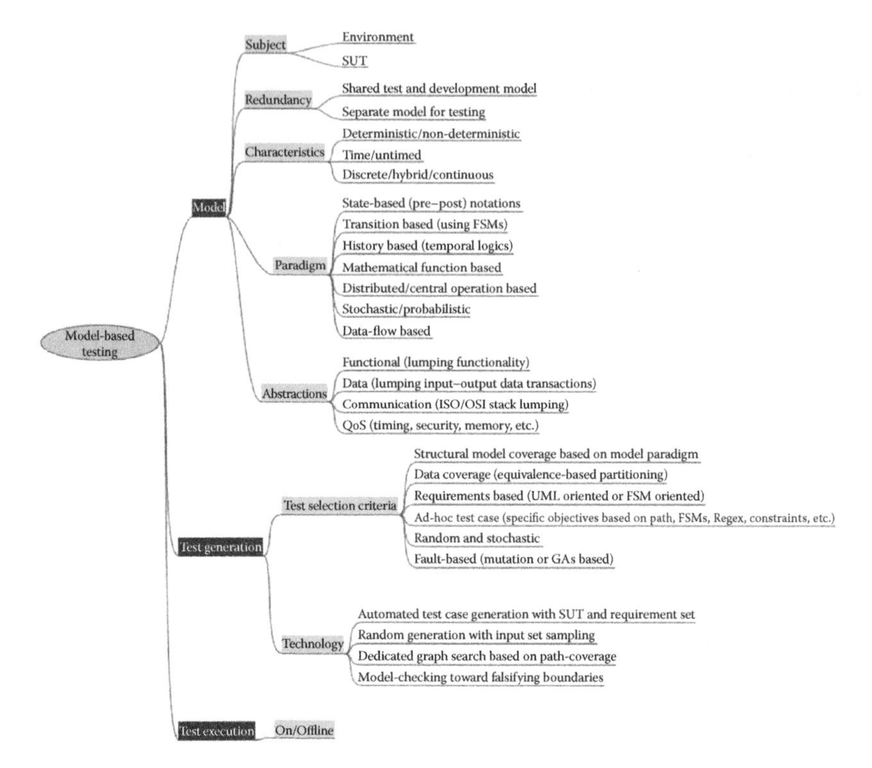

FIGURE 12.3 Summarizing model-based testing.

set. Modeling languages are used to specify the requirement set and generate test cases. UML is the most widely used and preferred means of such specifications. Williams (1999) was the first to present UML as a test-planning tool. However, he also concluded that the information collected is insufficient as it lacks pragmatic details and the diagrams must be augmented in order to be used by test programmers. Another approach (Offutt & Abdurazik, 1999) suggested was to build a standardized library, but then again it requires a collaborative effort that spans the entire domain industry.

Offutt et al. (1999; 2003) proposed techniques that adapt predefined state-based specifications to generate test cases from UML statecharts. This resulted in the development of UMLTEST—a test-data generation tool that was integrated with Rational Rose. A parallel effort was made by Vieiraet et al. (2000) using the same concept of UML statecharts, which resulted in the development of Design and Specification-Based Object-Oriented Testing (DAS-BOOT). The Java class to be tested was compared with the statechart specification of the class behavior, thereby defining the association between the code and the specification. Offutt et al. (2000; 2003) extended their system-level testing work to integration-level testing using UML collaboration diagrams. The message path coverage criterion was used to generate test cases from UML sequence diagrams. They concluded that, at the unit level, state charts were better than sequence charts, but at the integration level, it was vice versa.

Riebish et al. (2002) presented a procedure for the iterative software development process in generating test cases with sequence diagrams and use-cases as inputs for requirements engineering. They established that obtaining test cases systematically could help in documentation of software's usage and interactive behavior.

Another effort by Hartman et al. (2004) led to the development of a tool that integrates UML to automatically generate black-box conformance tests early in the development life cycle. For unit and integration testing, the authors derived tests from statechart and sequence diagrams, and for the system level, they used use-case and activity diagrams. The derived test cases were then executed using JUnit or the system test tool.

One more approach using Use-case was presented by Salem and Sumbramanian (2004). Use-cases were documented with precondition, postcondition, basic, and alternate flows and resulted in a traceability matrix. Indeed, this approach provides validation of the requirement set. The framework for model level testing of the behavioral UML model was proposed in another study by Toth et al. (2003). This process allowed different UML designs to be tested and design flaws be detected in the modeling phase of the development process. One similar detailed effort was done by Nebut et.al. (2006) where they employed UML use-case contracts as the starting point for construction of test cases. They enhanced use-cases with contracts (based on use-cases pre- and post-conditions) as they are defined by CockBurn (1997) and D'Souza and Wills (1998). Building up on the idea by Meyer's (1992) at the requirement level, they made these contracts executable by incorporating requirement-level logical expressions. Finally, they constructed a simulation model from these semi-formalized use-cases. The simulation model resulted in the extraction of relevant paths using coverage criteria. These paths are termed "test objectives." Each use-case is then described using a UML sequence diagram and results in test scenarios. Other approaches (Basanieri et al., 2002; Briand & Labiche, 2002) also propose to automatically generate test scenarios from use-cases and use-case scenarios. Testing through the use-case requirements can be summarized in Figure 12.4.

Automating methods to derive tests from fuzzy descriptions of the use-cases is a formidable task. The requirement-based testing techniques are based on formal methods that are difficult to maintain as well as rigorous, and only suitable for mission-critical applications (Bernot et al., 1991; Dick & Faivre, 1993; Legard et al., 2002; Tahat et al., 2001). Ryser, et al. (1998) suggested that for practical purposes, the testers need to focus on methods of the systematic test approach. Among varied efforts in proposing test cases (Kim et al., 1999; Offutt & Abdurazik, 1999), only a few address the system-level testing (Basanieri et al., 2002; Briand & Labiche, 2002; Frohlich & Link, 2000; Riebisch et al., 2002; Ryser & Glinz, 1999).

Model-based testing is a valuable methodology that helps test automation in conjunction with system development. Models allow testers to get more testing accomplished in shorter time. When integrated with model-based testing, model-based design development, supported by model continuity and experimental

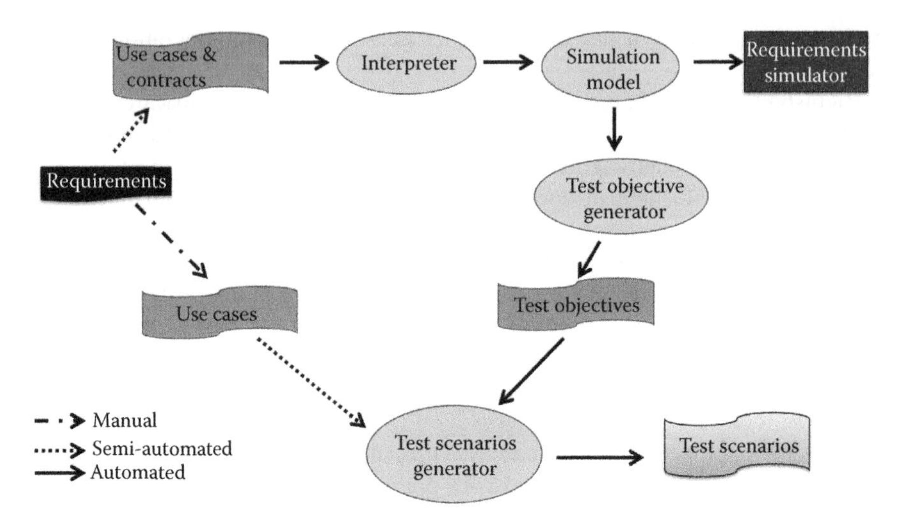

FIGURE 12.4 Test scenario generation based on requirement specifications.

frame designs, provides the best of all options. The next section presents these integrated ideas.

12.2.4 Architecture-Oriented Evaluation Methodologies

The next developmental progress in the area of system testing is in the area of Software Testing Environments (STEs) (Eickelmann & Richardson, 1996). An STE overcomes the deficiencies of manual testing through automating the test process and integrating testing tools to support a wide range of test capabilities. SAAM provides an established method for describing and analyzing SAs (Kazman et al., 1994). SAAM is defined as consisting of the following steps:

1. Characterize a canonical functional partition for the domain.
2. Create a graphical diagram of each system's structure using the SAAM graphical notation.
3. Allocate the functional partition onto each structural decomposition of the system.
4. Choose a set of quality attributes with which to assess the architectures.
5. Choose a set of concrete tasks that test the desired quality attributes.
6. Evaluate the degree to which each architecture provides support for each task, thus indicating satisfaction of the desired quality.

SAAM process creates three perspectives of the SA:

1. Canonical functional partition
2. System structure (structural decomposition)
3. Mapping of functionality with system structure

SAAM developed a STEP model specifically to evaluate the testing process for any STE. The canonical functional partition step describes six partitions for the testing process, as listed in Table 12.1.

However, the SAAM process is not complete in many respects and is a work in progress. It is a methodology and a framework through which various STEs could be evaluated as attempted by Eickelmann and Richardson (1996). The aim of analyzing the architecture is to predict the effects of the architecture, identify potential risks, and verify the quality requirements (Li & Henry, 1993). It is not possible to measure the quality attributes of the final system based on the SA design. Methods based on scenarios could be considered mature as they have been validated over the past several years, but development of the evaluation methods of attribute model-based architecture is still ongoing. SA analysis should reveal requirement conflicts and incomplete design descriptions from a particular stakeholder's perspective.

Various other contributions in the area of SA testing are available. Specification-based testing criteria are explored by Richardson and Wolf (1996). Bertolino and Inverardi (1996) discuss the advantages of using SA-level testing for reuse of tests

TABLE 12.1
Functional Partitioning of Testing Process

S.No.	Canonical Partition	Process Includes	Artifacts
1.	Test Execution	Execution of the instrumented source code and recording of execution traces	Test output results, test execution traces, and test status
2.	Test Development	Specification and implementation of a test configuration	Test oracles, test cases, and test adequacy criteria
3.	Test Failure Analysis	Behavior verification and documentation and analysis of pass/fail statistics	Pass/fail state and test failure reports
4.	Test Measurement	Test coverage measurement and analysis	Test coverage measure and test failure measures
5.	Test Management	Support for test artifact persistence, artifact relations persistence, and test execution state preservation. It requires a repository for test artifacts.	Updated active repository
6.	Test Planning	Development of master test plan, the feature of the system to be tested, and detailed test plans	Reports dealing with risk assessment issues, organizational training needs, required and available resources, comprehensive test strategy, resource and staffing requirements, roles and responsibility allocations, and overall schedule

and to test extra-functional properties. Harrold (1998, 2000) presents approaches toward using SAs for effective regression testing and testing as a whole. Richardson et al. (1998) present architecture-based integration testing that considers architecture testability and its simulation.

12.2.5 Discussion

Sections 12.2.1 through 12.2.4 lay out the evolution of testing methodologies corresponding to the growth of software systems that are termed as SAs. From simple software modules to SAs to architectural frameworks, the testing process has also grown from simple coverage tests to entire modeling and simulation frameworks focused toward testing the complete architectural frameworks. Classical testing methodologies in Section 12.2.1 appear no longer to be applicable to the SA domain. However, their use is recommended in supporting many functional aspects of the overall testing process for SAs. Model-based testing techniques covered in Section 12.2.2 that are pushed forward by the model-driven architecture paradigm have come up in recent years with an idea that "analyze before build." The testing process brought forward by this methodology now has another component to the original system code, which is the model. The model itself could be complete software in itself or just a graphical representation as in UML (described in Section 12.2.3). With the addition of one more software component in the software planning, design, and analysis stage comes whole arena of publications and testing of model-based systems and of models separately. Systems trying to run model-based analysis required a mechanism to execute the models. This resulted in the development of area called simulation-based testing which is defined as the testing of a model by a simulator.

Modeling and simulation, as independent as an area it could be, is practically linked to the development and analysis of systems, not necessarily software systems, before building them into the real world. The DEVS theory is built on the mathematical system theoretic foundation. The application of DEVS modeling and simulation in design and planning of software systems has been attempted only recently with the technology similar to that of MDA, known as model continuity (Hu & Zeigler, 2005). Advances in the latest DEVS M&S framework now allow it to construct entire SAs based on the enhanced MSVC paradigm as described in Chapter 7. GENETSCOPE System (Mittal et al., 2006) and ATC-Gen (Mak et al., 2010) provide enough evidence of the power of DEVS modeling in evaluation and construction of SA and systems.

UML has matured to great levels and has become an integral part of any model-based system design. Even the most complex and encompassing DoDAF lends itself to UML in description of its various artifacts. However, the bridge to develop an executable code from imprecise UML constructs is under research. Executable UML with the aid of the action semantic approach is one effort that brings to light this important gap of "Model to Code" directly. DEVS with its advanced model-continuity process provides this capability readily. Further, the process of developing executable architectures from object-oriented designs is in place (Wagenhals et al., 2002), but it is not being explored rigorously for testing SAs.

As capable as the DEVS M&S framework is, it is still not in the mainstream industrial software system design and planning. The ideal progress path now is the development of a mechanism to employ DEVS M&S with a UML-based developmental methodology. Chapter 11 has presented a mapping of UML with the DEVS elements that provides just the same.

This chapter surveys the literature on the topic of software-testing methodologies. As fuzzy as this topic becomes with respect to defining what is testing and what evaluation with respect to SA is, let alone the architecture framework, this chapter compiles some of the basic research ideas dealing with these two aspects. The problem of testing has become more complex as the complexity of the SA increased, and the most time-consuming task that has been researched widely is the generation of good test cases. Automated test-case generation has been the preferred choice, realizing the current state of automated code generation methodologies. UML as a modeling tool has been successful in meeting the demands of the testing process in a fragmented manner, but the testing framework as a whole is not mature enough to be included as a college text. The advantages of UML with the addition of executing the model provide promise to the model-based testing and analysis methodology. However, without a formal and scalable execution framework, this process is not foolproof. It is at this juncture that the DEVS theory, with its mathematical theoretic foundation, provides a mechanism to provide the basis to construct the model as well as a decoupled simulator that executes the model. The common denominator to construct SA is as important as the execution of model or the technology underlying the simulation methodology. DoDAF provides the needed guidelines to develop the net-centric architecture. The development of a testing framework involving DEVS, UML, and DoDAF is the proposed testing framework that takes into perspective the advances in UML, foundation of systems engineering in DEVS, and architectural framework principles in DoDAF.

We have seen that the area of software testing at best can be extended toward SAs. Software testing lacks the formal systems methodology to apply to architecture framework. As a reminder, an architecture framework enables creation of multiple architectures. A testing methodology that is designed for a framework lends itself automatically to any architecture that is created using that framework. Moving forward, we are looking at systems testing methodology for architecture frameworks, like DoDAF.

12.3 DODAF SPECIFICATIONS WITH SYSTEM ENTITY STRUCTURE

Now, having seen the shortcomings of DoDAF 1.0 in Chapter 11 and the augmented DoDAF descriptions such as OV-8 and OV-9 that are required for the M&S-based design, we will now turn to consider DoDAF as a documentation mechanism. We will build a metamodel for a complex system using System Entity Structure (SES) formalism and will associate the DoDAF documents to various elements of this metamodel. So far, the SES has been shown to provide a means of pigeonholing the various DoDAF views. The power of this representation, however, lies in the support it provides for deriving system behaviors that can be transferred in a semi-automated fashion to executable test federations. The systems view (SV) is further refined by explicitly adding messages as entities to it. For simplicity, components represent both the functions and their decomposition into services. The coupling associated with the components aspect specifies how messages are routed among the components. This information is what is required to automatically map the SV to a simulation model, that is, in this case, a test federation. To obtain such information, we develop a process for deriving it from specifications associated with the operational view (OV) and mappings between the OV elements and their realizations in the SV. If an operational node is engaged in an activity that requires a certain information exchange and the operational node is mapped to a component that executes the function implementing the activity, then this component must be observed to receive and send messages associated with that information exchange.

Information technology-based systems of the future will be increasingly complex with participants across the globe communicating through disparate channels. Interoperability is very much in question. Scalability and fault-tolerance issues have to be addressed. Capabilities have to be satisfied and reliability has to be ensured. Any large system that DoDAF specification documents intend to build has to realize these important facets of the architecture design. Modeling and simulation with its model-continuity principles is fast becoming an accepted method of evaluating design principles ensuring accountability to various components within the system. DoDAF has completely overlooked M&S as a possible means to evaluate design, capabilities, and planned expansion of current architectures. There is no provision for testing the constructed system, either in OV or in SV. The ability to configure systems for optimum performance is not allowed in the current DoDAF specification document.

DEVS-modeled systems are inherently object-oriented and DoDAF at the OV stage does not have full expressiveness to be transformed to an executable model. In the SES engineering approach, we have a hierarchical perspective representation that would enable DEVS to step into at various levels of resolutions. The four main perspectives are as follows:

1. Systems-based
2. Operation-based
3. Doctrine-based
4. Organization-based

DEVS Unified Process requires all four perspectives to be available in order for the system model be deployable. DoDAF 1.0, if enhanced with the new OV documents, does make the DoDAF a DEVS-compliant system. The SES is a high-level ontology framework targeted to modeling, simulation, systems design, and engineering, as described in detail in Chapter 10. Its expressive power, in both strength and limitation, derives from that domain of discourse. An SES is a formal structure governed by a small number of axioms that provide clarity and rigor to its models. The structure supports hierarchical and modular compositions allowing large complex structures to be built in stepwise fashion from smaller, simpler ones. Tools have

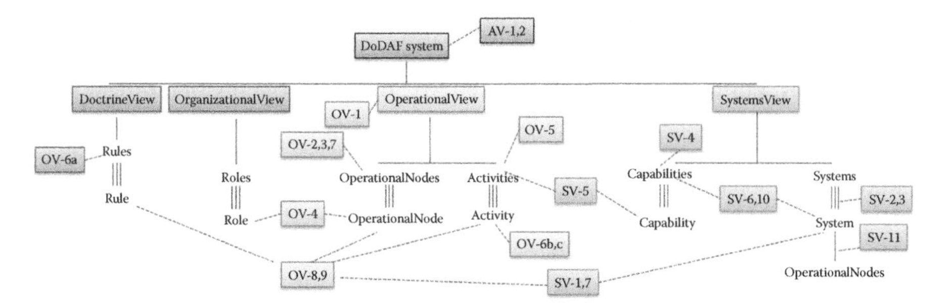

FIGURE 12.5 Representing DoDAF within the SES framework.

been developed to transform SESs back and forth to XML, allowing many operations to be specified in either SES directly or its XML/DSL guise. Further, the SES metamodel allows processing the model using an Abstract Syntax Tree (AST). The axioms and function-based semantics of the SES promote the pragmatic design and are easily understandable by data modelers. Together with the availability of appropriate tool support, this makes development of XML schema or AST transparent to the modeler. Finally, SES structures are compact relative to equivalent schema and automatically generate associated executable simulation models.

Figure 12.5 shows the various DoDAF views mapped into the SES framework. Operational and system perspectives are considered two different decompositions of the system under consideration. They are represented by corresponding nodes called aspects labeled by the names, *OperationalView* and *SystemView*, respectively. The *OperationalView* aspect has entities labeled *OperationalNode*s and *Activities.* The various OVs of DoDAF (other than OV-4) are easily interpreted as describing the entities and their interactions. Likewise, the *SystemView* aspect has entities that are associated with the functions and their interactions. One exception is SV-5, which is a relation between the functions of the *SystemView* and the activities of the *OperationalView*. SV-5 describes how the activities are implemented via executable functions supplied by the system. The other exceptions are SV-1 and SV-7 that relate connectivity information and performance parameters to the OV, respectively. To accommodate OV-4, we have added another aspect, the *OrganizationalView*, which represents the decomposition of the system into the roles played by participating personnel. Similarly, OV-6a required another aspect called *DoctrineView*, which describes doctrines or business rules.

12.4 MODELING AND SIMULATION AS APPLICABLE TO DoDAF TESTING

Any DoDAF-compliant architecture is essentially a system of system. It is highly likely that such architecture will be netcentric in an operational environment in future. The component systems within the architecture may have already been piecewise tested before deployment. So, what does DoDAF architecture testing mean?

A DoDAF-compliant operational system is put to use for executing DoD missions that spans multiple agencies. As a result, the focus is more on the execution of the mission. Let us call each mission a mission thread. The success and failure of a mission thread will ultimately decide the efficacy of the operational environment. Consequently, our testing efforts are geared toward testing the mission thread that takes a DoDAF system of system in a netcentric operational environment.

The testing plan is guided by the DEVS Unified Process and follows an iterative nature of development. The process begins with the mission thread requirements that are specified in a formal DoDAF representation. We have already discussed how DoDAF representation in UML or SES is DEVS executable. The formal DEVS-compliant models are preferred in DEVSML and stored in a netcentric repository for distributed execution. Now, given that we have an executable model of a mission thread, the test plan features the generation of test scenarios from mission thread specification. In DEVS jargon, this leads to specification of various experimental frames. Having DEVSML models gives us the capability to

1. Simulate the platform independent model
2. Generate a test suite automatedly

The DEVS test-suite model is designed to ensure that the required behavior as expressed in input–output pairs is correctly implemented when integrated in the system with temporal constraints. The Experimental Frame design built from the mission thread requirement interfaces between the executable model (as a simulation) and the Test-suite model. The validation of the Test-suite is undertaken in Experimental Frame. The results of the model execution and the adequacy of Test-suite model (which is also executed as a simulation) are compared and evaluated toward measures of conformance and performance. Analysis of the Experimental Frame simulations and the system Test-suite simulation results are compared and evaluated to determine the departure from the required behavior. This comparison is measured in a metric called Conformance measure. Ideally, the designed model has a 100% conformance with the Test-suite model. If the departure exceeds a given tolerance, the model is revised to increase the model-test conformance. All this assumes that the initial DoDAF specifications have been cast in stone.

However, the Conformance measure does not tell anything about the quality of the test case or a particular experimental frame. That is where the performance measure is helpful, which provides information about the relative performance of various system and operational parameters defined in the mission thread requirements. The analysis of these two measures is used to iteratively tune the model specifications and the Test-suite specification. Typically, however, the iterative process will also suggest new or modified specifications at the DoDAF level. The iterative loops can be seen in Figure 12.6. Finally, when the models conform to the system test specifications, the Test-suite presents the design and performance recommendations as the outcome of this data-centric process. The Model Repository serves as the basis of the component design based on model-continuity principles, and the Test-suite serves as the benchmark for performance evaluation and matching the technical specifications as developed in the Technical View (TV) DoDAF descriptions. Figure 12.6 shows these two iterative feedback loops as model tuning and Test-suite tuning.

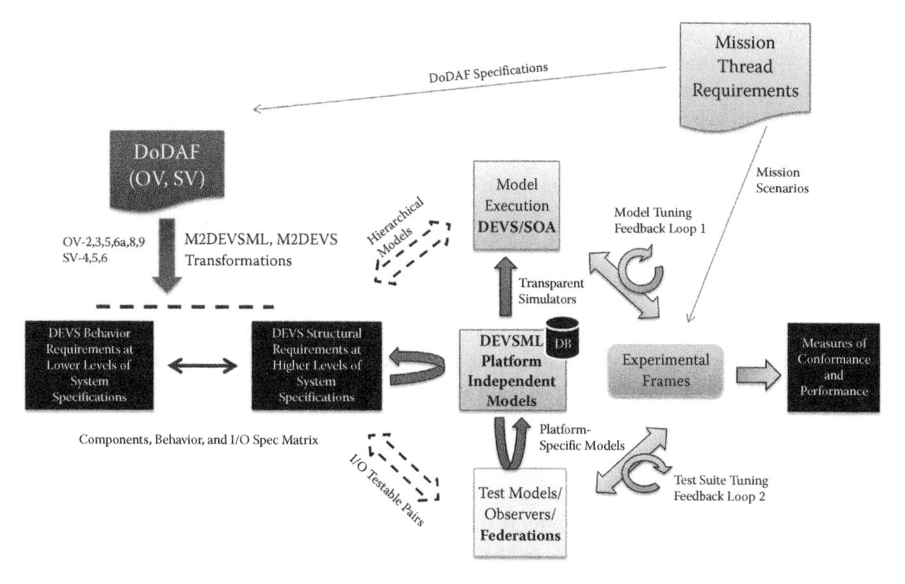

FIGURE 12.6 The lifecycle development process of the DUNIP-based DoDAF-to-DEVS system.

The iterative process is briefly summarized as follows:

1. The mission thread is specified in DoDAF-based DSL or any other formal means such as SES.
2. The Experimental Frame is specified from the mission thread requirements.
3. Using various model transformation technologies, the DoDAF-DSL models are transformed to DEVSML specification.
4. The DEVSML model generates the executable model and the Test-suite model.
5. The model and the Test-suite model are executed on a distributed netcentric platform.
6. Experimental Frame controls the execution and data analytics.
7. The conformance and performance measures are analyzed and validated with the mission thread requirements.
8. The model and the Test-suite model is tuned, based on the outcome of step 7.
9. Repeat steps 4 through 8.

12.4.1 NR-KPP and KIP

Net-Ready Key Performance Parameters (NR-KPP) are a key to measuring the readiness for transformation into a fully interoperable and secure netcentric warfare environment. As currently stated in the Joint Staff Guidance:

> The NR-KPP assesses information needs, information timeliness, information assurance (IA), and the net-ready attributes required for both the technical exchange

of information and the end-to-end operational effectiveness of that exchange. It consists of verifiable performance measures and the associated metrics required evaluating the timely, accurate and complete exchange and use of information to satisfy information needs of a given capability. It is composed of the following four pillars:

1. Compliance with the Net-Centric Operations and Warfare Reference model (NCOW-RM)
2. Compliance with applicable Global Information Grid (GIG) Key Interface Profiles (KIP)
3. Verification of compliance with DoD IA requirements; and
4. Supporting integrated architecture products required to assess information exchange and use for a given capability.

Integrated architectures are the most critical components of NR-KPP because they establish both the operational and the systems context for information exchange. They are developed and documented using the DoDAF, which provides templates for the 27 distinct views. Two other critical components in NR-KPP, namely, NCOW-RM and KIP, are in the evolving state. KIP tools help standardize and manage interfaces to networks, services, and communication pathways. Initial attempts to manage interfaces required identification of specific interfaces between each system, for which an operational interchange of information was specified. The netcentric vision requires a paradigm shift from one-to-one relationships to one-to-many relationships, and KIPs are the key profiles enabling that shift. The KIP consists of refined OVs and SVs, and interface control document/specification, an engineering management plan, a configuration management plan, a TV with an SV-TV bridge, and procedures for standards conformance and interoperability testing. The Joint Staff categorizes the key interfaces in four broad categories, namely, communications computing, enterprise services, network operations, and other explicit interfaces (in addition to 17 mentioned since first draft). From an evaluation perspective, as has been recognized, if the KIP is not defined, then it clearly cannot be tested. As a part of the overall test and evaluation (T&E) strategy, the NR-KPP T&E strategy is twofold:

1. Identify and gather sufficient data from developmental events to verify that the key subcomponents for the NR-KPP have been satisfied, and verify that the information and systems data exchanges can be accomplished.
2. Validate these exchanges to achieve the mission in an operational test environment.

The two steps mentioned in the overall strategy relate to the SV-6 document that specifies the system data-exchange matrix. The first step verifies the data exchange, while the second step validates it in the operational test environment (see Figure 12.7). The OV-8 and OV-9 documents lie in the operational test domain. As stated earlier, the current DoDAF OVs do not provide any

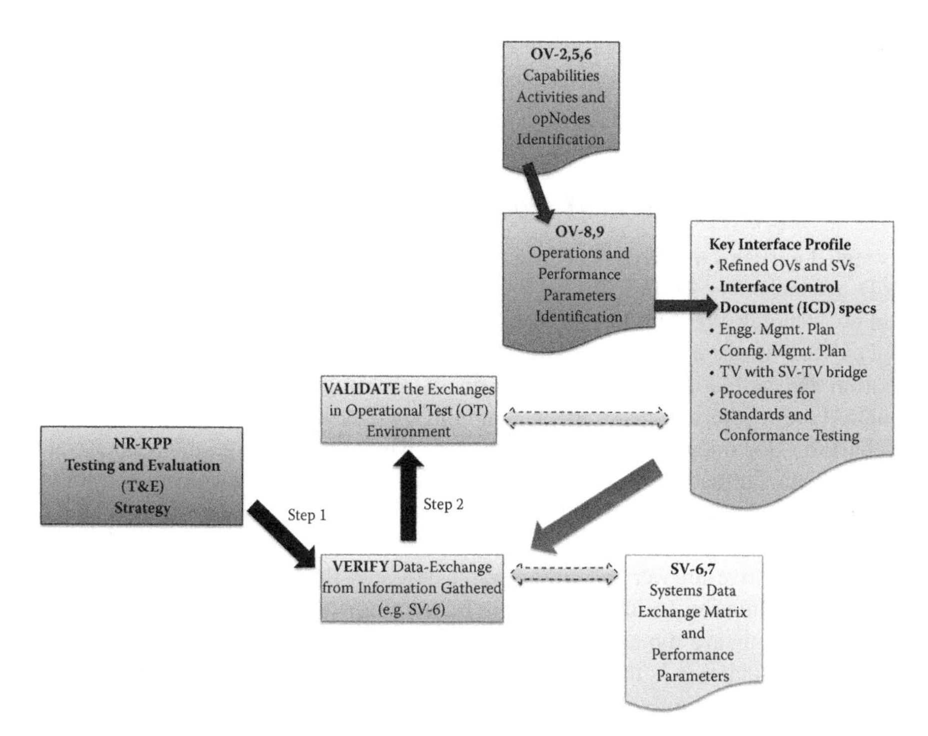

FIGURE 12.7 Role of OV-8 in construction of Key Interface Profile (KIP).

mechanism to facilitate M&S; these new OVs, in addition to specifying interfaces, are refined to identify these operational parameters for any particular activity component. As the KIP specifies an interface control document, development of OV-8 is expedited by the available operational interfaces, in relation to a particular capability, that can be mapped to specific system interfaces in relation to SV-6 and SV-7. Providing the mapping for SV-6 and SV-7 is outside the scope of this discussion. The objectives of proposing OV-8 and OV-9 have been to empower the DoDAF with M&S at the operational development stage and to show how different capabilities can be composed from various independent activities. Proving interface specification and allowing composition using this interface specification in OV-8 is a key element of this proposition. The refined OV-8 allows for specification of parameters, for both an activity and the composed capability that will be mapped to the NR-KPP with reference to any particular DoDAF architecture.

End-to-end interoperability is defined to include both the technical exchange of information and operational effectiveness of that exchange, as required for mission accomplishment (Joint Capabiliities Integration and Development, 2004). Accordingly, the strategy discussed next has the overarching goal of assessing end-to-end information exchange in critical joint mission threads that are stated within

DoDAF by system developers and users. Figure 12.7 provides an overview of the approach illustrating how it operationalizes the guidance for NR-KPP development and assessment.

12.5 DoDAF-COMPLIANT ARCHITECTURES

As stated earlier, DoDAF has been mandated as the common denominator across DoD agencies. The DoDAF-based architecture before deployment must be certified by agencies like Joint Interoperability Test Command for DoDAF compliance. DoDAF architecture is a complex set of documents with no mandated formal verification and validation processes. As a result, granting compliance certificates to DoDAF architecture appears to be a complicated task. While it is one thing to eyeball the documents, it is another thing to have an executable model that satisfies mission thread requirements. The key question is: Is an executable model of a would-be DoDAF-compliant architecture a mandatory requirement to receive certification?

An affirmative answer to this question enforces the use of the formal modeling and simulation framework that is founded on the general systems theory; DoDAF systems are dynamical systems in real world. Using DEVS and SES as the formal foundation is the understated outcome of such exploration. A negative answer to this question will develop a test-case generation strategy that interacts with a live system and not an executable model. Certainly, formal rigor is also needed here. The only difference is that an executable model is not required. The second approach was implemented in the ATC-Gen case study described in Mak et al. (2009).

The next question is: Given that an executable model could be generated from the formal DoDAF architecture specification, what is the minimal set of documents from the entire DoDAF set that is required for certification?

We have shown in this chapter and the previous chapter that an executable DoDAF 1.0 specification has two added OV documents, namely, OV-8 and OV-9, that are focused toward the M&S discipline. Not all documents are needed to construct these documents. However, an executable DoDAF specification, while essential, does not guarantee compliance either.

The last question is: How will the certification be executed?

The DoDAF being a static document set, with no underlying formal systems theory, is a complex multivariable decision space. We suggest that the certification process follows the steps:

1. *Request*: The initiation stage will require the community of interest (COI) requesting the certification to submit a minimal set of documents for an initial review to the certification coordinator. Table 12.2 shows the document set submitted to the coordinator.
2. *Evaluation*: The evaluation stage is initiated by the certification coordinator as a request for more information to the COI. This will require the document set that helps build the executable model of the DoDAF

TABLE 12.2
Initial DoDAF Document Set to Request DoDAF Compliance

DoDAF View	Description	Products Required to Support Certification by Coordinator
AV-1	Overview and Summary Information: Scope, purpose, intended users, environment depicted, analytic findings	• Mission thread descriptions • Overall MOEs (global performance levels that the collaboration must be able to accomplish, including workload levels and timing requirements) • These top-level MOEs will then be decomposed into MOPs that are elaborated and reconciled with those of COIs in OV-5 and OV-6 views.
AV-2	Integrated Dictionary: Data repository with definitions of all terms used in all products	• Pointers to data assets in the DoD Discovery Metadata Specification (DDMS) Registry that will be used in the exercise and that are not accounted for in any particular COI's contribution.
OV-2	Operational Node Connectivity Description	• Provide each operational node information as to which organization it will be assigned to, along with the classification level of this organization.
OV-3	Operational Information Exchange Matrix: Information exchanged among nodes with attributes	• Information exchange requirements as introduced by the NECC approach (or equivalent) • Common information exchange data model, if employed
OV-5	Operational Activity Diagram	• Activities performed by operational nodes and their hierarchical relationships.
SV-6	Data Exchange Matrix: Provides details of information exchanges among systems/services	• Any additional information that might be needed to enable Joint Interoperability Test Command to map OV-3 and SV-1 products (as outlined in Table 12.3) to SV-6 representation of Web service to Web–service interactions.

specification. Table 12.3 lists the document set. The certification agency will build the executable model and will conduct formal M&S-based testing and analysis.

3. *Conformance*: The conformance stage will highlight the gaps in the DoDAF operational model and will send a validation report to the COI along with various tolerance levels. Either the COI will submit a revision of DoDAF specs or the process terminates with no certification due to lack of the incomplete DoDAF specification.

TABLE 12.3
DoDAF Document Set to Develop an Executable Model by COI Participant

DoDAF View	Description	Products Required to Support Interoperability Certification Supplied by COI Participant
OV-4	Organizational Relationships	• Provide organization nodes and roles of personnel involved in exercise
OV-6b,c	Operational Event—Trace Description	• MOPs (performance levels that its services must be able to accomplish including workload levels and timing requirements) • Descriptions of human participation in operations must be specified at an appropriate level of resolution
OV-7	Logical Data Model Documentation of the data requirements and structural business rules	• Translations to/from its data schemas and the common information exchange data model, if employed, or alternatively • Translation to/from its data schemas and the COIs that it will interact with
OV-8	Activity-Component Mapping	• Tabular format of mapping of all the components responsible for executing activities listed in OV-5 and OV-6
OV-9	Rule-Component Mapping	• Tabular format of mapping of all the components and their activities, with rules specified in OV-6a
SV-1	Systems Interface Description: Identification of systems and system components and their interconnections within and between nodes	• Pointers in the DDMS registry to its data assets, including its XML schema, that will be used (to be accumulated in AV-2) • Web Service Description Language descriptions for consuming any Web services that it will provide or requires from sources external to the collaboration • Mapping from Information Exchange Requirements in OV-3 to specific Web-service definitions (in the form of a table)
SV-2	Communication Description	• Minimum network configuration as a basic assurance level
SV-4	Functionality Description	• Description of encapsulation of its web services—boundaries enclosing which web services are to be considered internal to it and those that are public (exposed to the test exercise)
SV-5	Operational Activity to Function Traceability Matrix	• An exchange table that provides translation or transformation of top-level MOEs at the systems level to the operational level (OV5,6b.c) MOPs

4. *Certification*: Once the conformance level has been reached, the certification agency will document the conformance, performance, and tolerance analyses of the submitted architecture and will request the remaining DoDAF documents that were not submitted earlier. Once submitted by COI, a DoDAF-compliant certification is issued and the architecture is entered in the DoD repository.

12.6 SUMMARY

DoDAF documents are static documents and M&S is not mandated therein. Testing a static document without a formal temporal behavior in an operational setting is impossible in such a case. This chapter discussed various testing methodologies that are prevalent in the industry and academia for software programs, models, architectures, and test-case generation mechanisms, but all fall short when it comes to a level above the architecture, that is, the architecture framework such as DoDAF. DEVS and SES, with the underlying formal systems theory, have proven to be a strong contender for developing model-based and simulation-based testing methodologies. DoDAF was cast in SES, and Chapter 11 has put DoDAF in DEVS to give it a formal representation. We have also discussed how testing is different for an architecture and an architecture framework. We have also highlighted the role of NR-KPP and KIP in determining the integration, interoperability, and executable nature of any DoDAF architecture. Finally, we have discussed various steps that may lead a DoDAF-architecture to a DoD certification.

REFERENCES

Avritzer, A., & Weyuker, E. (1995). The automatic generation of load test suites and the assessment of the resulting software. *IEEE Transactions on Software Engineering, 21*(9), 705–716.

Basanieri, F., Bertolino, A., & Marchetti, E. (2002). The cow suite approach to planning and deriving test suites in UML projects. *Proceedings of 5th International Conference on UML: Model Engineering Languages, Concepts and Tools* (pp. 383–397). Dresden, Germany.

Baudry, B., Fleurey, F., Jzquel, J., & Traon, Y. (2002). Genes and bacteria for automated test cases optimizations in the .NET environment. *Proceedings of the 13th International Symposium on Software Reliability Engineering* (pp. 195–206). Annapolis, MD.

Bernot, G., Gaudeland, M., & Marre, B. (1991). Software testing based on formal specifications: A theory and a tool. *Journal of Software Engineering, 6*(6), 387–405.

Bertolino, A., & Inverardi, P. (1996). Architecture-based software testing. *Proceedings of International Software Architecture Workshop*. San Francisco, CA.

Beydeda, S., & Gruhn, V. (2003). Test data generation based on binary search for class-level testing. *Book of Abstracts: ACS/IEEE International Conference on Computer Systems and Applications*. IEEE Computer Society.

Briand, L., & Labiche, Y. (2002). A UML-based approach to system testing. *Journal of Software and Systems Modeling, 1*(1), 10–42.

Adrion, W.R., Branstad, M.A., Cherniavsky, J.C. (1981). *Validation, Verification and Testing of Computer Software*. National Bureau of Standards, Center for Programming Science and Technology, United States.

Clasessen, K., & Hughes, J. (2000). QuickCheck: A lightweight tool for random testing of haskell programs. *SIGPLAN, 35*(9), 268–279.

CockBurn, A. (1997). Structuring use cases with goals. *Journal of Object Oriented Programming*, 35–40.

DeMillo, R., & Offutt, A. (1991). Constraint-based automatic test data generation. *IEEE Transactions on Software Engineering, 17*(9), 900–910.

Diaz, E., Tuva, J., & Blanco, R. (2003). Automated software testing using metaheuristic technique based on Tabu. *Proceedings of the 18th IEEE International Conference on Automated Software Engineering* (pp. 310–313). IEEE Computer Society.

Dick, J., & Faivre, A. (1993). Automating the generation and sequencing of test cases from model-based specifications. *Proceedings of International Symposium on Formal Methods.* Odense, Denmark.

D'Souza, D., & Wills, A. (1998). Interaction models: uses, case actions and collaborations. In *Objects, Components and Frameworks with UML: The Catalysis Approach*. Boston, MA: Addison-Wesley.

Duran, J., & Ntfasos, S. (1999). An evaluation of random testing. *Proceedings of the 2nd Conference on Computer Science and Engineering*. Montreal, QC, Canada.

Eickelmann, N., & Richardson, D. (1996). An evaluation of software test architectures. *Proceedings of 18th IEEE International Conference on Software Engineering*. Berlin, Germany.

Ferguson, R., & Korel, B. (1996). Generating test data for distributed software using the chaining approach. *Journal of Information and Software Technology, 38*(5), 343–353.

Frohlich, P., & Link, J. (2000). Automated test case generation from dynamic models. *Proceedings of the 14th European Conference on Object Oriented Programming*. Sophia Antipolis and Cannes, France.

Hamlet, D. (1994). Random testing. In J. Marciniak, Ed., *Encyclopedia of Software Engineering* (pp. 970–978). John Wiley & Sons. DOI: 10.1002/0471028959.

Harrold, M. (1998). Architecture-based regression testing of evolving systems. *International Workshop on the Role of Software Architecture in Testing and Analysis*. Clearwater Beach, FL.

Harrold, M. (2000). Testing: A roadmap. *Proceedings of ACM International Conference on Software Engineering: The Future of Software Engineering*. Limerick, Ireland.

Hartmann, J., Vieira, M., Foster, H., & Ruder, A. (2004). *UML-Based Test Generation and Execution. TAV21.* Berlin, Germany.

Hu, X., & Zeigler, B. (2005). Model continuity in the design of dynamic distributed real-time systems. *IEEE Transactions on Systems, Man and Cybernetics-Part A: Systems and Humans, 35*(6), 867–878.

Jamoussi, A. (1997). An automated tool for efficiently generating a massive number of random test cases. *Proceedings of the 2nd High-Assurance Systems Engineering Workshop. IEEE Computer Society.* Washington, DC., USA.

Jones, B., Sthamer, H., & Eyres, D. (1996). Automatic structural testing using genetic algorithm. *Software Engineering Journal, 11*(5), 299–306.

Juristo, N., Morano, A., & Vegas, S. (2004). Reviewing 25 Years of Testing Technique Experiments. *Empirical Software Engineering, 9*(1–2), 7–44.

Kazman, R., Bass, L., Abowd, G., & Webb, M. (1994). SAAM: A method for analyzing the properties of software architectures. *Proceedings of the Sixteenth International Conference on Software Engineering*. Italy.

Kim, Y., Honh, H., Cho, S., Bae, D., & Cha, S. (1999). Test cases generation from UML state diagrams. *IEEE Software, 146*(4), 187–192.

Korel, B. (1990). Automated software test data generation. *IEEE Transactions on Software Engineering, 16*(8), 870–879.

Korel, B., & Al-Yami, A. (1996). Assertion-oriented automated test data generation. *Proceedings of the 18th International Conference on Software Engineering*. IEEE Computer Society.

Legard, B., Peureux, F., & Utting, U. (2002). Automated boundary testing from Z and B. *Proceedings of the International Conference on Formal Methods*. Shanghai, China.

Li, W., & Henry, S. (1993). Object-oriented metrics that predict maintainability. *Journal of Systems and Software, 23*(2), 111–122.

Lin, J., & Yeh, P. (2001). Automatic test data generation for path testing using GAs. *Information Sciences: An International Journal, 131*(1–4), 47–64.

Lutsky, P. (2000). Information extraction from documents for automating software testing. *Journal of Artificial Intelligence in Engineering, 14*(1), 63–69.

Mak, E., Mittal, S., Ho, M.H., Nutaro, J.J. (2010). Automating link 16 testing using DEVS and XML. *Journal of Defense Modeling and Simulation, 7*(1), 39–62.

Meyer, B. (1992). Applying design by contract. *IEEE Computer, 25*(10), 40–51.

Michael, C., McGraw, G., Schatz, M., & Walton, C. (1997). Genetic algorithms for dynamic test data generation. *Proceedings of the 12th International Conference on Automated Software Engineering* (pp. 307–308). Incline Village, NV.

Mittal, S., Mak, E., & Nutaro, J. (2006). DEVS-based dynamic model reconfiguration and simulation control in the enhanced DoDAF design process. *Journal of Defense Modeling and Simulation, 3*(4), 239–267.

Myers, G. (1978). *The Art of Software Testing*. New York, NY: John Wiley & Sons.

Nebut, C., Fleurey, F., Le Traon, Y., & Jezequel, J.-M. (2006). Automatic test case generation: A use case driven approach. *IEEE Transactions on Software Engineering, 32*(3), 140–155.

Offutt, J., Liu, S. (1999). Generating test data from SOFL specifications. *Journal of Systems and Software, 49*(1), 49–62.

Offutt, J., & Abdurazik, A. (1999). Generating test cases from UML specification. *Proceedings of 2nd International Conference on the Unified Modeling Language* (pp. 416–429). Fort Collins, CO.

Offutt, J., & Abdurazik, A. (2000). Using UML collaboration diagrams for static checking and test generation. *Proceedings of the 3rd International Conference on UML*. York, UK.

Offutt, J., Liu, S., Abdurazik, A., & Ammann, P. (2003). Generating test data from state-based specifications. *Journal of Software Testing, 13*(1), 25–53.

Pargas, R., Harrold, M., & Peck, R. (1999). Test data generation using genetic algorithms. *Journal of Software Testing, Verification and Reliability, 9*(4), 263–282.

Ramamoorthy, C., Ho, S., & Chen, W. (1997). On the automated generation of program test data. *IEEE Transactions on Software Engineering, 2*(4), 293–300.

Richardson, D., & Wolf, A. (1996). Software testing at the architectural level. *Proceedings of Software Architecture Workshop*, ACM SIGSOFT.

Richardson, D., Stafford, J., & Wolf, A. (1998). *A Formal Approach to Architecture-based Software Testing*. Irvine: Technical Report, University of California.

Riebisch, M., Philippow, I., & Gotze, M. (2002). UML-based statistical test-case generation. *Revised papers from the Proceedings of International Conference on NetObjectDays on Objects, Components, Architectures, Services, and Applications for a Networked World*, 2591 (pp. 394–411), Erfurt, Germany.

Rothermel, G., Untch, R., & Chu, C. (2001). Prioritizing test cases for regression testing. *IEEE Transactions on Software Engineering, 17*(10), 929–948.

Ryser, J., Berner, S., & Glinz, M. (1998). *On the State of the Art in Requirement-based Validation and Test of Software*. Inst. Fur Informatic, University of Zurich.

Ryser, J., & Glinz, M. (1999). A scenario-based approach to validating and testing software systems using statecharts. *Proceedings of the 12th International Conference on Software and Systems Engineering and their Applications*. Paris, France.

Salem, A., & Subramanian, L. (2004). Utilizing UML use cases for testing requirements. *International Conference on Software Engineering Research and Practice*. Scotland, UK.

Tahat, L., Vaysbur, B., Koreland, B., & Bader, A. (2001). Requirement-based automated black-box test generation. *Proceedings of the 25th Annual International Computer Software and Applications Conference*. Chicago, IL.

Torkar, R. (2005). *A Literature Study of Software Testing and the Automated Aspects Thereof.* Sweden: S-CORE Scientific Report, University of Troliihattan/Uddevalla.

Toth, A., Varro, D., & Pararicca, A. (2003). Model level automatic test generation for UML state-charts. *Proceedings of 6th IEEE Workshop on Design and Diagnostics of Electronic Circuits and Systems*. Poznan, Poland.

Utting, M., Pretshner, A., & Legeard, B. (2006). A Taxonomy of Model-based Testing. *Working Paper Series ISSN 1170-487X.*

Vieira, M., Dias, M., & Richardson, D. (2000). Object-oriented specification-based testing using UML statechart diagrams. *Proceedings of the Workshop on Automated Program Analysis, Testing and Verification*. Limerick, Ireland.

Wagenhals, L., Haider, S., & Levis, A. (2002). Synthesizing executable models of object oriented architectures. *Workshop on Formal Methods Applied to Defense Systems.* Adelaide, Australia.

Whittaker, J. (2000). *What Is Software Testing? And Why It Is So Hard?* IEEE Software, *17*(1), 70–79.

Williams, C. (1999). Software testing and UML.*International Symposium on Software Reliability Engineering*. Boca Raton, FL.

Section III

Netcentric System of Systems

13 DEVS Standard

13.1 INTRODUCTION

Since the early 1970s, the modeling and simulation (M&S) community has tried to define different formalisms for varied systems specifications, even though different models of the source systems existed before their computational implementations. Examples are, for instance, systems described using differential equations in the continuous domain or models operating on a discrete time base.

The DEVS formalism was defined to bring coherence and to unify the field of discrete-event M&S with formal rigor and an underlying systems theoretical framework. Despite this coherence and unity, Discrete Event Systems Specification (DEVS) is still tightly linked to the simulation language implementations, like C++, C#, or Java. Today, there are numerous libraries and tools for expressing DEVS models across the globe, such as DEVSJAVA (ACIMS), xDEVS (Universidad Complutense de Madrid), DEVS/C++ (ACIMS), CD++ (Wainer, 2009), aDEVS (Nutaro, 2005), DEVS-Suite (Kim et al., 2009), and MicroSim (developed in this book). Although this proliferation of libraries shows the numerous advantages in doing DEVS M&S, its multiplicity presents a difficulty in sharing models and modelers still have to learn the programming language in which the simulator is implemented, and most of the time they remain tied to it. There exists a fragmentation in the field as a result of growing specialization of the knowledge.

Thus, a widely accepted framework is now more necessary than ever. This framework is needed not only to reach a universal specification of both models and simulators in a programming environment, but also to allow the system designer to solve fundamental issues such as validation, verification, and interoperation. A common and well-accepted framework is High Level Architecture (HLA) (U.S. Department of Defense, 2001). HLA focuses on interoperability of existing geographically dispersed M&S assets, and it was initially built focusing on defense applications. These days, however, users other than the military domain such as business enterprise and e-commerce are interested in distributed simulation applications as well. HLA does not address this migration of capabilities easily. In addition, it does not focus on how to solve the problem of creating models that are to be executed in a common simulation environment.

We focus on the fact that the DEVS theoretical foundations are, in principle, independent of the various programming languages and hardware platforms. Within a study group of the Simulation Interoperability Standards Organization (SISO), a standard is under development to support interoperability of DEVS models implemented on different platforms as well as with legacy simulations (see Figure 13.1). The goal of the study group is to find a core of the DEVS formalism that is suitable for standardization of activities at the level of

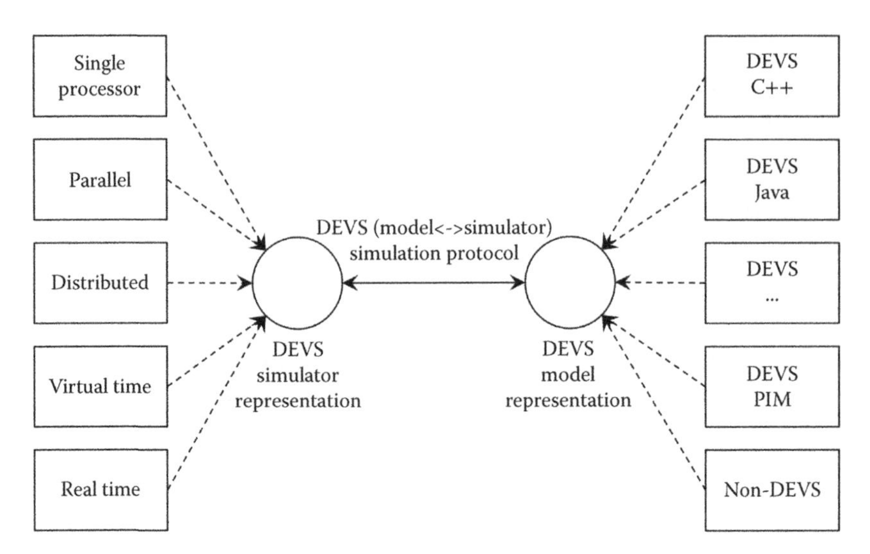

FIGURE 13.1 Conceptual Architecture of Standard (a).

modeling, thereby bridging the gap between existing simulation frameworks and modeling activities using a standard notation. In this part, we discuss fundamentals of various designs on how to achieve these goals through the standardization of DEVS models and their simulators.

One of the main issues to be addressed is a direct consequence of separating the model from the simulator. This results in multiple ways in which the same model can be simulated—all adhering to the abstract simulator specification. For instance, there are virtual-time simulators (where the simulator can skip from one event time to the next without traversing the intervening time interval) and real-time simulators (where time is interpreted as wall clock readings, so the simulator must wait for the interval for its next scheduled event to expire before handling the event). In addition to the different combinations of model type/simulation software, a standard allows for the use of different forms of distribution of model components (e.g., single processor vs. multiprocessor, and within the latter, conservative vs. optimistic time advance for virtual time as well as centralized vs. noncentralized time control in real-time execution).

The standard can also be independent of different implementation platforms, such as Windows versus UNIX; different programming languages, such as Java versus C++; and different networking and middleware frameworks such as .NET versus Apache. As we can see, such a standard can have multiple simulation scenarios. For example, a model may be simulated in virtual time and in real time in both a distributed and a nondistributed fashion.

In summary, the proliferation of DEVS-based M&S engines has brought the need to improve and standardize DEVS tools, facilitating the work of DEVS designers, independently of the programming language implementations or algorithmic code expressions used. To better understand the problem, let us consider that DEVS categorically separates the model and the simulator. In this regard, the DEVS formalism specifies the abstract simulation engine that correctly simulates DEVS atomic

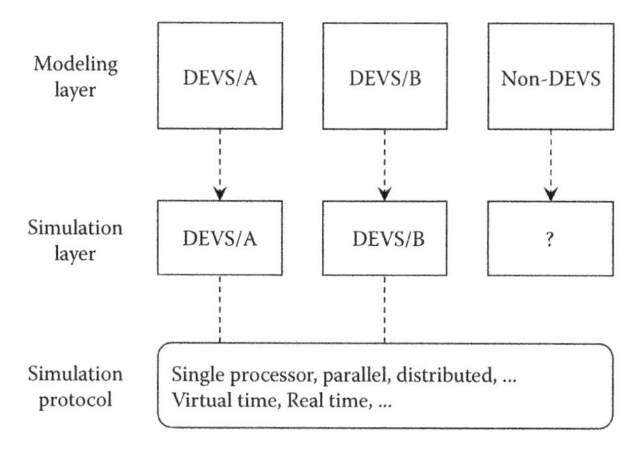

FIGURE 13.2 Conceptual Architecture of Standard (b).

and coupled models. Interpreted in a distributed simulation environment, the DEVS abstract simulator gives rise to a general protocol that has specific mechanisms for declaring who takes part in the simulation, that is, the federates. It also specifies how federates interact in an iterative cycle that controls how time advances, when federates exchange messages, and perform internal-state updating. A significant feature in comparison to simulation based on the HLA standard is that if the federates in simulation are DEVS compliant, then the simulation can be proved to be correct in the sense that the DEVS closure under coupling theorem guarantees a well-defined resulting structure and behavior.

In the next sections, we show how we can define a DEVS M&S standard platform. To this end, we transform the structure of the standard presented in Figure 13.1 into the structure presented in Figure 13.2, which will facilitate our definition.

We divide DEVS into three layers. The first one is the modeling layer, which represents the ways in which a model can be represented: using a specific DEVS M&S platform (DEVS/A or DEVS/B) in Figure 13.2, or using a Non-DEVS representation (e.g., a model definition using MATLAB®. The second layer is the simulation layer, which is strongly coupled to the modeling layer by the simulation relation. Every DEVS M&S platform has its own M&S layers. The last one is the simulation protocol, which resides as part of the simulation layer. Each simulator may implement one or several simulation protocols, like parallel or distributed simulations, simulation using real time or virtual time, and so on. Our purpose in this chapter is to show how to define a DEVS M&S standard platform (that we will name DEVS/* in the following) that is able to simulate its own models and models coming from other simulation platforms.

13.2 PLATFORM-INDEPENDENT MODELS AND DEVS STANDARDIZATION

Standardizing DEVS model representation as a platform-independent model (PIM) allows a model to run on any DEVS simulation environment. This is powerful in a sense that a model can be retrieved from a repository and run locally on a different tool other than the model's original, intended running environment. This is highly recommended to avoid the model run in a distributed simulation between remote environments for obvious performance reasons.

Different groups have used PIMs as a mechanism for interchanging model information. To do so, a PIM vocabulary needs to be defined, which is formally defined by a PIM schema against which every file written in the vocabulary can be automatically validated. This arrangement gives a PIM vocabulary several important advantages over platform-specific models (PSMs):

- Validation against a schema promotes stability of the standard.
- The schema can restrict data types.
- The schema can define key data to insure consistency in the models (including the validation of important properties in the data values; i.e., unicity or duplicity of use, when needed).
- The schema can be extended to include constraint types or simulator directives. Files that are validated under the original schema continue to validate under the extended one (though, of course, the reverse is not guaranteed).

Indeed, PIM libraries for validating models, defining keys, compressing files, and so on are provided by numerous PIM tools designed for manipulating and parsing the defined schema. It suffices to define our PIM vocabulary in the form of a schema that these tools can work with. This contrasts to ad hoc formats that require writing, debugging, and maintaining routines equivalent to these tools.

Figure 13.3 shows a DEVS/PIM standard definition framework. This figure extends the previous framework depicted in Figure 13.2, but including a DEVS/PIM definition as well as a hypothetical DEVS M&S standard platform (named DEVS/*).

In this section, we address the construction of the DEVS/PIM definition. This process starts from the development of a DEVS/PIM schema that includes the grammar for both the structure and behavior of DEVS models, that is, coupled and

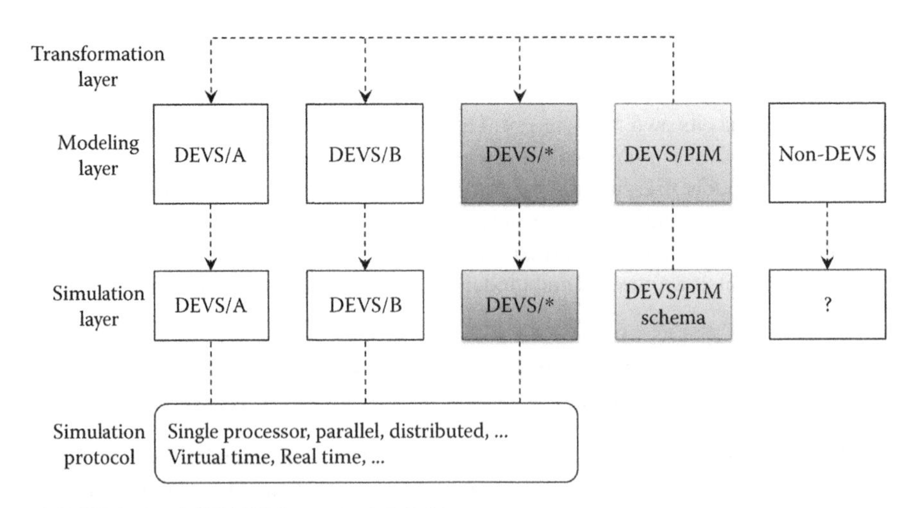

FIGURE 13.3 DEVS/PIM standard definition.

atomic models. When it is defined, the system designer is able to create DEVS models that may be validated against the earlier version of the schema. It forms a valid DEVS/PIM. Next, there are two possibilities. The first one is to transform the previous scenario into a PSM for a concrete simulation engine, such as DEVS/A, DEVS/B, or DEVS/*, as shown in Figure 13.3. The second one is the development of the DEVS/PIM simulator in a high-level programming language. Finally, a reverse engineering layer may be provided to come back from the PSMs to the PIMs. The reverse engineering is the hardest task to accomplish. In this case, DEVS/PIM must include a (complex) parser to obtain the corresponding DEVS/PIM scenario from a set of DEVS PSM files (atomic and coupled models) that compose the DEVS scenario. Thus, in some cases, other approaches are needed to facilitate the reverse engineering. In this chapter, we work neither on the DEVS/PIM simulator nor on the reverse engineering process, mainly because there are other approaches easier to develop that also implement a DEVS M&S standard platform. In fact, a full DEVS/PIM simulator does not exist in the literature yet. Table 13.1 shows different approaches that partially implement a DEVS/PIM standard (some of them presented in this book). Note that the modeling layer comprises both a structure and behavior model definition. All the proposals shown in Table 13.1 define a DEVS structure, whereas the behavior is completely defined in eUDEVS, DEVSML, and DML. Transformations from PIMs to PSMs are also defined in all the approaches. In addition, three PIM simulators have been developed for XFD-DEVS, DEVS/SCXML, and DEVSML 2.0. Finally, the reverse engineering has been partially applied in DEVS/Schema and DEVSML 1.0.

TABLE 13.1
Different DEVS/PIM Proposals of Standard

	Structure	Behavior	PIM→PSM	PIM Simulator	Reverse Engineering
DEVS/Schema (Risco-Martín et al., 2007)	✓	✓[a]	✓	✘	✓[b]
XFD-DEVS (Mittal et al., 2007)	✓	✓[a]	✓	✓	✘
DEVS/SCXML (Risco-Martín et al., 2007)	✓	✓[c]	✓	✓	✘
eUDEVS (Risco-Martín et al., 2009)	✓	✓	✓	✘	✘
DEVSML 1.0 (Mittal et al., 2007)	✓	✓	✓	✘	✓[b]
DEVSML 2.0 (Mittal et al., 2012)	✓	✓	✓	✓	✘
DML (Touraille et al., 2009)	✓	✓	✓	✘	✘

[a] Behavior limited by the corresponding schema definition.
[b] Implemented for two Java simulation platforms (DEVSJAVA and xDEVS).
[c] Behavior limited by the corresponding SCXML Schema definition.

As a result, there is not a fully functional DEVS standard to easily integrate all the different simulation engines developed and simultaneously define DEVS models. However, this should not surprise us, since this also happens in many software (as well as hardware) engineering fields. For example, in object-oriented programming languages, there are many Object-oriented programming (OOP) languages, such as Java, C#, or C++, and probably the most extended PIM specification for OOP software specification and definition is UML. There are also some integrated IDEs like NetBeans or Eclipse that allow the software engineer to program in several OOP languages. However, it is still difficult to build communications between, for example, C++ and Java applications, even when the full UML OOP model is provided.

In the rest of this chapter, we show a practical approach for the development of a DEVS M&S standard framework. By "standard," we mean a DEVS framework that, according to Figures 13.1 through 13.3, is able to model (1) systems coming from different DEVS platforms, and (2) systems coming from a non-DEVS framework. In this regard, we introduce the basis of a DEVS simulation framework that is able to simulate DEVS heterogeneous models.

13.3 MODELING LAYER

13.3.1 Standard DEVS Models Using XML Schemas

In the previous section, we provided the basic concept on how to unify different DEVS models coming from different libraries in order to reach a full DEVS interoperability at both the M&S levels. However, to accomplish a valid standardized framework, DEVS models (in particular, the behavior of these models) should be defined in a standard way. We explored different ways to reach a standard representation (UML, SysML, DEVSML, etc.) in this book. However, for the sake of clarity, we show in this section a way to represent DEVS models using XML schemas. Note that this example can be extended to more complex formalisms like UML state charts, mainly because these charts can be stored in XMI, which is an XML representation of a model behavior. A valid hypothesis is that if we are able to transform platform-independent DEVS models into platform-specific DEVS models (including structure and behavior), then any DEVS simulator will be able to simulate a DEVS/PIM. As a result, a DEVS/PIM simulator would have no tight platform binding, and any DEVS simulation engine (like DEVS/*) in Figure 13.3 could become a standard. In the following, we show one way to define DEVS/PIM atomic and coupled models via XML schemas.

13.3.1.1 Example DEVS/XML Schema

The XML schema that defines atomic DEVS models is called the Atomic DEVS-XML Schema. It is organized in a structure based on the same elements as DEVS formalism. These are inputs, outputs, states set, transition functions, output function, and time-advance function. Figure 13.4 shows the general structure of the Atomic DEVS-XML Schema (Risco-Martín et al., 2007).

The input and output specification of DEVS-XML is made by means of the port definition (see Figure 13.5). These ports have an internal structure formed by a group of signals: the ports may be simple (only one signal or message) or composed (a set of signals).

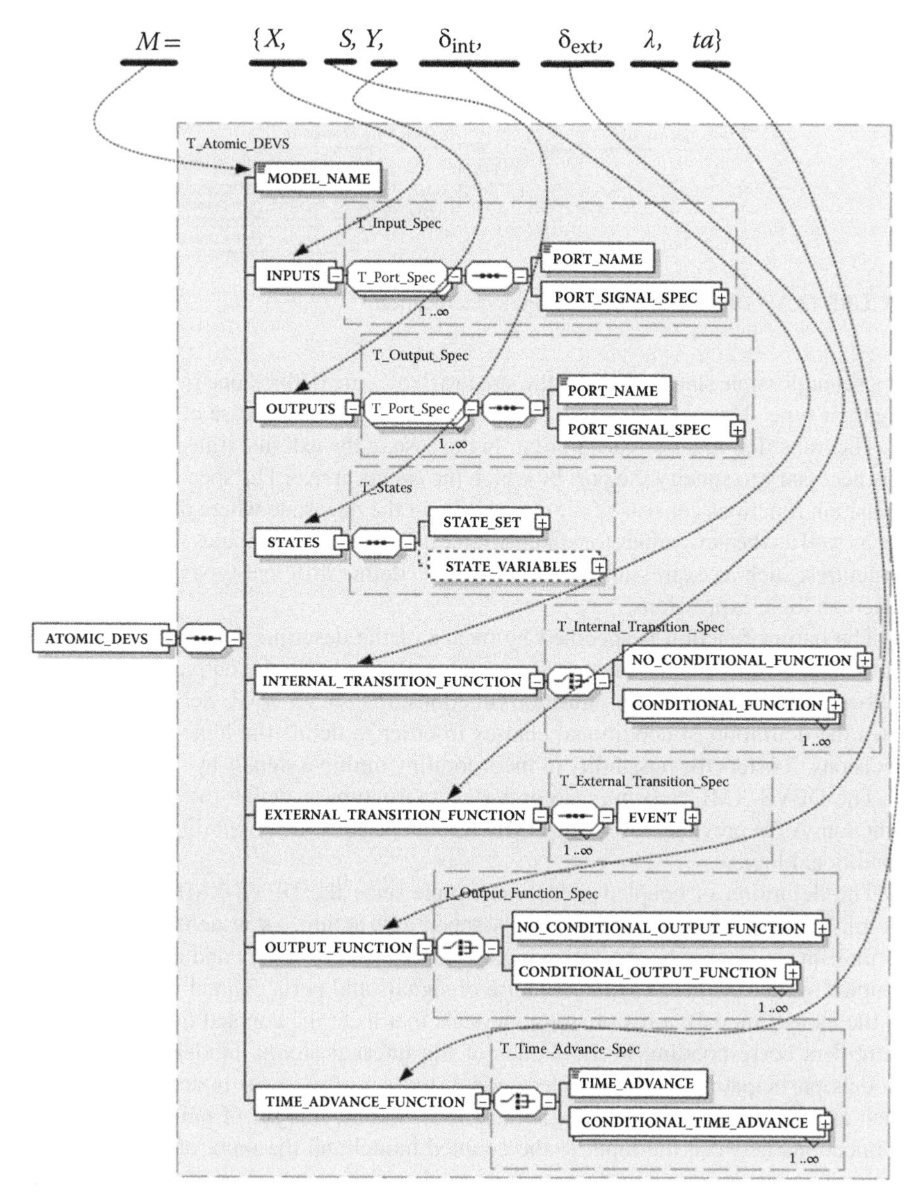

FIGURE 13.4 DEVS-XML Schema: Atomic model structure.

The DEVS-XML Schema incorporates the definition of signal types. These types are divided into two categories: elementary (integer, real, positive integer, etc.) and enumerated. This set of types can be extended within the DEVS-XML Schema to cover specific types needed.

The state specification is made by means of two structures: one for the state of the system and the other for the set of state variables. The state of the system is defined

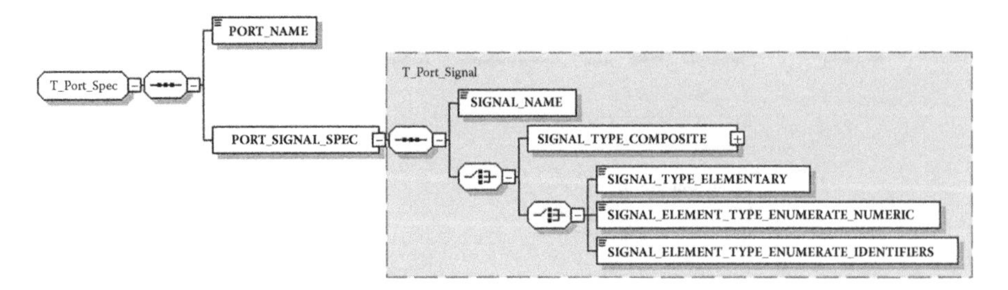

FIGURE 13.5 DEVS-XML Schema: Port specification.

as a set of possible states, whereas the state variables are defined one by one, specifying their type. These data types are the same explained for the case of signals.

The transition functions are similar. In the case of the external transition function, it is necessary to specify the port by which the events arrive. The specification of the transition functions consists of the description of the new state where the system will be, as well as the new values for state variables. This work contributes some behavior structures, such as expressions, that allow us to define different types of transitions based on logic expressions.

The output function is specified by means of the description of SEND clauses. These clauses provide the mechanism to define the path that the output signals must follow. As in the case of the transition functions, the DEVS-XML Schema incorporates the definition of conditional clauses in order to define the logic of the output decisions. It offers the possibility of incorporating multiple signals by multiple ports.

The DEVS-XML Schema also provides a structure to define the time-advance function. As in previous cases, the XML schema permits the possibility of including conditional logic.

The definition of coupled models is simple once the DEVS-XML Schema for atomic models has been defined. This schema structure, as is defined in DEVS formalism, is formed by the following elements: a set of inputs and outputs of the coupled system (with the same structure of signals and ports defined in the schema of the atomic model); a list of atomic models that form the coupled model; a list of identifiers corresponding to the names of the internal atomic models; the atomic models participating in the coupled model; the set of external input connections, such as a set of pairs (port/signal, atomic/port/signal; this set of pairs defines the connections between the input to the coupled model and the input of the elements within the coupled model); the set of external output connections, such as a set of pairs (atomic/port/signal, port/signal) that defines the connections between the output of the atomic models inside of the coupled model and the output of the coupled model); and the set of internal connections between atomic models, such as a set of pairs (atomic/port/signal, atomic/port/signal) that defines the connections among elements inside of the coupled model.

Figure 13.6 shows the XML schema for coupled DEVS models with the elements described above.

We now present an example of an atomic model developed using the XML schema described above. The atomic model shown in Listing 13.1 is the processor model.

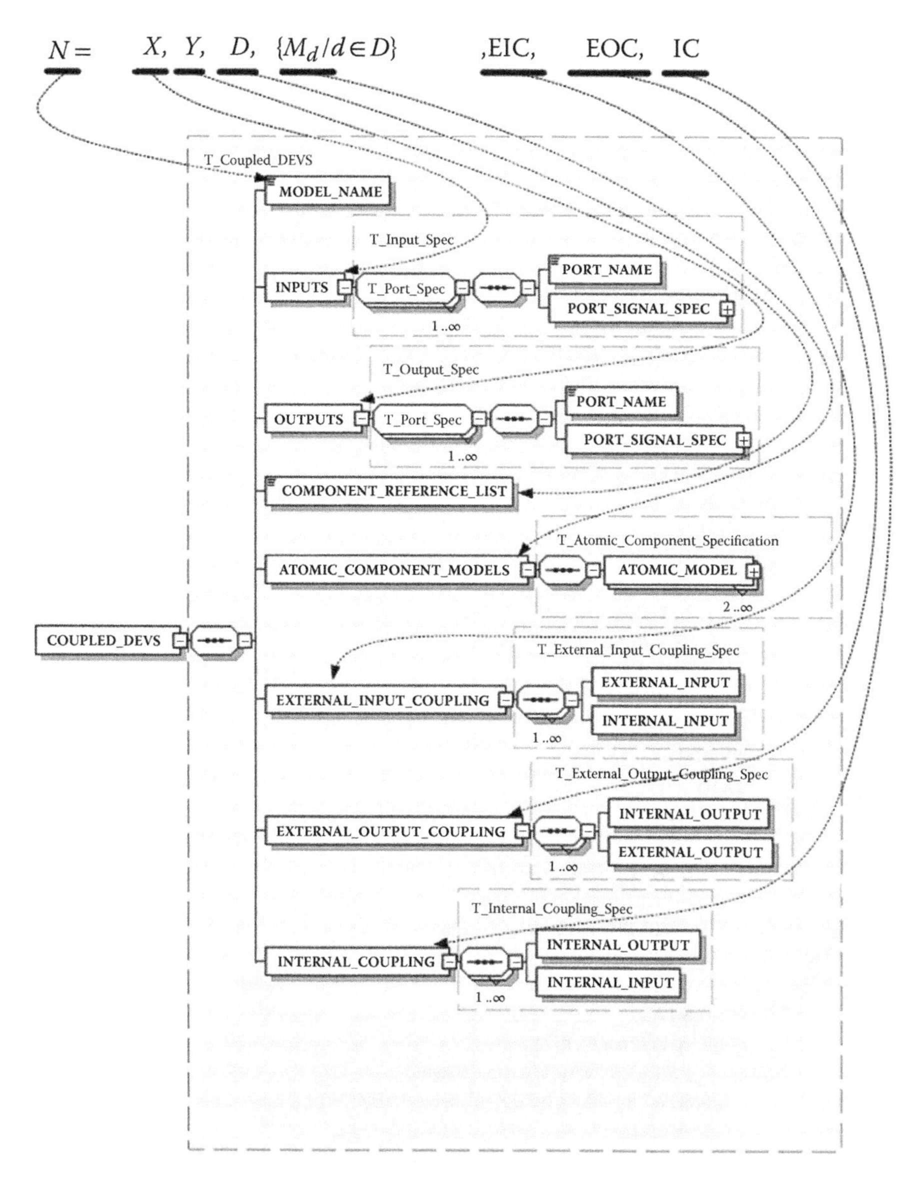

FIGURE 13.6 DEVS-XML Schema for coupled models.

13.3.2 DEVS-To-DEVS Interoperability

As stated above, there are many libraries for expressing DEVS models. All have efficient implementations for executing the DEVS protocol and provide advantages of object-oriented frameworks such as encapsulation, inheritance, and polymorphism. To simplify the notation, we use the term "DEVS/A" to denote a hypothetical DEVS library implemented in an "A" programming language and "DEVS/B"

LISTING 13.1 PROCESSOR.XML

```
<?xml version="1.0" encoding="UTF-8"?>
<ATOMIC_DEVS xmlns:xsi="http://www.w3.org/2001/
   XMLSchema-instance"
            xsi:noNamespaceSchemaLocation="D:\Atomic.xsd">
  <ATOMIC_MODEL_NAME>processor</ATOMIC_MODEL_NAME>
    <INPUTS>
      <PORT_NAME>inPortIn</PORT_NAME>
      <PORT_SIGNAL_SPEC>
             <SIGNAL_NAME>Job</SIGNAL_NAME>
             <SIGNAL_TYPE_ELEMENTARY>
              Integer
             </SIGNAL_TYPE_ELEMENTARY>
      </PORT_SIGNAL_SPEC>
      </INPUTS>
      <OUTPUTS>
      <PORT_NAME>oportOut</PORT_NAME>
      <PORT_SIGNAL_SPEC>
              <SIGNAL_NAME>Job</SIGNAL_NAME>
              <SIGNAL_TYPE_ELEMENTARY>
               Integer
              </SIGNAL_TYPE_ELEMENTARY>
      </PORT_SIGNAL_SPEC>
      </OUTPUTS>
      <STATES>
      <STATE_SET>
             <STATE_SET_NAME>phase</STATE_SET_NAME>
             <STATE_SET_VALUES>
              active passive
             </STATE_SET_VALUES>
      </STATE_SET>
      <STATE_VARIABLES>
             <STATE_VARIABLE_NAME>
              processingTime
             </STATE_VARIABLE_NAME>
             <STATE_VARIABLE_TYPE_ELEMENTARY>
              Real
             </STATE_VARIABLE_TYPE_ELEMENTARY>
             <STATE_VARIABLE_NAME>
              currentJob
             </STATE_VARIABLE_NAME>
             <STATE_VARIABLE_TYPE_ELEMENTARY>
              Integer
             </STATE_VARIABLE_TYPE_ELEMENTARY>
      </STATE_VARIABLES>
      </STATES>
```

```
    <INTERNAL_TRANSITION_FUNCTION>
    <NO_CONDITIONAL_FUNCTION>
           <NEW_STATE>passive</NEW_STATE>
           <STATE_VARIABLE_UPDATE
             State_Variable_Name="processingTime"
             State_Variable_Value="5"/>
    </NO_CONDITIONAL_FUNCTION>
    </INTERNAL_TRANSITION_FUNCTION>
    <EXTERNAL_TRANSITION_FUNCTION>
    <EVENT>
          <PORT Port_Name="inPortIn"/>
      <CONDITIONAL_FUNCTION>
          <STATE_CONDITION>phase==passive</STATE_CONDITION>
          <TRANSITION_FUNCTION>
                  <NEW_STATE>active</NEW_STATE>
                  <STATE_VARIABLE_UPDATE
                  State_Variable_ Name="processingTime"
                  State_Variable_Value="5"/>
          </TRANSITION_FUNCTION>
             </CONDITIONAL_FUNCTION>
             <CONDITIONAL_FUNCTION>
          <STATE_CONDITION>phase==active</STATE_CONDITION>
          <TRANSITION_FUNCTION>
            <NEW_STATE>active</NEW_STATE>
            <STATE_VARIABLE_UPDATE
              State_Variable_Name="processingTime"
              State_Variable_Value="sigma-e"/>
            <STATE_VARIABLE_UPDATE
              State_Variable_Name="currentJob"
              State_Variable_Value
               ="xSignalValue('in','Job')"/>
          </TRANSITION_FUNCTION>
            </CONDITIONAL_FUNCTION>
     </EVENT>
      </EXTERNAL_TRANSITION_FUNCTION>
      <OUTPUT_FUNCTION>
     <NO_CONDITIONAL_OUTPUT_FUNCTION>
      <SEND Port_Name="oportOut" Signal_Name="Job"
            Signal_Value="job"/>
    </NO_CONDITIONAL_OUTPUT_FUNCTION>
     </OUTPUT_FUNCTION>
     <TIME_ADVANCE_FUNCTION>
    <TIME_ADVANCE>sigma</TIME_ADVANCE>
     </TIME_ADVANCE_FUNCTION>
</ATOMIC_DEVS>
```

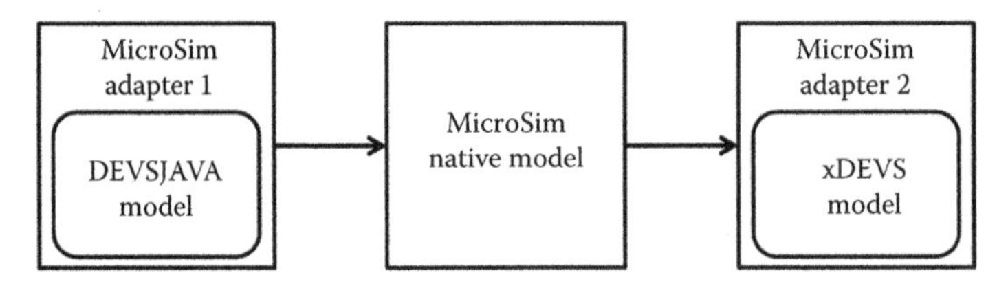

FIGURE 13.7 Heterogeneous DEVS/* model.

to denote a DEVS library implemented in a "B" programming language. Finally, we denote "DEVS/*" as a third DEVS simulation engine that will be the standard simulation engine.

First of all, to find an interoperable DEVS M&S framework, we must design compatibility between different DEVS simulation engines and DEVS/*. In other words, a model implemented using DEVS/A or DEVS/B should be able to be simulated in DEVS/*. In addition, a hybrid-coupled model implemented using all DEVS/A, DEVS/B, and DEVS/* models should be able to be simulated in DEVS/*. Finally, to reach a full standard, we should define a set of transformations to also simulate DEVS/PIMs. For the sake of clarity in the examples provided in this chapter, we use MicroSim as a DEVS/* interoperable framework. Indeed, we have extended our MicroSim engine to provide interoperability between different DEVS M&S frameworks.

The main modification to our current MicroSim architecture is that we must add some adapters of other DEVS simulation platforms to the modeling package. For instance, as Figure 13.7 depicts, a MicroSim atomic component can contain a DEVSJAVA atomic component or an xDEVS atomic component (as well as its own MicroSim native atomic component). With this implementation, MicroSim allows interoperability among different DEVS simulation engines.

Listing 13.2 shows how to build a DEVSJAVA adapter.

LISTING 13.2 ATOMICDEVSJAVA.JAVA (DEVSJAVA WRAPPER)

```
public class AtomicDevsJava extends Atomic {
  // DEVSJAVA atomic  model
  private genDevs.modeling.atomic atomic;
  // Input and output ports
  public HashMap<String, Port> input
                           = new HashMap<String, Port>();
  public HashMap<String, Port> output
                           = new HashMap<String, Port>();

  public AtomicDevsJava(genDevs.modeling.atomic atomic) {
    super(atomic.getName());
    this.atomic = atomic;
    this.atomic.initialize();
  }
```

```
  @Override
  public double ta() {
    double sigmaAux = atomic.ta();
    if (sigmaAux >=
        genDevs.modeling.DevsInterface.INFINITY) {
      return Constants.INFINITY;
    }
    return sigmaAux;
  }

  @Override
  public void deltint() {
    atomic.deltint();
  }
  @Override
  public void deltext(double e) {
    genDevs.modeling.message msg = buildMessage();
    atomic.deltext(e, msg);
  }

  @Override
  public void deltcon(double e) {
    genDevs.modeling.message msg = buildMessage();
    atomic.deltcon(e, msg);
  }

  @Override
  public void lambda() {
    genDevs.modeling.message msg = atomic.out();
    genDevs.modeling.ContentIteratorInterface itr =
                                  msg.mIterator();
    while (itr.hasNext()) {
      genDevs.modeling.ContentInterface devsJavaPort =
                                  itr.next();
      Port port = output.get(devsJavaPort.getPortName());
      if (port != null) {
        port.addValue(devsJavaPort.getValue());
      }
  }
}

  public genDevs.modeling.message buildMessage() {
    genDevs.modeling.message msg =
                                new genDevs.modeling.message();
    Iterator<String> itr = input.keySet().iterator();
    while (itr.hasNext()) {
      String portName = itr.next();
      Collection<Object> values =
```

```
                                    input.get(portName).getValues();
        for (Object value : values) {
          genDevs.modeling.content con =
          atomic.makeContent(portName, (GenCol.entity) value);
          msg.add(con);
        }
      }
      return msg;
    }
}
```

The *AtomicDevsJava* class has a DEVSJAVA atomic model (*atomicDevsJava*) that is used to invoke the different transition functions as well as the time advance and output functions. In both the external transition and output functions, the input and the output must be transformed to the MicroSim specification, respectively. A similar wrapper can be developed for the xDEVS simulation engine (Listing 13.3).

LISTING 13.3 ATOMICXDEVS.JAVA

```
public class AtomicXDevs extends Atomic {

  private xdevs.kernel.modeling.Atomic atomicXDevs;

  public AtomicXDevs(xdevs.kernel.modeling.Atomic
                     atomicXDevs) {
    super(atomicXDevs.getName());
    this.atomicXDevs = atomicXDevs;
  }

  @Override
  public double ta() {
    double sigmaAux = atomicXDevs.ta();
    if (sigmaAux >= xdevs.kernel.modeling.Atomic.INFINITY) {
      return Constants.INFINITY;
    }
    return sigmaAux;
  }

  @Override
  public void deltint() {
    atomicXDevs.deltint();
  }

  @Override
  public void deltext(double e) {
```

```
    buildInput();
    atomicXDevs.deltext(e);
  }

  @Override
  public void deltcon(double e) {
    buildInput();
    atomicXDevs.deltcon(e);
  }

  @Override
  public void lambda() {
    atomicXDevs.lambda();
    Collection<xdevs.kernel.modeling.PortInterface> ports =
                                 atomicXDevs.getOutports();
    for (xdevs.kernel.modeling.PortInterface portXDevs :
                                                  ports) {
      if (!portXDevs.isEmpty()) {
        Port port = outPorts.get(portXDevs.getName());
        if (portXDevs instanceof
                      xdevs.kernel.modeling.Port) {
          port.addValue(((xdevs.kernel.modeling.
                         Port) portXDevs).getValue());
        } else if (portXDevs instanceof
                   xdevs.kernel.modeling.PortComplex) {
          port.addValues(((xdevs.kernel.modeling
              .PortComplex) portXDevs).getValues());
        }
      }
    }
  }

  public void buildInput() {
    for (Port port : inPorts.values()) {
      xdevs.kernel.modeling.PortInterface portXDevs =
                      atomicXDevs.getInport(port.getName());
      boolean isComplexPort = (portXDevs instanceof
          xdevs.kernel.modeling.PortComplex) ? true : false;
      if (isComplexPort) {
        ((xdevs.kernel.modeling.PortComplex) portXDevs)
                              .addValues(port.getValues());
      } else {
        ((xdevs.kernel.modeling.Port) portXDevs)
                           .setValue(port.getSingleValue());
      }
    }
  }
}
```

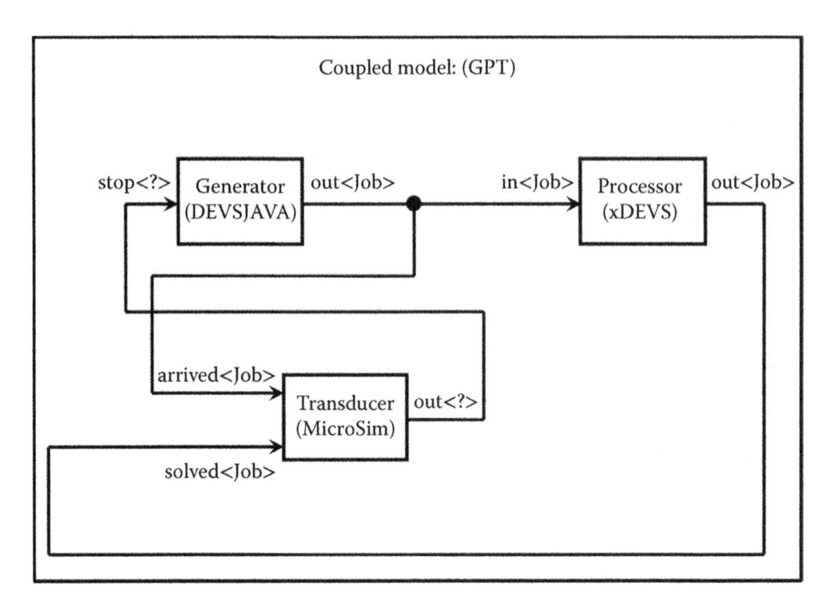

FIGURE 13.8 GPT-coupled model.

With these two adapters, we are able to implement heterogeneous DEVS models.

13.3.2.1 Example: Generator Processor Transducer Model

The Generator Processor Transducer (GPT) DEVS model is similar to the EFP model introduced in earlier chapters without the experimental frame-coupled model (see Figure 13.8). In this example, we compose a heterogeneous DEVS model using the extended MicroSim simulation engine.

DEVSJAVA only accept events derived from entity. As xDEVS and MicroSim accept any data type, we created a new class called *Job* in the MicroSim platform (Listing 13.4).

LISTING 13.4 JOB.JAVA

```
public class Job extends GenCol.entity {

  public String id;
  public double time;

  public Job(String name) {
    this.id = name;
    this.time = 0.0;
  }
}
```

Next, the *Generator* class is implemented as a DEVSJAVA atomic model (Listing 13.5).

LISTING 13.5 GENERATOR.JAVA (DEVSJAVA)

```
public class Generator extends genDevs.modeling.atomic {
  // Ports

  public static final String stop = "stop";
  public static final String out = "out";
  // State
  protected double period;
  protected int jobCounter;

  public Generator(String name, double period) {
    super(name);
    addInport(Generator.stop);
    addOutport(Generator.out);
    this.period = period;
    this.setSigma(period);
    jobCounter = 1;
  }

  @Override
  public void deltint() {
    jobCounter++;
    this.setSigma(period);

  }

  @Override
  public void deltext(double e, genDevs.modeling.message x) {
    Continue(e);
    if (!x.isEmpty()) {
      super.passivate();
    }
  }

  @Override
  public genDevs.modeling.message out() {
    Job job = new Job("" + jobCounter + "");
    genDevs.modeling.message m = new
      genDevs.modeling.message();
    m.add(makeContent(Generator.out, job));
    return m;
  }
}
```

The *Processor* class is implemented as a xDEVS atomic model (Listing 13.6).

The *Transducer* atomic model is the same as developed in previous chapters. Now, to build the GPT-coupled model, we use a MicroSim-coupled model (Listing 13.7).

As can be seen in Listing 13.8, the output of the simulation is exactly the same as the output provided for the EFP model.

LISTING 13.6 PROCESSOR.JAVA (XDEVS)

```
public class Processor extends xdevs.kernel.modeling.Atomic {
  // Ports
  protected xdevs.kernel.modeling.Port<Job> in =
                 new xdevs.kernel.modeling.Port<Job>("in");
  protected xdevs.kernel.modeling.Port<Job> out =
                 new xdevs.kernel.modeling.Port<Job>("out");
  // State
  protected Job currentJob;
  protected double processingTime;

  public Processor(String name, double processingTime) {
    super(name);
    super.addInport(in);
    super.addOutport(out);
    this.processingTime = processingTime;
    currentJob = null;
  }

  @Override
  public void deltint() {
    super.passivate();
    currentJob = null;
  }

  @Override
  public void deltext(double e) {
    super.resume(e);
    if (super.phaseIs("passive")) {
      Job job = in.getValue();
      currentJob = job;
      super.holdIn("active", processingTime);
    }
  }

  @Override
  public void lambda() {
    if (super.phaseIs("active")) {
      out.setValue(currentJob);
    }
  }
}
```

LISTING 13.7 GPT.JAVA

```
public class Gpt extends Coupled {

  public Gpt(String name) {
    super(name);
    AtomicDevsJava generator =
       new AtomicDevsJava(new Generator("Generator", 1));
    generator.addInPort(new Port<Job>(Job.class,
                                     Generator.stop));
    generator.addOutPort(new Port<Job>(Job.class,
                                      Generator.out));
    AtomicXDevs processor =
         new AtomicXDevs(new Processor("Processor", 2));
    processor.addInPort(new Port<Job>(Job.class, "in"));
    processor.addOutPort(new Port<Job>(Job.class, "out"));
    Transducer transducer = new Transducer("Transducer",
                                          20);
    super.addComponent(generator);
    super.addComponent(processor);
    super.addComponent(transducer);

    super.addCoupling(generator,
                      generator.getOutPort(Generator.out),
                      processor, processor.getInPort("in"));
    super.addCoupling(generator,
                      generator.getOutPort(Generator.out),
                      transducer, transducer.iportArrived);
    super.addCoupling(processor,
                      processor.getOutPort("out"),
                      transducer, transducer.iportSolved);
    super.addCoupling(transducer, transducer.oportOut,
                      generator,
                      generator.getInPort(Generator.stop));
  }

  public static void main(String args[]) {
    MicroSimLogger.setup(Level.INFO);
    Gpt gpt = new Gpt("GPT");
    Coordinator coordinator = new Coordinator(gpt);
    coordinator.initialize(0);
    coordinator.simulate(Long.MAX_VALUE);
  }
}
```

Now, we know how to adapt atomic models from other DEVS simulation platforms to MicroSim. Models coming from DEVS/C++ engines, for example, can also be adapted using JNI, but the implementation of JNIs is out of scope.

LISTING 13.8 GPT SIMULATION OUTPUT

```
[INFO|00:00:00.030]: Start job 1 @ t = 1.0
[INFO|00:00:00.065]: Start job 2 @ t = 2.0
[INFO|00:00:00.069]: Start job 3 @ t = 3.0
[INFO|00:00:00.071]: Finish job 1 @ t = 3.0
[INFO|00:00:00.075]: Start job 4 @ t = 4.0
[INFO|00:00:00.077]: Start job 5 @ t = 5.0
[INFO|00:00:00.080]: Finish job 3 @ t = 5.0
[INFO|00:00:00.083]: Start job 6 @ t = 6.0
[INFO|00:00:00.086]: Start job 7 @ t = 7.0
[INFO|00:00:00.088]: Finish job 5 @ t = 7.0
[INFO|00:00:00.091]: Start job 8 @ t = 8.0
[INFO|00:00:00.093]: Start job 9 @ t = 9.0
[INFO|00:00:00.102]: Finish job 7 @ t = 9.0
[INFO|00:00:00.105]: Start job 10 @ t = 10.0
[INFO|00:00:00.108]: Start job 11 @ t = 11.0
[INFO|00:00:00.110]: Finish job 9 @ t = 11.0
[INFO|00:00:00.116]: Start job 12 @ t = 12.0
[INFO|00:00:00.118]: Start job 13 @ t = 13.0
[INFO|00:00:00.121]: Finish job 11 @ t = 13.0
[INFO|00:00:00.124]: Start job 14 @ t = 14.0
[INFO|00:00:00.127]: Start job 15 @ t = 15.0
[INFO|00:00:00.129]: Finish job 13 @ t = 15.0
[INFO|00:00:00.132]: Start job 16 @ t = 16.0
[INFO|00:00:00.134]: Start job 17 @ t = 17.0
[INFO|00:00:00.144]: Finish job 15 @ t = 17.0
[INFO|00:00:00.146]: Start job 18 @ t = 18.0
[INFO|00:00:00.149]: Start job 19 @ t = 19.0
[INFO|00:00:00.151]: Finish job 17 @ t = 19.0
[INFO|00:00:00.154]: End time: 19.0
[INFO|00:00:00.164]: Jobs arrived : 19
[INFO|00:00:00.166]: Jobs solved : 9
[INFO|00:00:00.169]: Average TA = 2.0
[INFO|00:00:00.172]: Throughput = 0.47368421052631576
[INFO|00:00:00.175]: Start job 20 @ t = 19.0
[INFO|00:00:00.177]: Finish job 19 @ t = 20.0
```

However, there is an important issue here: How can we adapt coupled models? Due to *closure under coupling* of the DEVS formalism, we have an abstraction mechanism by which a coupled model can be executed like an atomic model. In contrast to the DEVS hierarchical modeling, where a coupled model is merely a container and has corresponding coupled simulators, we can consider an atomic model with the corresponding lowest level atomic simulator. This has been accomplished by implementing two new adapters (one for DEVSJAVA and other for xDEVS). The adapter takes each coupled component of the model and treats it as an atomic model (Listing 13.9).

LISTING 13.9 COUPLEDDEVSJAVA.JAVA (DEVSJAVA-COUPLED ADAPTER)

```
public class CoupledDevsJava extends
                             genDevs.modeling.atomic {

  private genDevs.simulation.CoordinatorInterface
        coordinator;

  public CoupledDevsJava(genDevs.modeling.digraph model) {
    super(model.getName());

    genDevs.modeling.couprel couplings = model.getCouprel();
    java.util.Iterator<?> itr = couplings.iterator();
    while (itr.hasNext()) {
      GenCol.Pair relation = (GenCol.Pair) itr.next();
      GenCol.Pair from = (GenCol.Pair) relation.getKey();
      String fromComponentName = (String) from.getKey();
      String fromPortName = (String) from.getValue();
      GenCol.Pair to = (GenCol.Pair) relation.getValue();
      String toComponentName = (String) to.getKey();
      String toPortName = (String) to.getValue();
      if (fromComponentName.equals(model.getName())) {
        this.addInport(fromPortName);
      } else if (toComponentName.equals(model.getName())) {
        this.addOutport(toPortName);
      }
    }
    coordinator = new genDevs.simulation.coordinator(model);
    coordinator.initialize();
  }

  @Override
  public void deltext(double e,
                      genDevs.modeling.message x) {
    coordinator.simInject(e, x);
  }

  @Override
  public void deltint() {
    coordinator.wrapDeltfunc(coordinator.getTN());
  }

  @Override
  public genDevs.modeling.message out() {
    coordinator.computeInputOutput(coordinator.getTN());
    return
    (genDevs.modeling.message) coordinator.getOutput();
  }
```

```
  @Override
  public double ta() {
    return coordinator.getTN() - coordinator.getTL();
  }
}
```

13.3.2.2 Example: The EFP Model

Now, we can implement the transducer using DEVSJAVA (Listing 13.10).

LISTING 13.10 TRANDUCER.JAVA (DEVSJAVA)

```
ppublic class Transducer extends genDevs.modeling.atomic {

  private static final Logger logger =
            Logger.getLogger(Transducer.class.getName());
  // Ports
  public static final String iportArrived = "arrived";
  public static final String iportSolved = "solved";
  public static final String oportOut = "out";
  // State
  protected LinkedList<Job> jobsArrived =
                                new LinkedList<Job>();
  protected LinkedList<Job> jobsSolved =
                                new LinkedList<Job>();
  protected double totalTa;
  protected double clock;

  public Transducer(String name, double observationTime) {
    super(name);
    super.addInport(iportArrived);
    super.addInport(iportSolved);
    super.addOutport(oportOut);
    totalTa = 0;
    clock = 0;
    super.holdIn("active", observationTime);
  }

  @Override
  public void deltint() {
    clock = clock + getSigma();
    double throughput;
    double avgTaTime;
    if (!jobsSolved.isEmpty()) {
      avgTaTime = totalTa / jobsSolved.size();
```

```
      if (clock > 0.0) {
        throughput = jobsSolved.size() / clock;
      } else {
        throughput = 0.0;
      }
    } else {
      avgTaTime = 0.0;
      throughput = 0.0;
    }
    logger.info("End time: " + clock);
    logger.info("Jobs arrived : " + jobsArrived.size());
    logger.info("Jobs solved : " + jobsSolved.size());
    logger.info("Average TA = " + avgTaTime);
    logger.info("Throughput = " + throughput);
    super.passivate();
  }

  @Override
  public void deltext(double e,
                        genDevs.modeling.message x) {
    Continue(e);
    clock = clock + e;
    Job job = null;
    if (!x.valuesOnPort(iportArrived).isEmpty()) {
      job = (Job) x.getValOnPort(iportArrived, 0);
      logger.info("Start job " + job.id
              + " @ t = " + clock);
      job.time = clock;
      jobsArrived.add(job);
    }
    if (!x.valuesOnPort(iportSolved).isEmpty()) {
      job = (Job) x.getValOnPort(iportSolved, 0);
      totalTa += (clock - job.time);
      logger.info("Finish job " + job.id
              + " @ t = " + clock);
      job.time = clock;
      jobsSolved.add(job);
    }
  }

  @Override
  public genDevs.modeling.message out() {
    Job job = new Job("null");
    genDevs.modeling.message m =
                      new genDevs.modeling.message();
    m.add(makeContent(Transducer.oportOut, job));
    return m;
  }
}
```

Now creating the experimental frame (Listing 13.11).

LISTING 13.11 EF.JAVA (DEVSJAVA)

```
public class Ef extends genDevs.modeling.digraph {
  // Ports

  public static final String in = "in";
  public static final String out = "out";

  public Ef(String name, double period,
            double observationTime) {
    super(name);
    addInport(Ef.in);
    addOutport(Ef.out);
    Generator generator =
              new Generator("Generator", period);
    super.add(generator);
    Transducer transducer =
              new Transducer("Transducer", observationTime);
    super.add(transducer);
    addCoupling(this, Ef.in,
                transducer, Transducer.iportSolved);
    addCoupling(generator, Generator.out, this, Ef.out);
    addCoupling(generator, Generator.out,
                transducer, Transducer.iportArrived);
    addCoupling(transducer, Transducer.oportOut,
                generator, Generator.stop);
  }
}
```

We can now, henceforth, compose new heterogeneous models using coupled models coming from different DEVS engines (Listing 13.12).

LISTING 13.12 EFP.JAVA (MICROSIM)

```
public class Efp extends Coupled {

  public Efp(String name) {
    super(name);
    AtomicDevsJava ef = new AtomicDevsJava(
                        new CoupledDevsJava(
                        new Ef("ExpFrame", 1, 20)));
    ef.addInPort(new Port<Job>(Job.class, Ef.in));
    ef.addOutPort(new Port<Job>(Job.class, Ef.out));
    addComponent(ef);
    AtomicXDevs processor = new AtomicXDevs(
                            new Processor("Processor", 2));
```

```
    processor.addInPort(new Port<Job>(Job.class, "in"));
    processor.addOutPort(new Port<Job>(Job.class, "out"));
    addComponent(processor);
    addCoupling(ef, ef.getOutPort(Ef.out),
              processor, processor.getInPort("in"));
    addCoupling(processor, processor.getOutPort("out"),
              ef, ef.getInPort(Ef.in));
  }

  public static void main(String args[]) {
    MicroSimLogger.setup(Level.INFO);
    Efp efp = new Efp("Coordinator");
    Coordinator coordinator = new Coordinator(efp);
    coordinator.initialize(0);
    coordinator.simulate(Long.MAX_VALUE);
  }
}
```

A new and simplified DEVS standard architecture can be built from Figure 13.3. We now address the non-DEVS models shown in Figure 13.9.

13.3.3 DEVS-to-Non-DEVS Interoperability

When implementing a DEVS-coupled model, some atomic models inside the conceptual DEVS scenario may be implemented using external libraries, for example, MATLAB m-functions. In this section, we describe DEVS simulation interoperability between two different DEVS model implementations: a DEVS-compliant model written in Java executed along with a non-DEVS model extended with MATLAB functionality.

Figure 13.10 depicts the integration methodology. We have a DEVS model interface, naturally implemented as a MicroSim atomic abstract class. Keeping track of the inheritance path in Figure 13.10, we move down from the DEVS model MicroSim

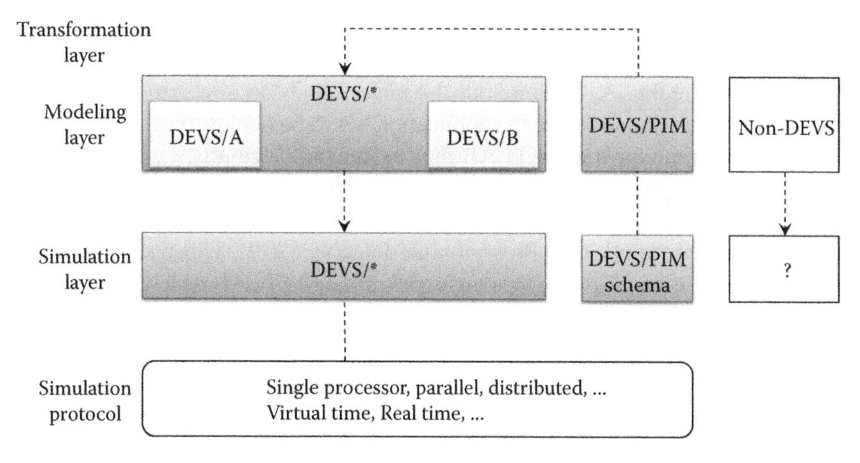

FIGURE 13.9 DEVS standard revisited.

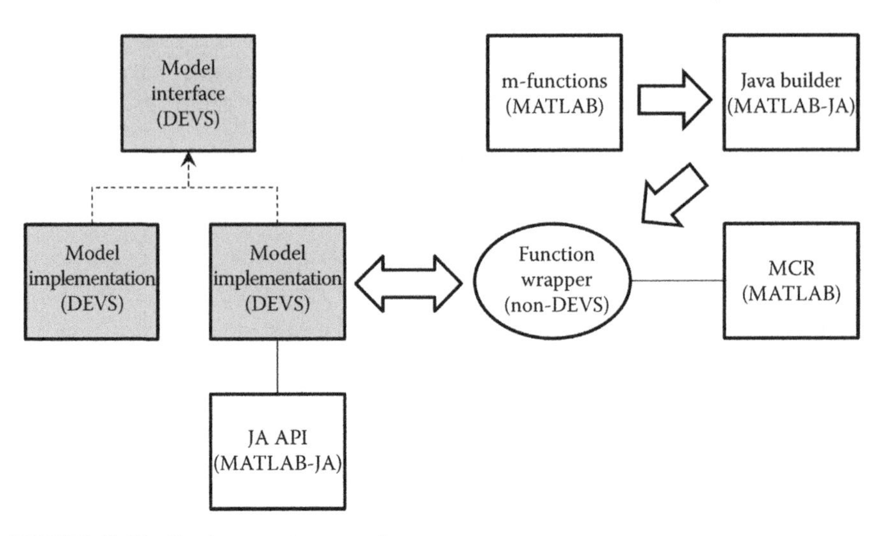

FIGURE 13.10 Implementation overview.

abstract class at the top to DEVS model implementations that ensure DEVS formalism specifications. These DEVS models can be implemented as full DEVS models or otherwise act as adapters for MATLAB integration. Let us now look in more detail the MATLAB integration technical requirements by focusing on MATLAB functionality abstraction.

Figure 13.10 illustrates the MATLAB Java wrapper function assembly process with MATLAB Builder support. The application tool MATLAB Java Builder allows the user to deploy MATLAB functions by creating MATLAB-based Java classes that can be used with a Java-application-building environment. This additional MATLAB tool creates the Java classes by encrypting MATLAB functions and generates a Java wrapper around them. The system integrates MATLAB computation capabilities to our DEVS model adapter as non-DEVS functionality. The DEVS model adapter implementation wraps as a MATLAB connector and requests the inclusion of the encrypted MATLAB functions translated as Java classes in order to take advantage of their computing capacity.

In addition, Figure 13.10 points out the major handicap concerning the system operability—the machine where the application has to be deployed needs MATLAB Component Runtime for the MATLAB files to function normally.

Next, we take a quick look at the technical procedures for MATLAB integration. To enable Java applications to exchange data with MATLAB methods they invoke, Java Builder provides an Application Programming Interface (API). This package provides a set of data conversion classes. Each class represents a MATLAB data type. As seen in Figure 13.10, MATLAB integration within the DEVS model adapter requires the linkage to the Java Builder API. This API assists the information exchange process management between a Java application environment and a MATLAB programming framework.

Summarizing from a global perspective, we developed a DEVS model component within xDEVS interface global guidelines implemented as a Java model. Alternatively, exploiting the MATLAB numerical computing environment, we generated a wrapper for the desired MATLAB functionality, managed by the specific model interface implementation.

We now have the capability of modeling and simulating atomic and coupled models that share the same semantics given the DEVS mathematical specification via the DEVS simulation environment incorporated in the MicroSim framework with dissimilar computing platforms. So far, the implementations of both Java-compliant models and MATLAB-core-based models that share a common DEVS interface have been presented. Now, the DEVS Simulator MicroSim that remains unchanged is able to simulate the formerly stated interoperable DEVS model implementations.

13.3.3.1 Barrel Filler Example

We describe the barrel filler model that embraces interoperation between DEVS and non-DEVS model implementations. Filling a barrel is a typical engineering process that involves both continuous and discrete event archetypes. This experiment is motivated by a simple example that can be found in Zeigler's book, but with a different unrelated purpose. Figure 13.11 shows the system structure: it has one continuous input port, *inflow*, and two output ports for observing the contents, one discrete output port, *barrel*, and one continuous port, *cout*. There is also a continuous state variable, *contents*. The derivative of the continuous state variable contents is established by the input value of inflow.

Barrels are sent out at the discrete output port barrel whenever the variable contents reaches the value 10. At that point, contents is reinitialized.

In order to build a model of a barrel filler, we implemented a coupled model that involves an integrator atomic model with an additional reset entrance and a simple model consisting of a threshold level detector.

Even though the integrator atomic model fulfills the Mealy Discrete Time Specified System (Mealy DTSS) formalism, it is shown (Zeigler et al., 2000) how these systems can be specified as DEVS models. A Mealy DTSS is represented as a DEVS in a manner similar to a memoryless function with the difference that the state is updated when it receives an input. The system will *passivate* until it collects an input. Next, it moves to the transition state, delivers the output, and in the internal

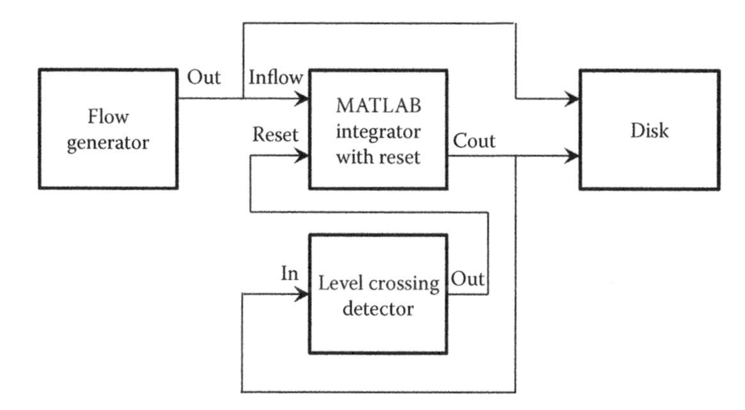

FIGURE 13.11 Barrel filler model.

transition function computes the next state and makes $\sigma = \infty$, waiting for the arrival of the new period. In the barrel filler sample, the integrator block comprising the state variable contents, the two input port inflow and reset and the output port cout, computes the next state, handling the integration function parameterized by the inflow value that is reliant on the reset entrance value within the internal transition.

$$\text{Integrator} = < X, S, Y, \delta_{\text{int}}, \delta_{\text{ext}}, \delta_{\text{con}}, \lambda >$$

where

$X = \{\text{inflow} \times \text{reset} | \text{inflow} \in \mathbb{R}, \text{reset} \in \{\text{true, false}\}\}$

$S = \{\text{contents} | \text{contents} \in \mathbb{R}\}$

$Y = \{\text{cout} | \text{cout} \in \mathbb{R}\}$

$\delta_{\text{ext}}(\text{contents}, e, \text{inflow})$

- $\sigma = \infty$
- Case reset
  - false : $\dfrac{\text{dcontents}}{\text{dt}} = \text{inflow}$
  - true:contents $= 0$

$\delta_{\text{con}}(s, \phi) = \delta_{\text{int}}(s)$

λ(contents, e, inflow)

cout = contents

$\text{ta}(s) := \sigma$

In contrast, the atomic model that embodies a threshold level detector (state event trigger) exemplifies DEVS formalism throughout a DEVS model. The level detector model only updates the internal event state variable to *true* and triggers the internal transition function ($s = 0$) whenever the input port value crosses the threshold level. Just before the internal transition passivates, the system delivers the state event value through the output port:

$$\text{Level} = < X, S, Y, \delta_{\text{int}}, \delta_{\text{ext}}, \delta_{\text{con}}, \lambda >$$

where

$X = \{\text{in} | \text{in} \in \mathbb{R}\}$

$S = \{\text{state event} \in \{\text{true, false}\}\}$

$Y = \{\text{out} \in \{\text{true, false}\}\}$

$\delta_{\text{ext}}(\text{state event}, e, \text{in})$

- if in ≥ 10 then
  - state event = true
  - $\sigma = 0$

δ_{int}(state event, e, in)

- $\sigma = 0$

$\delta_{\text{con}}(s, \phi) = \delta_{\text{int}}(s)$

λ(state event, e, in)

- out = state event
- state event = false

$\text{ta}(s) = \sigma$

At last, as illustrated on Figure 13.11, to complete the coupling, we link the output port cout of the integrator with the level detector input port and the output port out of the level detector component with the input port reset of the integrator module.

Now, the main goal is to segregate the integration duty to the MATLAB computing environment. In our first step, with MATLAB Java Builder JA support, we generated a Java wrapper that embraces the MATLAB integration function "ode45" as shown in Figure 13.12. With the MicroSim model interface as the skeleton, along with previous integrator and level detector DEVS model description, without leaving away MATLAB immense functionality, we are ready to simulate the coupled model example.

Figure 13.13 illustrates the model simulation. The X-axis posts the time elapsed, while the Y-axis presents the measure unit for the barrel filler inflow and the internal variable contents status value. The sine wave series pictures the inflow value at time t, while the other series illustrates the value of the internal variable contents at time $t + h$. Whenever the content reaches the tolerance limit level 10, the level detector component triggers a state event on the way to rearrange the integrator's internal initial value to 0. As can be seen in Figure 13.13, we attach a source generator to

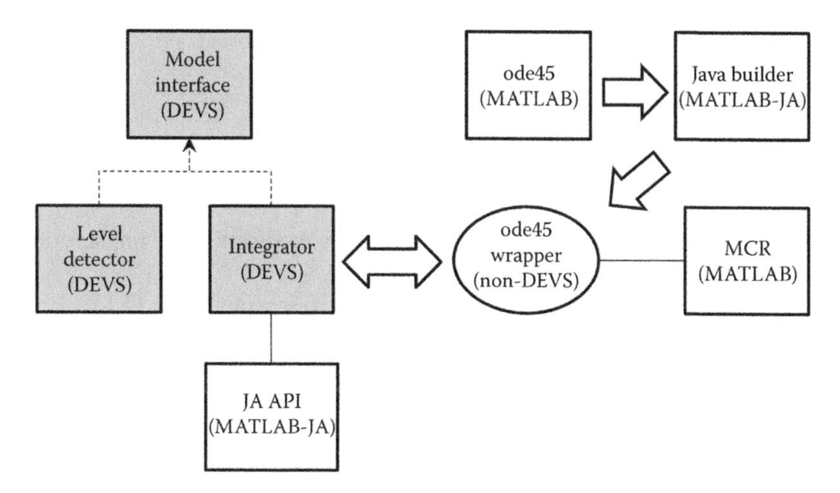

FIGURE 13.12 Barrel filler implementation.

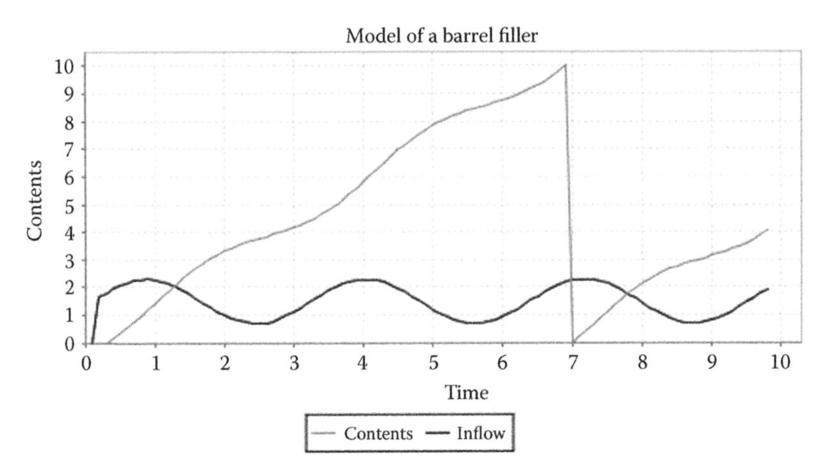

FIGURE 13.13 Execution of the barrel filler model.

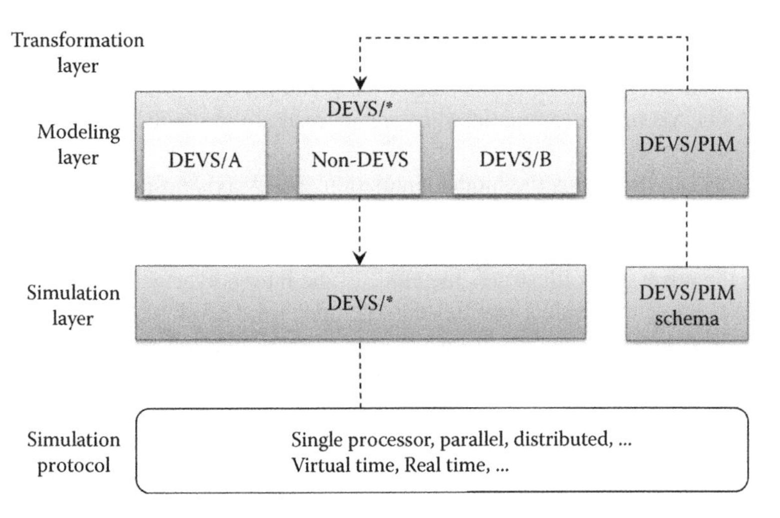

FIGURE 13.14 DEVS: Final architecture of the standard.

the model input that propagates the oscillation of a sine function and impacts on the deviation of the internal variable contents value.

The DEVS/ MATLAB integrator source code is shown in Listing 13.13.

In conclusion, the implementation of non-DEVS models using a DEVS/* standard platform requires ad hoc solutions; however, it is possible to create DEVS/* adapters. As a result, our final DEVS standard architecture is shown in Figure 13.14.

Next, we visit the simulation layer, mainly to consider some issues that can arise under this conceptual architecture.

LISTING 13.13 MATLABINTEGRATOR.JAVA

```
public class MatlabIntegrator extends Atomic {

  protected Port<Double> iportReset =
                                new Port<Double>("reset");
  protected Port<Double> iportIn = new Port<Double>("in");
  protected Port<Double> oportOut = new Port<Double>("out");

  protected double time;
  protected double x;
  protected double dx;
  protected double sampleTime;

  public MatlabIntegrator(String name, double sampleTime,
                          double x0, double t0) {
    super(name);

    this.x = x0;
    this.time = t0;
    this.sampleTime = sampleTime;

    super.addInPort(iportIn);
    super.addOutPort(oportOut);
    super.addInPort(iportReset);
  }

  public MatlabIntegrator() {
    this("MatlabIntegrator", 0.1, 0, 0);
  }

  @Override
  public void deltext(double e) {
    if(!iportReset.isEmpty()) {
      x = 0;
    }
    else {
      dx = iportIn.getSingleValue();
    }
    super.setSigma(sampleTime);
  }

  @Override
  public void deltint() {
    Double[] ut = {dx};
    Double[] xt = {x};
    // integration interval
    Double[] ts = {time, time + super.getSigma()};
    int method = 1;  // Default integration method
```

```
    Double[] xx = doIntegral(ts, xt, ut, method);
    x = xx[0].doubleValue();
    time = time + super.getSigma();

    super.passivate();
  }

  @Override
  public void lambda() {
    oportOut.addValue(x);
  }
  /**
   * @param funcion coming from Matlab
   * @param ts  time span [ti, tf]
   * @param x0  initial state vector: x(ti)
   * @param U   constant input vector
   * @param method integration method:
   *         {0,1:ode45, 2:ode23, 3:ode115, 4:ode15s}
   */
  public Double[] doIntegral(Double[] ts, Double[] X0,
                             Double[] U, int method) {
    // Create Matlab Object
    MWIntegrator theIntegrator = null;
    // Java object reference to be passed to component
    MWJavaObjectRef origRef = null;
    MWNumericArray X = null; // Integration result
    MWNumericArray T = null; // Integration instants
    // X(tf): state at the final time
    Double[] XF = new Double[X0.length];
    // result of the Matlab function
    Object[] result = null;
    try {
      theIntegrator = new MWIntegrator();
      origRef = new MWJavaObjectRef(this);
      if (U == null) {
        U = new Double[1];
        U[0] = 0.0;
      }
      result = theIntegrator.MWdoIntegral(2, origRef,
                     ts, X0, U, (double) method);
      T = (MWNumericArray) result[0];
      X = (MWNumericArray) result[1];

      int[] dim = X.getDimensions();
      int[] index = new int[2];
      index[0] = dim[0];
      for (index[1] = 1; index[1] <= dim[1]; index[1]++) {
        XF[index[1] - 1] = X.getDouble(index);
      }
```

```
            } catch (Exception e) {
              System.out.println("Exception: " + e.toString());
            } finally {
              MWArray.disposeArray(result);
              MWArray.disposeArray(T);
              MWArray.disposeArray(X);
              MWArray.disposeArray(origRef);
            }
            return XF;
        }
    }
```

13.4 SIMULATION LAYER

DEVS framework treats a model and its simulator as two distinct elements. The simulation protocol describes how a DEVS model should be simulated whether in a standalone fashion or in a coupled model. Such a protocol is implemented by a simulation algorithm implemented as an atomic simulator or a coordinator.

In order to reach a DEVS standard for the simulation layer, we focus on the implementation of a unique DEVS simulation architecture that we name "DEVS/*." Ideally, such architecture must allow the simulation of any DEVS/* model, as well as any other DEVS and non-DEVS models. Next, we tackle the standard's primary simulation issues through the GPT example.

A typical DEVS simulation consists of a loop that is repeatedly executing four functions (compute tN, compute output, propagate output, and apply transition function) in each of the three DEVS simulators involved in this process.

Figure 13.15 shows a simulation of a GPT DEVS model. In this example, the root GPT-coupled model has been implemented using the standard DEVS/* modeling framework. The generator model ("G") has been implemented using another DEVS modeling framework (called DEVS/A). The processor model ("P") has been developed using a third DEVS modeling platform (called DEVS/B). The transducer atomic model ("T") comes from a non-DEVS modeling library.

To support interoperability, the ideal DEVS/* standard framework must contain model adapters, as we have seen in previous sections. It will facilitate the simulation of a heterogeneous model like the one presented in Figure 13.13.

There have been many attempts to build standard DEVS simulation frameworks; one of them was presented in (Wainer, 2009), where a standard simulation interface was proposed.

As can be seen, the implementation of a standard simulation layer is not a hard task when the interoperability is solved in the modeling layer. However, if the software engineer wants to put different DEVS simulators in scene, a major problem appears.

Figure 13.16 shows an example. If our hypothetical DEVS/* adapts a DEVS/B simulator, we will find several incompatibilities in the message-passing protocol. Since different DEVS models use different message formats, we must also convert

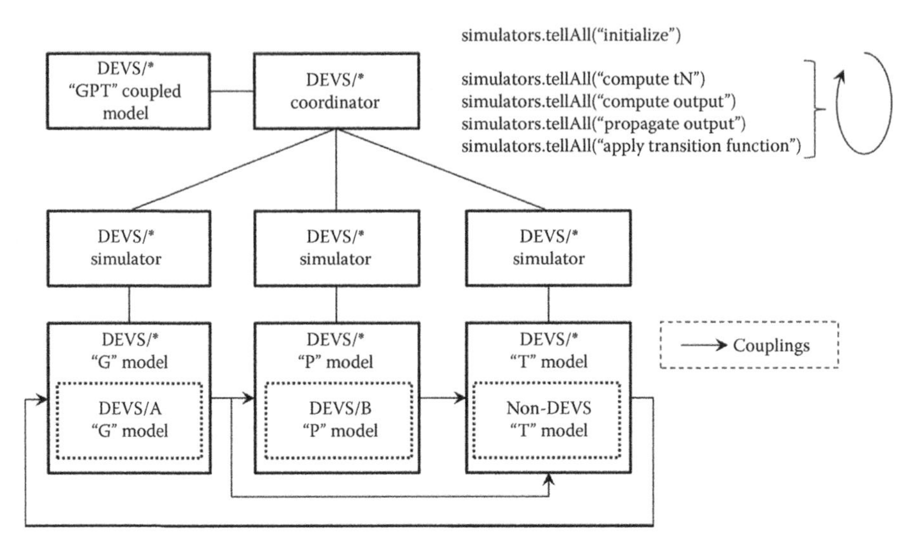

FIGURE 13.15 Federation of DEVS simulators in the GPT model simulation process.

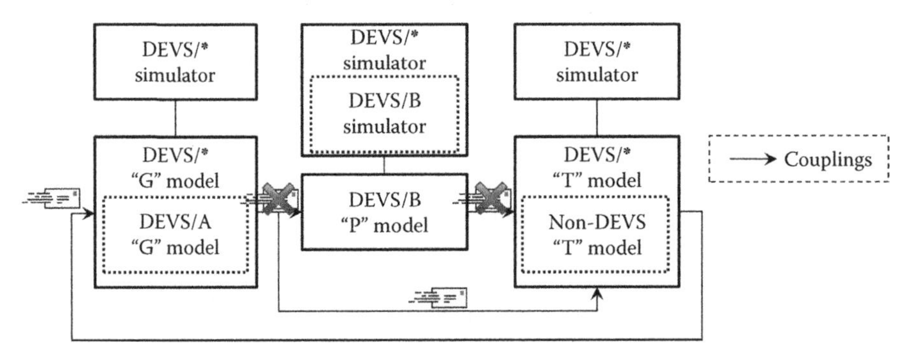

FIGURE 13.16 Interoperability at the simulation layer.

messages coming from DEVS/* to messages entering into DEVS/B models, as Figure 13.16 depicts. This problem also arises when the message-passing protocol is implemented in the simulation layer, as it occurs in the DEVSJAVA framework.

13.5 CONCLUSIONS

In this chapter, we have presented several ways to implement a DEVS standard protocol. To do so, we have divided the DEVS M&S formalism in two layers: modeling and simulation. In the modeling layer, we can provide interoperability between different DEVS M&S platforms using the well-known adapter pattern. In the simulation layer, the DEVS standard can be easily implemented using model adapters to provide a unified simulation layer. As we have seen, the simulation platform using different DEVS simulators is a hard task as it implies the translation of messages or, in other way, the implementation of a standard message format. Selecting XML as the message format is a natural way and the foundation of net-centric simulation transparency. We shall explore this option in chapters ahead.

REFERENCES

ACIMS—Software site. (n.d.). Retrieved from www.acims.arizona.edu/SOFTWARE/software.shtml. Last accessed: Apr. 2012

Nutaro, J.J., (2005): aDEVS - a Discrete EVent System simulator. Retrieved from www.ornl.gov/~1qn/adevs.last accessed: Apr. 2012

Kim, S., Sarjoughian, H. S., & Elamvazhuthi, V. (2009). DEVS-Suite: A component-based simulation tool for rapid experimentation and evaluation. *Spring Simulation Multi-*conference, San Diego, CA.

Mittal, S., Risco-Martín, J. L., & Zeigler, B. P. (2007). DEVSML: Automating DEVS execution over SOA towards transparent simulators. *SpringSim '07: Proceedings of the 2007 Spring Simulation Multiconference* (pp. 287–295). San Diego, CA: Society for Computer Simulation International.

Mittal, S., Zeigler, B., & Hwang, M. (2007). *XFD-DEVS: XML-Based Finite Deterministic DEVS*. Retrieved from duniptechnologies.com: www.duniptechnologies.com/research/xfddevs. Last accessed November 2012.

Mittal, S., & Douglass, S. (2012). DEVSML 2.0: The language and the stack. *DEVS Symposium*. Orlando, FL.

Risco-Martín, J. L., Mittal, S., Cruz, J. M., & Zeigler, B. P. (2009). eUDEVS: Executable UML using DEVS theory of modeling and simulation. *SIMULATION: Transactions of SCS*, *85*(11–12), 750–777. doi: http://dx.doi.org/10.1177/0037549709104727.

Risco-Martín, J. L., Mittal, S., López-Pérez, M. A., & Cruz, J. M. (2007). A W3C XML schema for DEVS scenarios. *SpringSim '07: Proceedings of the 2007 Spring Simulation Multiconference* (pp. 279–286). San Diego, CA: Society for Computer Simulation International.

Risco-Martín, J. L., Mittal, S., Zeigler, B. P., & Cruz, J. M. (2007). From UML state charts to DEVS state machines using XML. *MPM'07: Proceedings of the Workshop on Multi-Paradigm Modeling: Concepts and Tools at the 10th International Conference on Model-Driven Engineering Languages and Systems* (pp. 35–48). Nashville, TN.

Touraille, L., Traore, M. K., & Hill, D. R. C. (2009). A mark-up language for the storage, retrieval, sharing and interoperability of DEVS models. *Proceedings of the 2009 ACM/SCS Spring Simulation Multiconference*.

U.S. Department of Defense. (2001). *RTI 1.3-Next Generation Programmer's Guide, Version 4*.

Wainer, G. A. (2009). *Discrete-Event Modeling and Simulation: A Practitioner's Approach* (p. 520). CRC Press, Boca Raton, FL.

xDEVS. Retrieved from https://sourceforge.net/projects/xdevs.last accessed: Oct. 2012

Zeigler, B. P., Kim, T. G., & Praehofer, H. (2000). *Theory of Modeling and Simulation*. New York: Academic Press.

14 Architecture for DEVS/SOA

14.1 OVERVIEW

Discrete Event Systems Specification (DEVS) environments such as DEVSJAVA, DEVS-C++, and others (ACIMS) are embedded in object-oriented implementations; they support the goal of representing executable model architectures in an object-oriented representational language. As a mathematical formalism, DEVS is platform independent, and its implementations adhere to the DEVS protocol so that DEVS models easily translate from one form (e.g., C++) to another (e.g., Java) (Zeigler et al., 2000). Moreover, DEVS environments, such as DEVSJAVA, execute on commercial, off-the-shelf desktops or workstations and employ state-of-the-art libraries to produce graphical output that complies with industry and international standards. DEVS environments are typically open architectures that have been extended to execute on various middleware such as the Department of Defense's (DoD) High Level Architecture (HLA) standard, Common Object Request Broker Architecture (CORBA), Simple Object Access Protocol (SOAP), and others and can be readily interfaced to other engineering and simulation and modeling tools (ACIMS, n.d.; Cho et al., 2001; Hu & Zeigler, 2005; Wainer & Giambiasi, 2001; Zhang et al., 2005). Furthermore, DEVS operation over web middleware (SOAP) enables it to fully participate in the netcentric environment of the Global Information Grid/Service-Oriented Architecture (GIG/SOA) (Atkinson, 2004). As a result of recent advances, DEVS can support model continuity through a simulation-based development and testing life cycle (Hu & Zeigler, 2005). This means that the mapping of high-level requirement specifications to lower-level DEVS formalizations enables such specifications to be thoroughly tested in virtual simulation environments before being easily and consistently transitioned to operate in a real environment for further testing and fielding.

DEVS formalism categorically separates the model, the simulator, and the experimental frame. However, one of the major problems in this kind of mutually exclusively system is that the formalism implementation is itself limited by the underlying programming language. In other words, the model and the simulator exist in the same programming language. Consequently, legacy models as well as models that are available in one implementation are hard to translate from one language to another even though both the implementations are object oriented. Other constraints such as libraries inherent in C++ and Java are another source of bottleneck that prevents such interoperability.

We will show how a netcentric system can be developed on the DEVS/SOA distributed M&S framework. In addition to supporting SOA application development, the framework enables verification and validation testing of application. Before we embark on the details of this distributed SOA-based M&S framework, it is important to acknowledge the advantages and power provided by the DEVS M&S framework.

After the development of the World Wide Web (WWW), many efforts in the distributed simulation field have been made for modeling, executing simulation, and creating model libraries that can be assembled and executed over the WWW. By means of eXtensible Markup Language (XML) and web services technology, these efforts have entered upon a new phase. Today, cloud-based solutions and virtualization technologies are paving the way for Internet platform as a service. We already described DEVSML to facilitate model interoperability and several ways to simulate models coming from different DEVS simulation engines through adapter containers. This chapter describes an updated version a simulation framework called DEVS/SOA (Mittal et al., 2009a, b) that is implemented using latest web services standards such as JAX-WS 2.0. The central point resides in executing the simulator as a web service. Again, the notion of simulator here should be clear in terms of DEVS elements; that is, model and simulator are bound together by a simulation relation or protocol or a contract and both are independent of each other. The development of this kind of framework will help solve large-scale problems and guarantees interoperability among different networked systems and specifically DEVS-validated models. This chapter focuses on the overall approach, and the symmetrical SOA-based architecture that allows for DEVS-as-a-Service (DaaS) through a netcentric DEVS virtual machine (DEVSVM).

There have been a lot of efforts in the area of distributed simulation using parallelized DEVS formalism. Issues like "causal dependency" (Zeigler et al., 2000) and "synchronization problem" (Fujimoto, 1999) have been adequately dealt with solutions such as (1) restriction of global simulation clock until all the models are in sync or (2) rolling back the simulation of the model that has resulted in the causality error. Our chosen method of webcentric simulation does not address these problems as they fall in a different domain. In our proposed work, the simulation engine rests solely on the server. Consequently, the coordinator and the model simulators are always in sync.

Most of the existing webcentric simulation efforts consist of the following components:

1. *Application*: The top-level coupled model with (optional) integrated visualization
2. *Model partitioner*: Element that partitions the model into various smaller coupled models to be executed at a different remote location
3. *Model deployer*: Element that deployed the smaller partitioned models to different locations
4. *Model initializer*: Element that initializes the partitioned model and makes it ready for simulation
5. *Model simulator*: Element that coordinates with root coordinator about the execution of partitioned model execution

The simulator design is almost the same in all of the implementations and is derived directly from parallel DEVS formalism (Zeigler et al., 2000). There are, however, different methods to implement the former four components in the previous list. DEVS/grid (Seo et al., 2004) uses all the components mentioned in the preceding list. DEVS/P2P (Cheon et al., 2004) implements step 2 using hierarchical model partitioning on a cost-based metric. DEVS/RMI (Zhang et al., 2005) has a configuring engine that integrates the functionality of steps 1, 2, and 3 mentioned in the preceding list. DEVS/cluster (Kim & Kang, 2004) is a multithreaded distributed DEVS simulator built upon CORBA, which again is focused toward development of the simulation engine.

As stated earlier, these efforts have been in the area of using the parallel DEVS and implementing the simulator engine in the same language as that of the model. These efforts are in no way similar to what we envision in the DEVS Unified Process and the DEVSML 2.0 stack. Our work is focused toward interoperability at the application level, specifically, at the modeling level and hiding the simulator engine as a whole. Our vision and solution development is along the lines of model-as-a-service (MaaS), simulation-as-a-service (SimaaS), and ultimately, DaaS. We are focused toward taking XML just as a communication middleware, as used in SOAP, for existing DEVS models, but not as a complete solution in itself. We would like the user or designer to code the behavior in any of the programming languages, ideally a domain-specific language (DSL) of his choice and let the DEVSML 2.0 stack develop the transformations. The DEVS/SOA architecture is responsible for taking a DSL or a coupled DEVSML model, integrating code within their DSLs, and delivering us with an executable model that can be simulated on any DEVS platform. The user need not learn any new syntax or any new language, except the newly developed DSL. This kind of capability where the user can integrate his model from models stored in any web repository, whether it contained public models of legacy systems or proprietary standardized models, will provide more benefit to the industry as well as to the user, thereby truly realizing the model-reuse paradigm. Next, we describe the technical architecture of DEVS/SOA and its implementation in a netcentric environment. The realization of netcentric DEVS has the following pieces:

- *DEVSML 2.0*: the central concept
- Distributed simulation using SOA
- DEVSVM (both client and server)
- Design and development of netcentric systems with DEVS

In summary, our architecture is based on two layers that correspond to the client side and the server side of the DEVS/SOA layer (Figure 14.1). The client is provided with a list of servers that provide DaaS. The user selects some servers to distribute the simulation of his model. Then, if the server does not have a copy of the model, the model is uploaded and compiled at all the servers. The user selects the server that can act as the coordinator of the imminent simulation and it creates new simulators on other remote DaaS providers. Finally, the distributed simulation is executed and the results are sent back to the user.

Thus, from a user's perspective, the simulation process is done through the following three steps (Figure 14.1):

1. Write a DSL model and autotransform it to DEVS model using the developed M2DEVS or M2DEVSML transformations.
2. From a list of DaaS providers (through, e.g., Universal Description Discovery and Integration [UDDI]), select your simulation resources.
3. Run the simulation (upload, compile, and simulate) and wait for the results.

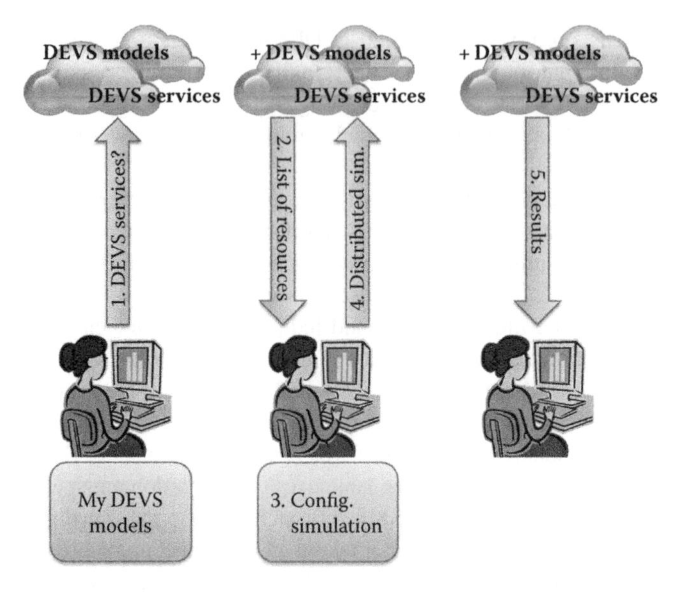

FIGURE 14.1 Execution of DEVS SOA-based M&S.

14.2 NETCENTRIC DEVS VIRTUAL MACHINE

The web service framework is essentially a client–server framework wherein a server provides the requested services to the clients. These services are nothing but computational code that is executed at the server's end with a valid return value. The mode of communication between the client and the server employs standard transport protocols like XML, Hypertext Transfer Protocol (HTTP), and SOAP. This standardized mode of communication provides interoperability between various services as the data, expressed in XML, is machine-readable.

To implement our DEVS/SOA framework, we have to go beyond this client–server paradigm, for this paradigm is not distributed in nature. Even though it operates on a network (Internet), it may not be distributed. We need to implement a distributed framework in a netcentric setup to have the capability of distributed M&S. The distributed DEVS protocol has two types of components: the coordinator that corresponds to a coupled model and the simulator that corresponds to an atomic model. These components need to be deployed at remote nodes in the same session, so that a distributed execution can take place.

In the current SOA framework, the server only acts as a provider of service and the client only acts as a consumer of service. Contrary to this functionality, the DEVS simulation components mentioned in Section 14.1 can be placed anywhere on the network. It is unavoidable that the same server can act as a provider and a consumer while executing DEVS simulation protocol. Consequently, the SOA that executes the DEVS simulation protocol is constructed such that the servers that provide DEVS service can play the role of both the coordinator and the simulator. As shown in Figure 14.1, step 2 provides a list of resources (servers) available on the Internet that provides DEVS simulation services. Once the list of servers is available to the user, the user assigns the role of coordinator to one of the servers and the rest of them become simulators. This fundamental concept of a DEVS node that acts as both a server and a client, defines the architecture of a netcentric DEVS Virtual Machine (DEVSVM). We provide more details on this assignment in Section 14.7.

During the execution of DEVS simulation protocol, each of the simulators makes calls to other simulators. Such calls are executed using the SOA framework. These simulators also coordinate with the coordinator using the same transport mechanism. As a result, the same simulator is invoking services from other simulators while providing services to other simulators or coordinators. This results in an architecture that is symmetrical by default; that is, it acts as both a service provider and a service consumer. The temporal role of a remote node is guided by the DEVS simulation protocol.

The DEVS simulation layer services are defined in a separate Web Service Description Language (WSDL) that implements this symmetrical execution. Further, in addition to the roles of simulator consumer and provider, the architecture allows the remote node to act as either a coordinator or a simulator. This assignment is made at step 3 in Figure 14.1 and is elaborated in sections 14.3–14.5.

Figure 14.2 depicts the architecture of DEVS/SOA. Starting at the bottom of Figure 14.2, the modeling layer of DEVS/SOA is built upon our MicroSim modeling layer; that is, all the DEVS/SOA repository is MicroSim compliant. To achieve distributed interoperability, we have developed two versions of MicroSim, one oriented to implement Java-based models and the other to implement .NET-based models. We showed in Chapter 13 that a DEVS M&S engine is able to simulate models coming from different DEVS simulation engines and also non-DEVS models if a proper wrapper is developed. Thus, the interoperability is also considered in DEVS/SOA mainly because MicroSim implements several DEVS adapters. Later, we show that this interoperability is trivially solved between Java and .NET models through the DEVS/SOA simulation layer.

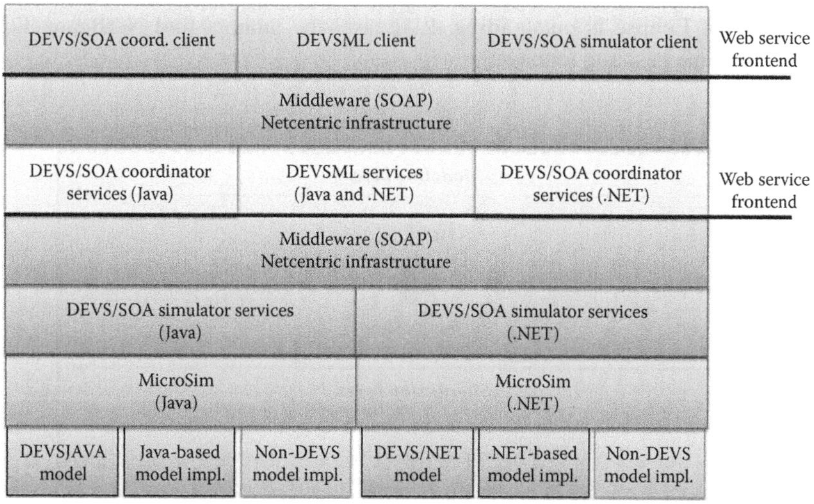

DEVS/SOA coord. client
DEVSML client
DEVS/SOA simulator client
Web service frontend
Middleware (SOAP)
Netcentric infrastructure
DEVS/SOA coordinator services (Java)
DEVSML services (Java and .NET)
DEVS/SOA coordinator services (.NET)
Web service frontend
Middleware (SOAP)
Netcentric infrastructure
DEVS/SOA simulator services (Java)
DEVS/SOA simulator services (.NET)
MicroSim (Java)
MicroSim (.NET)
DEVSJAVA model
Java-based model impl.
Non-DEVS model impl.
DEVS/NET model
.NET-based model impl.
Non-DEVS model impl.

FIGURE 14.2 Complete DEVS/SOA architecture for a netcentric DEVS Virtual Machine.

Following Figure 14.2, the next layer is the simulation layer. We have developed a DEVS/SOA simulator, where all the functionality is offered as web services. In addition, we have a DEVS/SOA coordinator that is also DEVSML compliant. As with the DEVS/SOA simulator, the DEVS/SOA coordinator is also available through web services. In addition, the DEVS/SOA coordinator uses other coordinators and simulators as web service references. Regarding DEVSML services, it allows the system engineer the possibility to upload DEVSML model and transform or compile it to a MicroSim model.

Coordinator, simulator, and DEVSML web services are available to develop client applications (top layer of Figure 14.2). Based on the capabilities of DEVS/SOA coordinators and simulators as well as the DEVSML integrated into DEVS/SOA, a basic DEVS/SOA client application consists of the following functionalities:

- Upload DEVSML model
- Compile DEVSML model to MicroSim model
- Simulate DEVS model (centralized and local, distributed and remote)
- Simulate real-time (RT) DEVS model (centralized and local, distributed and remote)

Figure 14.3 shows a simplified view of our DEVS/SOA architecture. Figure 14.3 shows how the modeling layer can be seen as a composition of DEVSML plus MicroSim, whereas the simulation layer is composed by DEVS/SOA coordinators and simulators. This scheme is used in the rest of this chapter. Web services are automatically created from DEVSML, DEVS/SOA coordinators, and DEVS/SOA simulators.

To implement this architecture, we have used superior Java Enterprise and .NET integration capabilities of NetBeans 7 and Microsoft Visual Studio 2010. It is our experience that NetBeans and Eclipse are complementary to each other and advancements in one of them leads the other. For example, the Xtext Eclipse EMF framework of Eclipse is outstanding. Likewise, the authors find NetBeans J2EE framework quite easier to manage.

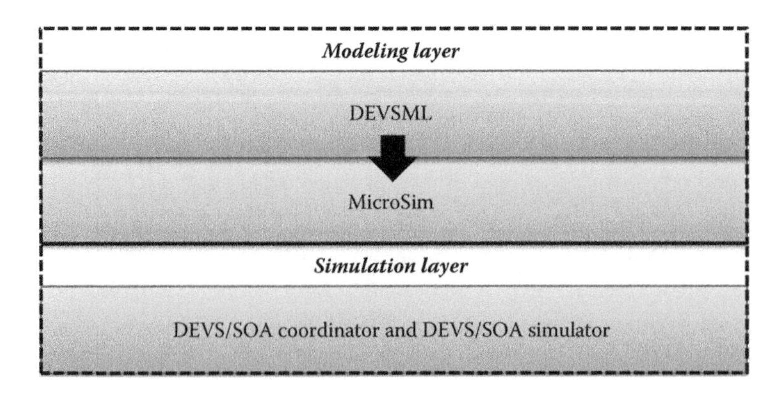

FIGURE 14.3 DEVS/SOA simplified architecture.

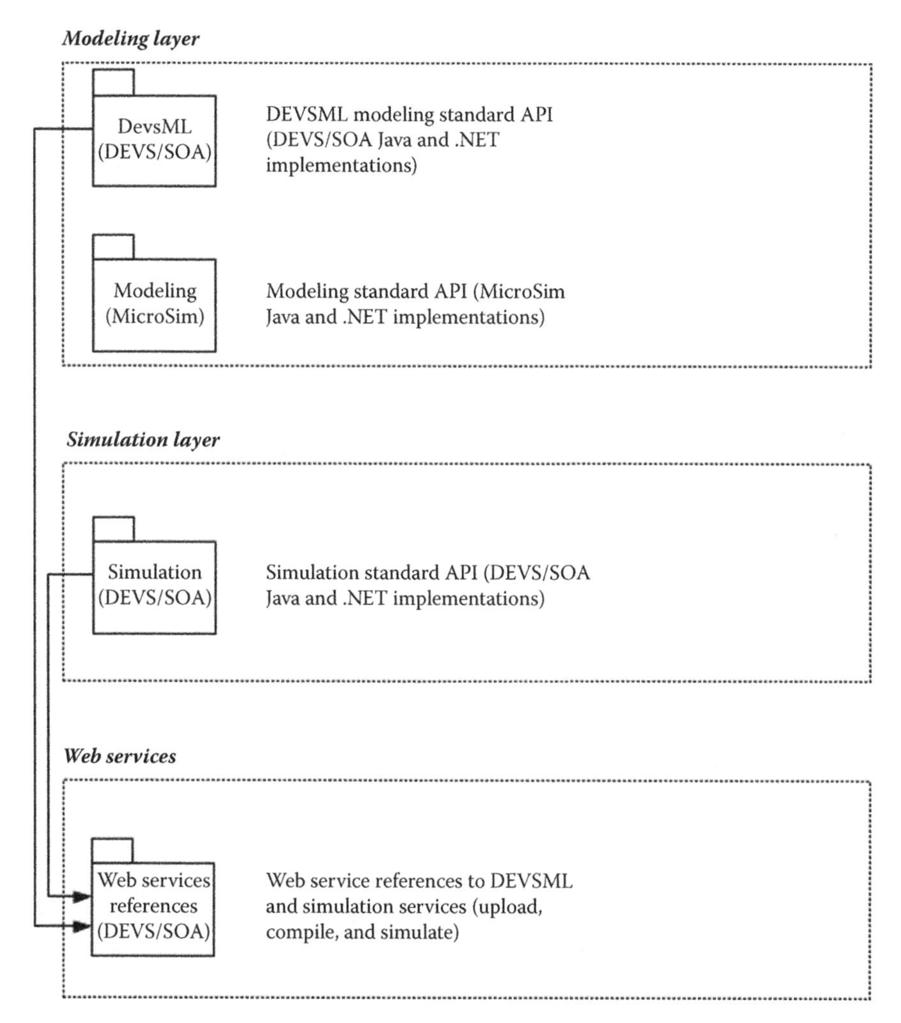

FIGURE 14.4 Server's package structure for DEVS/SOA.

The complete design of the architecture at server's end is as follows and as shown in Figure 14.4. All the software packages must be implemented in both Java and .NET environments.

Next, we briefly describe our modeling layer formed by DEVSML and MicroSim. In sections 14.3–14.6, we will describe the simulation architecture. Finally, we provide an example of a client application.

14.3 DEVSML PACKAGE

We have already introduced the DEVSML and employed this DEVS DSL in the DEVSML 2.0 stack. It highlighted the use of DSLs along with M2DEVS, M2DEVSML, and M2M transformation to lead these DSLs into the DEVS M&S domain. The DEVSML 2.0 stack is fully integrated into our DEVS/SOA architecture,

SoaDevsML
+upload(serverPackage : String, arrayOfFileContents : byte [], arrayOfFileNames : byte []) : void +compile(serverPackage : String, arrayOfFileNames : byte []) : void

FIGURE 14.5 DEVSML services.

providing two basic services: *upload* and *compile* DEVSML models. Thus, the user can upload DEVSML models to different servers, increasing the size of the initial repository with new models. These DEVSML models are automatically transformed and compiled to the target server (to DEVS/SOA Java or DEVS/SOA .NET models, or more specifically MicroSim/Java or MicroSim/.NET), and this operation is performed at the server as a service.

Figure 14.5 shows the set of services offered by the DEVSML package:

- *upload*: It is used to upload the model to the different servers. This service enables the user to take their DEVS models and upload the code physically from their machines to the designated DEVS/SOA server farm. This service receives (1) the package name, which is the folder where the model is saved at the server side; (2) the content of the DEVSML files, which is in fact the DEVS model implementation; and (3) the name of these files.
- *compile*: This service is used to transform and compile the uploaded DEVSML model files to the target architecture (MicroSim/Java or MicroSim/.NET). It receives (1) the package name, which is the server folder where the model was previously uploaded and (2) the file names of the DEVSML model.

14.4 MICROSIM PACKAGE

The modeling package constitutes the DEVS modeling library and is basically formed by two packages: the DEVSML modeling package and the MicroSim (both Java and .NET implementations) have been designed to wrap different DEVS libraries, such as DEVSJAVA, xDEVS, or DEVS/NET. Thus, DEVS models in different DEVS modeling DSLs can be incorporated. The MicroSim DEVS modeling package has already been described in this book. We provide some details about the MicroSim/.NET implementation in Chapter 15.

14.5 DEVS/SOA SIMULATION PACKAGE

14.5.1 Introduction

Web-based simulation requires the convergence of simulation methodology and WWW technology (mainly web service technology). The fundamental concept of web services is to integrate software application as services. Web services allow the applications to communicate with other applications using open standards. We are offering DEVS-based simulators as a web service, and they must have these standard technologies: communication protocol (SOAP), service description (WSDL), and service discovery (UDDI). The key difference between a parallel and distributed system that uses a single platform (and most likely the same programming language) and a netcentric system is the usage and specification of industry-wide standards. The standards define the contracts between different parties, and each of the components in a netcentric system is a stand-alone system, which is a first requirement in designing a system of system. In DEVS Unified Process, we are focused toward netcentricity.

Each DEVS/SOA server (Java or .NET) executes DEVS simulation protocol as a service. Thus, client applications can be developed to run distributed simulations. In these simulations, DEVS/SOA models are located in server repositories and model composition is performed by the client user through XML documents.

Figure 14.6 shows the framework of the proposed distributed simulation. The complete setup requires one or more servers that are capable of providing DEVS coordinator and DEVS simulator services. The capability to run the simulation service is provided by the server side design of the DEVS simulation protocol.

The simulation interface is built upon DEVS specification. As a result, there are also other DEVS standard functions published as services by DEVS/SOA, which can be used by the client application to improve functionality:

- Initialize simulator *i*
- Run transition in simulator *i*
- Run lambda function in simulator *i*
- Inject message to simulator *i*
- Get time of next event from simulator *i*
- Get time advance from simulator *i*
- Get console log from all the simulators
- Finalize simulation service

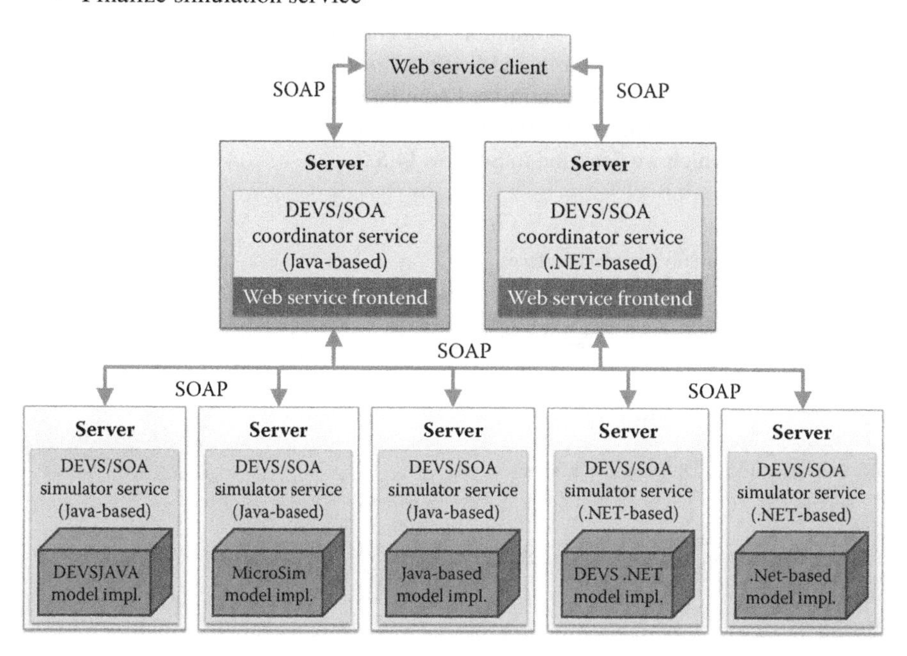

FIGURE 14.6 DEVS/SOA distributed simulation.

The explicit transition functions, namely, the internal transition function, the external transition function, and the confluent transition function, are abstracted to a single transition function that is made available as a service. The transition function that needs to be executed depends on the simulator implementation and is decided at runtime. For example, if the simulator implements the parallel DEVS formalism, it will choose among the internal transition, external transition, or confluent transition.

14.5.2 Message Serialization

The issue of message passing and model upload is done through serialization and SOA technologies. Figure 14.7 illustrates the message serialization process. When a component makes an external transition or executes the output function, the message received or emitted is serialized in the corresponding simulator and then sent to the coordinator through the simulation service. The coordinator stores the location of each simulation service, so it is able to request all the messages after each simulation step.

All the communication between the coordinator and the simulation services is done through SOAP over HTTP. The serialization is done through Java or .NET serialization utilities. In the RT version, each simulator knows each simulation service at its end (from coupling information). So the established logical communication is made possible by passing messages from simulation services to simulation services directly, without using the coordinator.

14.5.3 DEVS/SOA Metamodel

Figure 14.8 shows the Unified Modeling Language (UML) diagram of the DEVS/SOA simulation package. It contains simulators and coordinators, that is, *SoaSimulator, SoaCoordinator, SoaRTSimulator*, and *SoaRTCoordinator* classes. *SoaCoordinator* and *SoaSimulator* are used in centralized simulations, whereas the RT prefix indicates that the class is designed for RT simulation. Basically, RT and non-RT use the same interface, which we describe in Section 14.5.5.

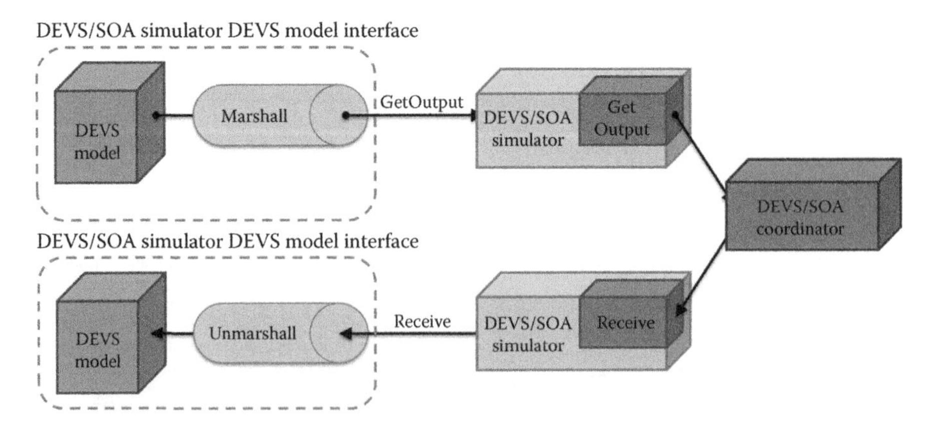

FIGURE 14.7 Message serialization through marshall and unmarshall operations.

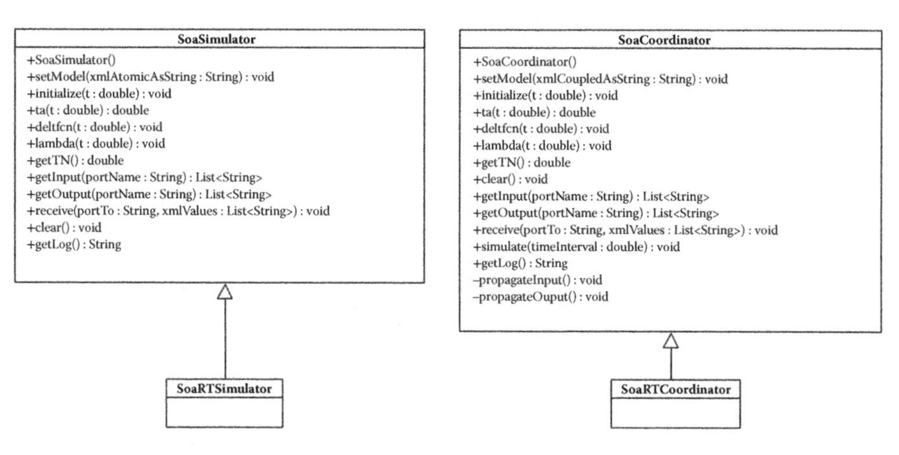

FIGURE 14.8 DEVS/SOA simulation package.

14.5.3.1 DEVS/SOA Coordinator

The main difference with other simulator platforms starts here. In both centralized and RT simulations, the coordinator receives an XML file containing the DEVS model to be simulated and the location of each of its components. The root coordinator is executed at the server selected by the user. The XML file (Listing 14.1) shows the initial configuration to run the Generator Processor Transducer (GPT) example as a DEVS/SOA model.

LISTING 14.1 DEVS/SOA XML CONFIGURATION FILE

```
<?xml version="1.0" encoding="UTF-8" standalone="yes"?>
<devs name="Gpt" host="192.168.1.2">
  <atomic name="gene"
          class="microsim.lib.examples.gpt.Generator"
          host="192.168.1.3"/>
  <atomic name="proc"
          class="microsim.lib.examples.gpt.Processor"
          host="192.168.1.2"/>
  <atomic name="tran"
          class="microsim.lib.examples.gpt.Transducer"
          host="192.168.1.5"/>
  <coupling componentFrom="gene" portFrom="out"
           componentTo="proc" portTo="in"/>
  <coupling componentFrom="gene" portFrom="out"
           componentTo="tran" portTo="arrived"/>
  <coupling componentFrom="proc" portFrom="out"
           componentTo="tran" portTo="solved"/>
  <coupling componentFrom="tran" portFrom="out"
           componentTo="gene" portTo="stop"/>
</devs>
```

The first component, labeled as "devs," will be simulated by a root DEVS/SOA coordinator, located at host "192.168.1.2". Next, each atomic model will be simulated by a DEVS/SOA simulator (and each coupled model by a DEVS/SOA coordinator). As a result, from the XML configuration file, the host "192.168.1.2" will manage one coordinator and one simulator, whereas the other two hosts will manage one simulator each.

Following is the description of the set of DEVS/SOA coordinator member functions:

- *setModel*: This service may receive an full XML configuration file as a string or a "coupled" XML node as part of the configuration file. In the first case, the DEVS/SOA coordinator will be the root coordinator, whereas in the second case, the DEVS/SOA coordinator will be in charge of managing one DEVS coupled model. This method also creates all the children of the root coordinator. Note that in this case, all the children are web service references.
- *initialize*: This web method initializes the last timed event and the next time event for the current DEVS/SOA coordinator and each of its children.
- *ta*: This is the time advance function, which receives the current time and computes the time of next event.
- *deltfcn*: This service receives the current simulation time. The service propagates the input from the external input connections in the simulated coupled model and executes an internal or external or confluent transition function in each child. It then updates the time of last event and the time of next event.
- *lambda*: This service executes the output function in each child (simulator or coordinator). At the end, this web method propagates the output generated using coupling information.
- *getTN*: This web method returns the time of next event.
- *clear*: This service empties all the ports in the entire DEVS model.
- *getInput*: This service returns the input stored in the component simulated at the corresponding port name, passed as an argument.
- *getOutput*: This service returns the output stored in the corresponding port name.
- *receive*: The web method stores the set of values received in the input port specified.
- *getLog*: This service returns the content of the log file generated for the current simulation.
- *simulate*: This is the main web method. After the DEVS/SOA coordinator has been initialized, this member receives the interval to run the simulation. In centralized simulation, this is virtual time. In RT simulation, this is wall clock time.
- *propagateInput and propagateOutput* are auxiliary member functions used in *deltfcn* and *lambda* to propagate input and output between connections respectively. These two member functions have not been declared as web methods.

In the most basic usage from a client application, the root DEVS/SOA coordinator is called through the *simulate* service reference, using a XML configuration file. Using this file, the coordinator receives the name of the root coupled model and a list of network IP addresses for each subcomponent (each subcomponent linked to a DEVS/SOA coordinator or simulator). Such list of IPs is used to invoke simulation services in other remote servers. In this way, the components of the model are distributed among N servers, where N is the length of that list. The coordinator stores the list of simulation or coordination services activated. In the case of centralized simulation, this list is used to propagate and to receive messages through the coupling protocol stored in the root coupled model. In addition, the coordinator stores the time of last event and the time of next event. In the case of RT simulations, instead of event times, the DEVS/SOA RT coordinator only knows the time in which the simulation must be stopped.

The task of the coordinator is to execute a typical DEVS loop over the distributed simulators. Both DEVS/SOA coordinators and simulators store their "state," that is, time of last event, simulators, coupling information, simulated component, and so on, using session variables.

14.5.3.2 DEVS/SOA Simulator

The DEVS/SOA simulator receives an "atomic" XML node of the configuration file shown in Listing 14.1. The member function and services provided by DEVS/SOA simulator (Figure 14.8) are enumerated as follows:

- *setModel*: This service receives an XML atomic node of the configuration file. It instantiates the associated atomic model and stores it in a session variable.
- *initialize*: This web method initializes the last timed event and the next time event for the current DEVS/SOA simulator.
- *ta*: This is the time advance function, which just executes the time advance method of the simulated atomic model.
- *deltfcn*: This service executes an internal or external or confluent transition function in the simulated model. After that, it updates the last timed event and the next time event.
- *lambda*: This service just executes the output function in the corresponding atomic model.
- *getTN*: This web method returns the current next time event.
- *clear*: This service just empties all the ports in the atomic model.
- *getInput*: This service returns the input stored in the component simulated at the corresponding port name, passed as argument.
- *getOutput*: This service returns the output stored in the corresponding port name.
- *receive*: The web method stores the set of values received in the specified input port.
- *getLog*: This service returns the content of the log file generated for the current simulator.

Next, we describe the difference between centralized and RT simulations.

14.5.4 Centralized Simulation

The centralized simulation is done through a central coordinator that is located at the main server. The coordinator creates *n* simulation services over the selected DaaS. Each simulation service simulates its corresponding model. Figure 14.9 shows the process for a distributed generator-processor DEVS/SOA model, where the coordinator and both the simulators can be located at different machines. Once the simulation starts, the coordinator executes the output function of the simulation services (in Figure 14.9, points 0 and 1). After that, the output is propagated and internal transitions occur. Propagating an output means that once the coordinator takes the serialized output from the simulation services (2 and 3), it is sent to other simulation services by means of coupling information (4 and 5). This information is known only by the coordinator as all messages must flow through the coordinator.

As it appears, the coordinator participates in all message passing and is the bottleneck. We designed distributed DEVS SOA protocol where the coupling information is downloaded to each of the models and coordinator is relieved of message passing. It is described in Section 14.5.5.

14.5.5 Real-Time Simulation

RT DEVS simulation is defined as the execution of DEVS simulation protocol in wall-clock time rather than logical virtual time. In this particular execution of DEVS, each of the simulators runs in an independent thread that is located on remote machines. The DEVS coordinator presides over the DEVS simulators and performs the job of starting and stopping the simulator threads. Each simulator executes the DEVS simulation protocol and schedules the next event based on the value of its own

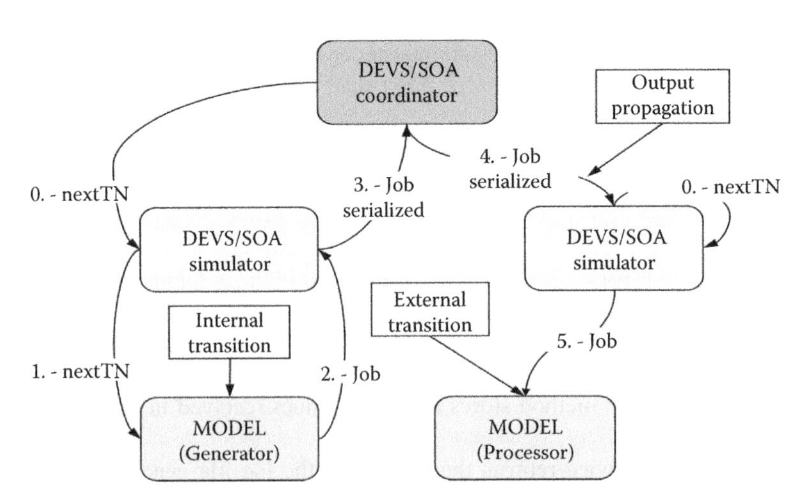

FIGURE 14.9 Centralized communication among services.

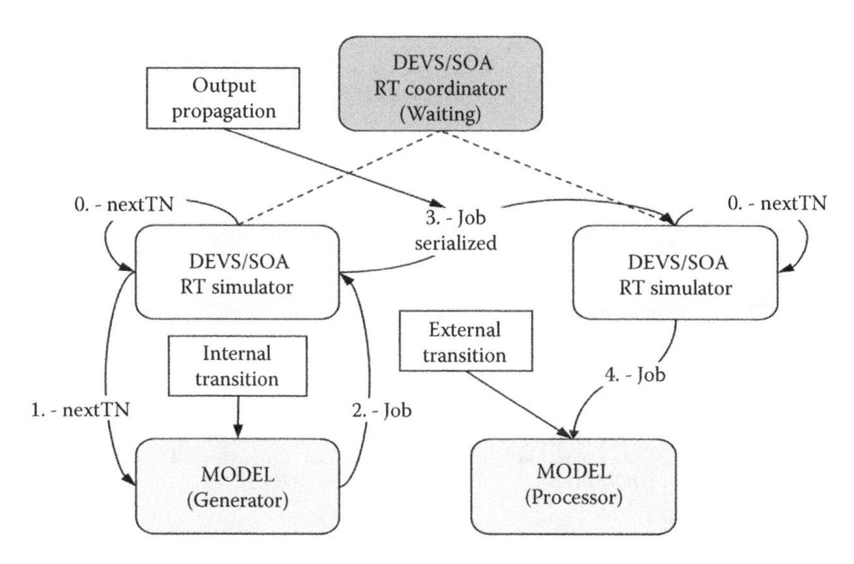

FIGURE 14.10 RT communication among services.

tN variable. It should be noted that the network delay in a RT simulation is a real network delay and the simulation is executed over a live network.

For the RT simulation, we have incorporated a DEVS/SOA RT coordinator and a DEVS/SOA RT simulator. All their functions are also published as web services, which are the same set of functions than in the centralized simulation.

The design is similar in many aspects, but instead of a central coordinator, all the simulation is observed by an RT coordinator without any intervention. Furthermore, each RT simulation service knows the coupling information, so the message passing is made directly from one simulation service to other simulation service. The RT coordinator is located at the main server. This coordinator creates n RT simulators over the Internet. Each simulator simulates its own component of the model. After that the coupling information is broken down (on a per-model basis) and sent to the corresponding RT simulator. Figure 14.10 illustrates the process. Once the simulation starts, the coordinator executes the simulate service and nothing else. The simulate service waits for internal or external transitions using RT (0). If an internal transition happens (1), the output is generated and propagated using the coupling information serializing and de-serializing messages (2, 3, and 4).

14.6 CROSS-PLATFORM DEVELOPMENT AND EXECUTION OVER DEVS/SOA

One other section that requires some description is the multiplatform simulation capability as provided by DEVS/SOA framework. It consists of realizing distributed simulation using different DEVS and non-DEVS models such as DEVSJAVA, DEVS-C++, MatLab®, and so on. This kind of interoperability is trivially solved when using both DEVS/SOA Java and DEVS/SOA.NET architectures. On one hand,

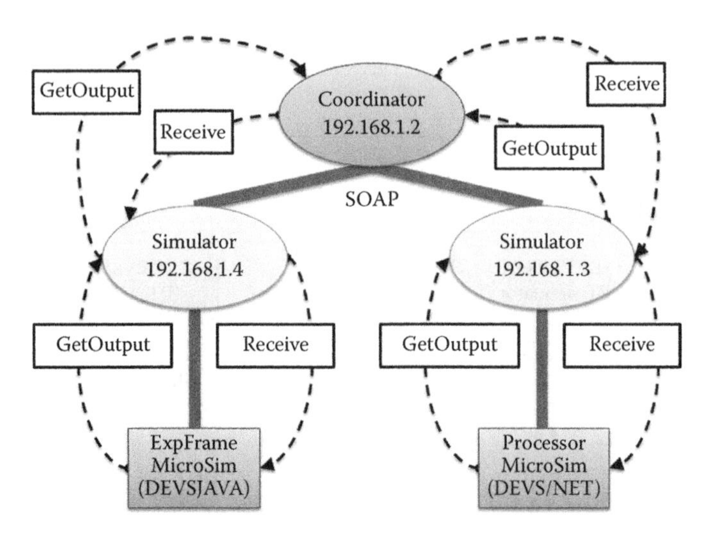

FIGURE 14.11 DEVS/SOA interoperability.

the interoperability at the modeling level of the same platform (either Java or .NET) is solved by MicroSim, as we showed in Chapter 13. On the other hand, the interoperability among different Java and .NET DEVS/SOA engines is solved in the message passing interface.

Figure 14.11 depicts the EF-P example and how DEVS/SOA allows us interoperability. In this example, the Experimental Frame is used as MicroSim/Java atomic model, which is wrapping the Experimental Frame coupled model implemented in DEVS/Java. On the other hand, the Processor atomic model is implemented using MicroSim/.NET. As can be seen, the interoperability between DEVS models in either Java or .NET is solved through the adaptation of different models to the MicroSim modeling package. The interoperability between DEVSJAVA and .NET simulators is trivially solved through the message passing interface. Messages passed are represented in XML and thus platform independent.

14.7 EXAMPLE: A CLIENT APPLICATION

All the services offered by DEVS/SOA can be used in a client application. The end user of DEVS/SOA must use web service references to build his own applications. We can only use DEVS/SOA coordinator services, and thus, just composition of preexisting models and simulation is allowed. However, using DEVSML services, the user can also upload his models.

This section describes an example of a client application to execute DEVS model over SOA framework using simulation-as-a-service (SaaS). In Chapter 15, we will see another example at lower level. For simplicity, we only use here a graphical user interface (GUI) to interact with DEVS/SOA Java servers.

The DEVS/SOA client takes the DEVS models package and through the dedicated servers hosting simulation services, it performs the following operations:

- Upload the models to specific IP locations
- Runtime compile at respective sites
- Simulate the coupled model
- Receive the simulation output at client's end

The DEVSV/SOA client, as shown in Figure 14.12, operates in the following sequential manner:

- The user selects the DEVS package folder at his machine.
- The top-level coupled model is selected as shown (JCASNum1 in Figure 14.12).
- Various available servers are selected (Figure 14.13). Any number of available servers can be selected (one at least).
- Clicking the button labeled "Assign Servers to Model Components," the user selects where it is going to simulate each of the coupled models, including the top-level one, that is, the main server where the coordinator will be created (Figure 14.13).
- The user then uploads the model by clicking the Upload button. The models are partitioned and distributed among the servers chosen in the previous point.
- The user then compiles the models at the server's end by clicking the Compile button.
- The user then clicks the Simulate button to run the simulation.
- The user sees the simulation results on the console, as shown in Figure 14.12.

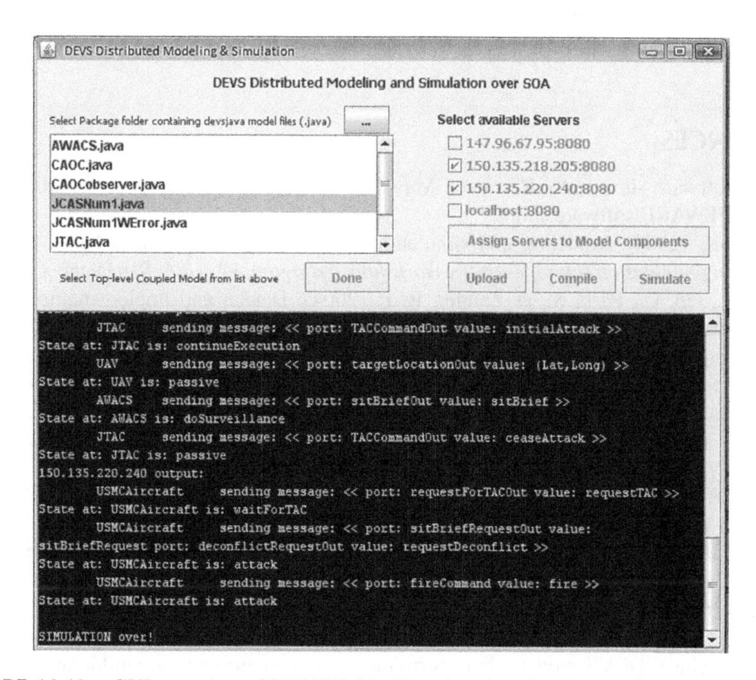

FIGURE 14.12 GUI snapshot of DEVS/SOA client hosting distributed simulation.

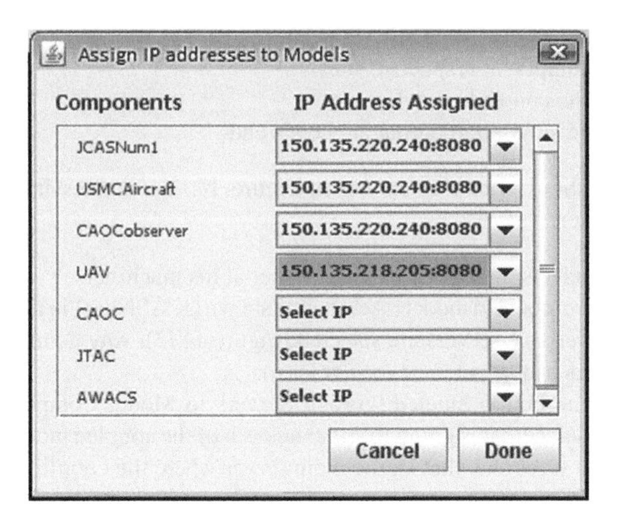

FIGURE 14.13 Server assignment to models.

14.8 SUMMARY

This chapter has described the software architecture of DEVS/SOA. It has described the interfaces, the design of the DEVS Virtual Machine, and the DEVS/SOA client that utilizes SimaaS leading to DaaS. It also highlighted multiplatform architecture inherent in MicroSim and DEVSML. In Chapter 15, we will see how the DEVSVM gets deployed in a local or remote environment.

REFERENCES

ACIMS Software site. (n.d.). Retrieved May 10, 2012 from http://www.acims.arizona.edu/SOFTWARE/software.shtml.

Atkinson, K. (2004). Modeling and simulation foundation for capabilities based planning. *Spring Simulation Interoperability Workshop, Arlington, VA, USA*. San Diego, CA, USA.

Cheon, S., Seo, C., Park, S., & Zeigler, B. P. (2004). Design and implementation of distributed DEVS simulation in a peer to peer networked system. *Advanced Simulation Technologies Conference*. Arlington, VA, USA.

Cho, Y. K., Zeigler, B. P., & Sarjoughian, H. S. (2001). Design and implementation of distributed real-time DEVS/CORBA. *2001 IEEE International Conference on Systems, Man and Cybernetics. e-Systems and e-Man for Cybernetics in Cyberspace (Cat. No.01CH37236)* (Vol. 5, pp. 3081–3086). Tucson, AZ. IEEE.

Fujimoto, R. M. (1999). *Parallel and Distribution Simulation Systems*. New York, NY, USA: John Wiley & Sons.

Hu, X., & Zeigler, B. P. (2005). Model continuity in the design of dynamic distributed real-time systems. *IEEE Transactions on Systems, Man, and Cybernetics - Part A: Systems and Humans*, *35*(6), 867–878.

Kim, K.-H., & W.-S. Kang. (2004). CORBA-based, multi-threaded distributed simulation of hierarchical DEVS models: Transforming model structure into a non-hierarchical one. *Computational Science and Its Applications* (pp. 167–176). LNCS (3046) - Springer.

Mittal, S., Risco-Martín, J. L., & Zeigler, B. P. (2009a). DEVS/SOA: A cross-platform framework for net-centric modeling and simulation in DEVS Unified Process, *SIMULATION*, *85*(7), 419–450.

Mittal, S., Zeigler, B. P., & Risco-Martín, J. L. (2009b) Implementation of a Formal Standard for Interoperability M&S/System of Systems Integration with DEVS/SOA, *International C2 Journal*, *3*(1). Available from: http://www.dodccrp.org/files/IC2J_v3n1_01_Mittal.pdf. Last accessed: November 2, 2012.

Seo, C., Park, S., Kim, B., Cheon, S., & Zeigler, B. (2004). Implementation of distributed high-performance DEVS simulation framework in the grid computing environment. *Advanced Simulation Technologies Conference.* Arlington, VA. SCS.

Wainer, G., & Giambiasi, N. (2001). Timed cell-DEVS: Modeling and simulation of cell spaces. *Discrete Event Modeling and Simulation Technologies* (pp. 187–214). Springer-Verlag New York, Inc. New York, NY, USA.

Zeigler, B. P., Praehofer, H., & Kim, T. G. (2000). *Theory of Modeling and Simulation: Integrating Discrete Event and Continuous Complex Dynamic Systems* (2nd ed.). Academic Press, San Diego, CA.

Zhang, M., Zeigler, B. P., & Hammonds, P. (2005). DEVS/RMI: An auto-adaptive and reconfigurable distributed simulation environment for engineering studies. *ITEA Journal* *27*(1): 49–60.

15 Model and Simulator Deployment in a Netcentric Environment

15.1 INTRODUCTION

In this chapter, we present how to develop and deploy our DEVS/SOA M&S framework. As described in the previous chapter, we are going to develop two DEVS/SOA platforms. The first one, the DEVS/SOA JAVA platform, is developed using NetBeans 7 and GlassFish as the main server (Oracle). NetBeans has been already introduced in Chapter 4. GlassFish is an open-source enterprise application server project started by Sun Microsystems for the Java EE platform and now sponsored by Oracle Corporation. Inside NetBeans, we are using the Java API for XML Web Services (JAX-WS), which is a Java programming language API for creating web services. It is part of the Java EE platform. Thus, the corresponding NetBeans plugin must be installed. The second one, DEVS/SOA.NET, is developed using Microsoft Visual Studio 2010. Microsoft Visual Studio is an Integrated Development Environment (IDE) from Microsoft. We can use it to develop console and graphical user interface applications, websites, web applications, and web services (MSDN).

Having both MicroSim/JAVA and MicroSim/.NET at the modeling layer and at the simulation layer, we will have a full interoperable and distributed framework for DEVS M&S. We start by revisiting the DEVS/SOA metamodel already presented in the previous chapter (Figure 15.1).

To implement both DEVS/SOA simulators and coordinators in a distributed fashion, we describe how to create the starting project in both platforms (JAVA and .NET). Next, we implement each of the functions shown in the UML diagram (Figure 15.1). Finally, we show how to deploy the resultant DEVS/SOA web applications and make them available in the cloud.

15.2 PROJECT PREPARATION

15.2.1 NetBeans

First of all, we create an empty project in NetBeans. Select File→New Project …, and in the next dialog box, select Web Application (see Figure 15.2).

Enter a project name (MicroSimServer, for instance) and save the project in the project folder (Figure 15.3).

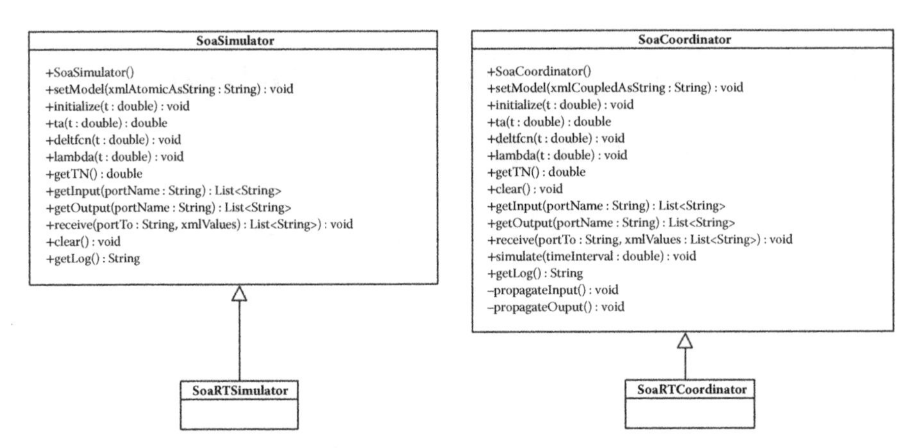

FIGURE 15.1 DEVS/SOA simulation package.

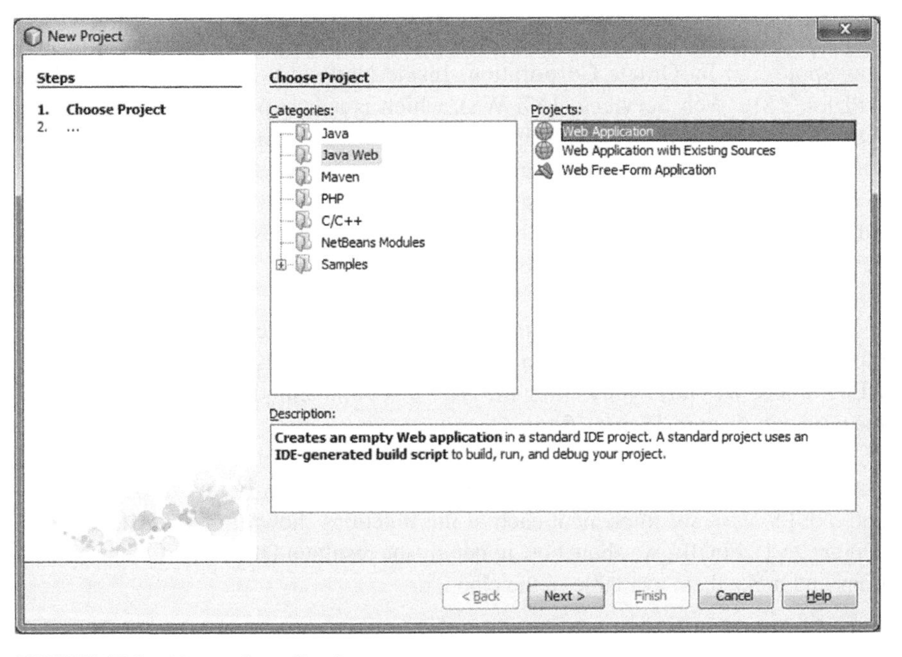

FIGURE 15.2 New web application.

Select the server where the web application will be deployed. At this point, we can install several servers if they are not already installed. In our case, we added the GlassFish server (Figure 15.4).

In the empty project, we create a package structure. Add two packages: "modeling" for DEVSML web services and "simulation" for DEVS/SOA web simulators (Figure 15.5).

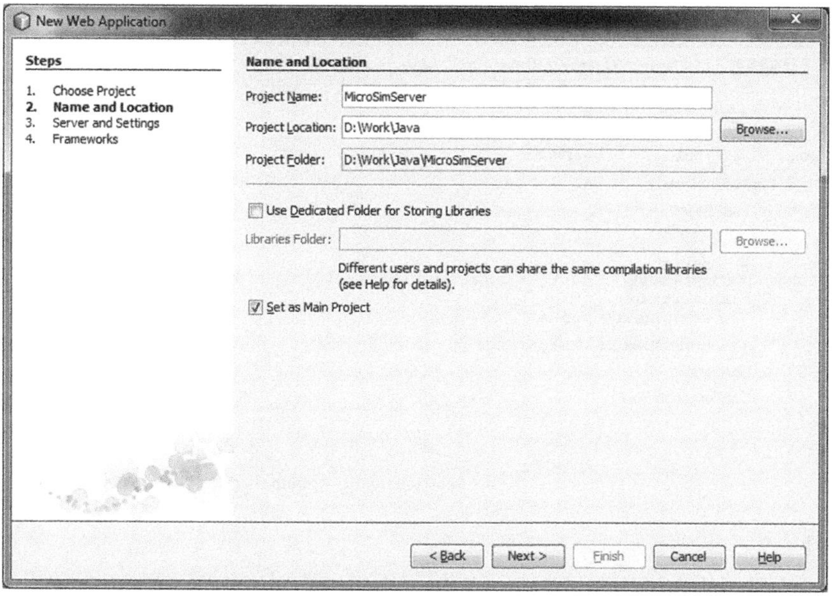

FIGURE 15.3 Project name: MicroSimServer.

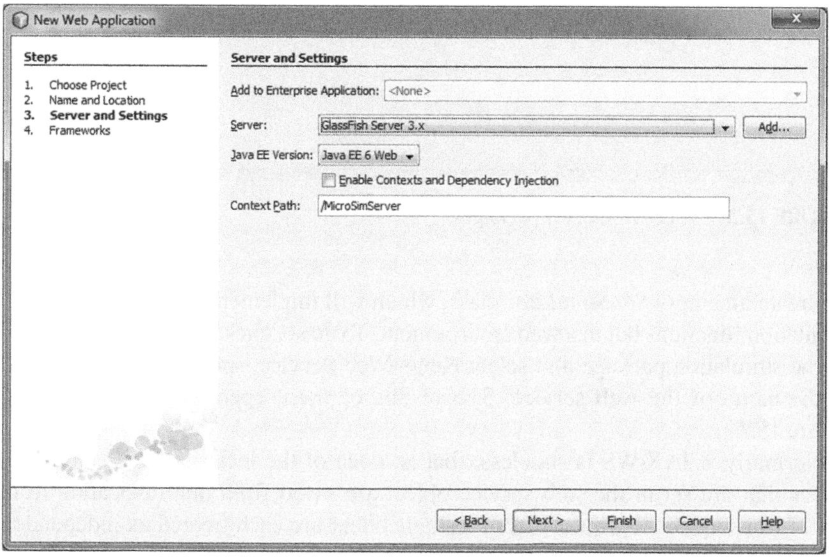

FIGURE 15.4 Server settings.

Since MicroSim/JAVA is needed to implement DEVS models, we create a *lib* folder in the MicroSimServer main directory and add the external libraries MicroSim.jar and devsjava.jar. Include these two external libraries in the MicroSimServer project (Figure 15.6).

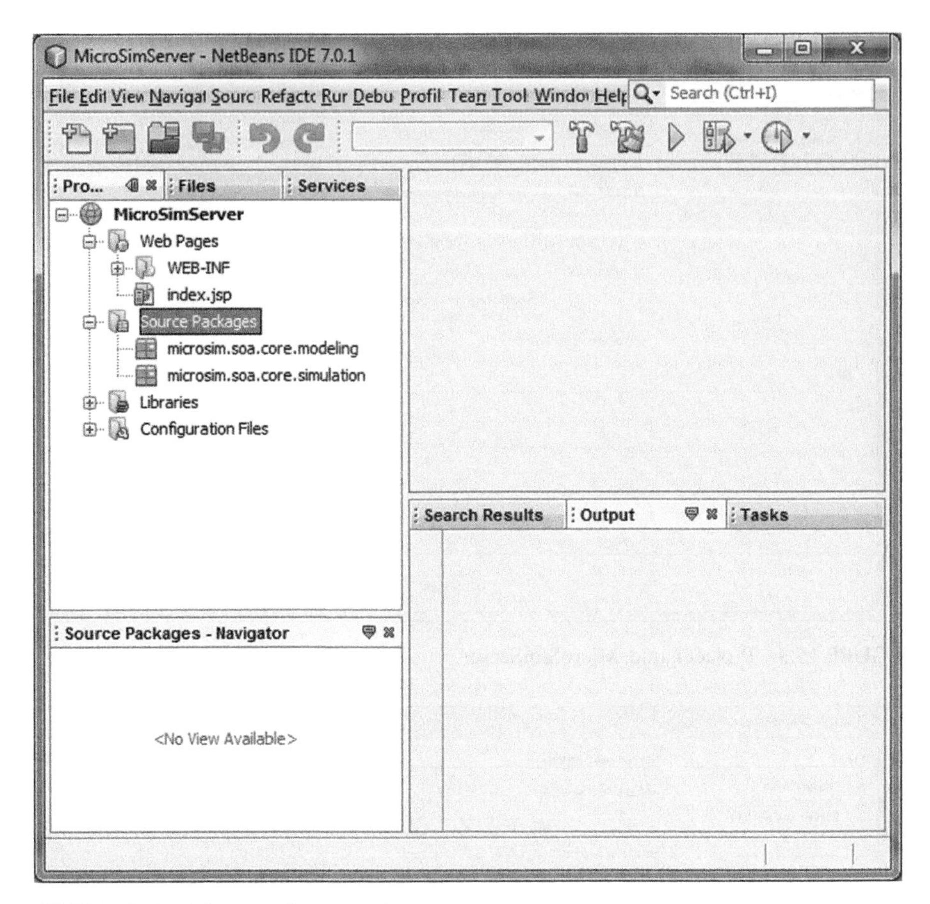

FIGURE 15.5 MicroSimServer packages.

Create an empty *SoaSimulator* class, which will implement all the typical DEVS simulation functions but in a web environment. To do so, click the right mouse button on the simulation package and select New→Web Service, specifying SoaSimulator as the name of the web service. The results of these operations can be seen in Figure 15.7.

Normally, a JAX-WS is stateless; that is, none of the local variables and object values that are set in the web service object are saved from one invocation to the next. Even sequential requests from a single client are each treated as independent, stateless invocations.

In our case, we want to save data on the service during one invocation and then use that data in the subsequent invocation. For example, the time of the last event must be saved between invocations at the remote simulator. This state variable (among others) can be maintained using HTTP sessions, and then returned to the client when necessary.

Enabling stateful support in a JAX-WS demands a minimal amount of coding on both the client and the server.

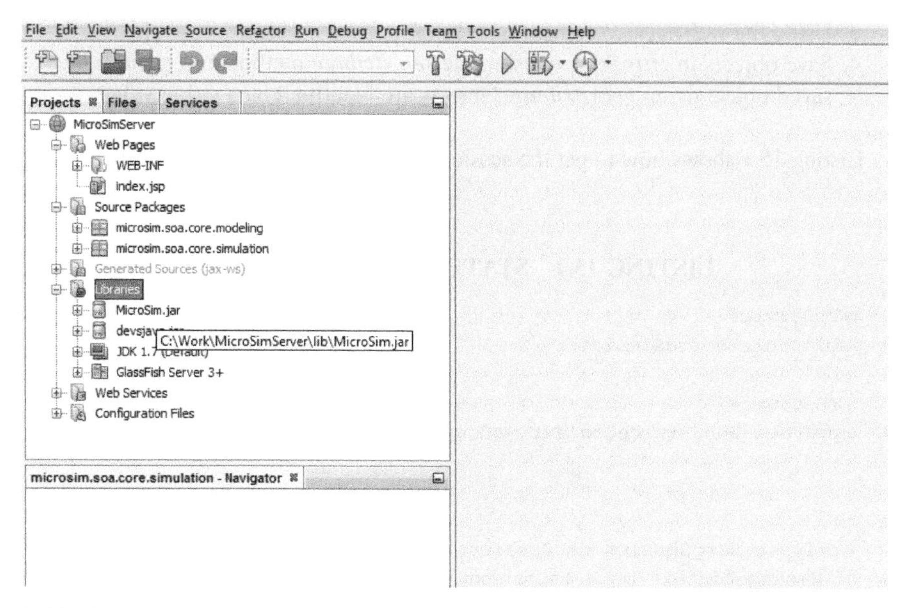

FIGURE 15.6 External libraries.

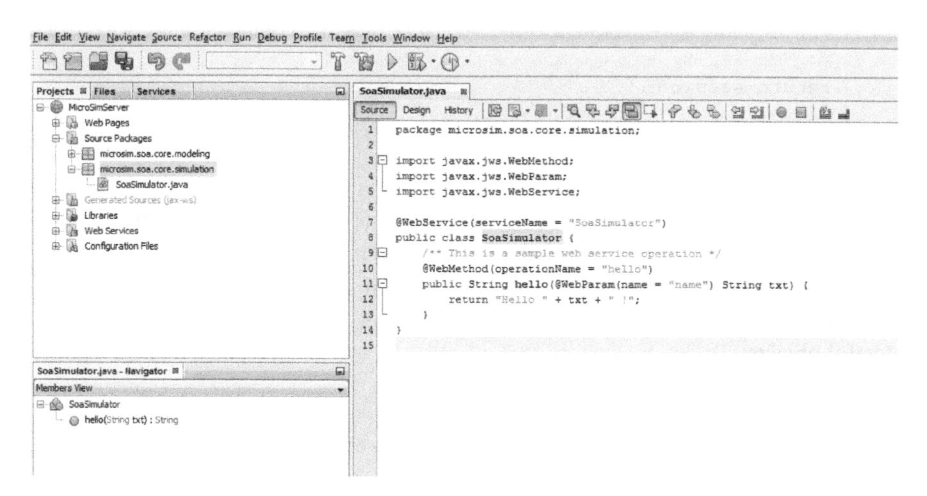

FIGURE 15.7 SoaSimulator web service.

The steps required on the server are as follows:

1. Add @Resource at the top of the web service.
2. Add a variable of type, *WebServiceContext*, that will have the context injected into it.

3. Using the web service context, get the *HttpSession* object.
4. Save objects in *HttpSession* using the *setAttribute* method and retrieve the saved object using *getAttribute*. Objects are identified by a string value.

Listing 15.1 shows how to get the session object using a *getSession()* function.

LISTING 15.1 STATEFUL WEB SERVICE

```
@WebService
public class SoaSimulator {

  @Resource
  private WebServiceContext wsContext;

  ...

  private HttpSession getSession() {
    MessageContext mc = wsContext.getMessageContext();
    HttpSession session =
        ((javax.servlet.http.HttpServletRequest) mc
       .get(MessageContext.SERVLET_REQUEST)).getSession();
    if (session == null) {
      throw new
           WebServiceException("No HTTP Session found");
    }
    return session;
  }
```

On the client side, we need to set the SESSION_MAINTAIN_PROPERTY in the request context. This tells the client to pass the HTTP cookies back to the server-side web service. The cookie contains a session ID that allows the server to match the web service invocation with the correct HttpSession, providing access to any saved stateful objects:

```
WebServiceClient client = (new WebServiClient()).
    getSoaSimulatorPort();
((BindingProvider)client).getRequestContext().
  put(BindingProvider.SESSION_MAINTAIN_PROPERTY, true);
```

15.2.2 Microsoft Visual Studio

Select File→New Project. In the following window, select ASP.NET empty web application using the same project name (MicroSimServer) and locating the project inside a solution (Net in our case) (Figure 15.8).

Once the project has been created, set the default namespace to microsim.soa (in Project→Properties) (Figure 15.9).

As in NetBeans, create two folders: one for the modeling layer and the other for the simulation layer. Click the right mouse button on the project name and select Add→New Folder (Figure 15.10).

Add the MicroSim/.NET library to have the MicroSim modeling layer available. Add the MicroSim/.NET project to the current solution (click the right mouse button on the solution and select Add Existing Project) (Figure 15.11).

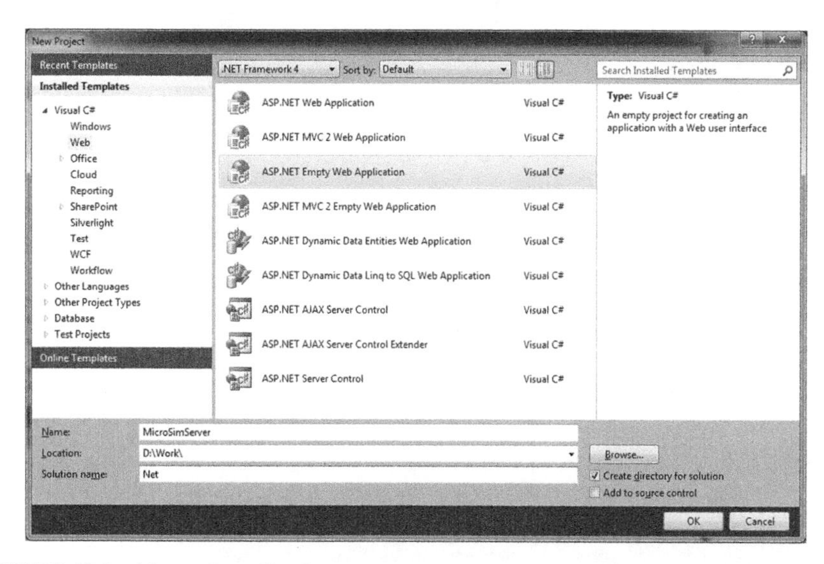

FIGURE 15.8 New web application.

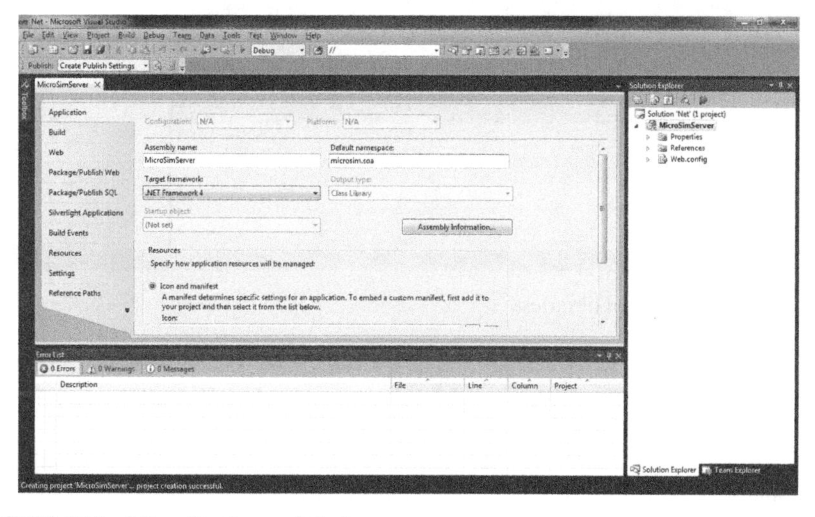

FIGURE 15.9 MicroSimServer default namespace.

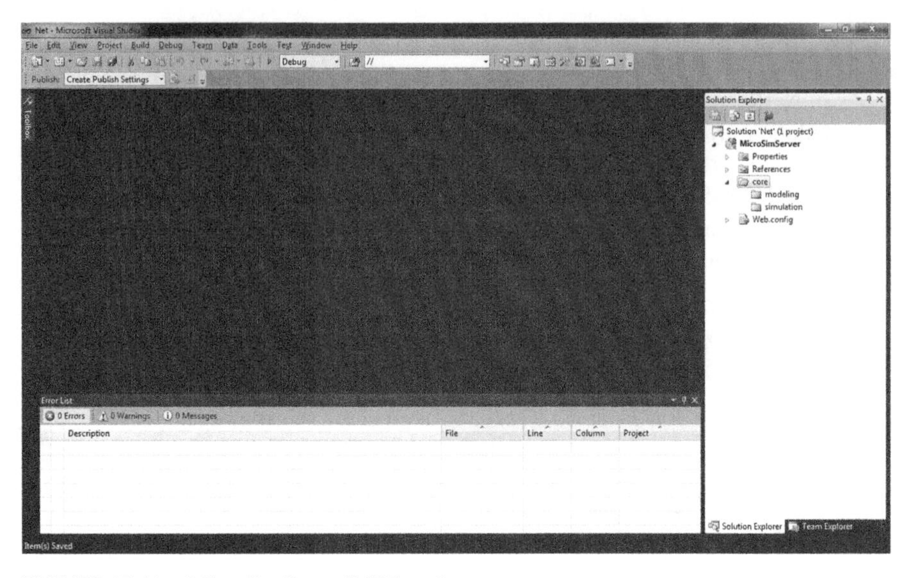

FIGURE 15.10 MicroSimServer/.NET packages.

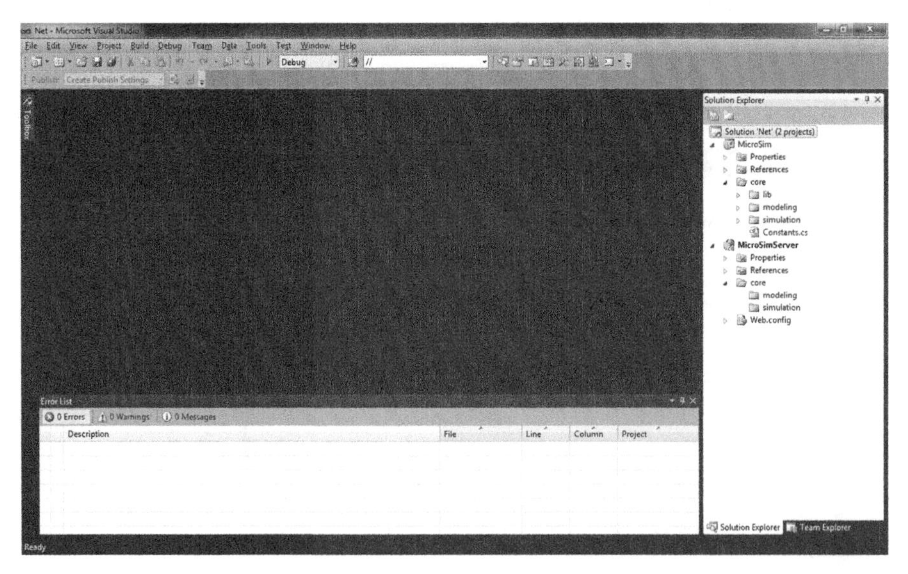

FIGURE 15.11 External libraries.

Click the right mouse button on the solution and select Properties, adding the solution dependencies (MicroSimServer depends on MicroSim) (Figure 15.12)

To start creating the web service, click the right mouse button on the simulation folder and select Add→Class. In the following window, select Web Service, and enter SoaSimulator.asmx as the name of the service (Figure 15.13).

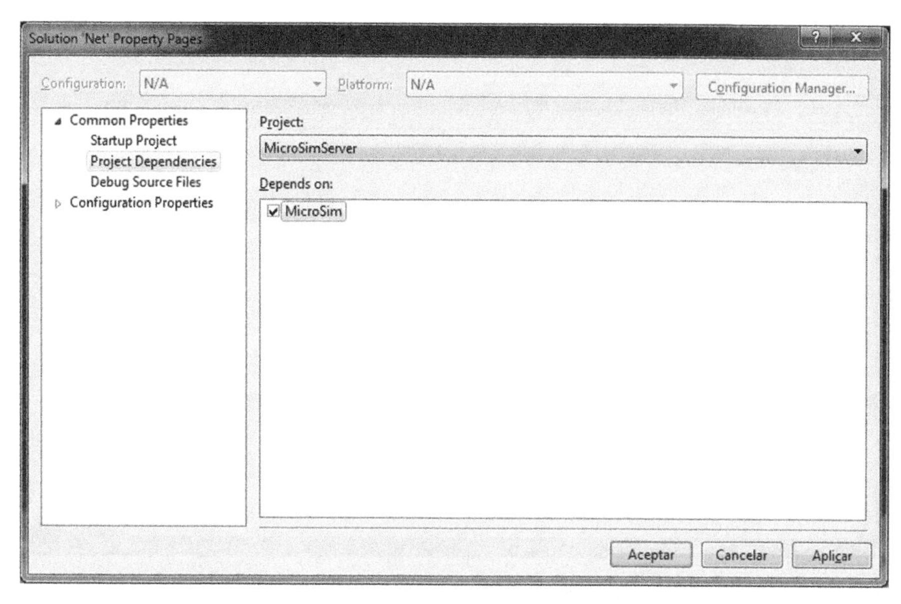

FIGURE 15.12 MicroSimServer/.NET dependencies.

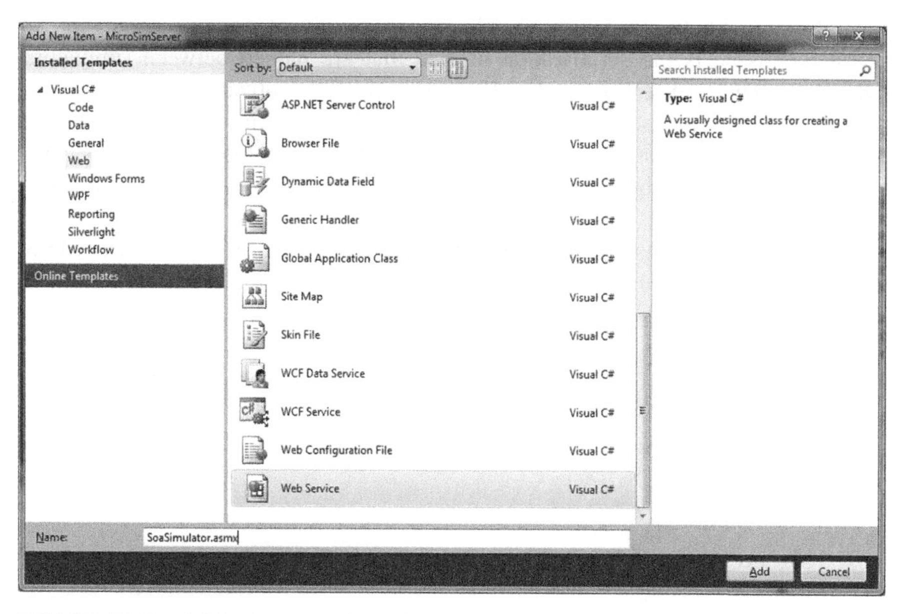

FIGURE 15.13 Adding a new web service.

Figure 15.14 shows the resultant class file. As can be seen, the notation is completely equivalent to the JAX-WS, making the task quite straightforward.

To use stateful web services, set the property *EnableSession* as true in every web method we need to save the state. Besides, Web.config (in server applications) and

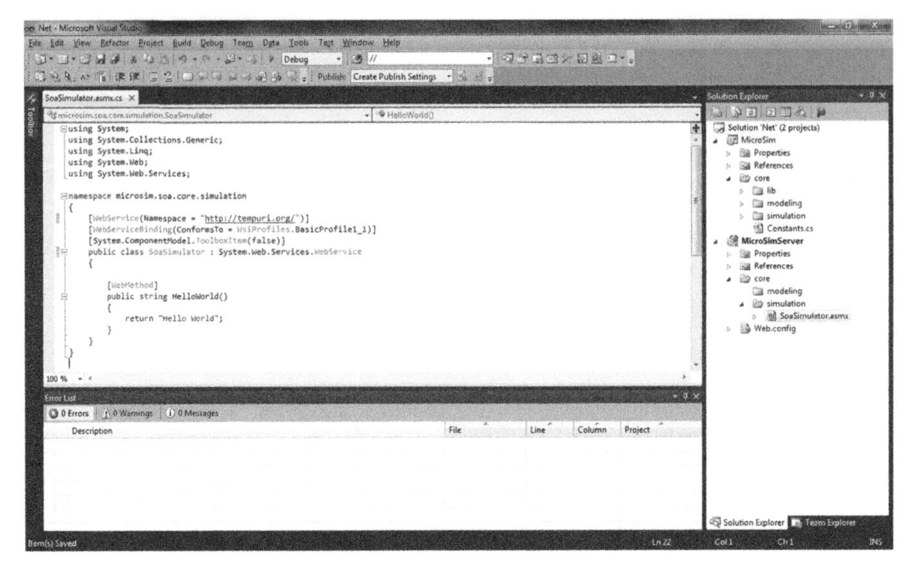

FIGURE 15.14 SoaSimulator web service.

app.config (in client applications) must be modified to allow cookies to store the session information between successive calls (parameter allowCookies = "true").

Another issue is the serialization of data that are passed between simulators (DEVS messages). JAX-WS uses JAXB to perform this serialization to XML. In the case of .NET, we can use the *XmlTextWriter* class. However, for simplicity, we have developed a JAXB class in our MicroSimServer/.NET web application. This class just encapsulates the procedure that implements the serialization and deserialization process (Listing 15.2).

Having both projects created, we now have a basis for the DEVS/SOA M&S framework. Next, we describe how both DEVS/SOA simulator and coordinator are developed in both JAVA and .NET platforms.

LISTING 15.2 JAXB.NET CLASS

```
namespace microsim.soa.core
{
  public class JAXB
  {
    private static String UTF8ByteArrayToString(Byte[]
                                              characters)
    {
      UTF8Encoding encoding = new UTF8Encoding();
      String constructedString =
                 encoding.GetString(characters);
```

```
        return (constructedString);
    }

    public static String marshal(Object value)
    {
      MemoryStream memoryStream = new MemoryStream();
      XmlSerializer serializer =
                        new XmlSerializer(value.GetType());
      XmlTextWriter xmlTextWriter =
              new XmlTextWriter(memoryStream, Encoding.UTF8);
      serializer.Serialize(xmlTextWriter, value);
      memoryStream = (MemoryStream)xmlTextWriter.BaseStream;
      String XmlizedString =
               UTF8ByteArrayToString(memoryStream.ToArray());
      xmlTextWriter.Close();
      return XmlizedString.Remove(0, 1);
    }

    public static Object unmarshal(String xmlAsString,
                                   Type valueType)
    {
      XmlSerializer serializer =
                        new XmlSerializer(valueType);
      StringReader reader = new StringReader(xmlAsString);
      return serializer.Deserialize(reader);
    }
  }
}
```

15.3 DEVS/SOA SIMULATOR

In this section, we describe how to develop each member function in a DEVS/SOA simulator service (see Figure 15.1).

15.3.1 DEVS Model Instantiation

A distributed DEVS/SOA simulation is performed using an XML configuration file. For example, a file for configuring a distributed simulation of the Generator Processor Transducer (GPT) DEVS model is shown in Listing 15.3.

Note that a simulator will be created for each DEVS atomic model in the XML configuration file. Thus, a DEVS/SOA simulator will receive, for example, the following entry:

```
<atomic name = "gene" class = "microsim.lib.examples.efp
  .Generator" host = "192.168.1.3"/>
```

LISTING 15.3 DEVS/SOA GPT XML CONFIGURATION FILE

```
<?xml version="1.0" encoding="UTF-8" standalone="yes"?>
<devs name="Gpt" host="192.168.1.2">
  <atomic name="gene"
          class="microsim.lib.examples.efp.Generator"
          host="192.168.1.3"/>
  <atomic name="proc"
          class="microsim.lib.examples.efp.Processor"
          host="192.168.1.2"/>
  <atomic name="tran"
          class="microsim.lib.examples.efp.Transducer"
          host="192.168.1.5"/>
  <coupling componentFrom="gene" portFrom="out"
            componentTo="proc" portTo="in"/>
  <coupling componentFrom="gene" portFrom="out"
            componentTo="tran" portTo="arrived"/>
  <coupling componentFrom="proc" portFrom="out"
            componentTo="tran" portTo="solved"/>
  <coupling componentFrom="tran" portFrom="out"
            componentTo="gene" portTo="stop"/>
</devs>
```

This implies that the DEVS/SOA simulator web service must instantiate generator to be used as a simulated DEVS model. The above atomic XML node is received by the simulator as a string. Thus, the *setModel* web method will extract both the name and the class of the atomic model and will create the corresponding object. This task is performed in both JAVA and .NET *SoaSimulator* classes (Listing 15.4).

Both the methods instantiate the proper atomic model. Besides, this method uses two session variables to store the name of the simulator (that will be same as the atomic model name) and the MicroSim atomic model.

LISTING 15.4 SOASIMULATOR::SETMODEL(STRING) WEB METHOD

JAVA

```
  @WebMethod
  public void setModel(String xmlAtomicAsString)
    throwsException {
    Document xmlDocument =
               DocumentBuilderFactory.newInstance()
              .newDocumentBuilder().parse(new
               InputSource(new StringReader
                 (xmlAtomicAsStr ing)));
    Element xmlAtomic = (Element) xmlDocument.getFirstChild();

    String nameAtomic = xmlAtomic.getAttribute("name");
```

```
    String clasAtomic = xmlAtomic.getAttribute("class");
    Atomic model =
          (Atomic) Class.forName(clasAtomic).newInstance();
    model.setName(nameAtomic);

    HttpSession session = getSession();
    session.setAttribute("name", nameAtomic);
    session.setAttribute("model", model);
  }
```

.NET

```
  [WebMethod(EnableSession = true)]
  public void setModel(String xmlAtomicAsString)
  {
    XmlDocument xmlDocument = new XmlDocument();
    xmlDocument.LoadXml(xmlAtomicAsString);
    XmlNode xmlAtomic =
          xmlDocument.GetElementsByTagName("atomic")
            .Item(0);
    String nameAtomic =
          xmlAtomic.Attributes.GetNamedItem("name").Value;
    String clasAtomic =
          xmlAtomic.Attributes.GetNamedItem("class").Value;
    Type classType = Type.GetType(clasAtomic);
    Object modelAux = Activator.CreateInstance(classType,
      true);
    Session["name"] = nameAtomic;
    Session["model"] = (Atomic)modelAux;
  }
```

15.3.2 Simulator Initialization and Time Advance

The initialization consists of assigning initial values to both the time of last event and the time of next event. The time of last event is directly assigned the value of the argument received. The time of next event is assigned the value of time of last event plus the result of the model's time advance function (Listing 15.5).

Both variables (tL and TN) are saved using session objects.

LISTING 15.5 SOASIMULATOR::INITIALIZE(DOUBLE) AND SOASIMULATOR::TA(DOUBLE) WEB METHODS

JAVA

```
  @WebMethod
  public void initialize(double t) {
    HttpSession session = getSession();
    session.setAttribute("tL", t);
    session.setAttribute("tN", t + ta(t));
  }
```

```
@WebMethod
public double ta(double t) {
  HttpSession session = getSession();
  Atomic model = (Atomic) session.getAttribute("model");
  return model.ta();
}
```

.NET

```
[WebMethod(EnableSession = true)]
public void initialize(double t)
{
  Session["tL"] = t;
  Session["tN"] = t + ta(t);
}

[WebMethod(EnableSession = true)]
public double ta(double t)
{
  Atomic model = (Atomic)Session["model"];
  return model.ta();
}
```

15.3.3 Transition Function

As we have already shown in Chapter 4, MicroSim simulators implement all the three transition functions in *deltfcn*. This also applies to our DEVS/SOA framework (Listing 15.6).

Once again, we make use of the session object to update tL and tN state variables.

LISTING 15.6 SOASIMULATOR::DELTFCN(DOUBLE) WEB METHOD

JAVA

```
@WebMethod
public void deltfcn(double t) {
  HttpSession session = getSession();
  Atomic model = (Atomic) session.getAttribute("model");
  double tL = (Double) session.getAttribute("tL");
  double tN = (Double) session.getAttribute("tN");
  boolean isInputEmpty = model.isInputEmpty();
  if (isInputEmpty && t != tN) {
    return;
  } else if (!isInputEmpty && t == tN) {
    double e = t - tL;
```

```
      model.setSigma(model.getSigma() - e);
      model.deltcon(e);
    } else if (isInputEmpty && t == tN) {
      model.deltint();
    } else if (!isInputEmpty && t != tN) {
      double e = t - tL;
      model.setSigma(model.getSigma() - e);
      model.deltext(e);
    }
    tL = t;
    tN = tL + model.ta();
    session.setAttribute("tL", tL);
    session.setAttribute("tN", tN);
  }
```

.NET

```
  [WebMethod(EnableSession = true)]
  public void deltfcn(double t)
  {
    Atomic model = (Atomic)Session["model"];
    double tL = (Double)Session["tL"];
    double tN = (Double)Session["tN"];
    bool isInputEmpty = model.isInputEmpty();
    if (isInputEmpty && t != tN)
    {
      return;
    }
    else if (!isInputEmpty && t == tN)
    {
      double e = t - tL;
      model.setSigma(model.getSigma() - e);
      model.deltcon(e);
    }
    else if (isInputEmpty && t == tN)
    {
      model.deltint();
    }
    else if (!isInputEmpty && t != tN)
    {
      double e = t - tL;
      model.setSigma(model.getSigma() - e);
      model.deltext(e);
    }
    tL = t;
    tN = tL + model.ta();
    Session["tL"] = tL;
    Session["tN"] = tN;
    }
```

15.3.4 Output Function

The output function calls the model's output function that is being simulated. The use of sessions is also crucial here (Listing 15.7).

LISTING 15.7 SOASIMULATOR::LAMBDA(DOUBLE) WEB METHOD

```
JAVA
  @WebMethod
  public void lambda(double t) {
    double tN = (Double) session.getAttribute("tN");
    Atomic model = (Atomic) session.getAttribute("model");
    if (t == tN) {
      model.lambda();
    }
  }

.NET
  [WebMethod(EnableSession = true)]
  public void lambda(double t)
  {
    double tN = (Double)Session["tN"];
    Atomic model = (Atomic)Session["model"];
    if (t == tN)
    {
      model.lambda();
    }
  }
```

15.3.5 Other Member Functions

A DEVS/SOA coordinator asks for the time of next event from another simulator during the execution of the simulation protocol. This is accomplished using the getter web method (Listing 15.8).

In addition, to pass messages from one simulator to another, we need to extract the data stored in all the ports of the attached model. Input and output data must

LISTING 15.8 SOASIMULATOR::GETTN() WEB METHOD

```
JAVA
  @WebMethod
  public double getTN() {
    HttpSession session = getSession();
    double tN = (Double) session.getAttribute("tN");
    return tN;
  }
```

```
.NET
  [WebMethod(EnableSession = true)]
  public double getTN()
  {
    double tN = (Double)Session["tN"];
    return tN;
  }
```

be available in XML format for serialization. Thus, these web methods are implemented taking the original data in model ports, and serializing them to XML using JAXB. Listing 15.9 shows how the data are stored in input ports.

LISTING 15.9 SOASIMULATOR::GETINPUT(STRING) WEB METHOD

```
JAVA
  @WebMethod
  public List<String> getInput(String portName) {
    HttpSession session = getSession();
    Atomic model = (Atomic) session.getAttribute("model");
    Port port = model.getInPort(portName);
    Collection values = port.getValues();
    LinkedList<String> xmlValues = new LinkedList<String>();
    for (Object value : values) {
      StringWriter writer = new StringWriter();
      Result result = new StreamResult(writer);
      JAXB.marshal(value, result);
      String xmlValue = writer.toString();
      xmlValues.add(xmlValue);
    }
    return xmlValues;
  }

.NET
  [WebMethod(EnableSession = true)]
  public List<String> getInput(String portName)
  {
    Atomic model = (Atomic)Session["model"];
    Port port = model.getInPort(portName);
    ICollection values = port.getValues();
    List<String> xmlValues = new List<String>();
    foreach (Object value in values)
    {
     xmlValues.Add(JAXB.marshal(value));
    }
    return xmlValues;
  }
```

Exercise 15.1

Implement the *SoaSimulator::getOutput* web method.

Additionally, a DEVS/SOA simulator receives messages from other simulators. We implement another web method called *receive*. This web method takes the serialized message and adds the deserialized data to the corresponding port of the model (Listing 15.10).

LISTING 15.10 SOASIMULATOR::RECEIVE(STRING, LIST<STRING>) WEB METHOD

JAVA

```
@WebMethod
public void receive(String portTo,
    List<String> xmlValues) {
  HttpSession session = getSession();
  Atomic model = (Atomic) session.
    getAttribute("model");
  for (String xmlValue : xmlValues) {
    StringReader reader = new StringReader(xmlValue);
    Object value = JAXB.unmarshal(reader,
      model.getInPort(portTo).getValueType());
    model.getInPort(portTo).addValue(value);
  }
}
```

.NET

```
[WebMethod(EnableSession = true)]
public void receive(String portTo,
    List<String> xmlValues)
{
  Atomic model = (Atomic)Session["model"];
  Port port = model.getInPort(portTo);
  foreach (String xmlValue in xmlValues)
  {
    Object value = JAXB.unmarshal(xmlValue,
                                  port.getValueType());
    port.addValue(value);
  }
}
```

Finally, we create a web method clear to erase all the data stored in ports (Listing 15.11). This function is used at the end of each DEVS simulation step.

15.4 CREATING WEB SERVICE REFERENCES

A DEVS/SOA coordinator will use several web service references, one for each child (simulator or coordinator for a coupled subcomponent model) in the simulation scenario. In the GPT XML configuration file shown in Listing 15.3, for example, the root

LISTING 15.11 SOASIMULATOR::CLEAR() WEB METHOD

JAVA

```
@WebMethod
public void clear() {
  HttpSession session = getSession();
  Atomic model = (Atomic) session.getAttribute("model");
  Collection<Port> ports = null;

  ports = model.getInPorts();
  for (Port port : ports) {
    port.clear();
  }

  ports = model.getOutPorts();
  for (Port port : ports) {
    port.clear();
  }
}
```

.NET

```
[WebMethod(EnableSession = true)]
public void clear()
{
  Atomic model = (Atomic)Session["model"];
  ICollection<Port> ports = null;

  ports = model.getInPorts();
  foreach (Port port in ports)
  {
    port.clear();
  }

  ports = model.getOutPorts();
  foreach (Port port in ports)
  {
    port.clear();
  }
}
```

coordinator has three web service references. In more complex (hierarchical) scenarios, a DEVS/SOA coordinator can use other DEVS/SOA coordinator references (in fact, it occurs each time we have more than one coupled model in our complete DEVS model). Next, we describe how to create a web service reference in both platforms.

15.4.1 NetBeans

With MicroSimServer opened and Glassfish running, click the right mouse button on the project name and select New→Web Service Client. Enter a package name (e.g., microsim.soa.client.simulator) and select Browse … in Project (Figure 15.15).

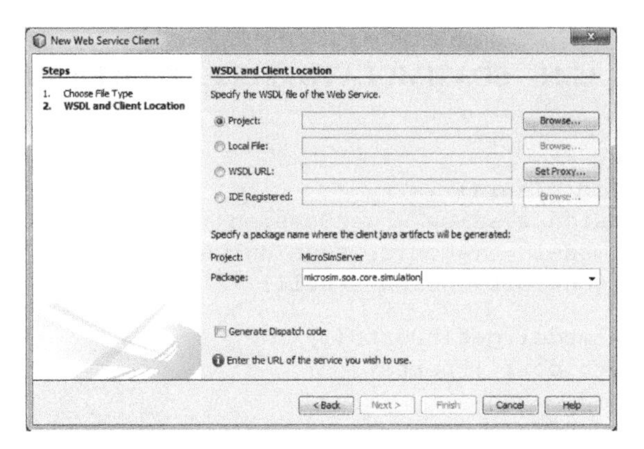

FIGURE 15.15 New web service client.

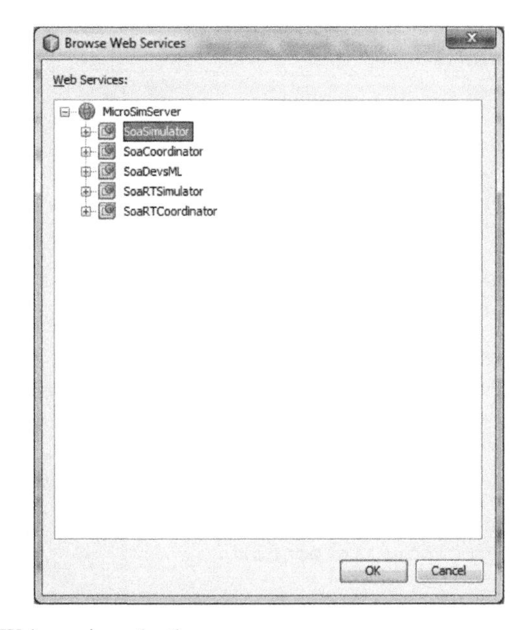

FIGURE 15.16 Web service selection.

Select *SoaSimulator* web service (Figure 15.16).

Next, the client code is automatically generated. This code will be used in the corresponding DEVS/SOA coordinator web service.

15.4.2 Microsoft Visual Studio

Click the right mouse button on the MicroSimServer project and select Add Service Reference… (Figure 15.17).

Click on Discover→Services in Solutions and select SoaSimulator with a namespace (e.g., SoaSimulatorService) (Figure 15.18).

Now, we are ready to implement our DEVS/SOA coordinator web service.

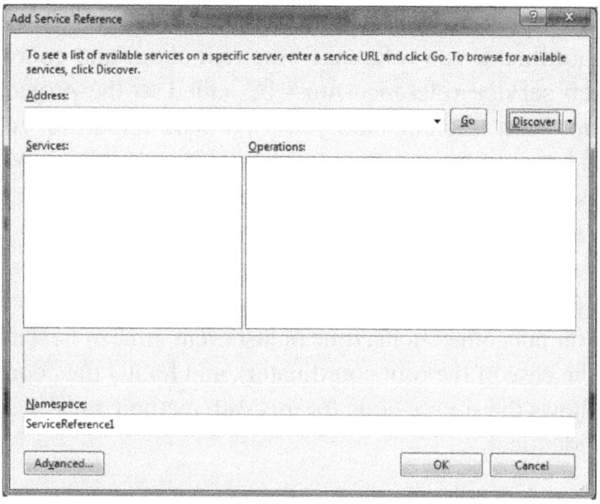

FIGURE 15.17 New web service reference.

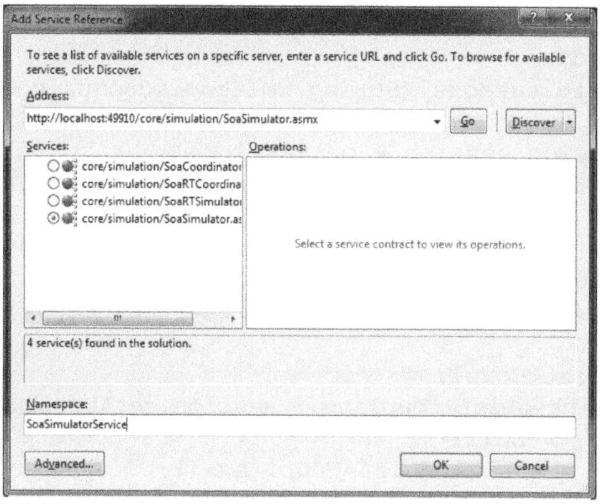

FIGURE 15.18 Web service selection.

15.5 DEVS/SOA COORDINATOR

15.5.1 DEVS Coupled Model Instantiation

Similar to the *SoaSimulator setModel* web method, the root DEVS/SOA coordinator receives the full XML configuration file. On the contrary, a non-root DEVS/SOA coordinator will receive a `<coupled>` XML node. Thus, we must consider both possibilities in the implementation of the *setModel* web method. In the first case, the coordinator will not instantiate a root coupled model; that is, the attached coupled model is NULL. In the second case, the coordinator will instantiate the corresponding coupled model defined in the XML configuration file. In both cases, the coordinator must visit all the immediate children in the XML configuration file, creating coordinator and/or simulator service references.

The *SoaCoordinator::setModel* function is the most complex web method in DEVS/SOA. A coordinator can reference other simulator and coordinator services. A coordinator web service reference must be added to the project. However, the coordinator has not been implemented yet. This is an important issue. The way to fix the problem is by creating an empty coordinator web service with all the member functions presented in Figure 15.1. Next, create the web service reference and, finally, continue inserting the body of all these member functions.

The *setModel* web method also creates the initial state of the current coordinator, that is, simulators (or coordinators) managed, internal connections, external input connections, external output connections, time of last event, time of next event, the coupled model (except in the case of the root coordinator), and finally the coordinator's name.

Listing 15.12 shows the source code for this web method, only in JAVA. The .NET version is almost identical.

LISTING 15.12 SOACOORDINATOR::SET MODEL(STRING) WEB METHOD

JAVA

```
@WebMethod
public void setModel(String xmlCoupledAsString) throws
                                                Exception {
  HttpSession session = getSession();
  HashMap<String, Object> simulators =
                     (HashMap<String, Object>) session
                    .getAttribute("simulators");
  if (simulators == null) {
    simulators = new HashMap<String, Object>();
  }
  LinkedList<Coupling> eic =
       (LinkedList<Coupling>) session.getAttribute("eic");
  if (eic == null) {
    eic = new LinkedList<Coupling>();
  }
  // The same here with eoc and ic
  // ...

  Document xmlDocument =
             DocumentBuilderFactory.newInstance()
             .newDocumentBuilder()
             .parse(new InputSource(new
             StringReader(xmlCoupledAsString)));
  Element xmlCoupled =
                   (Element) xmlDocument.getFirstChild();
  String nameCoupled = xmlCoupled.getAttribute("name");
  session.setAttribute("name", nameCoupled);
  if (xmlCoupled.getNodeName().equals("devs")) {
     session.setAttribute("model", null);
  } else if (xmlCoupled.getNodeName().equals("coupled")) {
```

```
        String clasCoupled = xmlCoupled.getAttribute("class");
        Coupled coupled =
            (Coupled) Class.forName(clasCoupled).newInstance();
        coupled.setName(nameCoupled);
        session.setAttribute("model", coupled);
      }
      // Components
      NodeList xmlComponentList = xmlCoupled.getChildNodes();
      for (int i = 0; i < xmlComponentList.getLength(); ++i) {
        Node xmlComponentAsNode = xmlComponentList.item(i);
        if (xmlComponentAsNode.getNodeType() ==
                                        Node.ELEMENT_NODE) {

          Element xmlComponent = (Element) xmlComponentAsNode;
          String nodeName = xmlComponent.getNodeName();
          if (nodeName.equals("coupled") ||
              nodeName.equals("atomic")) {

            StringWriter writer = new StringWriter();
            Result result = new StreamResult(writer);
            TransformerFactory.newInstance().newTransformer()
              .transform(new DOMSource(xmlComponent), result);
            String xmlComponentAsString = writer.toString();

            String nameComponent =
                   xmlComponent.getAttribute("name");
            String hostComponent =
                   xmlComponent.getAttribute("host");

            if (nodeName.equals("coupled")) {
              microsim.services.coordinator
             .SoaCoordinator coordinator = (new
              microsim.services.coordinator
             .SoaCoordinatorService(new
              URL(hostComponent)).getSoaCoordinatorPort();
              ((BindingProvider) coordinator)
              .getRequestContext().put(BindingProvider
              .SESSION_MAINTAIN_PROPERTY, true);
              coordinator.setModel(xmlComponentAsString);
              simulators.put(nameComponent, coordinator);
            } else if (nodeName.equals("atomic")) {
              microsim.services.simulator.SoaSimulator
               simulator = (new microsim.services.simulator
              .SoaSimulatorService(new
               URL(hostComponent)).getSoaSimulatorPort();
              ((BindingProvider) simulator)
              .getRequestContext()
              .put(BindingProvider.SESSION_MAINTAIN_PROPERTY,
                true);
```

```
                simulator.setModel(xmlComponentAsString);
                simulators.put(nameComponent, simulator);
            }
        } else if (nodeName.equals("coupling")) {
            String cFrom =
                    xmlComponent.getAttribute("componentFrom");
            String pFrom =
                    xmlComponent.getAttribute("portFrom");
            String cTo =
                    xmlComponent.getAttribute("componentTo");
            String pTo = xmlComponent.getAttribute("portTo");
            Coupling c = new Coupling(cFrom, pFrom, cTo, pTo);
            if (cFrom.equals(nameCoupled)) {
              eic.add(c);
            } else if (cTo.equals(nameCoupled)) {
              eoc.add(c);
            } else {
              ic.add(c);
            }
        }
      }
    }
    session.setAttribute("simulators", simulators);
    session.setAttribute("eic", eic);
    session.setAttribute("ic", ic);
    session.setAttribute("eoc", eoc);
}
```

Exercise 15.2

Implement the DevsSoaCoordinator::setModel(String) web method in C#.

15.5.2 Coordinator Initialization and Time Advance

In most of the web methods in a coordinator, the DEVS simulation functions are quite similar to the ones seen in Chapter 4. The main difference with respect to a simulator is that the coordinator must update its own state and call to update to the immediate descendant in the model hierarchy. This task is being performed by invoking the corresponding web methods in these descendants.

Thus, the initialization function initializes the descendants and then the coordinator updates its state (tL and tN) (Listing 15.13).

LISTING 15.13 SOACOORDINATOR::INITIALIZE (DOUBLE) WEB METHOD

JAVA

```
  @WebMethod
  public void initialize(double t) {
```

```
    HttpSession session = getSession();
    HashMap<String, Object> simulators =
        (HashMap<String, Object>) session
                        .getAttribute("simulators");
    for (Object simulator : simulators.values()) {
      if (simulator instanceof microsim.services
                        .simulator.SoaSimulator) {
        ((microsim.services.simulator
        .SoaSimulator) simulator).initialize(t);
      } else if (simulator instanceof microsim.services
                        .coordinator.SoaCoordinator) {
        ((microsim.services.coordinator
        .SoaCoordinator) simulator).initialize(t);
      }
    }
    session.setAttribute("tL", t);
    session.setAttribute("tN", t + ta(t));
  }
```

.NET

```
  [WebMethod(EnableSession = true)]
  public void initialize(double t)
  {
    Dictionary<String, Object> simulators =
          (Dictionary<String, Object>)Session["simulators"];
    foreach (Object simulator in simulators.Values)
    {
      if (simulator is
          SoaSimulatorService.SoaSimulatorServiceSoapClient)
      {
        ((SoaSimulatorService
        .SoaSimulatorServiceSoapClient)simulator)
        .initialize(t);
      }
      else if (simulator is
      SoaCoordinatorService.SoaCoordinatorServiceSoapClient)
      {
        ((SoaCoordinatorService
        .SoaCoordinatorServiceSoapClient)simulator)
        .initialize(t);
      }
    }
    Session["tL"] = t;
    Session["tN"] = t + ta(t);
  }
```

The time advance function is also the same as described in Chapter 4 for the regular DEVS coordinator. Here we can see the utility of the *getTN* web method (Listing 15.14).

LISTING 15.14 SOACOORDINATOR::TA(DOUBLE) WEB METHOD

JAVA

```
@WebMethod
public double ta(double t) {
  double tn = Double.POSITIVE_INFINITY;
  double tnAux = Double.POSITIVE_INFINITY;
  HttpSession session = getSession();
  HashMap<String, Object> simulators =
   (HashMap<String, Object>)
    session.getAttribute("simulators");
  for (Object simulator : simulators.values()) {
    if (simulator instanceof microsim.services
                           .simulator.SoaSimulator) {
      tnAux = ((microsim.services.simulator
             .SoaSimulator) simulator).getTN();
    } else if (simulator instanceof microsim.services
                         .coordinator.SoaCoordinator) {
      tnAux = ((microsim.services.coordinator
             .SoaCoordinator) simulator).getTN();
    }
    if (tnAux < tn) {
      tn = tnAux;
    }
  }
  return tn - t;
}
```

.NET

```
[WebMethod(EnableSession = true)]
public double ta(double t)
{
  double tn = Double.PositiveInfinity;
  double tnAux = Double.PositiveInfinity;
  Dictionary<String, Object> simulators =
     (Dictionary<String, Object>)Session["simulators"];
  foreach (Object simulator in simulators.Values)
  {
    if (simulator is SoaSimulatorService
           .SoaSimulatorServiceSoapClient)
    {
      tnAux = ((SoaSimulatorService
           .SoaSimulatorServiceSoapClient)
```

```
                    simulator).getTN();
        }
        else if (simulator is SoaCoordinatorService
                    .SoaCoordinatorServiceSoapClient)
        {
          tnAux = ((SoaCoordinatorService
                    .SoaCoordinatorServiceSoapClient)simulator)
                    .getTN();
        }
        if (tnAux < tn)
        {
          tn = tnAux;
        }
      }
      return tn - t;
    }
```

Exercise 15.3

Implement the SoaCoordinator::getTN() member function.

15.5.3 Transition Function

In the transition function, the coordinator must first propagate the data stored at the input ports of its coupled model (using the external input connections). It will be performed recursively among coordinators, because, after that, we call the transition function of all the descendants.

Both tL and tN state variables are updated (Listing 15.15).

LISTING 15.15 SOACOORDINATOR::DELTFCN(DOUBLE) WEB METHOD

JAVA

```
  @WebMethod
  public void deltfcn(double t) {
    propagateInput();
    HttpSession session = getSession();
    HashMap<String, Object> simulators =
                    (HashMap<String, Object>)
                    session.getAttribute("simulators");
    for (Object simulator : simulators.values()) {
      if (simulator instanceof microsim.services
                                .simulator.SoaSimulator) {
        ((microsim.services.simulator
        .SoaSimulator) simulator).deltfcn(t);
      } else if (simulator instanceof microsim.services
                            .coordinator.SoaCoordinator) {
```

```
        ((microsim.services.coordinator
        .SoaCoordinator) simulator).deltfcn(t);
      }
    }
    double tL = t;
    double tN = tL + ta(t);
    session.setAttribute("tL", tL);
    session.setAttribute("tN", tN);
  }
```

.NET

```
  [WebMethod(EnableSession = true)]
  public void deltfcn(double t)
  {
    propagateInput();
    Dictionary<String, Object> simulators =
          (Dictionary<String, Object>)Session["simulators"];
    foreach (Object simulator in simulators.Values)
    {
      if (simulator is SoaSimulatorService
                    .SoaSimulatorServiceSoapClient)
      {
        ((SoaSimulatorService
             .SoaSimulatorServiceSoapClient)
        simulator).deltfcn(t);
      }
      else if (simulator is SoaCoordinatorService
             .SoaCoordinatorServiceSoapClient)
      {
        ((SoaCoordinatorService
        .SoaCoordinatorServiceSoapClient)
        simulator).deltfcn(t);
      }
    }
    double tL = t;
    double tN = tL + ta(t);
    Session["tL"] = tL;
    Session["tN"] = tN;
  }
```

To propagate the input, we iterate over the set of external input connections. For each port at the origin, in the coupled model, we take all the data stored and we copy them to the destination port. To perform this task, we must (1) transform the data stored in the origin port (in the coupled model) to XML, (2) move the data to the simulator (or coordinator) at destination, and (3) deserialize the data and inject them in the corresponding atomic or coupled model at destination. Data propagation in XML from web service to web service is the most important point to allow us interoperability between different DEVS/SOA platforms (Listing 15.16).

LISTING 15.16 SOACOORDINATOR::PROPAGATEINPUT() MEMBER FUNCTION

JAVA

```
private void propagateInput() {
  HttpSession session = getSession();
  HashMap<String, Object> simulators =
                      (HashMap<String, Object>)
                      session.getAttribute("simulators");
    LinkedList<Coupling> eic =
       (LinkedList<Coupling>) session.getAttribute("eic");
    for (Coupling c : eic) {
      List<String> xmlValues =
                   this.getInput(c.getPortFrom());
      if (!xmlValues.isEmpty()) {
        Object simulatorTo =
               simulators.get(c.getComponentTo());
        if (simulatorTo instanceof microsim.soa.services
                                   .simulator.SoaSimulator) {
          ((microsim.soa.services.simulator
          .SoaSimulator) simulatorTo)
          .receive(c.getPortTo(), xmlValues);
        } else if (simulatorTo instanceof microsim.soa
                  .services.coordinator.SoaCoordinator) {
          ((microsim.soa.services
          .coordinator.SoaCoordinator)
          simulatorTo).receive(c.getPortTo(), xmlValues);
        }
      }
    }
  }
```

.NET

```
  private void propagateInput()
  {
    Dictionary<String, Object> simulators =
      (Dictionary<String, Object>)Session["simulators"];
    List<Coupling> eic = (List<Coupling>)Session["eic"];
    foreach (Coupling c in eic)
    {
      Object simulatorFrom =
             simulators[c.getComponentFrom()];
      List<String> xmlValues =
                   this.getInput(c.getPortFrom());
      if (xmlValues.Count > 0)
      {
        Object simulatorTo = simulators[c.getComponentTo()];
        if (simulatorTo is SoaSimulatorService
                           .SoaSimulatorServiceSoapClient)
        {
```

```
          SoaSimulatorService.ArrayOfString soapValues =
                    new SoaSimulatorService.ArrayOfString();
          soapValues.AddRange(xmlValues);
          ((SoaSimulatorService
          .SoaSimulatorServiceSoapClient)
          simulatorTo).receive(c.getPortTo(), soapValues);
        }
        else if (simulatorTo is SoaCoordinatorService
                        .SoaCoordinatorServiceSoapClient)
        {
          SoaCoordinatorService.ArrayOfString soapValues =
                  new SoaCoordinatorService.ArrayOfString();
          soapValues.AddRange(xmlValues);
          ((SoaCoordinatorService
          .SoaCoordinatorServiceSoapClient)
          simulatorTo).receive(c.getPortTo(), soapValues);
        }
      }
    }
  }
```

15.5.4 Output Function

The output function just executes the output function of all the descendants. After that, the coordinator propagates this output through the connections in the coupled model, also stored in the coordinator in the ic, eic, and eoc session variables. The output propagation is quite similar to the input propagation, but, obviously, using the internal connections and the external output connections (Listing 15.17).

LISTING 15.17 SOACOORDINATOR::LAMBDA(DOUBLE) WEB METHOD

JAVA

```
  @WebMethod
  public void lambda(double t) {
    HttpSession session = getSession();
    HashMap<String, Object> simulators =
                          (HashMap<String, Object>)
                          session.getAttribute("simulators");
    for (Object simulator : simulators.values()) {
      if (simulator instanceof microsim.services
                            .simulator.SoaSimulator) {
        ((microsim.services.simulator.SoaSimulator)
          simulator).lambda(t);
      } else if (simulator instanceof microsim.services
                          .coordinator.SoaCoordinator) {
        ((microsim.services.coordinator.SoaCoordinator)
```

```
            simulator).lambda(t);
        }
      }
      propagateOutput();
    }
```

```
.NET
  [WebMethod(EnableSession = true)]
  public void lambda(double t)
  {
    Dictionary<String, Object> simulators =
          (Dictionary<String, Object>)Session["simulators"];
    foreach (Object simulator in simulators.Values)
    {
      if (simulator is SoaSimulatorService
                       .SoaSimulatorServiceSoapClient)
      {
        ((SoaSimulatorService.SoaSimulatorServiceSoapClient
          )simulator).lambda(t);
      }
      else if (simulator is SoaCoordinatorService
                             .SoaCoordinatorServiceSoapClient)
      {
        ((SoaCoordinatorService
        .SoaCoordinatorServiceSoapClient)
        simulator).lambda(t);
      }
    }
    propagateOutput();
  }
```

Exercise 15.4

Implement the SoaCoordinator::propagateOutput() member function.

15.5.5 Other Member Functions

As in the DEVS/SOA simulator, create a web method to clear all the data stored in ports (Listing 15.18).

LISTING 15.18 SOACOORDINATOR::CLEAR() WEB METHOD

```
JAVA
  @WebMethod
  public void clear() {
    HttpSession session = getSession();
    Coupled model = (Coupled) session.getAttribute("model");
    if (model != null) {
      Collection<Port> ports = null;
```

```
      ports = model.getInPorts();
      for (Port port : ports) {
        port.clear();
      }
      ports = model.getOutPorts();
      for (Port port : ports) {
        port.clear();
      }
    }
    HashMap<String, Object> simulators =
                          (HashMap<String, Object>)
                          session.getAttribute("simulators");
    for (Object simulator : simulators.values()) {
      if (simulator instanceof microsim.services
                                .simulator.SoaSimulator) {
        ((microsim.services.simulator.SoaSimulator)
          simulator).clear();
      } else if (simulator instanceof microsim.services
                            .coordinator.SoaCoordinator) {
        ((microsim.services.coordinator.SoaCoordinator)
          simulator).clear();
      }
    }
  }
```

.NET

```
  [WebMethod(EnableSession = true)]
  public void clear()
  {
    Coupled model = (Coupled)Session["model"];
    if (model != null)
    {
      ICollection<Port> ports = null;
      ports = model.getInPorts();
      foreach (Port port in ports)
      {
        port.clear();
      }
      ports = model.getOutPorts();
      foreach (Port port in ports)
      {
        port.clear();
      }
    }
    Dictionary<String, Object> simulators =
       (Dictionary<String, Object>)Session["simulators"];
    foreach (Object simulator in simulators.Values)
    {
      if (simulator is SoaSimulatorService
                       .SoaSimulatorServiceSoapClient)
```

```
      {
        ((SoaSimulatorService
        .SoaSimulatorServiceSoapClient)simulator).clear();
      }
      else if (simulator is SoaCoordinatorService
                              .SoaCoordinatorServiceSoapClient)
      {
        ((SoaCoordinatorService
        .SoaCoordinatorServiceSoapClient)simulator).clear();
      }
    }
  }
```

We also need *getInput* and *getOutput* web methods to propagate input and output data. The implementation is almost the same as that in our DEVS/SOA simulator (Listing 15.19).

LISTING 15.19 SOACOORDINATOR::GETINPUT(STRING) WEB METHOD

JAVA

```
  @WebMethod
  public List<String> getInput(String portName) {
    HttpSession session = getSession();
    Coupled model = (Coupled) session.getAttribute("model");
    Port port = model.getInPort(portName);
    Collection values = port.getValues();
    LinkedList<String> xmlValues = new LinkedList<String>();
    for (Object value : values) {
      StringWriter writer = new StringWriter();
      Result result = new StreamResult(writer);
      JAXB.marshal(value, result);
      String xmlValue = writer.toString();
      xmlValues.add(xmlValue);
    }
    return xmlValues;
  }
```

.NET

```
  [WebMethod(EnableSession = true)]
  public List<String> getInput(String portName)
  {
    Coupled model = (Coupled)Session["model"];
    Port port = model.getInPort(portName);
    ICollection values = port.getValues();
    List<String> xmlValues = new List<String>();
    foreach (Object value in values)
    {
```

```
        xmlValues.Add(JAXB.marshal(value));
    }
    return xmlValues;
}
```

Exercise 15.5

Implement the List<String> SoaCoordinator::getOutput(String) web method.

Finally, we need a function to receive XML data from other DEVS/SOA coordinator or simulator services. However, there is an important issue here. In coupled models, messages propagate from an output port to an output port. Thus, instead of the *getInPort* member function of the coupled model, we must use the function called *getPort*, which will give the input or output port with the specified name (Listing 15.20).

LISTING 15.20 SOACOORDINATOR::RECEIVE(STRING, LIST<STRING>) WEB METHOD

JAVA

```
@WebMethod
public void receive(String portTo,
                    List<String> xmlValues) {
    HttpSession session = getSession();
    Coupled model = (Coupled) session.getAttribute("model");
    for (String xmlValue : xmlValues) {
        StringReader reader = new StringReader(xmlValue);
        Object value = JAXB.unmarshal(reader,
                    model.getPort(portTo).getValueType());
        model.getPort(portTo).addValue(value);
    }
}
```

.NET

```
[WebMethod(EnableSession = true)]
public void receive(String portTo, List<String> xmlValues)
{
    Coupled model = (Coupled)Session["model"];
    Port port = model.getPort(portTo);
    foreach (String xmlValue in xmlValues)
    {
        Object value = JAXB.unmarshal(xmlValue,
                                port.getValueType());
        port.addValue(value);
    }
}
```

Exercise 15.6

Add the *Component::getPort(String)* member function to MicroSim.

15.5.6 Simulation Function

At this point, a control function is created that executes DEVS/SOA distributed simulations. The body of this function is quite similar to the sequential version. The only difference is that we use session variables to store tL and tN state variables (Listing 15.21).

LISTING 15.21 SOACOORDINATOR::SIMULATE(DOUBLE) WEB METHOD

JAVA

```
@WebMethod
public void simulate(double timeInterval) {
  HttpSession session = getSession();
  double t = (Double) session.getAttribute("tN");
  double tF = t + timeInterval;
  while (t < Constants.INFINITY && t < tF) {
    lambda(t);
    deltfcn(t);
    clear();
    t = (Double) session.getAttribute("tN");
  }
}
```

.NET

```
[WebMethod(EnableSession = true)]
public void simulate(double timeInterval)
{
  double t = (Double)Session["tN"];
  double tF = t + timeInterval;
  while (t < Constants.INFINITY && t < tF)
  {
    lambda(t);
    deltfcn(t);
    clear();
    t = (Double)Session["tN"];
  }
}
```

Until now, we have not seen the implementation of the *getLog()* web method. There are many ways to implement the logger in DEVS/SOA. The first one is redirecting our initial logger mechanism to a string. This is the most elegant way, since we will not lose all the logs performed by our original DEVS models. However, it would demand a high modification in our current source code. Another mechanism consists of managing session variables (string) storing all the information needed to follow the simulation. In this case, the implementation is straightforward, but we will lose all the information logged by the DEVS models (for now stored in the server logs). We encourage the readers to implement their own *getLog* web method.

Exercise 15.7

Implement both *SoaSimulator* and *SoaCoordinator getLog* web methods.

Project 15.1

Implement the DEVS/SOA real-time simulation package.

Hint: The implementation is straightforward if we pay attention to the following four points.

- Start supposing a flattened XML configuration file, that is, only one coordinator and *n* simulators.
- DEVS/SOA simulators have their own set of external output connections. In this case, messages are directly sent from simulator to simulator, without using the root coordinator.
- The initialization time is the wall clock time.
- The root coordinator just observes the scenario; that is, the service is alive in the interval of the simulation time defined by the user.

15.6 WEB APPLICATION DEPLOYMENT

There are two ways to deploy the DEVS/SOA web application (called MicroSimServer). One is from the IDE (NetBeans or Visual Studio in our case), and the other is sending the web application to the target server (mainly using the server frontend). We will mainly use the first one and briefly describe the second one as follows. Each time we build a web application in NetBeans, a .WAR file containing all our source code and libraries is created (typically in the dist folder). This .WAR file can be uploaded to the target server using a server frontend (see Figure 15.19, where the `MicroSimServer.war` file is being installed in the server).

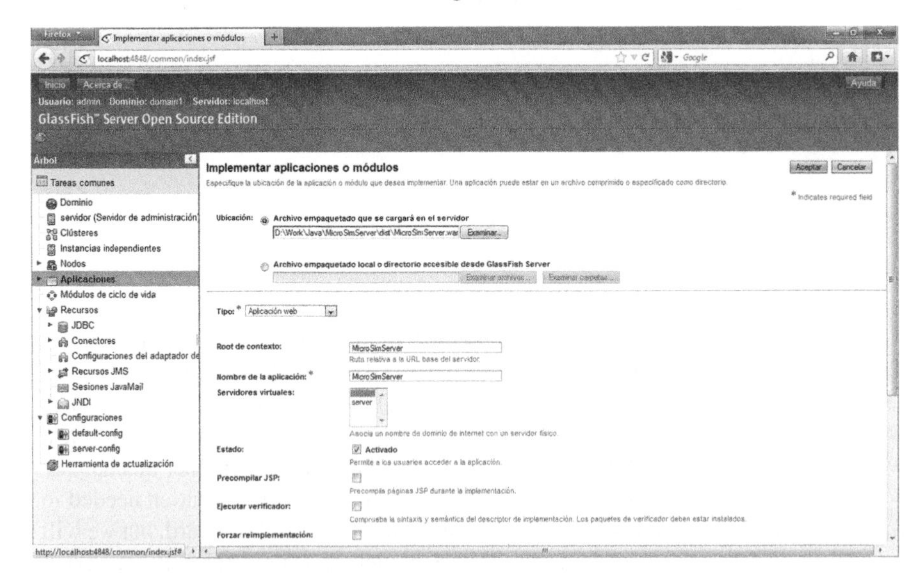

FIGURE 15.19 Web service application deployment.

Using NetBeans IDE, clean and build the current application and deploy the application by clicking with the right mouse button on the project name and selecting Deploy. The server will start and the web application will be available for client applications. Using Microsoft Visual Studio, the web application can be installed in an external server (see Figure 15.20), or we can just run a client application in the same solution and the IDE will automatically deploy our web application if there are dependencies between them.

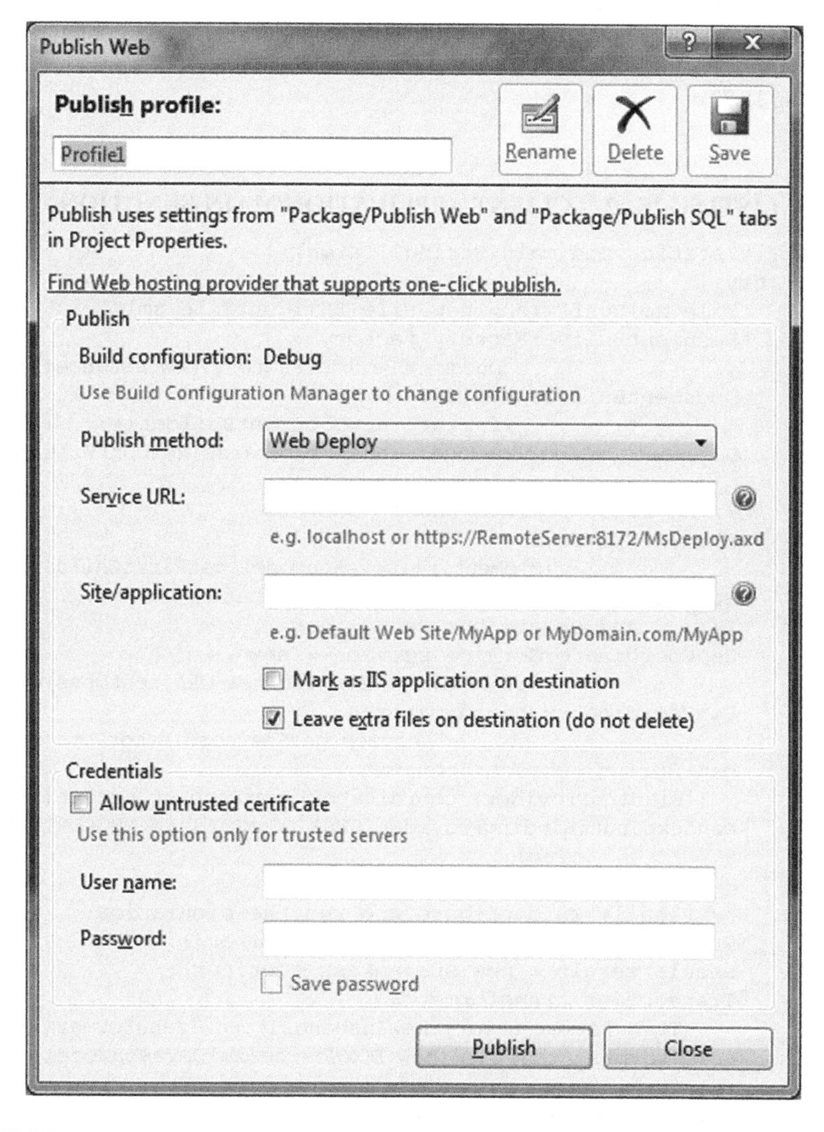

FIGURE 15.20 Installing a web application in an external server (click with the right mouse button on the project name and select Publish…).

15.7 CLIENT APPLICATION

15.7.1 NetBeans

To create a client application in NetBeans, open the MicroSimServer project with NetBeans and, in addition, create an empty Java project called MicroSimClient. Add a web service reference to *SoaCoordinator* as we did in the previous section. If we have both DEVS/SOA JAVA and .NET web applications deployed, a web service reference can be created for a DEVS/SOA JAVA coordinator or for a DEVS/SOA .NET coordinator, bearing no impact on the simulation execution.

Having the web service reference, a main function that reads XML configuration files and executes the simulation web method in the coordinator client is devised (Listing 15.22).

LISTING 15.22 CLIENT APPLICATION/MAIN FUNCTION

```
public static void main(String[] args) {
    try {
      File xmlConfFile = new File("XmlConfFile.xml");
      DocumentBuilderFactory factory =
                       DocumentBuilderFactory.newInstance();
      DocumentBuilder builder =
                       factory.newDocumentBuilder();
      Document xmlDevsSoaModel = builder.parse(xmlConfFile);

      // We first compute the I. address
      Element xmlDevs =
                   (Element)xmlDevsSoaModel.getFirstChild();
      String hostDevs = xmlDevs.getAttribute("host");
      // ... And create the web client
      SoaCoordinatorService service = new
                   SoaCoordinatorService(new URL(hostDevs));
      SoaCoordinator coordinator =
                        service.getSoaCoordinatorPort();
      Map<String, Object> context =
        ((BindingProvider) coordinator).getRequestContext();
      context.put(BindingProvider.SESSION_MAINTAIN_PROPERTY,
                 true);

      // Finally we configure and run the simulation:
      StringWriter writer = new StringWriter();
      Result result = new StreamResult(writer);
      Transformer transformer =
          TransformerFactory.newInstance().newTransformer();
      transformer.transform(new DOMSource(xmlDevsSoaModel),
                          result);
      String xmlModelAsString = writer.toString();
      coordinator.setModel(xmlModelAsString);
```

```
            coordinator.initialize(0);
            coordinator.simulate(30.0);
            String log = coordinator.getLog();
            System.out.println(log);
        } catch (Exception ex) {
            Logger.getLogger(MicroSimClientConsole.class
            .getName()).log(Level.SEVERE, null, ex);
        }
    }
```

The main function first reads the devs XML node in the configuration file to localize the DEVS/SOA root coordinator address.

15.7.2 Microsoft Visual Studio

The procedure to create a client application using Microsoft Visual Studio is the same. Create an empty console project. Create a web service reference to *SoaCoordinator* as we did in previous sections. Finally, create a main program, quite similar to the one developed in NetBeans (Listing 15.23).

LISTING 15.23 CLIENT APPLICATION/MAIN FUNCTION

```
static void Main(string[] args)
  {
    XmlDocument xmlDevsSoaModel = new XmlDocument();
    xmlDevsSoaModel.Load("XmlConfFile.xml");
    XmlNode xmlDevs =
       xmlDevsSoaModel.GetElementsByTagName("devs").Item(0);
    String hostDevs =
              xmlDevs.Attributes.GetNamedItem("host").Value;
    SoaCoordinatorServiceSoapClient coordinator =
              new SoaCoordinatorServiceSoapClient(hostDevs);
    coordinator.setModel(xmlDevs.OuterXml);
    coordinator.initialize(0);
    coordinator.simulate(30);
    String log = coordinator.getLog();
    System.Console.WriteLine(log);
    System.Console.ReadKey();
  }
```

15.8 EXAMPLES

In this section, we describe two different examples to show the capabilities of our DEVS/SOA framework. The first one is the GPT example executed over different DEVS/SOA JAVA servers. The second one is the EF-P example, where the Experimental Frame is located over DEVS/SOA JAVA servers and the Processor is simulated in a DEVS/SOA.NET server. In these two cases, all the models are being implemented using MicroSim/Java or MicroSim/.Net. With the second example, we prove interoperability between DEVS/SOA simulation frameworks. Note that the interoperability between different DEVS modeling frameworks has been already described in MicroSim (see Chapter 13).

The first issue to solve is the data propagation. The XML representation of data must be consistent among different DEVS/SOA simulation platforms. This can be done using different directives in both NetBeans and Visual Studio IDEs. Next, we show how to define a Job (the data type used in the two following examples) with the purpose that the serialization and deseralization give the same result in different platforms (Listing 15.24).

LISTING 15.24 JOB SERIALIZATION CONFIGURATION

JAVA

```
@XmlRootElement(name = "job")
@XmlAccessorType(XmlAccessType.FIELD)
public class Job {

   @XmlAttribute
   protected String id;
   @XmlAttribute
   protected double time;

   // Constructors, getters and setters goes here:
   // ...
}
```

.NET

```
[XmlRoot("job")]
public class Job
{
   public String id;
   public Double time;

   // Constructors, getters and setters goes here:
   // ...
}
```

15.8.1 GPT Example

We first show the complete XML configuration file to simulate the GPT example in a DEVS/SOA JAVA distributed environment (Listing 15.25).

As can be seen, we must enter the full URL of the WSDL of each DEVS/SOA coordinator or simulator. In this case, we are distributing the GPT simulation over a small local network of two servers (192.168.1.37 and 192.168.1.38).

In the client application (NetBeans or Visual Studio version), we must change the path of the XML configuration file to the XML given in Listing 15.25. After running the distributed simulation, we obtain the following result (Listing 15.26).

LISTING 15.25 GPTJAVA_XML CONFIGURATION FILE

```
<?xml version="1.0" encoding="UTF-8"?>
<devs name="Gpt"
      host="http://192.168.1.37:8080/MicroSimServer
            /SoaCoordinatorService?wsdl">
  <atomic name="gene"
          class="microsim.soa.lib.examples.efp.Generator"
          host="http:// 192.168.1.38:8080/MicroSimServer
                /SoaSimulatorService?wsdl"/>
  <atomic name="proc"
          class="microsim.soa.lib.examples.efp.Processor"
          host="http:// 192.168.1.37:8080/MicroSimServer
                /SoaSimulatorService?wsdl"/>
  <atomic name="tran"
          class="microsim.soa.lib.examples.efp.Transducer"
          host="http:// 192.168.1.37:8080/MicroSimServer
                /SoaSimulatorService?wsdl"/>
  <coupling componentFrom="gene" portFrom="out"
            componentTo="proc" portTo="in"/>
  <coupling componentFrom="gene" portFrom="out"
            componentTo="tran" portTo="arrived"/>
  <coupling componentFrom="proc" portFrom="out"
            componentTo="tran" portTo="solved"/>
  <coupling componentFrom="tran" portFrom="out"
            componentTo="gene" portTo="stop"/>
</devs>
```

LISTING 15.26 GPT SIMULATION RESULTS

```
INFO: Start job 1 @ t = 1.0
INFO: Start job 2 @ t = 2.0
INFO: Start job 3 @ t = 3.0
INFO: Finish job 1 @ t = 3.0
INFO: Start job 4 @ t = 4.0
INFO: Start job 5 @ t = 5.0
INFO: Finish job 3 @ t = 5.0
INFO: Start job 6 @ t = 6.0
INFO: Start job 7 @ t = 7.0
INFO: Finish job 5 @ t = 7.0
INFO: Start job 8 @ t = 8.0
INFO: Start job 9 @ t = 9.0
INFO: Finish job 7 @ t = 9.0
INFO: Start job 10 @ t = 10.0
INFO: Start job 11 @ t = 11.0
INFO: Finish job 9 @ t = 11.0
INFO: Start job 12 @ t = 12.0
```

```
INFO: Start job 13 @ t = 13.0
INFO: Finish job 11 @ t = 13.0
INFO: Start job 14 @ t = 14.0
INFO: Start job 15 @ t = 15.0
INFO: Finish job 13 @ t = 15.0
INFO: Start job 16 @ t = 16.0
INFO: Start job 17 @ t = 17.0
INFO: Finish job 15 @ t = 17.0
INFO: Start job 18 @ t = 18.0
INFO: Start job 19 @ t = 19.0
INFO: Finish job 17 @ t = 19.0
INFO: End time: 19.0
INFO: Jobs arrived : 19
INFO: Jobs solved : 9
INFO: Average TA = 11.0
INFO: Throughput = 0.47368421052631576
INFO: Start job 20 @ t = 19.0
INFO: Finish job 19 @ t = 20.0
```

15.8.2 Experimental Frame-Processor in DEVS/SOA JAVA and DEVS/SOA.NET

In this example, the Experimental Frame will be simulated by DEVS/SOA JAVA, whereas the Processor will be simulated by DEVS/SOA.NET. The root coordinator will also be executed in DEVS/SOA JAVA.

We have a major issue to address here. All the web references generated depend on the DEVS/SOA platform. Thus, if we use the following configuration file in Listing 15.27, the MicroSimClient implemented in either DEVS/SOA JAVA or DEVS/SOA .NET crashes.

LISTING 15.27 WRONG XML CONFIGURATION FILE

```
<?xml version="1.0" encoding="UTF-8"?>
<devs name="Efp"
      host="http://192.168.1.37:8080/MicroSimServer
           /SoaCoordinatorService?wsdl">
  <coupled name="ef"
           class="microsim.soa.lib.examples.efp.Ef"
           host="http://192.168.1.37:8080/MicroSimServer
                /SoaCoordinatorService?wsdl">
    <atomic name="gene"
            class="microsim.soa.lib.examples.efp.Generator"
            host="http://192.168.1.37:8080/MicroSimServer
                 /SoaSimulatorService?wsdl"/>
    <atomic name="tran"
```

```
          class="microsim.soa.lib.examples.efp.
             Transducer"
          host="http://192.168.1.37:8080/MicroSimServer
                /SoaSimulatorService?wsdl"/>
    <coupling componentFrom="ef" portFrom="in"
              componentTo="tran" portTo="solved"/>
    <coupling componentFrom="gene" portFrom="out"
              componentTo="ef" portTo="out"/>
    <coupling componentFrom="gene" portFrom="out"
              componentTo="tran" portTo="arrived"/>
    <coupling componentFrom="tran" portFrom="out"
              componentTo="gene" portTo="stop"/>
  </coupled>
  <atomic name="proc"
   class="microsim.core.lib.examples.efp.Processor,
     MicroSim"
   host="http://192.168.1.37:49910/core/simulation
        /SoaSimulator.asmx?WSDL"/>
  <coupling componentFrom="ef" portFrom="out"
            componentTo="proc" portTo="in"/>
  <coupling componentFrom="proc" portFrom="out"
            componentTo="ef" portTo="in"/>
</devs>
```

This is mainly because the WSDL files are not identical. There are many ways to fix this issue. One, for example, is to incorporate two web references more in each MicroSimServer project. Regarding DEVS/SOA JAVA, we must add a web reference to the DEVS/SOA.NET coordinator and another one to the DEVS/SOA.NET simulator. With respect to DEVS/SOA.NET, we must proceed in the same way.

Project 15.2

Add cross web service references to both DEVS/SOA projects in order to fix this issue.

Hint: In the case of DEVS/SOA JAVA, we must add a web service reference to http://localhost:49910/core/simulation/SoaSimulator.asmx?WSDL. This service reference must be added to a package named (for example) microsim.soa.net .services.simulator, we will have the client side of our DEVS/SOA.NET simulator. Next, we must modify several web services in the SoaSimulator class. We provide here two examples: SoaCoordinator ::setModel and ::initialize web methods (Listing 15.28 and Listing 15.29).

As can be seen in Listing 15.30, we must add a platform definition to the XML configuration file. Thus, having performed these tasks, we may now redefine our XML configuration file:

After running the XML configuration file in MicroSimClient/Java, we obtain the following results from the Transducer (Listing 15.31).

LISTING 15.28 SOACOORDINATOR::SETMODEL MODIFICATIONS

```
// …
  String nameComponent = xmlComponent.getAttribute("name");
  String hostComponent = xmlComponent.getAttribute("host");
  String platformCompo =
                    xmlComponent.getAttribute("platform");
// …
  } else if (nodeName.equals("atomic")) {
    if (platformCompo.equals("devs.soa.java")) {
      microsim.soa.services.simulator.SoaSimulator
        simulator = (new microsim.soa.services.simulator
             .SoaSimulatorService(new URL(hostComponent)))
             .getSoaSimulatorPort();
      ((BindingProvider) simulator).getRequestContext()
      .put(BindingProvider.SESSION_MAINTAIN_PROPERTY, true);
      simulator.setModel(xmlComponentAsString);
      simulators.put(nameComponent, simulator);
    } else if (platformCompo.equals("devs.soa.net")) {
      microsim.soa.net.services.simulator
        .SoaSimulatorServiceSoap simulator = (new
          microsim.soa.net.services.simulator
          .SoaSimulatorService(new URL(hostComponent)))
          .getSoaSimulatorServiceSoap();
      ((BindingProvider) simulator).getRequestContext()
      .put(BindingProvider.SESSION_MAINTAIN_PROPERTY, true);
      simulator.setModel(xmlComponentAsString);
      simulators.put(nameComponent, simulator);
    }
  }
```

LISTING 15.29 SOACOORDINATOR::INITIALIZE MODIFICATIONS

```
 @WebMethod
  public void initialize(double t) {
    // …
    for (Object simulator : simulators.values()) {
      if (simulator instanceof microsim.soa.services
                           .simulator.SoaSimulator) {
        ((microsim.soa.services.simulator.SoaSimulator)
          simulator).initialize(t);
      } else if (simulator instanceof microsim.soa.net
            .services.simulator.SoaSimulatorServiceSoap) {
        ((microsim.soa.net.services.simulator
        .SoaSimulatorServiceSoap) simulator).initialize(t);
```

```
        } else if (simulator instanceof microsim.soa.services
                                      .coordinator.SoaCoordinator) {
          ((microsim.soa.services.coordinator.SoaCoordinator)
           simulator).initialize(t);
        }
      }
      // ...
    }
```

LISTING 15.30 CORRECT EFP XML CONFIGURATION FILE

```
<?xml version="1.0" encoding="UTF-8"?>
<devs name="Efp"
      host="http://192.168.1.37:8080/MicroSimServer
           /SoaCoordinatorService?wsdl">
  <coupled name="ef" platform="devs.soa.java"
           class="microsim.soa.lib.examples.efp.Ef"
           host="http://192.168.1.37:8080/MicroSimServer
                 /SoaCoordinatorService?wsdl">
    <atomic name="gene" platform="devs.soa.java"
            class="microsim.soa.lib.examples.efp.Generator"
            host="http://192.168.1.37:8080/MicroSimServer
                  /SoaSimulatorService?wsdl"/>
    <atomic name="tran" platform="devs.soa.java"
            class="microsim.soa.lib.examples.efp.Transducer"
            host="http://192.168.1.37:8080/MicroSimServer
                  /SoaSimulatorService?wsdl"/>
    <coupling componentFrom="ef" portFrom="in"
              componentTo="tran" portTo="solved"/>
    <coupling componentFrom="gene" portFrom="out"
              componentTo="ef" portTo="out"/>
    <coupling componentFrom="gene" portFrom="out"
              componentTo="tran" portTo="arrived"/>
    <coupling componentFrom="tran" portFrom="out"
              componentTo="gene" portTo="stop"/>
  </coupled>
  <atomic name="proc" platform="devs.soa.net"
   class="microsim.core.lib.examples.efp.Processor,
    MicroSim"
   host="http://192.168.1.38:49910/core/simulation
         /SoaSimulator.asmx?WSDL"/>
  <coupling componentFrom="ef" portFrom="out"
            componentTo="proc" portTo="in"/>
  <coupling componentFrom="proc" portFrom="out"
            componentTo="ef" portTo="in"/>
</devs>
```

LISTING 15.31 DEVS/SOA EFP SIMULATION RESULTS

```
INFO: Start job 1 @ t = 1.0
INFO: Start job 2 @ t = 2.0
INFO: Start job 3 @ t = 3.0
INFO: Finish job null @ t = 3.0
INFO: Start job 4 @ t = 4.0
INFO: Start job 5 @ t = 5.0
INFO: Finish job null @ t = 5.0
INFO: Start job 6 @ t = 6.0
INFO: Start job 7 @ t = 7.0
INFO: Finish job null @ t = 7.0
INFO: Start job 8 @ t = 8.0
INFO: Start job 9 @ t = 9.0
INFO: Finish job null @ t = 9.0
INFO: Start job 10 @ t = 10.0
INFO: Start job 11 @ t = 11.0
INFO: Finish job null @ t = 11.0
INFO: Start job 12 @ t = 12.0
INFO: Start job 13 @ t = 13.0
INFO: Finish job null @ t = 13.0
INFO: Start job 14 @ t = 14.0
INFO: Start job 15 @ t = 15.0
INFO: Finish job null @ t = 15.0
INFO: Start job 16 @ t = 16.0
INFO: Start job 17 @ t = 17.0
INFO: Finish job null @ t = 17.0
INFO: Start job 18 @ t = 18.0
INFO: Start job 19 @ t = 19.0
INFO: Finish job null @ t = 19.0
INFO: End time: 19.0
INFO: Jobs arrived : 19
INFO: Jobs solved : 9
INFO: Average TA = 11.0
INFO: Throughput = 0.47368421052631576
INFO: Start job 20 @ t = 19.0
INFO: Finish job null @ t = 20.0
```

Exercise 15.8

Implement a DEVS/SOA GPT simulation where the Generator is defined in DEVS/JAVA, Processor in MicroSim/Java, and Transducer in MicroSim/.Net.

REFERENCES

GlassFish. (n.d.). Retrieved November 02, 2012, from http://glassfish.java.net/.

Java API for XML Web Services (JAX-WS). (n.d.). Retrieved November 02, 2012, from http://jax-ws.java.net/.

Microsoft Visual Studio. (n.d.). Retrieved November 02, 2012, from www.microsoft.com/visualstudio.

16 Netcentric System of Systems with DEVS-Based Event-Driven Architectures

16.1 INTRODUCTION

Industry and government are spending extensively to transform their business processes and governance to Service Oriented Architecture (SOA) implementations for efficient information reuse, integration, collaboration, and cost sharing. SOA enables orchestrating web services to execute such processes. For example, the Department of Defense's (DoD) grand vision is the Global Information Grid (GIG) that is founded on the SOA infrastructure. As illustrated in Figure 16.1, the SOA infrastructure is to be based on a small set of capabilities known as Core Enterprise Services (CES), whose use is mandated to enable interoperability and increased information sharing within and across mission areas, such as the warfighter domain, business processes, defense intelligence, and so on (CIO, 2007). Net-Centric Enterprise Services (NCES) is the DoD's implementation of its data strategy over the GIG. NCES provides SOA infrastructure capabilities such as service and metadata registries, service discovery, user authentication, machine-to-machine messaging, service management, orchestration, and service governance.

However, composing/orchestrating web services in a process workflow (i.e. mission thread in the DoD domain) is currently bounded by the Business Process Modeling Notation (BPMN)/Business Process Execution Language (BPEL) technologies. Moreover, there are few methodologies to support such composition/orchestration. Furthermore, BPMN and BPEL are not integrated in a robust manner, and different proprietary BPMN diagrams from commercial tools fail to deliver the same BPEL translations. Today, these two technologies are mostly used by executives and managers who devise process flows without touching the technological aspects. With so much resting on SOA, its reliability and analysis must be rigorously considered. The BPMN/BPEL combination neither has any grounding in system theoretical principles nor can be used in designing netcentric systems based on SOA in its current state (Mittal, 2011).

A netcentric system is a System of Systems (SoS) by nature. A common defining attribute of an SoS that critically differentiates it from a single monolithic system is interoperability, or lack thereof, among the constituent disparate systems. The plethora of perspectives on SoS problems evident in the literature suggests that interoperability may take the form of integration of constituent systems (e.g., element A

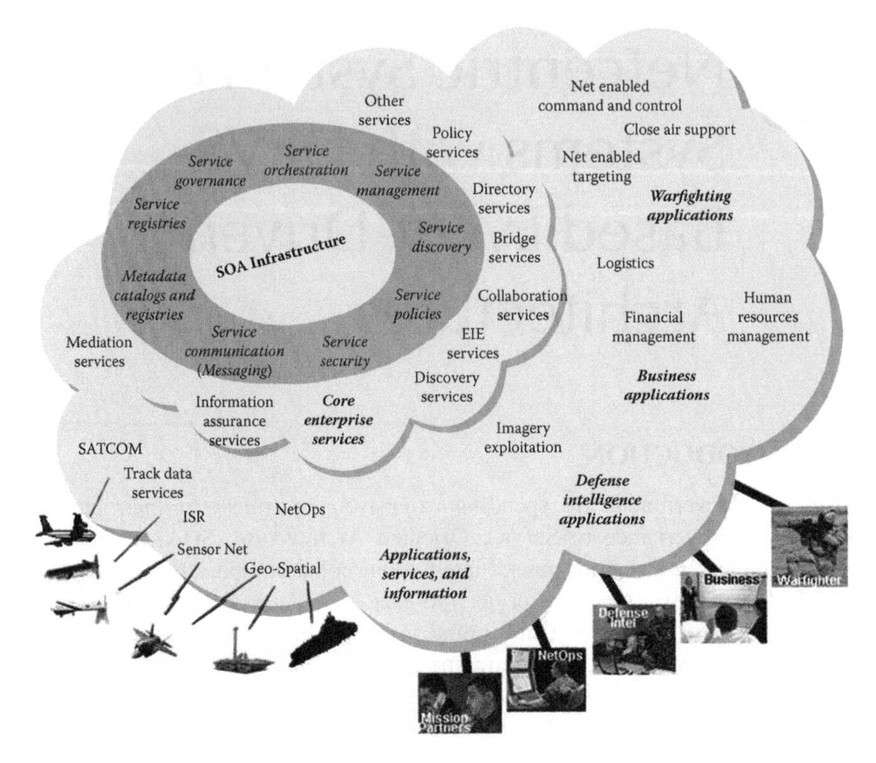

FIGURE 16.1 Core enterprise services in Global Information Grid (CIO, 2007).

is hierarchically superior to element B) or interoperation of constituent systems (e.g., two or more independent elements or systems with no identified hierarchy) (Morganwalp & Sage, 2004; Sage, 2007; Sage & Cuppan, 2001).

Sage (2007) drew the parallel between viewing the construction of SoS as federation of systems and the federation that is supported by the High Level Architecture (HLA), an IEEE standard fostered by the DoD to enable composition of simulations (Dahmann et al., 1998; Sarjoughian & Zeigler, 2000). As illustrated in Figure 16.2, HLA is a network middleware layer that supports message exchanges among simulations, called federates, in a neutral format.* However, experience with HLA has been disappointing and forced acknowledging the difference between enabling heterogeneous simulations to exchange data (the so-called technical interoperability) and the desired outcome of exchanging meaningful data so that coherent interaction among federates takes place (the so-called substantive interoperability) (Yilmaz & Oren, 2004). Tolk introduced the Levels of Conceptual Interoperability Model (LCIM), which identified seven levels of interoperability among participating systems (Tolk & Muguira, 2003). These levels can be viewed as a refinement of the *operational* interoperability type, which is one of the three defined by Dimario (2006). The operational type concerns linkages between systems in their interactions

* HLA also provides a range of services to support execution of simulations.

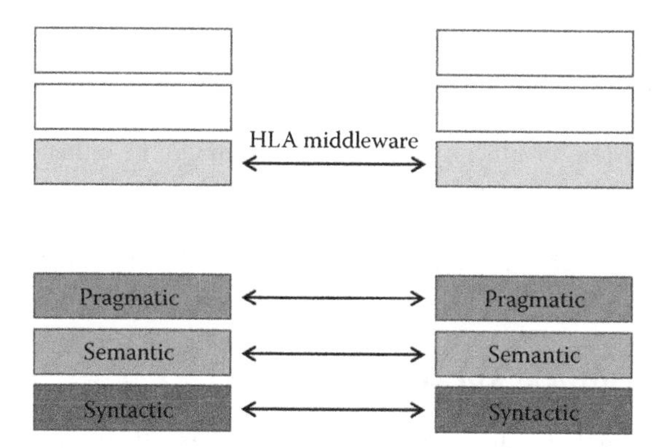

FIGURE 16.2 Interoperability levels in distributed simulation.

TABLE 16.1
Linguistic Levels of Interoperability

Linguistic Level	A Collaboration of Systems or Services Interoperates at This Level If
Pragmatic: how information in messages is used	The receiver reacts to the message in a manner that the sender intends (assuming nonhostility in the collaboration).
Semantic: shared understanding of meaning of messages	The receiver assigns the same meaning as the sender did to the message.
Syntactic: common rules governing composition and transmitting of messages	The consumer is able to receive and parse the sender's message.

with one another, with the environment, and with users. The other types apply to the context in which systems are constructed and acquired. They are *constructive*–relating to linkages between organizations responsible for system construction—and *programmatic*—linkages between program offices to manage system acquisition.

Subsequently, Zeigler and Hammonds (2007) and Mittal et al. (2008) mapped the LCIM into three linguistically inspired levels: *syntactic, semantic,* and *pragmatic.* The levels are summarized in Table 16.1.

In an SoS, systems and/or subsystems often interact with each other because of interoperability and overall integration of the SoS. These interactions are achieved by efficient communication among the systems using either peer-to-peer communication or a central coordinator in a given SoS. Since the systems within an SoS are operationally independent, interactions among systems are generally asynchronous in nature. A simple yet robust solution to handle such asynchronous interactions (specifically, receiving messages) is to throw an event at the receiving end to capture the messages from single or multiple systems. In a netcentric SoS, such systems are integrated using an XML-based middleware and interact through service interfaces. A system architecture where the constituent systems communicate using events is called an Event-Driven Architecture (EDA). It is at the event level where pragmatics, semantics, and syntactic levels converge to provide interoperability.

In this chapter, we will describe the foundation to develop a DEVS-based EDA. A DEVS system is composed of events and components/systems that produce and consume those events. An event is any change in state that merits attention from the originating system or other systems. The system can be either a simple atomic black box that performs a single task, or a complex SoS that receives the event and delegates it to one of its subcomponents. We will employ DEVS to a netcentric system deployed using web services. SOA is composed of web services and is a great enabler to EDA.

16.2 EVENT-DRIVEN ARCHITECTURE

An EDA is a system that is driven by events. So, what exactly is an event? Intuitively, an event happens when something occurs. This occurrence of event affects the causal agent or other agents or both. An agent here can be a complete system in itself and the discipline can be anything, namely, business, defense, and security. While the notion of an event is quite intuitive, a computational meaning is warranted at this point.

Consider the following example, taken from Taylor et al. (2009: p. 15):

> If you exceed your allowance of minutes of cellular phone time, your wireless carrier bills you for the overage. In EDA terms, the state of your minutes goes from 'Under to Over', and that change in state triggers the billing of the overage charge. The value of your account balance changes as you exceed the minutes allowance. The shift from minutes = under to minutes = over is an event.

Taylor et al. (2009) discuss the notion of an event with respect to information contained in the event description. An event has three levels of description. At the first level, the event occurs; that is, a notification happens. At the second level, the event is defined, that is, what constitutes to mark the event occurrence. At the third level, there is precise detail about additional information about the event. In the example above, "change in minutes" event occurs, "minutes = over" is defined, and "overage amount = 10 minutes"—these are the three levels of description.

This description is analogous to the pragmatic, semantic, and syntactic levels of interoperability. At the pragmatic level, the cellular billing system is concerned only with the "change in minutes" event. This pragmatic event is usable in one of the cellular billing system's process flows. At the semantic level, the billing system adds additional information to what it means to "change in minutes" by defining *minutes = over.* At the syntactic level, where the system internal dynamics are implemented, semantics are translated to system's internal logic *overage amount = 10 minutes.* To interoperate, such interoperability is not only desired but also a necessity. Interoperation at pragmatic, semantic, and syntactic levels helps design events that convey complete meaning.

An EDA is composed of the following five functional components:

1. Event producer
2. Event consumer/listener
3. Event processor
4. Event reaction
5. Messaging backbone

An event producer generates a pragmatic event and puts it on the messaging backbone. The event the consumer picks up is the pragmatic event that is usable by the event consumer. The consumer forwards it to the event processor for action on the event that generates either an automated or a human reaction in the form of another event that may or may not be put on the messaging backbone. In an EDA, the interoperability at the syntactic level between a producer, a consumer, and the processor is taken for granted, due to the adoption of Open standards and selection of XML as the preferred mode of communication.

An EDA is of two types:

1. *Explicit EDA*: When the producer and the consumer are tightly coupled and the communication is point-to-point. Messaging may be synchronous and the backbone has point-to-point queues.
2. *Implicit EDA*: When the producer simply publishes to the messaging backbone and is unaware of the consumer. An implicit EDA is a completely decoupled system with agents or system interfacing only with the messaging backbone. Messaging is asynchronous and is of publish/subscribe in nature. Producers publish a specific event type, and the consumers subscribe to a particular event type.

An EDA is granular at the event level and is decentralized. EDA uses a commonly accessible messaging backbone, such as an Enterprise Service Bus (ESB) as well as adapters or middleware to transport messages/events. In an EDA, the event processing is of three types:

1. *Simple event processing*: The producer generates an event, and the consumer accepts it and ultimately forwards it to the processor that may or may not generate a reaction. In this case, it is a single occurrence of an event. Each event is processed exclusively.
2. *Event stream processing*: The event processors receive multiple events from event producers (via listeners), but act only when certain criteria are met at the producer level. Here, time series is implemented and events may be ignored as they arrive in a continuous stream. A single event is not an indication of a reaction. Multiple events in a sequence, on meeting specific criteria, may generate a reaction.
3. *Complex event processing* (CEP): The processors attend to multiple event streams on different time scales, thereby logically correlating them into multiple meaningful reactions. At this level, CEP almost symbolizes pattern matching on information sets as addressed in artificial intelligence (AI) literature. An EDA that implements CEP is inherently complex, dynamic, and implicit.

The more advanced forms of the EDA employ CEP and are loosely coupled by definition and by design. An EDA component can act 'portray all' or a subset of the functionalities such as a producer, consumer, processor, or reactionary agent. The other important aspect is that any EDA component must be modular; that is, it has defined interfaces. The EDA implemented on an SOA has modularity by default, as the components take the notion of web services, described by the Web Service Description Language (WSDL).

EDAs are driven by system's extensibility where events are powered by business needs. EDA component functionalities can be designed at any level of resolution, that is, from low-level instrumentation details to high-level contextual and abstract system response. Most of the EDA components are stateless; that is, the implemented EDA functionalities are stateless. The state of the system is carried with the event details. Lack of persistence in EDA components makes them highly agile and reusable. This has been the primary design considerations driving the evolution of EDAs.

Summarizing, the EDA, at the most basic level, is a collection of modular rules implemented as EDA functional components interacting through events in an interoperable distributed netcentric decentralized environment, encapsulating system's state or context within the events. The reader is encouraged to review Taylor et al. (2009) and Etzion and Niblett (2011) for in-depth EDA consideration.

16.3 NETCENTRIC DEVS SYSTEMS AS EVENT-DRIVEN ARCHITECTURES

The functional components of an EDA can be represented effectively as discrete-event models. In discrete-event modeling, events are generated at random time intervals as opposed to some predetermined time intervals seen commonly in discrete-time systems. More specifically, the state change of a discrete-event system happens only upon arrival (or generation) of an event, not necessarily at equally spaced time intervals. To this end, a discrete-event model is a feasible approach in simulating the SoS framework and its interaction. Several discrete-event simulation engines are available that can be used in simulating an interaction in a heterogeneous mixture of independent systems. The advantages of DEVS are its effective mathematical representation, its integration with SOA systems and its support to distributed simulations using a middleware such as the DoD's HLA.

EDAs differ from DEVS M&S framework in the following aspects:

1. Event description is implicit in DEVS, whereas it is central to any EDA-design.
   DEVS components interact through messages. These messages are explicit events at the receiving end, and DEVS deals with them formally in the external transition function. The notion of implicit event is handled in the internal transition function. Indeed, having interoperability at all the three levels—pragmatic, semantic, and syntactic—encapsulated within the event description is the preferred means to formally address the interoperability issue. DEVS messages need to be extended from programmatic objects as message structures to multilevel event descriptions as elaborated in the EDA literature.
2. For agility, the EDA functional components are stateless as far as possible. Event producers, listeners, and processors are suggested to be stateless, and the system's state is encapsulated in the event structure and is communicated along with the event. It is to be noted that the EDA functional components are not a complete I/O system per DEVS levels of system specification

at level 3, but operate at the I/O behavior level at 0, 1 and 2. At level 3, DEVS formalism groups various I/O behaviors into abstract states and the EDA components are designed to not have state descriptions.

3. The EDA has a common communication backbone.
   The functional components of an EDA communicate using a message bus such as an ESB. DEVS systems have explicit coupling relations with other DEVS components. Netcentric DEVS has a DEVS virtual machine that interoperates across network. The notion of DEVS bus is not new and has been addressed in literature.
4. The EDA is not hierarchical in a system structure.
   The hierarchy in an EDA is encoded in the event description. The event structure itself contains information about an implicit hierarchical system. The event processors receive events at a particular system level and have the capacity to communicate a different level of detail to the next consumer. In DEVS, hierarchy is explicit as defined in coupled models. Any DEVS coupled system is "closed under coupling," implying that, theoretically, there exists an atomic DEVS that can specify the exact same behavior as the coupled DEVS.

The parallels between EDA and DEVS can be best understood by mapping with DEVS levels of system specification, as shown in Table 16.2. EDA processes events in three ways, namely, simple events, event streams, and complex events. Table 16.2 maps the event processing levels with DEVS levels of system specification.

TABLE 16.2
Mapping EDA Constructs to DEVS Levels of System Specifications

Level	Name	DEVS System Specifications	EDA
4	Coupled systems	System built by several component systems that are coupled together	Does not exist. There is no containment to specify hierarchy. Event producers and consumers are agnostic of each other and are connected through a common messaging backbone.
3	I/O system	System with state and state transitions to generate behavior	Does not exist.
2	I/O function	Collection of input/output pairs partitioned according to the initial state the system is in when the input is applied	Complex event processing
1	I/O behavior	Collection of input/output pairs constituting the allowed behavior of the system from an external black box view	Event stream processing
0	I/O frame	Input and output variables and ports together with allowed values	Simple event processing

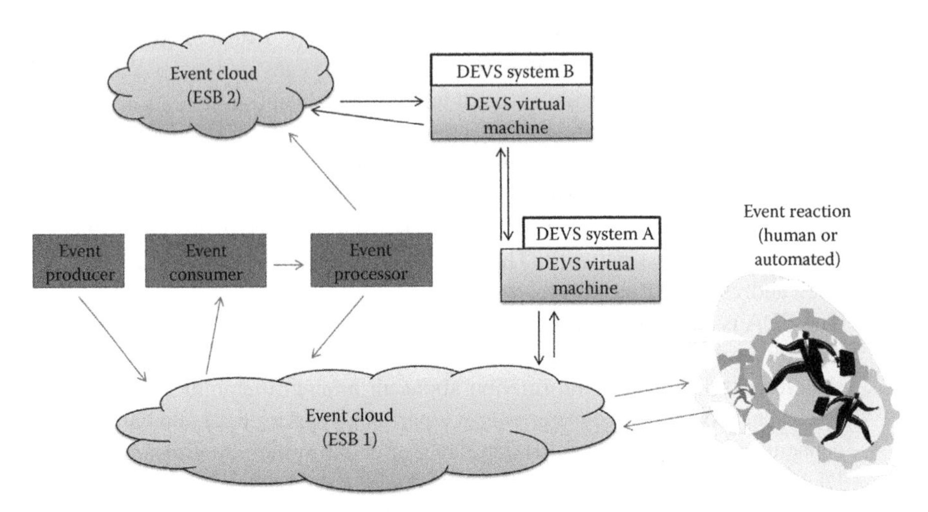

FIGURE 16.3 EDA functional components with netcentric DEVS systems.

Let us now look at the integration of an EDA with DEVS M&S framework. An EDA built on DEVS atomics and coupled components adds structure to the underlying system, thereby managing the state and hierarchical components that are abstracted behind the formal event descriptions. The EDA is real-time by default because there is no framework to manage time. With the DEVS M&S framework underneath the EDA, the EDA can be used for large-scale simulation systems that can run in logical time. A DEVS atomic component encapsulates all of the event-processing concepts and can act as producer, consumer, and processor at the same time. It displays complex behavior wherein the producing, processing, and consuming events are interlaced on a time basis. These events can be formally analyzed, with mathematical rigor. DEVS messages, transformed to formal event structures, indeed enhance the interoperability of a DEVS system at various levels. With a DEVS virtual machine, any DEVS component can be executed locally or as a component to a larger netcentric system communicating to other DEVS virtual machines through a transparent netcentric infrastructure. Figure 16.3 shows an EDA and DEVS system interfacing through a common messaging backbone and interacting with other netcentric messaging backbones. DEVS virtual machines communicate with each other as per the DEVSML 2.0 stack. The DEVS virtual machines can also put message events in the event cloud for the EDA to act on them.

The last piece of integrating a DEVS system with an SOA is the wrapping up of a web service in a DEVS atomic component. The next section describes the implementation of a DEVS service wrapper agent.

16.4 ABSTRACT DEVS SERVICE WRAPPER AGENT

As a crucial part of a DEVS netcentric system, we designed an abstract DEVS service agent to link DEVS models with web services and to generate statistics regarding remote method calls and response times. Figure 16.4 depicts the concept. When the operation is processed, this atomic model calculates the round trip time (RTT) taken

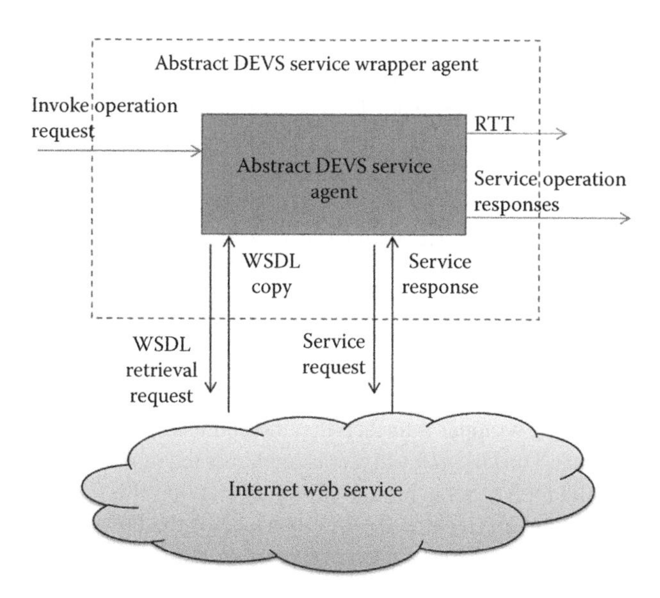

FIGURE 16.4 Schematic showing the architecture of DEVS agent service model.

by such operations and directs both the RTT and the received response from the web service to the DEVS agent's outport. The DEVS service wrapper agent needs to be configured by means of the following:

- URL of the web service
- Package name of the generated entities for web service operations
- Name of the operations offered by the web service
- Parameters needed by these operations

This information is specified in the WSDL document. To restrain the user from extracting this information by hand, we have implemented a wrapper that automatically generates the DEVS web service consumer for a live web service. Such automation is provided by the code generation framework Xtext. Thus, given a WSDL address, our framework is able to generate the corresponding DEVS service agent.

The operation of the DEVS service wrapper agent has three steps:

1. Retrieve and parse the WSDL, stored locally or through a live URL. This capability is implemented by the interface *IServiceWsdl* (Listing 16.1). The WSDL is parsed using the *wsdl4j.jar* library that enumerates all the service operations along with complex message types that the operations receive and respond back with. This step is executed when the user saves the agent DSL file, ending in `.wsd`. The package structure of the generated artifacts resembles the namespace information in the WSDL. Alternatively, the end user is given an option to provide the package name.
2. Create a skeleton around WSDL operations and encapsulate it within a DEVS state machine. The DEVS agent's interface abstracts all the

DEVS-specific details and provides only four functions that are related to initializing the factory, service invocation, service response, and unmarshalling behavior. The DEVS agent behavior is specified as *IWsdlDevsAgent* interface (Listing 16.2). The first step may be embedded in the second step. If it is embedded, the end user is given an option to regenerate the Java artifacts at runtime. The agent, during the initialization phase, parses the URL using *IServiceWsdl* and creates inports and outports based on the service operations. Each operation gets an individual inport and an outport. WSDL also contains information about the complex data type that each operation takes as an argument and sends as a result. The inport and outport are configured to match the corresponding service interface configuration.

3. Create a modular wrapper interface that the end user can use to invoke service operations. The *IWsdlDevsAgent* is implemented in *WsdlDevsAgent* that formalizes the DEVS behavior and allows overriding the *IWsdlDevsAgent* interface (Listing 16.3). Once the I/O interface of the DEVS service agent is initialized, placeholders for each of the operations, in the internal and the external transition functions, are generated in the agent that is derived from *WsdlDevsAgent*.

Let us introduce an example URL here.

LISTING 16.1 INTERFACE FOR RETRIEVING AND CONSTRUCTING SERVICE WRAPPER

```
public interface IServiceWsdl {

        public abstract void makeClient();
        public abstract WsdlCollection getStore();
        public abstract String getPackagename();
        public abstract String getWsdlloc();

}
```

LISTING 16.2 DEVS WSDL-AGENT WRAPPER INTERFACE

```
public interface IWsdlDevsAgent {
        public void initializeFactory();
        public void invokeOperation(String operation);
        public void sendResponseMessage(message m);
        public void unmarshallResponseMessage()  throws
           JAXBException;
}
```

LISTING 16.3 DEVS SERVICE WRAPPER AGENT

```
public abstract class WsdlDevsAgent extends ViewableAtomic
  implements IWsdlDevsAgent{
      protected WsdlCollection store = null;
      protected IServiceWsdl serviceWsdl = null;
      protected String entitiesPackageName = null;
      protected double rtt = 0;
      …

      public WsdlDevsAgent(String name, String wsdlLoc,
        String packageName, boolean generate){

            super(name);

            …

            serviceWsdl = new ServiceWsdl(wsdlLoc,
               packageName, generate);
            initializeFactory();
            initializeWsdlElems();

            if(store == null) return;
            initializeJAXBContext(entitiesPackageName);

            for (String op : store.getOpInputs().keySet()) {
               for (String input : store.getOpInputs().
                 get(op)) {
                   addInport("in" + input);
                   addTestInput("in" + input, new
                     entity(input));
                   }
            }
            addInport("inResult");

            for (String op : store.getOpOutputs().keySet()) {
               for (String output : store.getOpOutputs().get(op))
               {
                   addOutport("out" + output);
               }
            }
            addOutport("RTT");
      }

      public void initializeWsdlElems(){
            if(store == null){
                  System.out.println(getName()+" WSDL Service
                  not initialized!");
            }
```

```
        store = serviceWsdl.getStore();
        entitiesPackageName = serviceWsdl.
          getPackagename();
        System.out.println(store);
}

...

public void initialize() {
        passivate();
}

public void deltint() {
        if (phaseIs("invoked")) {
                double start = System.
                  currentTimeMillis();
                invokeOperation(currentOp);
                double end = System.currentTimeMillis();
                rtt = (end - start) / 1000;
                setSigma(rtt);
                getSimulator().simInject(rtt,"inResult",
                        new entity());
        } else
                passivate();
}

public void deltext(double e, message x) {
        Continue(e);

        for (String op : store.getOpInputs().keySet()) {
         for (String input : store.getOpInputs().
           get(op)) {
                if (phaseIs("passive") &&
                        somethingOnPort(x, "in"+input)) {
                                currentOp = input;
                                holdIn("invoked", 0);
                        }
                else if (phaseIs("invoked")
                        && somethingOnPort(x,
                          "inResult")) {
                        try {
                           unmarshallResponseMessage();
                        } catch (JAXBException e1) {
                           e1.printStackTrace();
                        }
                        holdIn("done", rtt);
                        break;
                        }
                }
        }
}
```

```
        public message out() {
                message m = new message();
                if (phaseIs("done")) {
                        try {
                                sendResponseMessage(m);
                        } catch (Exception e) {
                                e.printStackTrace();
                        }
                        m.add(makeContent("RTT", new
                          doubleEnt(rtt)));
                }
                return m;
        }
}
```

Exercise 16.1

Draw the state machine for DEVS service wrapper agent in Listing 16.3.

16.4.1 Example

For illustration purposes, let us consider a publicly available WSDL for a stock quote web service at www.restfulwebservices.net. The WSDL is provided in Listing 16.4. The *StockQuoteService* has two operations:

1. *GetStockQuote*
2. *GetWorldMajorIndices*

LISTING 16.4 WSDL FOR STOCKQUOTESERVICE AT WWW.RESTFULWEBSERVICES.NET

```
<wsdl:definitions name="StockQuoteService"
  targetNamespace="http://www.restfulwebservices.net/
  ServiceContracts/2008/01">
    <wsdl:types>
      <xsd:schema targetNamespace="http://www
        .restfulwebservices.net/ServiceContracts/
        2008/01/Imports">
           <xsd:import schemaLocation="http://www
             .restfulwebservices.net/wcf/
             StockQuoteService.svc?xsd=xsd0"
             namespace="http://www.restfulwebservices.net/
             ServiceContracts/2008/01"/>
           <xsd:import schemaLocation="http://www
             .restfulwebservices.net/wcf/
             StockQuoteService.svc?xsd=xsd3"
```

```
                namespace="http://GOTLServices.FaultContracts/
                2008/01"/>
              <xsd:import schemaLocation="http://www
                .restfulwebservices.net/wcf/
                StockQuoteService.svc?xsd=xsd1"
                namespace="http://schemas.microsoft.com/
                2003/10/Serialization/"/>
              <xsd:import schemaLocation="http://www
                .restfulwebservices.net/wcf/
                StockQuoteService.svc?xsd=xsd2"
                namespace="http://www.restfulwebservices.net/
                DataContracts/2008/01"/>
        </xsd:schema>
    </wsdl:types>
    <wsdl:message
        name="IStockQuoteService_GetStockQuote_InputMessage">
        <wsdl:part name="parameters" element="tns:
        GetStockQuote"/>
    </wsdl:message>
    <wsdl:message
        name="IStockQuoteService_GetStockQuote_OutputMessage">
          <wsdl:part name="parameters" element="tns:
          GetStockQuoteResponse"/>
    </wsdl:message>
    <wsdl:message
        name="IStockQuoteService_GetStockQuote_
          DefaultFaultContractFault_FaultMessage">
            <wsdl:part name="detail" element="q1:DefaultFault
              Contract"/>
    </wsdl:message>
    <wsdl:message
        name="IStockQuoteService_GetWorldMajorIndices_
          InputMessage"><wsdl:part name="parameters"
          element= "tns:GetWorldMajorIndices"/>
    </wsdl:message>
    <wsdl:message
        name="IStockQuoteService_GetWorldMajorIndices_
          OutputMessage"><wsdl:part name="parameters"
          element="tns:GetWorldMajorIndicesResponse"/>
    </wsdl:message>
    <wsdl:message
        name="IStockQuoteService_GetWorldMajorIndices_
          DefaultFaultContractFault_FaultMessage">
            <wsdl:part name="detail" element="q2:
              DefaultFault Contract"/>
    </wsdl:message>
    <wsdl:portType name="IStockQuoteService">
          <wsdl:operation name="GetStockQuote">
```

```
          <wsdl:input wsaw:Action="GetStockQuote"
            message="tns:IStockQuoteService_
            GetStockQuote_InputMessage"/>
          <wsdl:output wsaw:Action="http://www
            .restfulwebservices.net/ServiceContracts/
            2008/01/IStockQuoteService/
            GetStockQuoteResponse" message="tns:IStockQ
            uoteService_GetStockQuote_OutputMessage"/>
          <wsdl:fault wsaw:Action="http://www
            .restfulwebservices.net/ServiceContracts/
            2008/01/IStockQuoteService/
            GetStockQuoteDefaultFaultContractFault" name=
            "DefaultFaultContractFault" message=
            "tns:IStockQuoteService_GetStockQuote_
            DefaultFaultContractFault_FaultMessage"/>
      </wsdl:operation>
      <wsdl:operation name="GetWorldMajorIndices">
          <wsdl:input wsaw:Action="GetWorldMajorIndices"
            message="tns:IStockQuoteService_
            GetWorldMajorIndices_InputMessage"/>
          <wsdl:output
          wsaw:Action="http://www.restfulwebservices.net/
            ServiceContracts/2008/01/IStockQuoteService/
            GetWorldMajorIndicesResponse"
            message="tns:IStockQuoteService_
            GetWorldMajorIndices_OutputMessage"/>
          <wsdl:fault
          wsaw:Action="http://www.restfulwebservices.net/
            ServiceContracts/2008/01/
            IStockQuoteService/GetWorldMajorIndices
            DefaultFaultContractFault"
            name="DefaultFaultContractFault"
            message="tns:IStockQuoteService_
            GetWorldMajorIndices_DefaultFault
            ContractFault_FaultMessage"/>
      </wsdl:operation>
    </wsdl:portType>
  <wsdl:binding name="BasicHttpBinding_IStockQuoteService"
     type="tns:IStockQuoteService">
     <soap:binding
     transport="http://schemas.xmlsoap.org/soap/http"/>
      <wsdl:operation name="GetStockQuote">
          <soap:operation soapAction="GetStockQuote"
          style="document"/>
          <wsdl:input>
               <soap:body use="literal"/>
          </wsdl:input>
```

```
                <wsdl:output>
                    <soap:body use="literal"/>
                </wsdl:output>
                <wsdl:fault name="DefaultFaultContractFault">
                    <soap:fault name="DefaultFaultContractFault"
                      use="literal"/>
                </wsdl:fault>
            </wsdl:operation>
            <wsdl:operation name="GetWorldMajorIndices">
                <soap:operation soapAction="GetWorldMajorIndices"
                style="document"/>
               <wsdl:input>
                   <soap:body use="literal"/>
               </wsdl:input>
               <wsdl:output>
                   <soap:body use="literal"/>
               </wsdl:output>
               <wsdl:fault name="DefaultFaultContractFault">
                 <soap:fault name="DefaultFaultContractFault"
                 use="literal"/>
               </wsdl:fault>
            </wsdl:operation>
        </wsdl:binding>
        <wsdl:service name="StockQuoteService">
          <wsdl:port name="BasicHttpBinding_IStockQuoteService"
          binding="tns:BasicHttpBinding_IStockQuoteService">
          <soap:address
          location="http://www.restfulwebservices.net/wcf/
            StockQuoteService.svc"/>
          </wsdl:port>
        </wsdl:service>
    </wsdl:definitions>
```

With each operation, there is an associated complex message type as input and output. Refer to Listing 16.4 for detailed description.

When the WSDL is specified in the Xtext *.wsd* editor, and saved as a file, the implemented code generation mechanism generates a web service client in the target namespace, as can be seen in the top window of Figure 16.5. As always, the DSL comes with *package* and *import* constructs to organize the developed agents. In the left portion of the Eclipse workbench (Figure 16.5), the client code is generated in the *stockQuote.entities* package. In addition to that, there is *StockQuoteAgent.java* generated in the "agents" package. The class *StockQuoteAgent* extends *WsdlDevsAgent*. The complete code of *StockQuoteAgent* is shown in Listing 16.5. It provides placeholders for the user to invoke the operation and attend to the service response. From this point onward, the end user can execute the *StockQuoteAgent* in DEVSJAVA SimViewer by running this file as an application. Figure 16.6 shows an operational *StockQuoteAgent* in SimViewer.

FIGURE 16.5 WSDL-based DSL that autogenerates a DEVS wrapper agent.

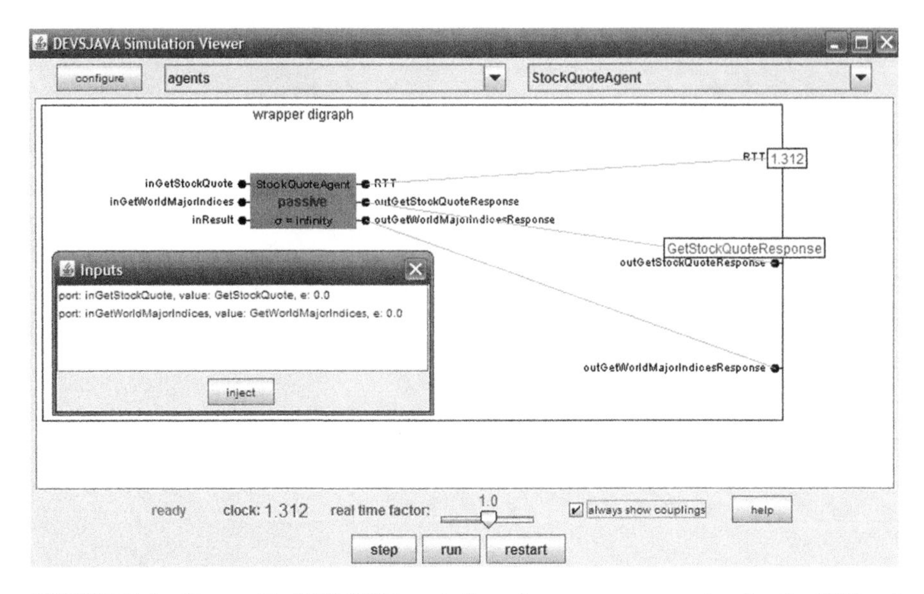

FIGURE 16.6 Executable DEVSJAVA code from the wrapper agent, showing the RTT and service response from the StockQuoteService.

LISTING 16.5 AUTOGENERATED STOCKQUOTE DEVS WRAPPER AGENT

```
public class StockQuoteAgent extends WsdlDevsAgent{

      protected ObjectFactory factory;

      public StockQuoteAgent(){
             this("http://www.restfulwebservices.net/wcf/
                    StockQuoteService.svc?wsdl",
                    "stockQuote.entities", false);
      }
      public StockQuoteAgent(String wsdlLoc, String
        packageName,
             boolean generate) {
             super("StockQuoteAgent", wsdlLoc,
                    packageName, generate);
             this.wsdlLoc = wsdlLoc;
             this.packageName = packageName;
             this.generate = generate;
      }

      @Override
      public void initializeFactory(){
             factory = new stockQuote.entities.
                 ObjectFactory();
      }
```

```
    @Override
    public void unmarshallResponseMessage() throws
      JAXBException{
           Object object = unmarshaller.unmarshal(doc);
           if(object instanceof GetStockQuoteResponse){
                   //add your code here
           }
           if(object instanceof GetWorldMajorIndices
             Response){
                   //add your code here
           }
    }

    @Override
    public void sendResponseMessage(message m){
           if(currentOp.equalsIgnoreCase("getStock
             Quote")){
                   //add your code here
           }
    if(currentOp.equalsIgnoreCase("getWorldMajor
      Indices")){
                   //add your code here
           }
    }

    @Override
    public void invokeOperation(String operationName) {
           IStockQuoteService service =
            new StockQuoteService().
            getBasicHttpBindingIStockQuoteService();

           if(operationName.equalsIgnoreCase("getStock
            Quote")){
                   //add your code here
           }

           if(operationName.equalsIgnoreCase(
                   "getWorldMajorIndices")){
                   //add your code here
           }
           printDoc();
    }

    public static void main(String... strings) {
           new SimView();
    }
}
```

However, the user still has to code specific pieces to invoke the operations and use the service response message. A sample implementation of *GetStockQuote* operation is shown in Listing 16.6. Stock quote for ticker symbol AAPL is requested. The execution can be seen in the bottom part of Figure 16.5. The visual execution of the DEVS agent can be seen in Figure 16.6, which shows the RTT of 1.312 seconds and the response received from the *StockQuoteService.* At this point, the agent is completely operational for *GetStockQuoteService* invocation and can be integrated with other DEVS systems.

LISTING 16.6 AUGMENTED STOCKQUOTE DEVS WRAPPER AGENT WITH USER CODE

```
public void unmarshallResponseMessage() throws JAXBException{
        Object object = unmarshaller.unmarshal(doc);
        if(object instanceof GetStockQuoteResponse){
                GetStockQuoteResponse response =
                        (GetStockQuoteResponse) object;
                StockQuote sq = response.
                  getGetStockQuoteResult()
                        .getValue();

                System.out.println( sq.getName().getValue()+
                        sq.getSymbol().getValue());
                }
        if(object instanceof GetWorldMajorIndicesResponse){
                        //add your code here
        }
}

public void sendResponseMessage(message m){
        if(currentOp.equalsIgnoreCase("getStockQuote")){
                m.add(makeContent("outGetStockQuoteResponse",
                        new entity("GetStockQuoteResponse")));
                }
        if(currentOp.equalsIgnoreCase("getWorldMajor
          Indices")){
                        //add your code here
        }
}

@Override
public void invokeOperation(String operationName) {

        IStockQuoteService service =
                new StockQuoteService().
                getBasicHttpBindingIStockQuoteService();
```

```
        try {
          if(operationName.equalsIgnoreCase("getStock
            Quote")){
              StockQuote sq = service.getStockQuote("AAPL");
              System.out.println( sq.getSymbol().getValue()
                + " "+
                                sq.getLast().getValue());

              JAXBElement<StockQuote> result = factory.
                createGetStockQuoteResponseGetStockQuote
                  Result(sq);

              GetStockQuoteResponse response = factory.
                createGetStockQuoteResponse();

              response.setGetStockQuoteResult(result);

              try {
                     marshaller.setProperty(
                      Marshaller.JAXB_FORMATTED_OUTPUT,
                      Boolean.TRUE);
                     doc = dbf.newDocumentBuilder().
                       newDocument();

                     marshaller.marshal(response, doc);

                      } catch (JAXBException ex) {
                     ex.printStackTrace();
              } catch (ParserConfigurationException e) {
                     e.printStackTrace();
              }
          }

         if(operationName.equalsIgnoreCase(
              "getWorldMajorIndices")){
                     //add your code here
         }

         printDoc();

       }
       catch
       (IStockQuoteServiceGetStockQuoteDefaultFault
         ContractFaultFaultMessage e) {
            e.printStackTrace();
       }
  }
```

The approach to develop a DEVS service wrapper agent is one of the two approaches. We have discussed a client-based approach. The other formal approach is through the development of handlers as described in the JAX-WS 2.0 specification. These handlers intercept the service requests and service response before the service invocation happens. These handlers are of two types: protocol handlers and logical handlers. Many handlers can be chained together to generate events before the actual service is invoked. These handlers act on the web service and can be deployed at both the client and the server sides. Although the server-side approach is easier to implement, it may not be a preferred solution when one does not have control over the service administration. The only choice left with us is to develop a client based on WSDL specifications. For more information about the design of handlers, please refer JAX-WS 2.0/2.1 specifications.

Exercise 16.2

Revisit the Experimental Frame Processor (EFP) example and replace the *Processor* with *StockQuoteAgent*. Modify the *Generator* to Generate jobs that contain ticker symbols. Plot RTT and compare it with results from the *Transducer*.

Exercise 16.3

Create SES for Exercise 16.2 with *StockQuoteAgent* as a specialization of the *Processor* entity.

Exercise 16.4

Run the EFP example developed in Exercise 16.2 on a netcentric DEVS virtual machine.

16.5 DISTRIBUTED MULTILEVEL TEST FEDERATIONS

The prime motivation of applying DEVS system theoretical principles to these emerging netcentric systems comes from an editorial by Carstairs (2005) that demands an M&S framework at higher levels of system specifications where systems of systems interact together using the netcentric platform. At this level, model interoperability is one of the major concerns. The motivation for this work stems from this need of model interoperability and characteristics of netcentric systems that are easier to simulate, test, and deploy with an underlying foundation of systems engineering principles. DEVS, which is known to be a component-based system, based on the formal systems theoretical framework is the preferred means. Table 16.3 outlines how it could provide solutions to the challenges in netcentric design and evaluation.

A test instrumentation system should provide a minimally intrusive test capability to support rigorous, ongoing, repeatable, and consistent testing and evaluation (T&E). Requirements for such a test implementation system (Mittal et al., 2008) include the ability to

- Deploy agents to interface with SoS component systems in specified assignments
- Enable agents to exchange information and coordinate their behaviors to achieve specified Experimental Frame data processing

TABLE 16.3
Desired M&S Capability for T&E in GIG/SOA as Addressed by DEVS Unified Process

Desired M&S Capability for Test and Evaluation (T&E)	Solutions Provided by DEVS Unified Process
Support for DoDAF need for executable architectures using M&S such as mission-based testing for GIG/SOA	DUNIP provides a methodology and an SOA infrastructure for integrated development and testing.
Interoperability and cross-platform M&S using GIG/SOA	Simulation architecture is layered to accomplish the technology migration. Model integrated computing employed in the DEVSML 2.0 stack provides platform neutrality. The DEVS virtual machine in a netcentric environment allows integration with an SOA-based EDA.
Automated test generation and deployment in a distributed simulation	Model is separated from the act of simulation. Model constructed offline can join a running live simulation. DEVS virtual machines provide this runtime capability. Automated test observers generation is integral to the DUNIP methodology.
Test artifact continuity and traceability through phases of systems development	Provide rapid prototyping by employing a spiral development process. The model *is* the deployed software.
Real-time observation and control of the test environment	Provide dynamic variable structure component modeling with Experimental Frame to enable control and reconfiguration of simulation.

- Respond in real-time to queries for test results while testing is still in progress
- Provide real-time alerts when conditions are detected that would invalidate results or otherwise indicate that intervention is required
- Centrally collect and process test results on demand, periodically, and/or at termination of testing
- Support consistent transfer and reuse of test cases/configurations from past test events to future test events, enabling life cycle tracking of SoS performance
- Enable rapid development of new test cases and configurations to keep up with the reduced SoS development times expected to characterize the reusable web service-based development supported on the GIG/SOA

Many of these requirements are not achievable with current data collection and testing. Instrumentation and automation are needed to meet these requirements.

A DEVS distributed federation is a DEVS coupled model whose components reside on different network nodes and whose coupling is implemented through the middleware connectivity characteristic of the environment (e.g., SOAP for GIG/SOA). The federation models are executed by DEVS simulator nodes that provide the time and data exchange coordination as specified in the DEVS abstract simulator protocol.

As discussed earlier in Chapter 2, in the general concept of EF, the generator sends inputs to the SoS under test (SUT), the transducer collects SUT outputs and develops statistical summaries, and the acceptor monitors SUT observables and decides about continuation or termination of the experiment (Mak et al., 2008). Since the SoS is composed of system components, the EF is distributed among SoS components. The event producers and consumers may be completely agnostic, though connected together by an event cloud in an EDA. With DEVS, the producers and consumers are connected together through the DEVS virtual machine. Each component may be coupled to an EF consisting of some subset of generator, acceptor, and transducer components. As mentioned, in addition, an observer couples the EF to the component using an interface provided by the integration infrastructure. We refer to the DEVS model that consists of the observer and EF as a test agent. In the case of DEVS service wrapper agent, the agent itself is an observer as well as a consumer and a producer. When coupled with EF, as in Exercise 16.2, the resulting DEVS system is a test agent.

SOA provides technologically feasible realization of the concept. As discussed earlier, the DEVS/SOA infrastructure enables DEVS models, and test agents in particular, to be deployed to the network nodes of interest. In this incarnation, the network inputs sent by EF generators are SOAP messages sent to other EFs as destinations; transducers record the arrival of messages and extract the data in their fields, while acceptors decide whether the gathered data indicate continuation or termination is in order.

Since EFs are implemented as DEVS models, distributed EFs are implemented as DEVS models, or agents as we have called them, residing on network nodes. Such a federation, illustrated in Figures 16.7 and 16.8, consists of DEVS simulators executing within the DEVS netcentric virtual machines on the nodes exchanging messages and obeying time relationships under the rules contained within their hosted DEVS models.

The linguistic levels of interoperability (Zeigler & Hammonds, 2007; Mittal et al., 2008) provide a basis for further structuring the test instrumentation system. In the following sections, we discuss the implementation of test federations that simultaneously operate at the syntactic, semantic, and pragmatic levels (Figure 16.9).

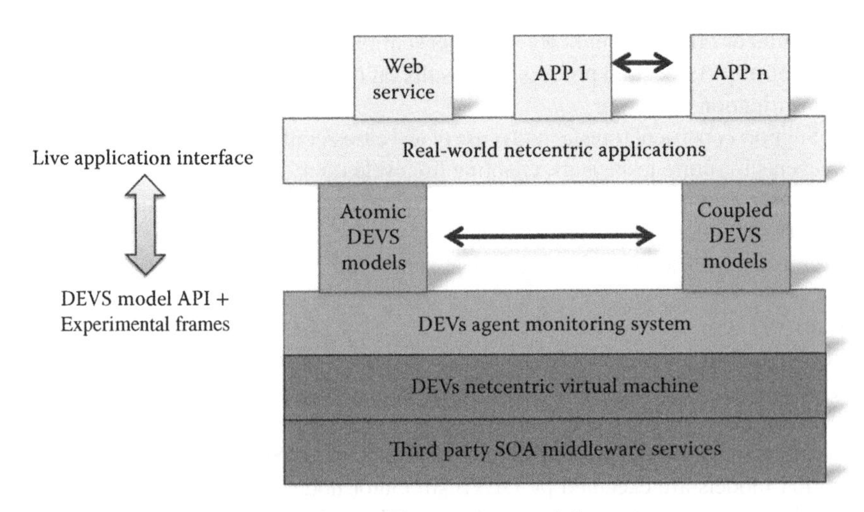

FIGURE 16.7 Deploying experimental frame agents and observers.

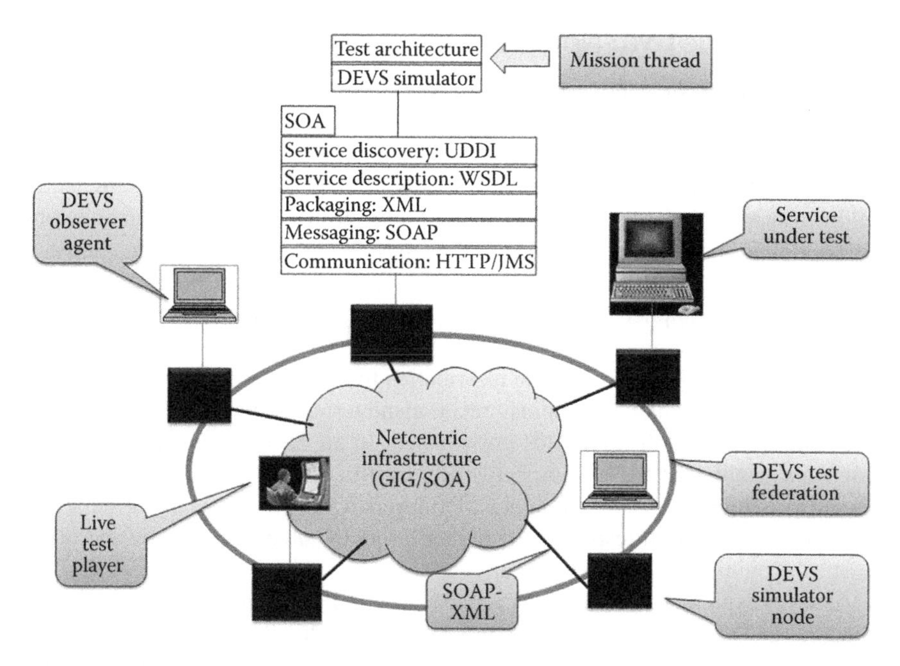

FIGURE 16.8 DEVS test federation in GIG/SOA environment.

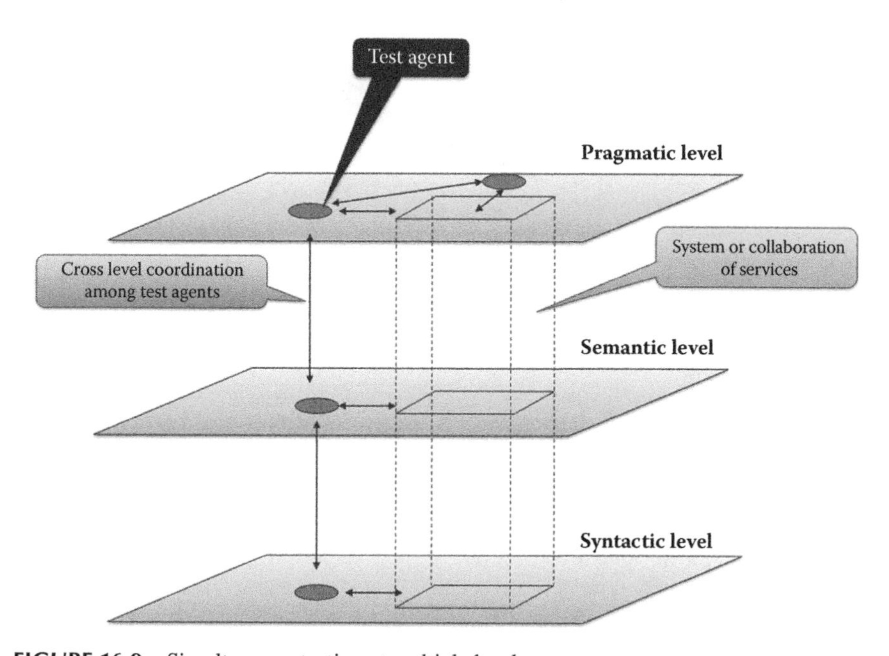

FIGURE 16.9 Simultaneous testing at multiple levels.

16.5.1 Syntactic Level: Network Health Monitoring

From the syntactic perspective, testing involves assessing whether the infrastructure can support the speed and accuracy needed for higher level exchange of information carried by multimedia data types, individually and in combination. We now consider this as a requirement to continually assess whether the network is sufficiently healthy to support the ongoing collaboration. Figure 16.10 illustrates the architecture that is implied by the use of subordinate probes. Nodal generator agents activate probes to meet the health-monitoring quality of service (QoS) thresholds determined from information supplied by the higher layer test agents, namely, the objectives of the higher layer tests.

Probes return statistics and alarm information to the transducers/acceptors at the DEVS health layer, which in turn may recommend termination of the experiment at the test layer when QoS thresholds are violated. In an EF for real-time evaluation of network health, the SUT is the network infrastructure (OSI layers 1 through 5) that supports higher session and application layers. QoS measures are at the levels required for meaningful testing at the higher layers to gather transit time and other statistics, providing QoS measurements.

Messages expressed in XML and carried by a SOAP middleware are directly generated by the DEVS generators and consumed by the DEVS transducers/acceptors. Such messages experience the network latencies and congestion conditions experienced by messages exchanged by the higher level web servers/clients. Under certain QoS conditions, however, video streamed and other data-typed packets may experience conditions different from the SOAP-borne messages. For these, we need to execute lower layer monitoring under the control of the nodal EFs.

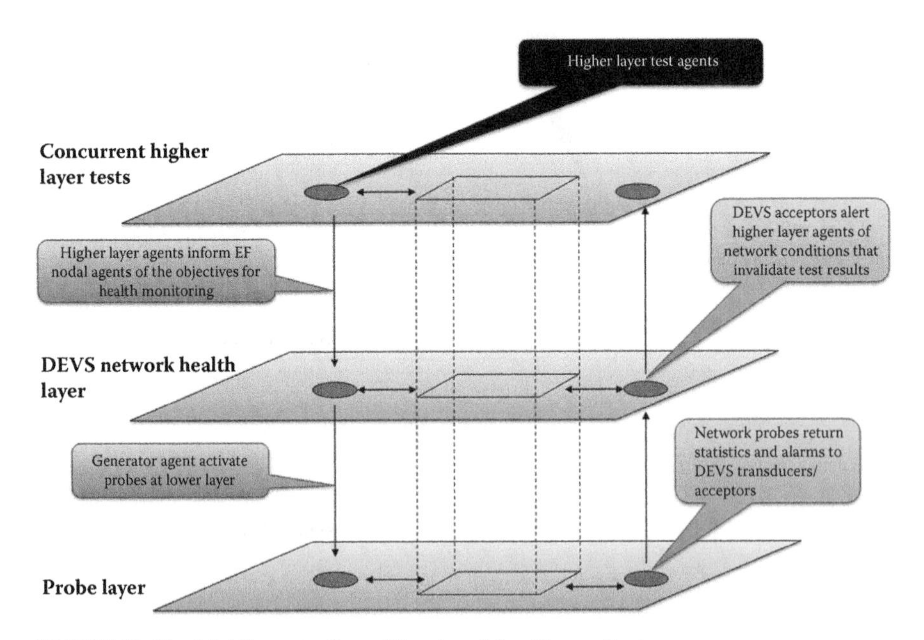

FIGURE 16.10 Multilayer testing with network health monitoring.

The collection of agent EFs has the objective of assessing the health of the network relative to the QoS that it is providing for the concurrent higher level tests. Thus, such a distributed EF is informed by the nature of the concurrent test for which it monitors network health. For example, if a higher level test involves exchanges of a limited subset of media data types (e.g., text and audio), then the lower layer distributed EFs need to monitor only the subset of types.

16.5.2 Semantic Level: Information Exchange in Collaborations

Mission threads consist of sequences of discrete information exchanges. A collaboration service supports such exchanges by enabling collaborators to employ a variety of media, such as text, audio, and video, in various combinations. For example, a drawing accompanied by a voice explanation involves both graphical and audio media data. Further, the service supports establishing producer/consumer relationships. For example, the graphical/audio combination might be directed to one or more participants interested in that particular item. From a multilevel perspective, testing of such exchanges involves pragmatic, semantic, and syntactic aspects. From the pragmatic point of view, the ultimate worth of an exchange is how well it contributes to the successful and timely completion of a mission thread. From the semantic perspective, the measures of performance involve the speed and accuracy with which an information item, such as a graphical/audio combination, is sent from producer to consumer. Accuracy may be measured by comparing the received item to the sent item using appropriate metrics. For example, is the received graphic/audio combination within an acceptable distance from the transmitted combination, where distance might be measured by pixel matching in the case of graphics and frequency matching in the case of audio? To automate this kind of comparison, metrics must be chosen that are both discriminative and quick to compute. Furthermore, if translation is involved, the meaning of the item must be preserved as discussed above. In addition, the delay involved in sending an item from sender to receiver must be within limits set by human psychology and physiology. Such limits are more stringent where exchanges are contingent on immediately prior ones as in a conversation. Instrumentation of such tests is similar to that at the syntactic level that was discussed earlier, with the understanding that the complexity of testing for accuracy and speed is of a higher order at the semantic level.

16.5.3 Pragmatic Level: Mission Thread Testing

A test federation observes an orchestration of web services to verify the message flow among participants and adheres to information exchange requirements. A mission thread is a series of activities executed by operational nodes and employing the information-processing functions of web services. Test agents watch messages sent and received by the services that host the participating operational nodes. Depending on the mode of testing, the test architecture may, or may not, have knowledge of the driving mission thread under test. If a mission thread is being executed and thread knowledge is available, testing can do a lot more than if it does not.

With knowledge of the thread being executed, DEVS test agents can be aware of the current activity of the operational nodes they are observing. This enables an agent to focus more efficiently on a smaller set of messages that are likely to provide test opportunities.

16.5.4 Measuring Success in Mission Thread Executions

The ultimate test of effectiveness of an integration infrastructure is its ability to support successful outcomes of mission thread executions. To measure such effectiveness, the test instrumentation system must be informed about the events and messages to expect during an execution, including those that provide evidence of success or failure, and must be able to detect and track these events and messages throughout the execution.

16.5.5 Measuring in Context

It is often said that the success of a mission depends on the right information arriving to the right place at the right time. Thus, an obvious measure of performance is the ability of an integration infrastructure to measure the extent to which the right information is delivered to the right place at the right time. This in turn places requirements on the Test Instrumentation System that it is able to gather the information needed to make such judgments. Much more than the simple ability to determine mission success or failure, such a capability would provide diagnostic capability to determine whether the outcome was due to failure in information flow and, if so, just what information was not received at the right time by which consumer.

Figure 16.11 graphically illustrates a formulation of the issue at hand. Given that a consumer is engaged in a particular activity for some period, there is a time window

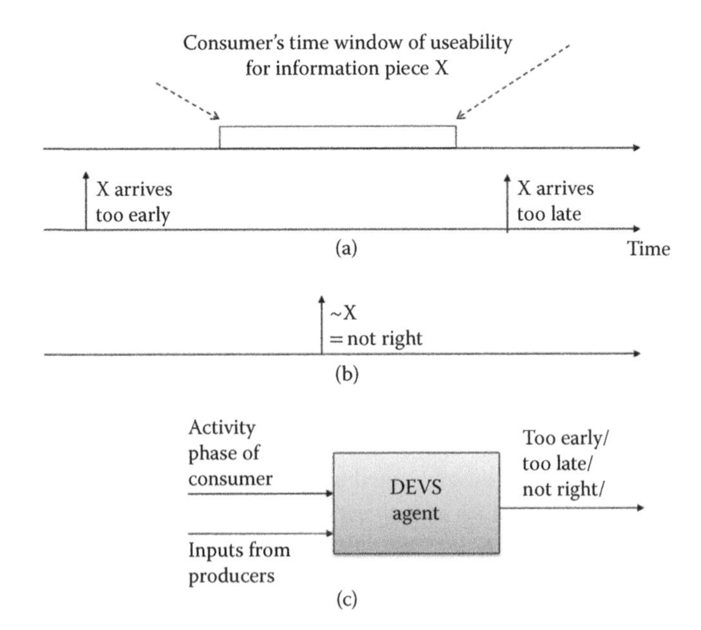

FIGURE 16.11 Instrumentation for the right information at the right place at the right time.

during which it expects, as well as can exploit in its processing, some piece of information denoted by X. As in Exhibit 16.11a, if X arrives before this window opens up, it may be too early and cannot be used when needed; likewise, if X arrives after the window has closed, it is too late for the information to be effectively used. Further, as shown in Exhibit 16.11b, if the wrong information, say ~X, arrives during the window of opportunity, it may be ignored by the consumer at best, or cause a back up of messages that clog the system at worst. As in Exhibit 16.11c, to make such determinations, a test agent has to know the current activity phase of the consumer, and has to be informed of the expected information X and its time window of usability.

Implementing such test capabilities requires not only the right test instrumentation and protocol but also the necessary back up analysis capabilities to provide the information items needed.

16.6 DISCUSSION

SOA has come a long way, and many of the businesses are seriously considering migration of their IT systems toward SOAs. The DoD's initiative toward migration of GIG/SOA and NCES requires reliability and robustness, not only in the execution but in the design and analysis phase as well. Web service orchestration is not just a research issue but also a more practical issue for which there is dire need. Further, SOA must be taken as another instance of system engineering for which there must be a laid out engineering process. Modeling and simulation provides the needed edge. Lack of methodologies to support design and analysis of such orchestration (except BPEL-related efforts) cost millions in failure. This chapter has illustrated that DEVS formalism can be used to build and analyze complex netcentric systems and workflows.

This chapter describes the integration of SOA and EDA with DEVS netcentric capabilities. After the development of the World Wide Web, many efforts in the distributed simulation field have been made for modeling, executing simulation, and creating model libraries that can be assembled and executed over the World Wide Web. By means of XML and web services technology, these efforts have entered upon a new phase. The DEVSML 2.0 stack allows the deployment of DEVS virtual machines across SOA and provides interoperability through the DEVS middleware. The central point resides in executing the simulator as a web service. The development of these kinds of frameworks will help solve large-scale problems and interoperability among different networked systems and specifically DEVS-validated models. This chapter has focused on the overall integration approach, and the symmetrical SOA-based architecture that allows for DEVS execution as a simulation SOA.

We have shown how a web service can be encapsulated into a DEVS atomic model and can be put toward a coupled DEVS system with other live web services and other DEVS models.

Further, based on the capabilities of the DEVS netcentric virtual machine, we have a foundation for the following:

1. Agent-implemented test instrumentation
2. Netcentric execution using simulation services

3. Distributed multilevel test federations
4. Analysis that can help optimally tune the instrumentation to provide confident scalability predictions
5. Mission thread testing and data gathering
   a. Definition and implementation of military-relevant mission threads enable constructing and/or validating models of user activity
   b. Comparison with current commercial testing tools shows that by replicating such models in large numbers it will be possible to produce more reliable load models than offered by the conventional use of scripts

We have taken the challenge of constructing netcentric systems as one of designing an infrastructure to integrate existing web services as components, each with its own structure and behavior with DEVS agents and interoperate with or without an EDA. We have discussed EDA and how a DEVS-based EDA gives formal rigor to an EDA. We have also recommended that DEVS refine the notion of an event structure in a more formal manner so that it can be integrated with an existing EDA, thereby facilitating the interoperability at pragmatic, semantic, and syntactic levels.

A netcentric system is analogous to an SoS where hierarchical coupled models could be created within the 'closure under composition' system theoretic principles. Various workflows can be integrated together using a component-based design. The netcentric system can be specified in many available frameworks, such as DoDAF, SES, BPMN/BPEL, and UML, or by using an integrative systems engineering-based framework such as DEVS.

We have established the capability to develop a live workflow example with a complete DEVS interface. In this role, DUNIP acts as a full netcentric production environment. Being DEVS enabled, it is also executable as an SUT model toward various verification and validation analysis that can be performed by coupling this SUT with other DEVS test models. Last but not the least, the developed DEVS system can be executed by both real and virtual users to the advantage of various performance and evaluation studies. As components comprising SoS are designed and analyzed, their integration and communication is the most critical part that must be addressed by the employed SoS M&S framework. We discussed the DoD's GIG as providing an integration infrastructure for SoS in the context of constructing collaborations of web services using SOA.

REFERENCES

Carstairs, D. (2005). Wanted: A New Test Approach for Military Net-Centric Operations. Guest Editorial. *ITEA Journal, 26*(3), 7–9.

CIO, DoD. (June 2007). *Department of Defense Architectural Vision, Version 1.0.* Retrieved from www.defenselink.mil/cio-nii/docs/GIGArchVision.pdf.

Dahmann, J., Kuhl, F., & Weatherly, R. (1998). Standards for simulation: as simple as possible but not simpler: the high level architecture for simulation. *Transactions of SCS, 71*(6), 378–387.

DiMario, M. (2006). System of Systems Interoperability Types and Characteristics in Joint Command and Control. *IEEE/SMC International Conference on System of Systems Engineering.* Los Angeles, CA.

Etzion, O., & Niblett, P. (2011). *Event Processing in Action.* Stamford, CT: Manning Publishers.

Mak, E., Mittal, S., Ho, M. H., & Nutaro, J. J. (2010). Automating Link 16 testing using DEVS and XML. *Journal of Defense Modeling and Simulation, 7*(1), 39–62.

Mittal, S. (2011). Agile netcentric systems with DEVS unified process. In A. Tolk, L. Jain (Eds), *Intelligence-based Systems Engineering*, Intelligent Systems Reference Library, Vol. 10, 159–199, Springer Berlin Heidelberg.

Mittal, S., Zeigler, B. P., Martín, J. L. R., Sahin, F., & Jamshidi, M. (2008). Modeling and Simulation for System of Systems Engineering. In M. Jamshidi (Ed.), *System of Systems Engineering for 21st Century.* John Wiley & Sons, Hoboken NJ, USA. doi: 10.1002/9780470403501.ch5.

Morganwalp, J., & Sage, A. (2004). Enterprise architecture measures of effectiveness. *International Journal of Technology, Policy and Management, 4*, 81–94.

Sage, A. (2007). From Engineering a System to Engineering an Integrated System Family, From Systems Engineering to System of Systems Engineering. *IEEE International Conference on System of Systems Engineering.* San Antonio, TX.

Sage, A., & Cuppan, C. (2001). On the systems engineering and management of system of systems and federation of systems. *Information Knowledge Systems Management, 2*, 325–345.

Sarjoughian, H., & Zeigler, B. (2000). DEVS and HLA: complimentary paradigms for M&S? *Transactions of SCS, 17*(4), 187–197.

Taylor, H., Yochem, A., Phillips, L., & Martinez, F. (2009). *Event Driven Architectures: How SOA Enables Real-Time Enterprise.* Boston, MA: Addison-Wesley.

Tolk, A., & Muguira, J. (2003) The Levels of Conceptual Interoperability Model (LCIM). *Proceedings of the Fall Simulation Interoperability Workshop*, Orlando, FL.

Yilmaz, L., & Oren, T. (2004). A Conceptual Model for Reusable Simulations within a Model-Simulator-Context Framework. *Conceptual Models Conference.* Genoa, Italy.

Zeigler, B., & Hammonds, P. (2007). *Modeling & Simulation-Based Data Engineering: Introducing Pragmatics into Ontologies for Net-Centric Information Exchange.* Academic Press, Burlington, MA.

17 Metamodeling in Department of Defense Architecture Framework (Version 2.0)

We have seen in Chapters 11 and 12 how extended Department of Defense Architecture Framework (Version 1.0) (DoDAF 1.0) with two new operational views, OV-8 and OV-9, provides an executable nature to the static DoDAF documents. We have highlighted some of the gaps and augmented the existing DoDAF views with information relevant to the executable aspect of the DoDAF specification. DoDAF 1.0 came out in late 2003 and has been revised twice since then. DoDAF 1.5 was released in mid-2007. DoDAF 1.5 kept the framework largely the same, but added a netcentric spin to it by including service oriented architecture (SOA) concepts in System views. As SOA technologies matured in the last decade, the focus shifted toward the data aspect of these architecture frameworks. Keeping pace with the changing landscape, DoDAF 2.0 was released in May 2009 (U.S. Department of Defense, 2009). DoDAF 2.0 is built on the foundation of DoDAF 1.5, and it introduces a totally new way to construct the architectures using the model-driven architecture (MDA) paradigm. This chapter provides an overview of DoDAF 2.0 and describes how DEVS and a System Entity Structure (SES) can build executable DoDAF models.

17.1 INTRODUCTION

DoDAF 2.0 replaces the Core Architecture Data Model (CADM) used in earlier DoDAF versions and proposes the DoDAF Metamodel 2 (DM2). The DM2 provides a high-level view of the data elements that are normally collected, organized, and maintained in an architectural document. In this manner, the DM2 supports the exchange and reuse of architectural information and data among various DoD agencies, Joint capability areas (JCAs), components, program boundaries, and coalition partners, thus facilitating the understanding and implementation of interoperability of processes and systems. The DM2 to a large extent solves the interoperability at semantic and syntactic levels. A subset of the semantic foundation of the DM2 Conceptual Data Model is provided in Table 17.1. For the entire list, see Vol. 1, DoDAF 2.0.

The "views" in earlier versions of DoDAF have been named as "viewpoints." Earlier versions of DoDAF suggested the use of all DoDAF products within the views. DoDAF 2.0 is "fit-for-purpose" and the products are now called DoDAF-described models. Not all the viewpoints and DoDAF-described models have to be created. The focus is more on the data sharing and development as the necessary ingredients

TABLE 17.1
High-Level Semantic Constructs in DoDAF 2.0 Conceptual Data Model

Concept	Description
Activity	Work, not specific to a single organization, weapon system, or individual that transforms input (resources) to output (resources) or changes their state.
Service	A mechanism to enable access to a set of one or more capabilities, where the access is provided using a prescribed interface and is exercised in consistency with the constraints and policies as specified by the service description. The mechanism is the performer. The capabilities accessed are resources—information, data, materiel, performers, and geopolitical extents.
Capability	The ability to achieve a desired effect under specified (performance) standards and conditions through combinations of ways and means (activities and resources) to perform a set of activities.
Desired effect	The result, outcome, or consequence of an action (activity).
Resource	Data, information, performers, materiel, or personnel types that are produced or consumed.
Performer	Any entity—human, automated, or any aggregation of human and/or automated—that performs an activity and provides a capability.
Condition	The state of an environment or situation in which a performer performs.
Materiel	Equipment, apparatus, or supplies that are of interest, without distinction as to its application for administrative or combat purposes.
Person type	A category of persons defined by the role or roles they share that are relevant to the architecture.
Measure	The magnitude of some attribute of an individual.
Measure type	A category of measures.
Organization	A specific real-world assemblage of people and other resources organized for an on-going purpose.
Rule	A principle or condition that governs behavior; a prescribed guide for conduct or action.
Standard	A formal agreement documenting generally accepted specifications or criteria for products, processes, procedures, policies, systems, and/or personnel.
System	A functionally, physically, and/or behaviorally related group of regularly interacting or interdependent elements.
Information	The state of something of interest that materializes—in any medium or form—and is communicated or received.
Data	Representation of information in a formalized manner suitable for communication, interpretation, or processing by humans or by automatic means.
Constraint	The range of permissible states of an object.
Agreement	A consent among parties regarding the terms and conditions of activities that said parties participate in.
Location	A point or extent in space that may be referred to physically or logically.

for architecture development. If an activity model is created, the necessary set of data for the activity model is required. Alternatively, the activity has been componentized. This addresses one of the primary gaps we highlighted in DoDAF 1.0 and is adequately addressed in the DoDAF 2.0 design approach.

DoDAF organizes the DoDAF-described model in a set of viewpoints, as listed in Table 17.2 and shown in Figure 17.1.

TABLE 17.2
Overview of DoDAF 2.0 Viewpoints

Viewpoint	Description
All	Overarching aspects of architecture context that relate to all viewpoints.
Capability	Capability requirements, the delivery timing, and the deployed capability.
Data and information	Articulates data relationships and alignment structures in the architecture content for the capability and operational requirements, systems engineering processes, and systems and services.
Operational	Includes operational scenarios, activities, and requirements that support capabilities.
Project	Describes relationships between the activity and capability requirements and the various projects that are being implemented. It details dependency among capability and operational requirements, system engineering processes, system design, and service design.
Services	Design for solutions that provide support for operational and capability functions by articulating performers, activities, services, and their exchanges.
Standards	Applicable operational, business, technical, and industry policies, standards, guidance, constraints, and forecasts that apply to capability and activity requirements, system engineering processes, and systems and services.
Systems	Design for solutions that provide support for operational and capability functions articulating systems, their composition, interconnectivity, and context.

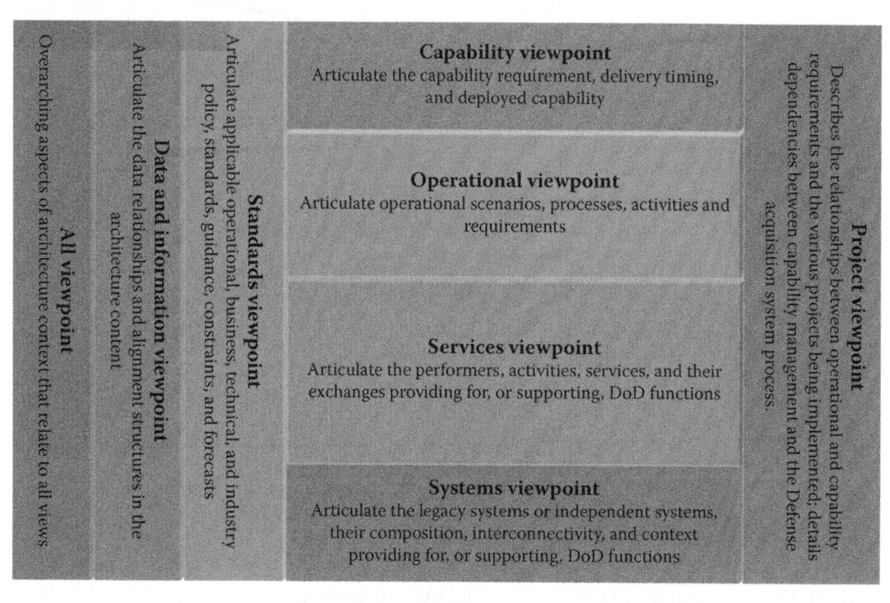

FIGURE 17.1 DoDAF 2.0 viewpoints. Reproduced from DoDAF 2.0 (2009).

The viewpoints categorize the latest DoDAF-described models as follows:

- The All, Operational, Systems, and Standards viewpoints borrowed from earlier All, Operational, System, and Technical views had their models organized based on the Conceptual Data Model for better understanding. The Systems and Service view in DoDAF 1.5 has been separated into System and Services Viewpoint to bring more focus toward netcentric or Service-oriented implementation in Services Viewpoint.
- All models of data (conceptual, logical, and physical) have been placed in the Data and Information Viewpoint rather than spread across the Operational and System views in earlier DoDAF.
- Systems Viewpoint has been enhanced to included legacy system descriptions.
- Standards Viewpoint has been enhanced to include business, commercial, and doctrinal standards, along with the technical standards for systems and services.
- Operational Viewpoint has been enhanced to include rules and constraints for any function (business, intelligence, warfighting, etc.).
- Two new Viewpoints—Project Viewpoint and Capability Viewpoint—have been added because of the feedback from the acquisition community and the emphasis within the Department on Capability Portfolio Management.

The evolution of DoDAF 2.0 viewpoints is shown in Figure 17.2.

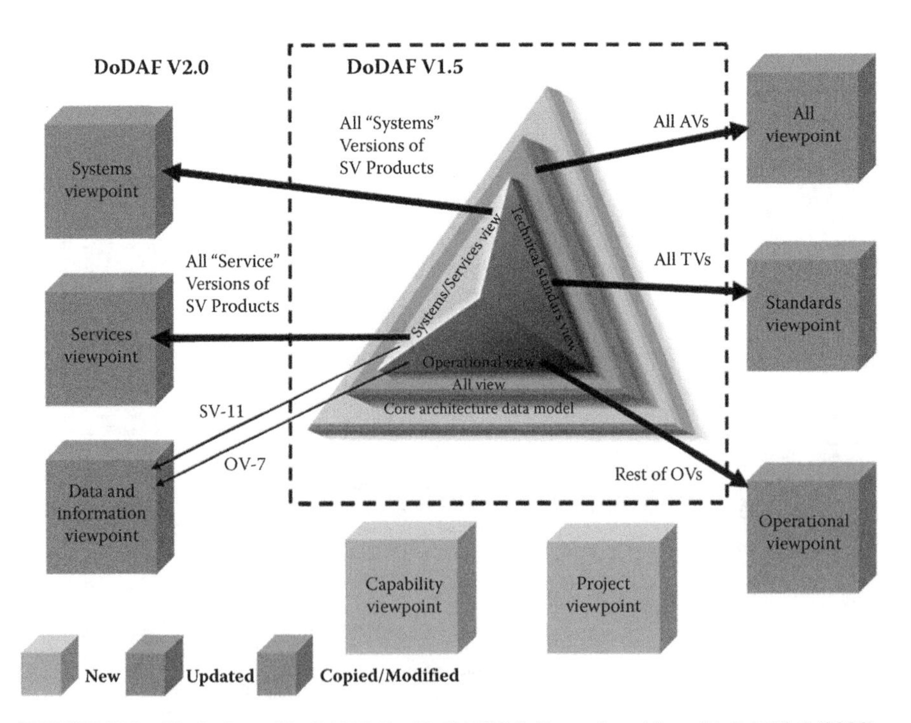

FIGURE 17.2 Evolution of DoDAF 1.5 to DoDAF 2.0. Reproduced from DoDAF 2.0 (2009).

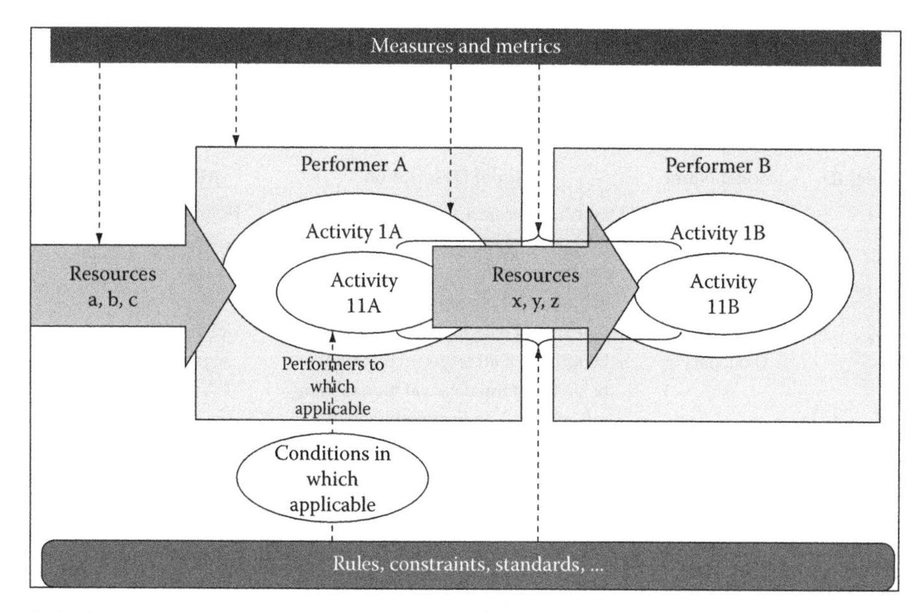

FIGURE 17.3 Top-level conceptual execution of a DoDAF scenario. Reproduced from DoDAF 2.0 (2009).

Based on the concepts described in Table 17.1, a typical DoDAF-designed model is executable in the manner shown in Figure 17.3. Activities are central to any DoDAF 2.0 scenario. A set of activities that may be organized hierarchically form a Capability. Activities can be implemented as services and use other service resources. Activities are performed by performers that use resources. Activities are constrained by rules, constraints, standards, and so on, in a given architectural description and have measures and metrics. Activities bring a change in the state of resources that triggers further activities in the DoDAF scenario.

17.2 DODAF 2.0 VIEWPOINTS AS RELEVANT TO BUILDING AN EXECUTABLE ARCHITECTURE

The reader is encouraged to refer to the original DoDAF 2.0 description document for details about various DoDAF-described models in each of the viewpoints. In this section, we look at only a subset of those models that aid in the development of an executable architecture (EA).

17.2.1 All Viewpoint

All Viewpoint (AV) DoDAF-described models provide an overview of the architectural effort including things such as the scope, context, rules, constraints, assumptions, and the derived vocabulary that pertains to the architectural description. It serves as a planning guide and is a structured text product. It consists of two models. Table 17.3 summarizes AV. This view also guides the EA's scope.

TABLE 17.3
All Viewpoint DoDAF-Described Model Relevant to an Executable Architecture Design

Model ID	Model Name	Model Description	M&S Relevance
AV-1	Overview and Summary Information	Describes a project's visions, goals, objectives, plans, activities, events, conditions, measures, effects (outcomes), and produced objects	Helps define the scope of the EA design
AV-2	Integrated Dictionary	Architectural data repository with definitions of all terms used throughout the architectural data and presentations. It provides taxonomy to all the other viewpoints.	Helps preserve the semantics used in EA development

17.2.2 Capability Viewpoint

Capability models describe the taxonomy, description, and evolution of a set of capabilities. This viewpoint is used by capability portfolio managers to capture the increasingly complex relationships between interdependent projects and capabilities. The CV DoDAF-described models are used to provide the strategic perspective and context for other architectural information. It consists of the models listed in Table 17.4.

17.2.3 Data and Information Viewpoint

DoDAF 2.0 incorporates three levels of abstraction that correlate to different levels associated with most data models developed in support of the operations or business. The considered levels (Table 17.5) are correspondingly translated to specific DoDAF-described models. These three levels can be effectively translated to the pragmatic, semantic, and syntactic levels.

17.2.4 Operational Viewpoint

DoDAF-described models in the OV describe the tasks and activities, operational elements, and resource flow exchanges required to conduct operations. A pure operational model is materiel independent. However, constraints may be introduced based on changing materiel requirements that impact the operational activity. The models in the OV are listed in Table 17.6.

17.2.5 Services Viewpoint

The DoDAF-described Service models associate Service resources to the operational and capability requirements. The Service resources support operational activities and facilitate exchange of information. Table 17.7 enumerates the service models.

TABLE 17.4
Capability Viewpoint DoDAF-Described Model Relevant to an Executable Architecture Design

Model ID	Model Name	Model Description	M&S Relevance
CV-1	Vision	Overall vision provides strategic context for the capabilities described at a high-level scope.	Helps define the scope and experimental frame
CV-2	Capability Taxonomy	Hierarchy of capabilities that are specified and referenced throughout the architectural description.	Model organization and containment
CV-4	Capability Dependencies	Dependencies between the planned capabilities and the definition of logical groupings of capabilities.	Couplings and containment between various capabilities
CV-5	Capability to Organizational Development Mapping	Planned solution for the capability phase in terms of performers, and locations and their associated concepts. It addresses deployment issues for capabilities.	Additional constraints to capability specification
CV-6	Capability to Operational Activities Mapping	A mapping between the capabilities required and the operational activities that those capabilities support for auditing and capability requirements traceability.	Associations, containment, and requirements traceability
CV-7	Capability to Services Mapping	A mapping between the capabilities and the services that these capabilities enable for auditing and capability requirements traceability.	Association and requirements traceability

Service resource flows incorporate human elements as types of performers—organizations and personnel types.

17.2.6 Standards Viewpoint

The standards model is a set of rules governing the arrangement, interaction, and interdependence of parts of the architectural description. Its purpose is to ensure that a solution satisfies a specified set of operational or capability requirements.

TABLE 17.5
Data and Information Viewpoint DoDAF-Described Model Relevant to an Executable Architecture Design

Model ID	Model Name	Model Description	M&S Relevance
DV-1	Conceptual Data Model	High-level data concepts, data requirements, and their relationships.	Pragmatic interoperability
DV-2	Logical Data Model	Documentation of the data requirements and structural business process (activity) rules. Semantics and data quality are introduced at this point.	Semantic interoperability
DV-3	Physical Data Model	Physical implementation format of the DV-2 entities, and so on. This is a direct transformation from semantic to syntactic levels of interoperability.	Syntactic interoperability

TABLE 17.6
Operational Viewpoint DoDAF-Described Model Relevant to an Executable Architecture Design

Model ID	Model Name	Model Description	M&S Relevance
OV-1	High-Level Operational Concept Graphic	High-level graphical/textual description of the operational concept	EA activity scope
OV-2	Operational Resource Flow Description	Description of resource flows exchanged between operational activities	EA events and message types
OV-3	Operational Resource Flow Matrix	Description of resources exchanged and the relevant attributes of the exchange	EA events and message structures
OV-4	Organizational Relationships Chart	The organizational context, role, or other relationships between organizations	EA association with organizations; additional constraints impacting activities
OV-5a	Operational Activity Decomposition Tree	Capabilities and activities organized in a hierarchical structure	Hierarchy and containment in association with capabilities
OV-5b	Operational Activity Model	Context of capabilities and activities and their relationships among activities, inputs, and outputs; cost, performance, or other pertinent information	Modularity of activities; defined I/O and other constraints

TABLE 17.6 (*Continued*)
Operational Viewpoint DoDAF-Described Model Relevant to an Executable Architecture Design

Model ID	Model Name	Model Description	M&S Relevance
OV-6a	Operational Rules Model	Identifies business rules that constrain activities	Business rules constraining or enabling the activities
OV-6b	State Transition Description	Identifies business process (activity) responses to events (usually, very short activities)	State transition behavior of modular activities
OV-6c	Event-Trace Description	Traces actions in a scenario or sequence of events	Event and I/O trajectories for requirements traceability

TABLE 17.7
Services Viewpoint DoDAF-Described Model Relevant to an Executable Architecture Design

Model ID	Model Name	Model Description	M&S Relevance
SvcV-1	Services Context Description	Identification of service, service items, and their interconnections	EA services scope, containment, and hierarchy
SvcV-2	Services Resource Flow Description	Description of resource flow exchanged between services	Service interface message types associated with services
SvcV-3a	Systems-Services Matrix	Relationships between systems and services in a given architectural description	Association of services with systems for containment, interoperability, and requirements traceability
SvcV-3b	Services-Services Matrix	Relationships among service in a given architectural description	Couplings, containment, interoperability, and association with other services
SvcV-4	Services Functionality Description	Functions performed by services and the service data flows among service functions (activities)	Operations within a service; decomposition of a service; encapsulation, containment, and exchanged message types

(Continued)

TABLE 17.7 (*Continued*)
Services Viewpoint DoDAF-Described Model Relevant to an Executable Architecture Design

Model ID	Model Name	Model Description	M&S Relevance
SvcV-5	Operational Activity to Service Traceability Matrix	Mapping of services (activities) back to operational activities (activities)	Association and containment
SvcV-6	Services Resource Flow Matrix	Provides details about the resource flow elements being exchanged between services and the attributes of that exchange	Message structures at the modular service interface
SvcV-7	Services Measures Matrix	Measures (metrics) of services model for the appropriate timeframe(s)	Metrics for services performance
SvcV-10a	Services Rules Model	Constraints imposed on service functionality	Behavior specification of services
SvcV-10b	Service State Transition Description	Responses of services to events	Behavior specification of services
SvcV-10c	Services Event-Trace Description	Service-specific refinements of critical event sequences described in operational viewpoint	Event trajectories at the I/O level. Behavior specification of services

It captures the doctrinal, operational, business, technical, or industry implementation guidelines upon which engineering specifications are based. This collection of constraints and specification overlay on the individual specification of activities, capabilities, services, and systems when they are reused within a particular architecture description. Table 17.8 lists the Standards model.

17.2.7 System Viewpoint

The DoDAF-described model of the System Viewpoint (SV) describes systems and interconnections providing for, or supporting, DoD functions. The system models associate system resources to the operational and capability requirements. These system resources support the operational activities and facilitate the exchange of information. However, the system models are in DoDAF 2.0 in support of legacy systems. As architectures are updated, they should transform from systems to services and utilize the models within the Services Viewpoint. The relationship between the SV and modeling and simulation (M&S) areas is provided in Table 17.9.

TABLE 17.8
Standards Viewpoint DoDAF-Described Model Relevant to an Executable Architecture Design

Model ID	Model Name	Model Description	M&S Relevance
StdV-1	Standards Profile	Listing of standards that apply to solution elements	Helps define the scope of EA design
StdV-2	Standards Forecast	Description of emerging standards and potential impact on current solution, with a set of timeframes	Helps create the experimental frame

TABLE 17.9
System Viewpoint DoDAF-Described Model Relevant to an Executable Architecture Design

Model ID	Model Name	Model Description	M&S Relevance
SV-1	System Interface Description	Identification of systems, system items, and their interconnections	EA services scope, containment, and hierarchy
SV-2	Systems Resource Flow Description	Description of resource flow exchanged between systems	System interface message types associated with systems
SV-3	Systems-Systems Matrix	Relationships between systems in a given architectural description	Association between systems for containment, interoperability, and requirements traceability
SV-4	Systems Functionality Description	Functions (activities) performed by systems and the system data flows among service functions (activities)	Operations within a system; decomposition of a system; encapsulation, containment, and exchanged message types
SV-5a	Operational Activity to Systems Function Traceability Matrix	Mapping of system functions (activities) back to operational activities (activities)	Association and containment
SV-5b	Operational Activity to Systems Traceability Matrix	Mapping of systems back to capabilities or operational activities (activities)	Association and containment

(Continued)

TABLE 17.9 (*Continued*)
System Viewpoint DoDAF-Described Model Relevant to an Executable Architecture Design

Model ID	Model Name	Model Description	M&S Relevance
SV-6	Systems Resource Flow Matrix	Provides details about the resource flow elements being exchanged between systems and the attributes of that exchange	Message structures at the system interface
SV-7	Systems Measures Matrix	Measures (metrics) of systems model for the appropriate timeframe(s)	Metrics for system performance
SV-10a	Systems Rules Model	Constraints imposed on systems functionality	Behavior specification of systems
SV-10b	System State Transition Description	Responses of systems to events	Behavior specification of systems
SV-10c	System Event-Trace Description	System-specific refinements of critical event sequences described in operational viewpoint	Event trajectories at the I/O level. Behavior specification of systems

17.3 DODAF 2.0 METAMODEL IN SES ONTOLOGY

Since the approach of DoDAF 2.0 is oriented toward data sharing and integration, the objective is facilitated by a group of DoDAF-described metamodels. This goes back to the subject of domain-specific modeling where a domain is specified using a set of constructs and semantics and a description language. We have seen in the earlier two sections an overview of the semantics and various DoDAF-described model constructs that help define a metamodel. While there is no one-to-one mapping between the DoDAF-described models and the semantics associated with the data model, a third component, the DoDAF conceptual model, is available that helps define mapping between the two. The DoDAF DM2 Conceptual Model is specified as a set of metamodels:

- Goals
- Capability
- Activities
- Performer
- Services
- Resource flows
- Information and data
- Project
- Training/skill/education
- Rules

- Measures
- Locations

For a better understanding, these metamodels can be grouped into categories (Table 17.10).

Figure 17.4 shows the relationships between the DM2 metamodels overlaid with architectural interrogatives.

The DM2 metamodels build on the foundation from the International Defense Enterprise Architecture Specification (IDEAS), from which all DoDAF concepts inherit several important properties, such as the concepts of the individual, types, tuples, whole-part, temporal whole-part, super-subtype, and interface. This binds the DoDAF domain concepts to a formal ontology amenable to mathematical analyses and computational representation toward an EA. For a detailed description on

TABLE 17.10
DM2 Metamodel Categories

Category	Metamodels in DM2
Goals and desired effects	Goals, Capabilities
Actual mission configurations	Activities, Performers, Services, Resource flows, Information and Data
Means (by which end-items are put in place)	Projects, and Training/Skill/Education
Characteristics of end-items	Rules, Measures, Locations

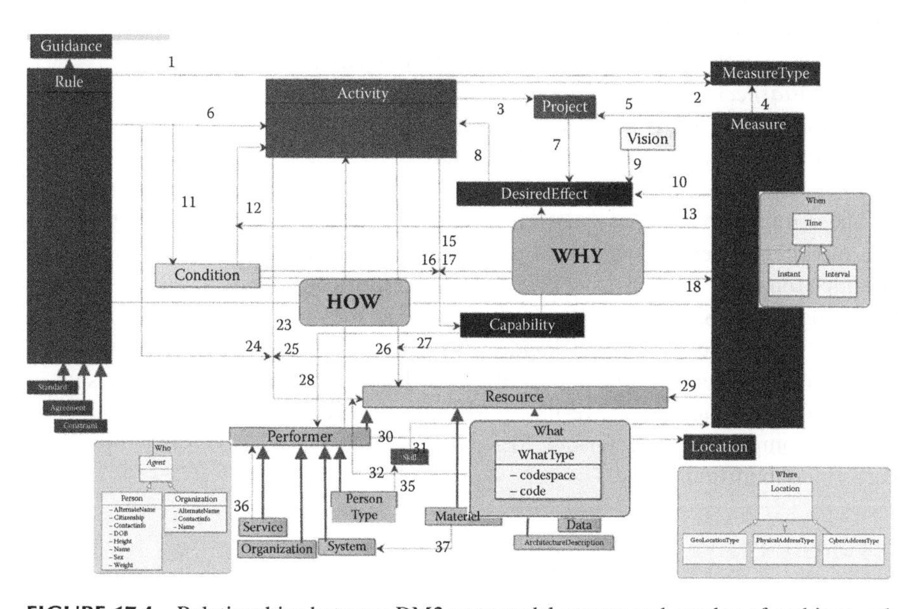

FIGURE 17.4 Relationships between DM2 metamodel groups and overlay of architectural interrogatives.

IDEAS usage in DoDAF, refer to the DoDAF 2.0 description (U.S. Department of Defense, 2009).

Similarly, the DoDAF domain concepts can be represented in a different ontology, such as an SES. The motivation here is that an SES is a high-level knowledge representation framework and it seamlessly lends itself to a formal M&S framework such as DEVS. Next, we look at each of the DM2 metamodels, as expressed in SESs. The SESs described ahead are in no manner complete or DoDAF 2.0 compliant. DoDAF compliance in this case is defined as incorporating every entity and relationships specified in the DoDAF 2.0 specification document. They are meant for illustrative purposes.

17.3.1 Performer

A performer is a class of entities that describe "who" in the architecture development process. The "how," tasks, activities, and processes (a set of activities), is assigned to the performers. Performers are allocated to organizations, personnel, and mechanization. Rules, location, and measures are then applicable specifically to a particular performer type. Figure 17.5 shows relations of performers with other DM2 metamodels.

Figure 17.5 is represented in the SES ontology (Figure 17.6). In terms of the natural language SES, Figure 17.6 can be read as follows:

- *Aspect*: A performer *is made of* PerformerRules, PeformerConditions, Location, and Activities.
- *Specialization*: A performer *can be* OrganizationType, PersonType, System, Service, and Ports.
- *Multiaspect*: A performer *is made of many* IndividualPerformers.
- *Specialization*: An IndividualPerformer *can be* an Organization.
- *Multiaspect*: A System *is made of many* Materiels.
- *Multiaspect*: A System *is made of many* PersonTypes.
- *Multiaspect*: A PersonType *is made of many* Skills.
- *Aspect*: A Skill *is made of* SkillMeasures.

The computational representation of the performer SES is shown in Figure 17.7. The metaPerformer package builds on the entities defined in other packages and provides top-level aspects, specializations, and multiaspects to reuse the domain knowledge. The SES editor (in the center) describes the Performer specialization and multiaspects. The package explorer shows the generated Java code as the file is being edited and the entities are being updated. The outline shows the hierarchical view of the resulting SES. As an exercise, the reader is encouraged to complete the Performer SES with elaboration of PerformerRules and PerformerConditions.

Exercise 17.1

Complete the performer SES by adding details for PerformerRules and PerformerConditions. Create separate packages for organizing the SES.

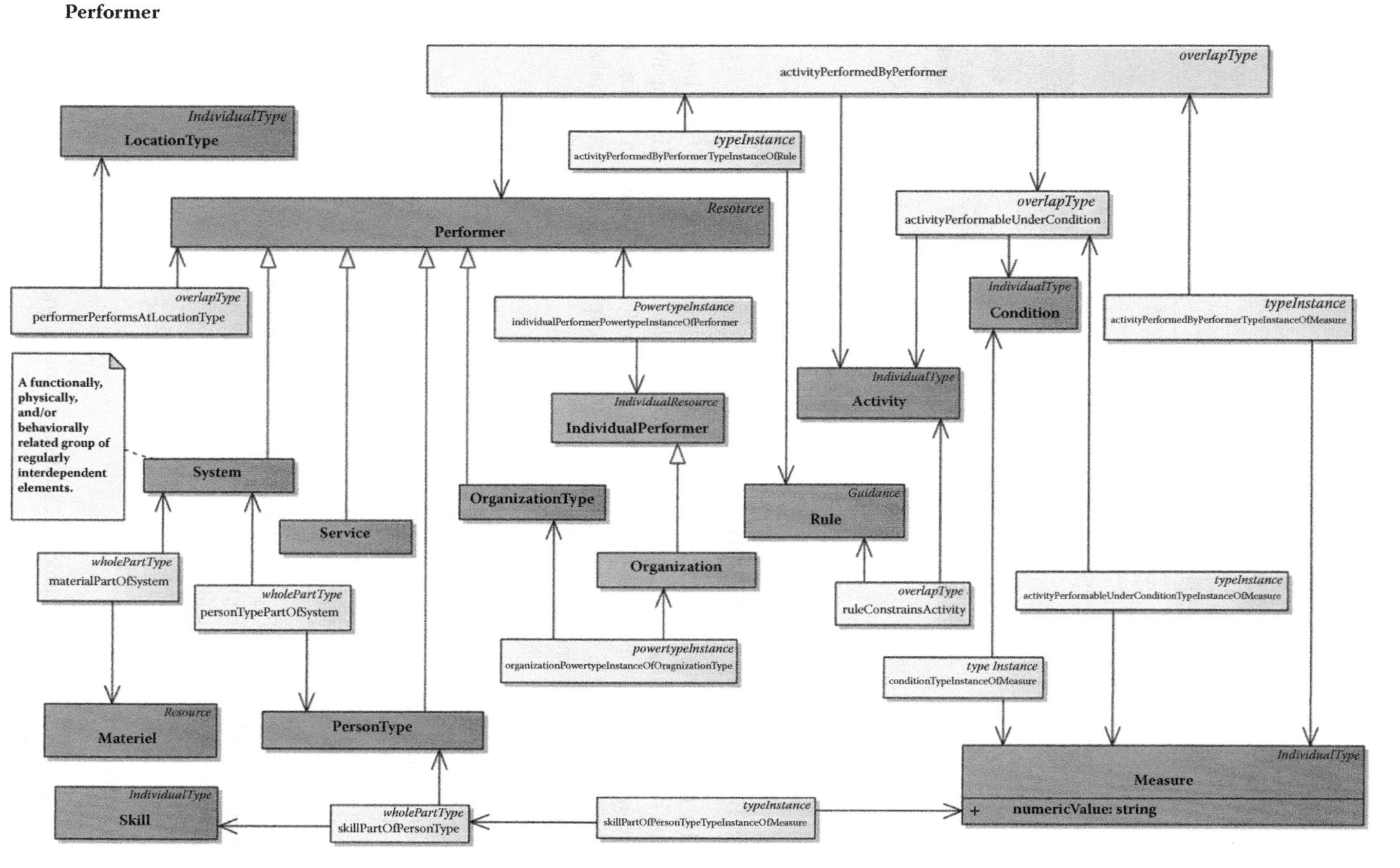

FIGURE 17.5 DoDAF metamodel for performer. Reproduced from DoDAF 2.0 specifications (2009).

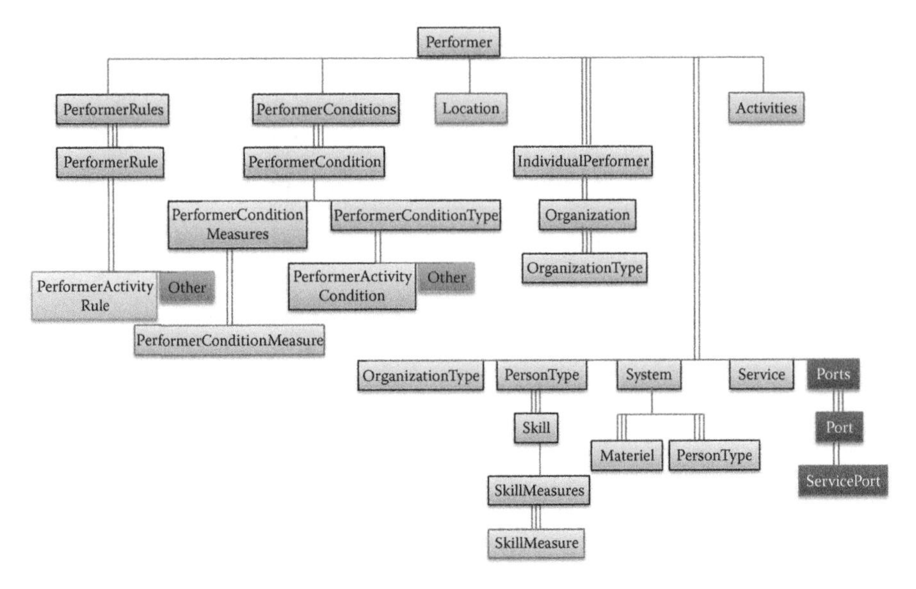

FIGURE 17.6 DoDAF metamodel for performer expressed in SESs.

17.3.2 Capability and Activities

The Capability metamodel provides information on the collection and integration of activities that combine to respond to a specific requirement. The Capability metamodel and its relationship with other metamodels are shown in Figure 17.8. The capability SES is shown in Figure 17.9. The capability SES incorporates the activity DM2 metamodel as well.

Exercise 17.2

Computationally represent the capability SES in Figure 17.9 in the SES editor. Use the earlier defined entities and, if required, enhance their description. Report any inconsistencies with the SES description. Refine the SES description by refactoring multiaspects and/or specializations.

Similarly, Project (Figure 17.10), Resource (Figure 17.11), Information and Data (Figure 17.12), Rules (Figure 17.13), and Measures (Figure 17.14) DM2 metamodels are expressed in graphical SESs.

Exercise 17.3

Represent a Project graphical SES into a computational SES. Refactor the earlier defined SESs.

FIGURE 17.7 Computational representation of the performer SES from Figure 17.6.

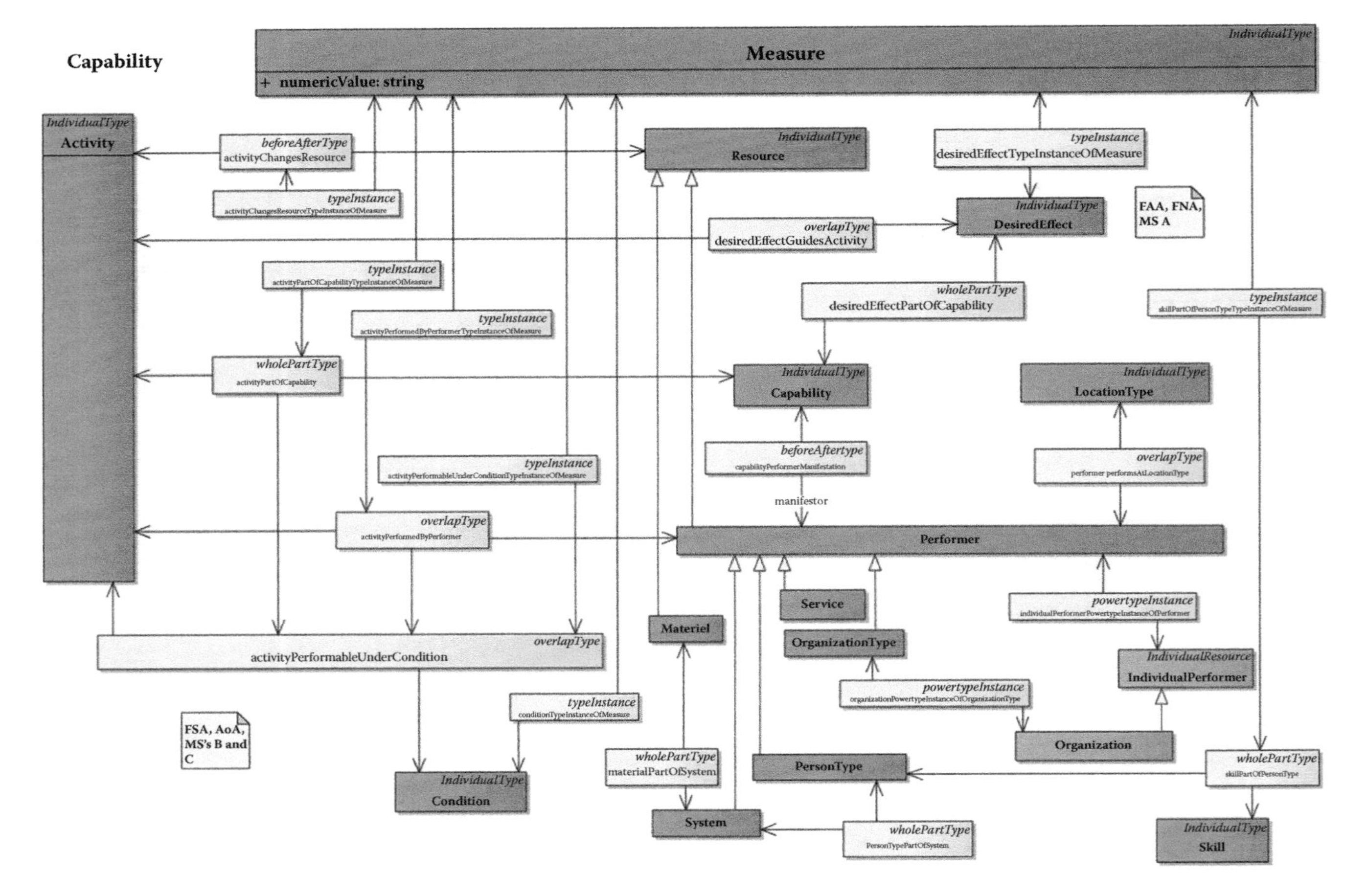

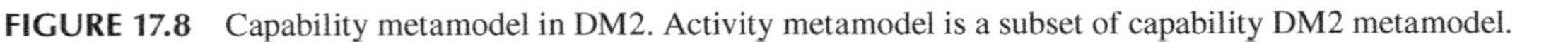
FIGURE 17.8 Capability metamodel in DM2. Activity metamodel is a subset of capability DM2 metamodel.

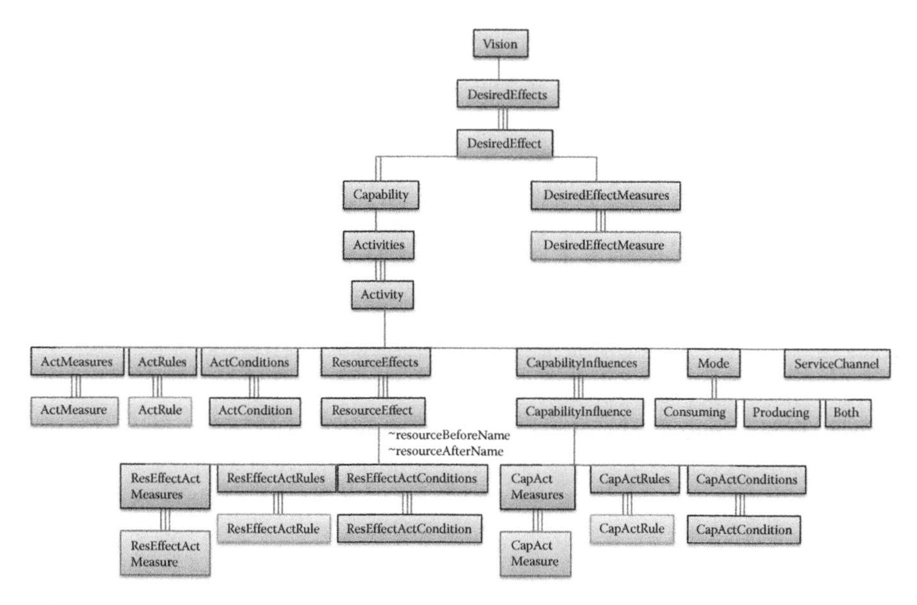

FIGURE 17.9 Capability SES based on the Capability DM2 metamodel in Figure 17.8.

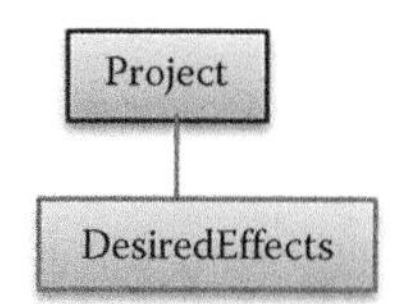

FIGURE 17.10 Project SES based on the Project DM2 description.

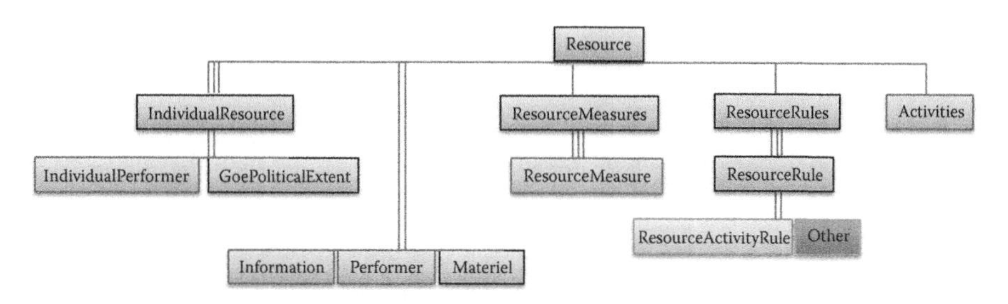

FIGURE 17.11 Resource SES based on the Resource flow DM2 description.

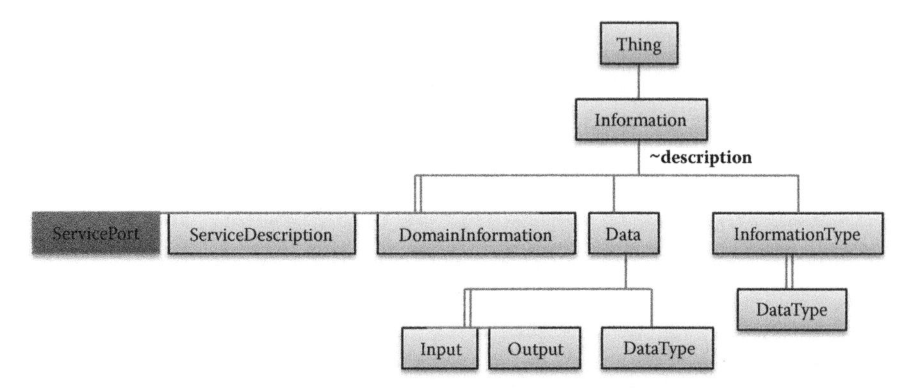

FIGURE 17.12 Thing SES based on Information and the Data DM2 description.

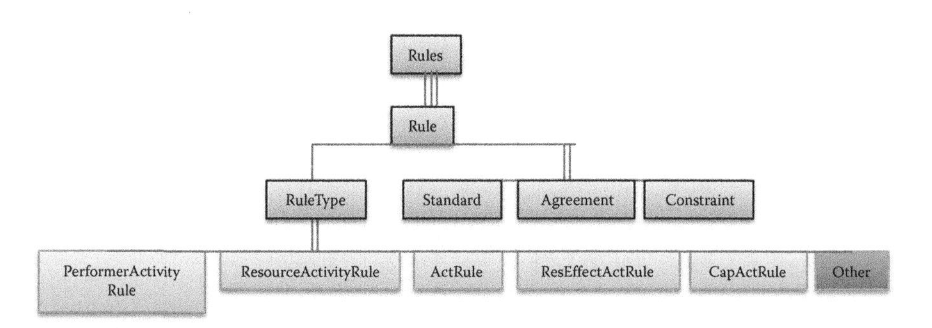

FIGURE 17.13 Rules SES based on the Rules DM2 description.

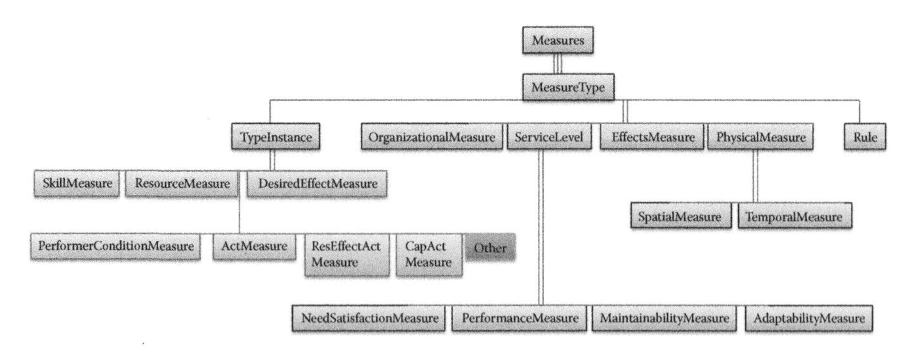

FIGURE 17.14 Measures SES based on the Measures DM2 description.

Exercise 17.4

Represent a Resource graphical SES into a computational SES. Refactor the earlier defined SESs.

Exercise 17.5

Represent Information and Data graphical SESs into computational SESs. Refactor the earlier defined SESs.

Exercise 17.6

Represent Rules and Measure graphical SESs into computational SESs. Refactor the earlier defined SESs.

17.4 DISCUSSION

We have provided an overview of the DoDAF 2.0 specification as given in the original DoDAF 2.0 specification document (U.S. Department of Defense, 2009) and analyzed it from an EA perspective. DoDAF 2.0 has totally redefined the architecture development approach with the latest version. The new approach has addressed many of our identified shortcomings in DoDAF 1.0 and is a step in the right direction. DoDAF 2.0 has also embraced the model-driven-engineering (MDE) paradigm by specifying various metamodels that provide semantic anchoring to various DoDAF domain concepts. Moving to data sharing and integration capabilities warrants interoperability at all three levels, namely, pragmatic, semantic, and syntactic. To achieve pragmatic interoperability, the kind of interoperability used in mission thread analyses, the interoperability at the lower two layers must exist. DoDAF 2.0 provides the foundation to achieve semantic and syntactic interoperability.

DoDAF and all its versions are requirements specification documents. The path to EA for advanced analytics and dynamic decision-making is not clear yet. However, the metamodeling approach certainly puts DoDAF in the area of model-driven analyses and the accompanying perspective that "everything is a model," rather than "everything is an object." The DM2 metamodel can be computationally represented in various ontologies, such as IDEAS and SES. We have illustrated how DoDAF domain concepts can be considered toward development of a DSL that is anchored to an M&S-based ontology such as the SES. Figure 17.15 describes the DSL use and how automated transformations can lead to an executable code.

Two types of analyses can be conducted on DoDAF-described models—static and dynamic analyses. In static analysis, the analysis is based on the data extracted from the architecture description. This kind of analysis can be either requirements-driven analysis or a historical data analysis for determining correlations and other trends. Dynamic analyses are based on the executable version of the architecture, typically in a simulation environment. The simulation is performed to gather predications

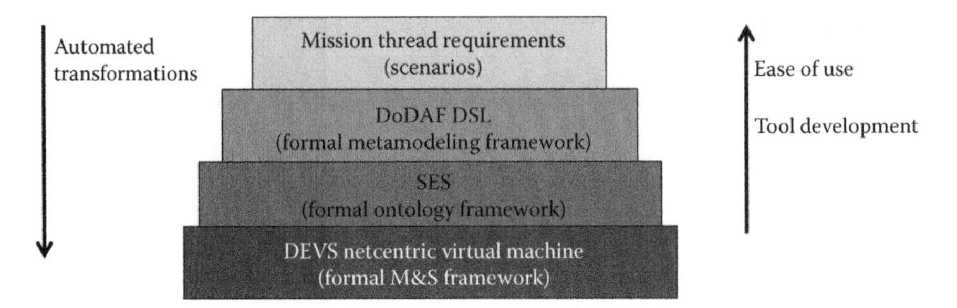

FIGURE 17.15 DoDAF-described models interfacing with the DEVS M&S infrastructure.

about the architecture performance and capabilities. Further, DoDAF suggests five principles of architectural analysis that may be applied to both the static and the dynamic analyses at different states of operations:

1. *Information consistency*: Data bounded and specified by metadata (e.g., schema). Semantic and syntactic interoperability is achieved when data and its derived forms are consistent through architecture description.
2. *Data completeness*: This refers to the availability of the required data in the description of a particular model that helps transformations to yield meaningful static and dynamic analyses.
3. *Transformation*: This refers to the ease with which data at syntactic levels are transformed to other required formats or possibly other toolsets, preserving semantics.
4. *Iteration*: The entire architecture development process should be iterative and agile.
5. *Lack of ambiguity*: This again refers to the semantic interoperability of a particular architecture description.

For a systems engineering perspective, the analysis is also done in two ways—structural and behavioral. The structural analysis is similar to the static analyses. The behavioral analysis is similar to the dynamic analysis in which simulation is used. While static and structural analyses entirely reside in the abstract modeling domain, the dynamic and behavioral analyses take the models to the simulation domain. This transition to an executable model warrants additional information such as time-based behavior because simulation is mostly time based (in DoDAF mission threads).

Figure 17.16 summarizes the DoDAF-suggested analyses with a systems M&S perspective. At the intersection of the axis, the static and structural analyses are in the modeling and metamodeling domain. Transformation to simulation infrastructure is fairly easy as there is not much dynamic behavior that needs to be analyzed. As more dynamism, conditional rules, and time based behavior are added, capabilities of metamodeling, transformation, and simulation framework are challenged to provide the needed analyses.

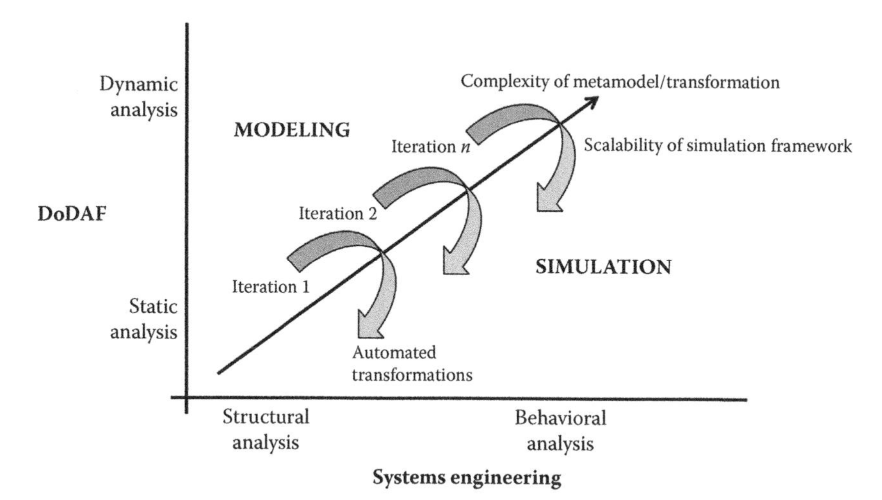

FIGURE 17.16 Modeling and simulation in DoDAF and systems engineering analysis.

REFERENCES

U.S. Department of Defense. (May 2009). *DoD Architecture Framework Version 2.0, Volume 1, Introduction, Overview, and Concepts; Manager's Guide*. Washington, DC.

U.S. Department of Defense. (May 2009). *DoD Architecture Framework Version 2.0: Volume 2, Architecture and Data Models, Architect's Guide*. Washington, DC.

U.S. Department of Defense. (May 2009). *DoD Architecture Framework 2.0: Volume 3, DoDAF Meta-model, Physical Exchange Specification*. Washington, DC.

Section IV

Case Studies

18 Joint Close Air Support *Designing from Informal Scenarios*

A Joint Mission Thread (JMT) is an operational and technical description of the end-to-end set of activities and systems that accomplishes the execution of a Joint mission (6212.01E, 2008). JMTs clarify joint operational and technical requirements to improve interoperability and integration. They provide an operational and technical context for objective joint analysis, assessment, testing, and training. JMTs establish common standards to verify operational and technical effectiveness into exchanges, thereby enabling improved cross-portfolio analysis of Joint and coalition capabilities (Behre, 2009). A JMT is developed in a three-tier process.

Tier 1: High-level generic data description with a totally reusable architecture-based information set. The information set is jointly provided and is specified in DoD-mandated architecture frameworks like DoDAF 2.0.

Tier 2: Various JMT strands are developed, wherein information represents specific documentation required to answer a particular question or satisfy a requirement or solve a problem. The strand is jointly integrated based on capabilities of the participated agencies resulting in refined DoDAF specifications.

Tier 3: This level provides the systems engineering level of detail to Tier 2 and contains enough rigor to inform Testing and Evaluation (T&E) and M&S communities. This step results in coordinated implementation and is jointly orchestrated. This step assumes that the DoDAF description lends itself to an executable architecture.

The Joint Close Air Support (JCAS) is a JMT that jointly operates between the ground forces, close air support from a fighter aircraft (U.S. Air Force, U.S. Navy, etc.), Joint Terminal Attack Controller (JTAC), mission headquarters, and many other cross-directory agencies and coalition partners. A typical JCAS scenario is shown in Figure 18.1 and is expressed in plain English below (Mittal, 2007; Zeigler, 2007).

JCAS JMT Operational Scenario #1

A. Special Operations Force (SOF) (Air Force Special Operations Command [AFSOC] and Naval Special Warfare [NSW]) JTAC working with Operational Detachment-Alpha is tasked to request immediate close air support on a stationary mechanized target in a mountainous terrain. A Predator unmanned aerial vehicle (UAV) is on station for support.
B. SOF JTAC contacts Airborne Warning and Control System (AWACS) with a request. AWACS passes the request to Special Operations Liaison Element (SOLE) in the Combine Air Operations Center (CAOC).

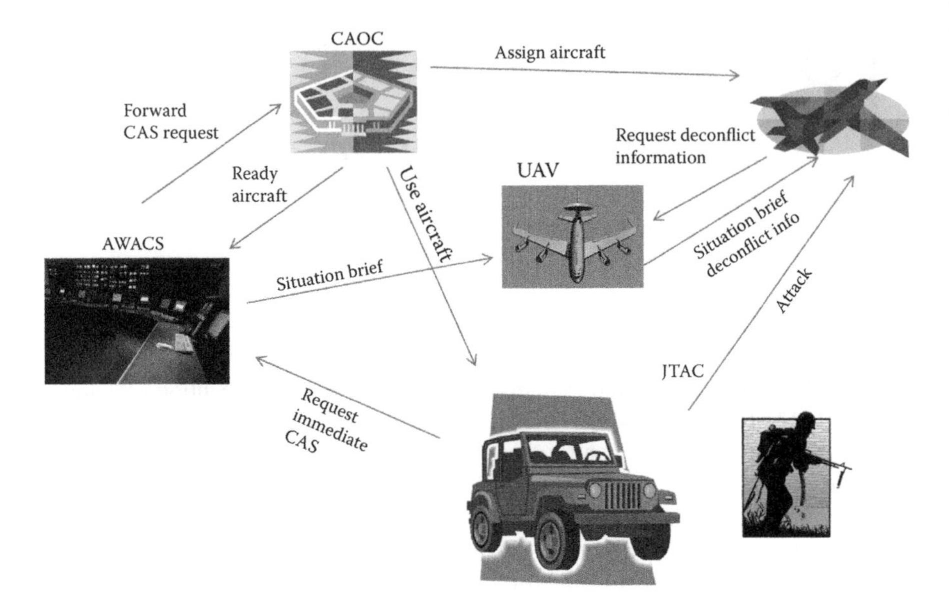

FIGURE 18.1 JCAS concept.

C. The Joint Special Operations Task Force (JSOFT) approves the request and the CAOC assigns a section of USMC F/A-18Ds, F-15Es, and a single B-1B. Ordnance consists of 20mm, Joint Direct Attack Munitions (JDAMs), and laser-guided bombs (LGBs).
D. The aircraft get situational brief from the AWACS aircraft while in route and then switch to SOF JTAC for terminal attack control and deconfliction from orbiting the UAV. A 9-Line brief will be given to each section/single aircraft. JTAC will continue to execute CAS missions until all weapons are expended.

Exercise 18.1

Develop a DoDAF description from the JCAS JMT Operational Scenario #1.

18.1 DESIGNING THE JCAS SYSTEM MODEL WITH DEVS SYSTEMS ENGINEERING APPROACH

Cleary, the requirements do not lend themselves to an executable model. The first task is to refine the requirements. Based on DEVS systems engineering, we need to know the structure and behavior of the components as well as the message exchanges. Developing the behavior requirements model as per the DEVS systems specification is based on the following steps:

1. *Identification*: What are the components and messages exchanged?
2. *Isolation*: How do you modularize components? What are the component interfaces?

3. *Assignment*: What is the I/O function? What are timing values? Is the information available?
4. *Abstraction*: What is the state space for each of the components? Are events clear?
5. *Refinement*: Does the component design match with reality and requirements?
6. *Communication*: Can a coupled model be created from components?

Each of these steps is analogous to a single iteration; after each step, the original requirement is eyeballed to ensure that the iteration makes sense and that any new information added or new assumptions made does not violate the behavioral specifications. Let us look at each of the steps in detail as we develop the JCAS JMT DEVS model. After the first five steps, it becomes very easy to put it into the DEVSML DSL.

18.1.1 Identification

The first step in analyzing such operational requirements is to figure out the entities (including messages), the atomic components exhibiting behavior, and the coupled components that exhibit communication exchanges through couplings. The first step is fairly easy, but it takes some design and analytical skills to figure out the behavior of each of the atomic components. Table 18.1 shows the identified components

TABLE 18.1
Identification of Components and Information Exchange

Operational Scenario Requirements	Identified Entities	Entity Type (Component/Message)
A	JTAC	Component
	Immediate CAS request	Message
	Target location	Message
	UAV	Component
B	AWACS	Component
	CAS request	Message
	CAOC	Component
	SOLE	Subcomponent of CAOC
C	Use aircraft	Message
	Assign USMC	Message
	USMC aircraft	Component
	Ordinance	Subcomponent of USMC aircraft
D	Situation brief	Message
	AWACS	Component
	Request TAC	Message
	Deconflict UAV	Message
	UAV	Component
	Execute CAS mission	Message
	Start attack	Message
	Cease attack	Message

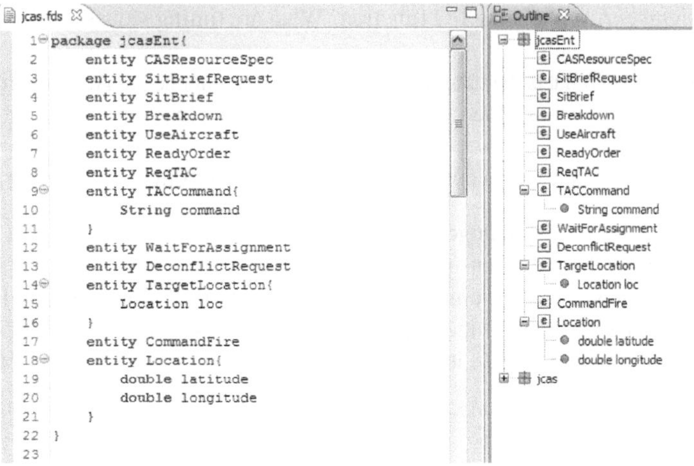

```
jcas.fds
1  package jcasEnt{
2      entity CASResourceSpec
3      entity SitBriefRequest
4      entity SitBrief
5      entity Breakdown
6      entity UseAircraft
7      entity ReadyOrder
8      entity ReqTAC
9      entity TACCommand{
10         String command
11     }
12     entity WaitForAssignment
13     entity DeconflictRequest
14     entity TargetLocation{
15         Location loc
16     }
17     entity CommandFire
18     entity Location{
19         double latitude
20         double longitude
21     }
22 }
23
```

```
Outline
jcasEnt
    CASResourceSpec
    SitBriefRequest
    SitBrief
    Breakdown
    UseAircraft
    ReadyOrder
    ReqTAC
    TACCommand
        String command
    WaitForAssignment
    DeconflictRequest
    TargetLocation
        Location loc
    CommandFire
    Location
        double latitude
        double longitude
jcas
```

FIGURE 18.2 Message entities for the JCAS model.

and messages. The analysis at this stage can very well be attempted using UML Sequence and Collaboration diagrams. Let us proceed with the DEVS design of the JCAS system.

So, the first cut gave us the following top-level components:

1. JTAC
2. UAV
3. CAOC
4. USMC aircraft
5. AWACS

The entities are ready to be put up in the DEVS modeling language constructs. Figure 18.2 shows the message entities encapsulated in a package, as they get refined in the iterations that ensue after further analysis of JCAS JMT Operational Scenario #1.

Exercise 18.2

Develop the UML sequence diagram based on information in Table 18.1.

18.1.2 Isolation

This step requires looking at each of the components in isolation and to modularize, basically to figure out what comes in and what goes out. This corresponds to Level 0 and Level 1 of the DEVS systems hierarchy, and we aim to find out the I/O pairs and ultimately the I/O mapping function. We start by observing the flow of messages first that will map directly to the external-event machine of the atomic components.

Just a reminder that a coupled system is a collection of atomic components and a behavior of a coupled DEVS is equivalent to an atomic model per the closure-under-coupling principle. Consequently, the five design steps mentioned earlier are applicable to atomic components specifically. For simplicity, we are taking the identified five components to be atomic and ignoring the subcomponent structure of the CAOC and USMC aircraft. Moving to the second iteration on analyzing JCAS Operational Scenario #1:

1. JTAC (from A, B, and D)
   a. Requests immediate CAS and waits for assignment
   b. Contacts AWACS with CAS request
   c. Receives approval to use aircraft from CAOC
   d. Sends TAC request on assignment of aircraft
   e. Sends TAC command to aircraft to prepare for attack
   f. Sends TAC command to aircraft to cease attack
2. AWACS (from B and D)
   a. Receives request to forward CAS ResourceSpec to CAOC
   b. Sends CAS request to CAOC
   c. Sends situational brief to USMC aircraft
3. UAV (from D)
   a. Sends target location on receiving deconflict request from JTAC
4. CAOC (from B and C)
   a. Allocates resources on receiving CAS request
   b. Sends use aircraft as approve request to JTAC and ready to USMC aircraft
5. USMC aircraft (from D)
   a. While flying, receives ready from CAOC
   b. While flying, receives situational brief from AWACS and starts en route
   c. Receives TAC command from JTAC to prepare for attack
   d. Receives target location from UAV on sending deconflict request
   e. Sends command to fire on receiving target location
   f. Receives situation brief from AWACS while preparing for attack
   g. Receives TAC command from JTAC to cease attack

Exercise 18.3

Refine the UML sequence diagram developed in Exercise 18.2.

18.1.3 Assignment

Once isolated, this step develops an I/O mapping table and the associated time advance between the input received and output generated. From this point onward, we are in the realm of assumptions and into the requirement elicitation process with the customer. It is at this step that many incomplete behavioral specifications will surface, and while we will make assumptions, it is strongly advised to get back to the customer or real-world data to figure out the timing values. This step generates Table 8.2.

Clearly, we have made some assumptions with respect to time relationships between the inputs and outputs, as shown in the last column of Table 18.2. For basic processing, we have assigned 1.0 time units. For example, the attack is conducted for the duration of 1000.0 time units and so forth. The objective is to be as reasonable as possible and refer to real-world data and subject matter experts (SMEs). This step enabled us to view an atomic component with defined input/output interfaces and their time-based trajectories.

TABLE 18.2
I/O Mapping Table for JCAS Atomic Components

Atomic Component	Input Message	Output Message	Time Advance/ Comments	Assumption
JTAC				
		CASResourceSpec	0.0	N
	UseAircraft			
	RequestTAC			
		TACCommand (initiateAttack)	10.0	Y
		TACCommand (ceaseAttack)	1000.0	Y
AWACS				
	CASResourceSpec	CASResourceSpec	1.0	Y
	SitBriefRequest	SitBrief	10.0	Y
UAV				
	DeconflictRequest	TargetLocation	1.0	Y
CAOC				
	CASResourceSpec	UseAircraft	1.0	Y
		ReadyOrder	1.0	Y
USMCaircraft				
	ReadyOrder			
	SitBrief		2000.0	Y
		RequestTAC	100.0	Y
	TACCommand (initiateAttack)	ReqDeconflict	0.0	
		SitBriefReq	0.0	N
	TargetLoc			
		Fire		
	TACCommand (ceaseAttack)			

18.1.4 Abstraction

Once we have identified all the message exchanges with respect to each of the components, the next step is the most crucial for identifying different states of the components. This is where the art of abstraction comes into play, and there is no set recipe to illustrate. What is critical is the sequence of events (input and output) and the associated time advances in each of the identified states within a component. Most of the time, the timing sequences provided as events are taken from a real-world scenario and it is easy to divide each state with the corresponding time advance. In our case, such timing is not provided, so we made reasonable assumptions. One other critical aspect worth mentioning is the situation where a component may be simultaneously receiving a message and, in the same instant, sending out a message. In that situation, it is important to separate these two events into separate events with a zero time state between the sequences. You can design any number of states to specify the behavior as long as the notion of an event is clear. For example, receiving a message qualifies as an external event, sending a message qualifies as an output event, and meaningful value change of a variable may qualify as an internal event. Such consideration will help build the state machine with an optimum number of states. The process begins with each identified event associated with a state, such that a DEVS state transition will either make the component receptive to the next input event or send an output at the expiry of time advance for the current state.

Table 18.3 adds more information and more assumptions. For example, if there is no correlation present for the sequence from generated output to receiving input, we make an assumption that after the generation of an output, the components anticipate an input and will wait indefinitely until that particular input is received. This can be seen in JTAC states, namely, waitForAssignment and waitForTACRequest, with a time advance of infinity. Once all the inputs and outputs have been mapped to specific states, it is good practice to add a closure state such as passive if the final state is not available from the behavioral specifications. In AWACS, the first I/O pair, CASResourceSpec :: CASResourceSpec, which translates to passing on the CASResourceSpec to CAOC, has been broken down into separate states, namely, *passive* and *passOnCASRequest*. This was necessary as the event sequence was first input and then output, and the input event without any prior correlation has to begin with time advance infinity. Consequently, the time advance for *passive* is infinity. UAV and CAOC are self-descriptive. In the USMC aircraft, the component continues to fly while it receives two inputs. Consequently, the state *flying* has time advance (infinity) as there is no prior correlation. If it is not enroute, then it turns enroute. During the *flying* phase, it receives either the situational brief *SitBrief* or *ReadyOrder* to request the TAC command. This kind of ambiguity will be best dealt with in the next refinement step. However, two separate states will help disambiguate the situation. Consequently, *headToTarget* and *requesTAC* are defined. The other very important transition in this component is the phase *prepForAttack* with a time advance of 100.0. As can be seen from Table 18.3, when the component is prepping for an attack, it receives the target location within the 100.0 units of time, and on

TABLE 18.3
Abstract States with Time Advances in JCAS Model Components

Atomic Component	Abstract State	Time Advance	Input Message	Output Message
JTAC				
	reqImmediateCAS	0.0		CASResourceSpec
	waitForAssignment	Infinity	UseAircraft	
	waitForTACRequest	Infinity	RequestTAC	
	provideTAC	10.0		TACCommand (initiateAttack)
	continueExec	1000.0		TACCommand (ceaseAttack)
	Passive	Infinity		
AWACS				
	Passive	Infinity	CASResourceSpec	
	passOnCASRequest	1.0		CASResourceSpec
	doSurveillance	Infinity	SitBriefRequest	
	provideSitBrief	10.0		SitBrief
	doSurveillance	Infinity		
UAV				
	Passive	Infinity	DeconflictRequest	
	provideDeconflict	1.0		TargetLocation
	Passive	Infinity		
CAOC				
	Passive	Infinity	CASResourceSpec	
	allocateResources	1.0		UseAircraft ReadyOrder
	Passive	Infinity		
USMC aircraft				
	Flying	Infinity	ReadyOrder	
	Flying	Infinity	SitBrief	
	headToTarget	2000.0		
	requestTAC	100.0		RequestTAC
	waitForTAC	Infinity	TACCommand (initiateAttack)	
	requestDeconflict	0.0		ReqDeconflict SitBriefReq
	prepForAttack	100.0	TargetLoc	
	Attack	0.0		Fire
	Firing	Infinity	TACCommand (ceaseAttack)	
	Flying	Infinity		

expiry of this period, if it is a valid message, it generates an output. Otherwise, it falls back to its previous state with the same time advance. Here we are also exploiting the notion of the fallback state in the internal event transition function as described in Chapter 5.

18.1.5 Refinement

Finally, there is the identification of state variables that hold information as the component goes about executing its behavior and reuse the information during its life cycle. Again, there is no recipe for variable identification, but a look at the requirements surely gives you an idea of what piece of information needs to be stored in the component's variable set so that as the component undergoes a state change, it can carry forward that information and use it for its advantage. For example, in the USMC aircraft, the component may check for validity of the target location (such as non-null) before beginning to fire. It waits for 100 time units to get a valid target location, failure of which results in defaulting on the attack with no firing. These kinds of checks and validations are entirely requirement-driven and form a part of the experimental frame design. Nevertheless, we have included these just for illustration purposes.

Another major component of this step is classifying the information in the previous step as init, deltext, reschedule, continue, deltint, and outfn so that they are easily put in the DEVSML DSL. The last column of Table 18.4 shows the classification.

Exercise 18.4

Design an experimental frame of the JCAS model.

18.1.6 Communication

This step provides information about the communicating components. Now that we know from our isolation step the interface boundary of each of the components, it is fairly straightforward to develop the connectivity chart (Table 18.5).

At this juncture of design, we have a basis of putting the information into DEVS DSL and further refinements are done with the help of the DEVSML editor. For example, code generation for the variables and logical expression are done in situ while DEVSML models are encoded.

Next, we show the designed component in the DEVSML text editor, its DEVS outline, and a pictorial representation of the state machine.

Exercise 18.5

Revise DoDAF OV-5, 6b,c with the new refined specifications and assumptions made during the DEVS design process.

TABLE 18.4
Classification to DEVS Elements

Atomic Component	Abstract State	Time Advance	Input Message	Output Message	DEVS
JTAC					
	reqImmediateCAS	0.0		CASResourceSpec	init + deltint + outfn
	waitForAssignment	Infinity	UseAircraft		deltext
	waitForTACRequest	Infinity	RequestTAC		deltext
	provideTAC	10.0		TACCommand (initiateAttack)	deltint + outfn
	continueExec	1000.0		TACCommand (ceaseAttack)	deltint + outfn
	Passive	Infinity			deltint
AWACS					
	Passive	Infinity	CASResourceSpec		init + deltext
	passOnCASRequest	1.0		CASResourceSpec	deltint + outfn
	doSurveillance	Infinity	SitBriefRequest		deltext
	provideSitBrief	10.0		SitBrief	deltint + outfn
	doSurveillance	Infinity			deltint
UAV					
	passive	infinity	DeconflictRequest		init + deltext
	provideDeconflict	1.0		TargetLocation	deltint + outfn
	Passive	Infinity			deltint

CAOC					
	Passive	Infinity	CASResourceSpec		init + deltext
	allocateResources	1.0		UseAircraft	deltint + outfn
	Passive	Infinity		ReadyOrder	deltint
USMC aircraft					
	Flying	Infinity	ReadyOrder		init + deltext
	Flying	Infinity	SitBrief		deltext
	headToTarget	2000.0			Deltint + reschedule
	requestTAC	100.0		RequestTAC	deltint + outfn
	waitForTAC	Infinity	TACCommand (initiateAttack)	SitBriefReq	deltext
	requestDeconflict	0.0		ReqDeconflict	deltint + outfn
	prepForAttack	100.0	TargetLoc	SitBriefReg	Deltext + continue
	Attack	0.0		Fire	deltint + outfn
	Firing	Infinity	TACCommand (ceaseAttack)		deltext
	Flying	Infinity			deltint

TABLE 18.5
Communication by Message Types

Source Component	Destination Component	Message Type
JTAC	USMC aircraft	TACCommand
JTAC	AWACS	CASResourceSpec
AWACS	CAOC	CASResourceSpec
AWACS	USMC aircraft	SitBrief
CAOC	JTAC	UseAircraft
CAOC	USMCaircraft	ReadyOrder
UAV	USMCaircraft	TargetLocation
USMC aircraft	JTAC	ReqTAC
USMC aircraft	AWACS	SitBriefReq
USMC aircraft	UAV	DeconflictReq
USMC aircraft	OUT	CommandFire

18.2 JCAS JMT MODEL IN DEVSML

This section presents the DEVS state machines and corresponding DEVSML specifications.

18.2.1 JTAC

Figure 18.3 shows the JTAC DEVS machine, and Figure 18.4 shows the DEVSML specifications.

18.2.2 AWACS

Figure 18.5 shows the AWACS DEVS machine, and Figure 18.6 shows the DEVSML specifications.

18.2.3 UAV

Figure 18.7 shows the UAV DEVS machine, and Figure 18.8 shows the DEVSML specifications.

18.2.4 CAOC

Figure 18.9 shows the CAOC DEVS machine, and Figure 18.10 shows the DEVSML specifications.

18.2.5 USMC Aircraft

Figure 18.11 shows the USMC Aircraft DEVS machine, and Figure 18.12 shows the DEVSML specifications. Figure 18.13 shows the DEVSML outline description.

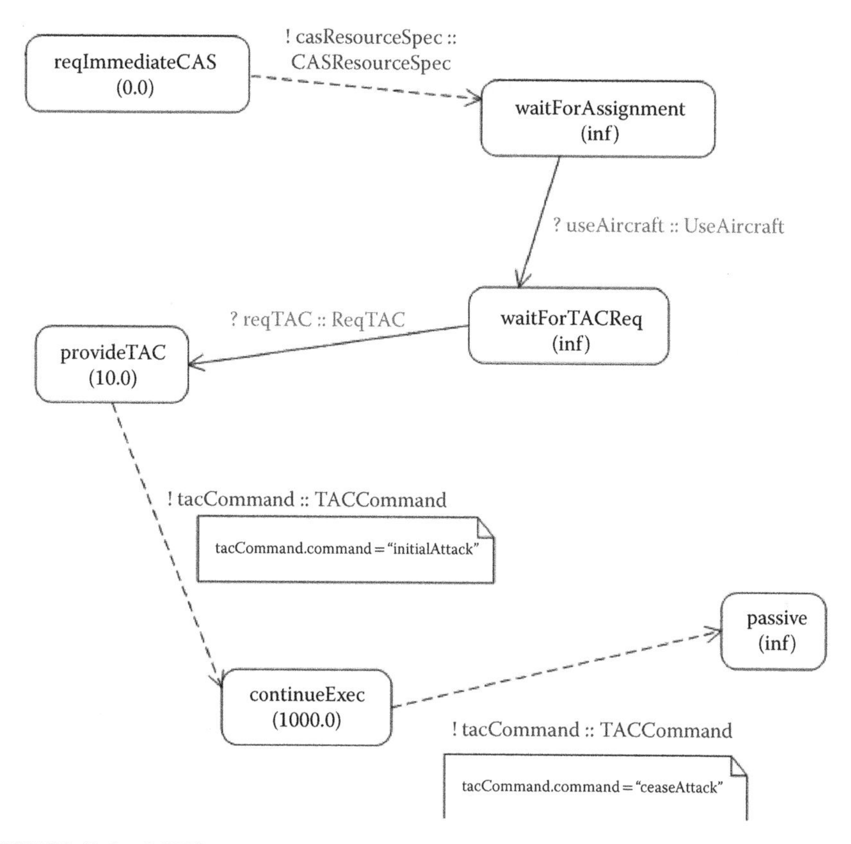

FIGURE 18.3 DEVS state machine for JTAC.

```
*jcas.fds
143 atomic JTAC{
144     vars{}
145     interfaceIO{
146         input ReqTAC reqTAC
147         input UseAircraft useAircraft
148         output CASResourceSpec casResourceSpec
149         output WaitForAssignment waitAssignment
150         output TACCommand tacCommand
151     }
152     state-time-advance{
153         reqImmediateCAS 0.0
154         waitForAssignment infinity
155         continueExec 1000.0
156         provideTAC 10.0
157         waitForTACReq infinity
158     }
159     state-machine{
160         start in reqImmediateCAS
161         deltint ( S: reqImmediateCAS) => S"? waitForAssignment
162         deltint ( S: provideTAC ) => S"? continueExec
163         deltext ( S: waitForTACReq, X: [ reqTAC ]) => S"? provideTAC
164         deltext ( S: waitForAssignment , X: [ useAircraft ]) => S"? waitForTACReq
165         outfn ( S: reqImmediateCAS) => Y: [ casResourceSpec ]
166         outfn ( S: provideTAC ) => Y: [ tacCommand ]{
167             "tacCommand.command = \"initialAttack\";"
168         }
169         outfn ( S: continueExec ) => Y: [ tacCommand ]{
170             "tacCommand.command = \"ceaseAttack\";"
171         }
172     }
173 }
174
175
176
177
```

```
Outline
jcasEnt
jcas
  jcasEnt.*
  JCASSystem
  USMCAircraft
  UAV
  JTAC
    Variables
    Inputs
      ReqTAC :: reqTAC
      UseAircraft :: useAircraft
    Outputs
      CASResourceSpec :: casResourceSpec
      WaitForAssignment :: waitAssignment
      TACCommand :: tacCommand
    State-Time-Advances
      reqImmediateCAS : 0.0
      waitForAssignment : Inf
      continueExec : 1000.0
      provideTAC : 10.0
      waitForTACReq : Inf
    DEVS State Machine
      start in : reqImmediateCAS
      Internal-Transitions
        reqImmediateCAS -> waitForAssignment
        provideTAC -> continueExec
      External-Transitions
        waitForTACReq [ reqTAC ] -> provideTAC
        waitForAssignment [ useAircraft ] -> waitForTACReq
      Output-Functions
        reqImmediateCAS-> [ casResourceSpec ]
        provideTAC-> [ tacCommand ]
        continueExec-> [ tacCommand ]
  CAOC
  AWACS
```

FIGURE 18.4 DEVSML specification for JTAC.

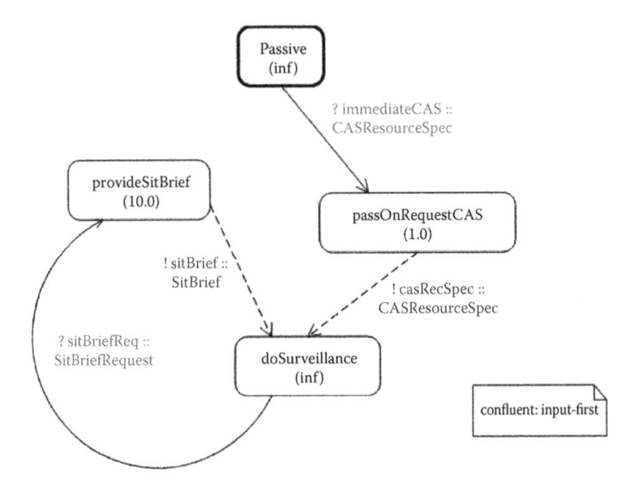

FIGURE 18.5 DEVS state machine for AWACS.

```
atomic AWACS{
    vars{}
    interfaceIO{
        input CASResourceSpec immideateCas
        output CASResourceSpec casRecSpec
        input SitBriefRequest sitBriefReq
        output SitBrief sitBrief
    }
    state-time-advance{
        passive infinity
        provideSitBrief 10.0
        doSurveillance infinity
        passOnRequestCAS 1.0
    }
    state-machine{
        start in passive
        deltint (S: passOnRequestCAS ) => S"? doSurveillance
        deltint (S: provideSitBrief ) => S"? doSurveillance
        deltext ( S: passive , X: [ immideateCas]) => S"? passOnRequestCAS
        deltext (S: doSurveillance, X: [ sitBriefReq ]) => S"? provideSitBrief
        outfn ( S: passOnRequestCAS ) => Y: [ casRecSpec ]
        outfn  ( S: provideSitBrief ) => Y: [ sitBrief]
        confluent input-first
    }
}
```

FIGURE 18.6 DEVSML specification for AWACS.

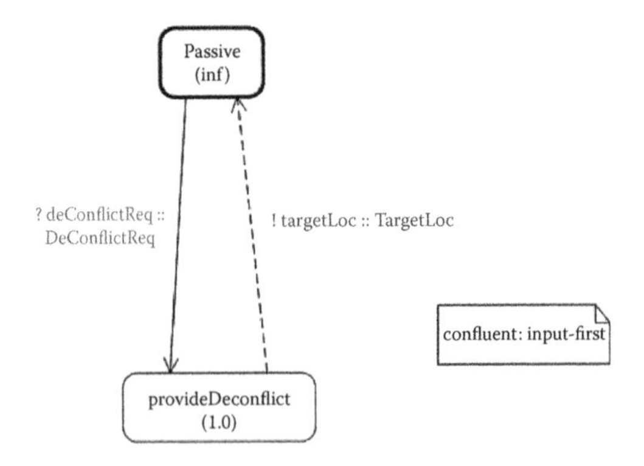

FIGURE 18.7 DEVS state machine for UAV.

```
*jcas.fds
131
132
133   atomic UAV{
134       vars{}
135       interfaceIO{
136           input DeconflictRequest deconflictReq
137           output TargetLocation targetLoc
138       }
139       state-time-advance{
140           provideDeconflict 1.0
141           passive infinity
142       }
143       state-machine {
144           start in  passive
145           deltext ( S: passive , X: [ deconflictReq ]) => S"? provideDeconflict
146           outfn ( S: provideDeconflict )  => Y: [ targetLoc ]
147           confluent input-first
148       }
149   }
150
151
152
153
154
155
```

```
Outline
jcasEnt
jcas
    jcasEnt.*
    JCASSystem
    USMCAircraft
    UAV
        Variables
        Inputs
            DeconflictRequest :: deconflictReq
        Outputs
            TargetLocation :: targetLoc
        State-Time-Advances
            provideDeconflict : 1.0
            passive : Inf
        DEVS State Machine
            start in : passive
            Internal-Transitions
            External-Transitions
                passive [ deconflictReq ] -> provideDeconflict
            Output-Functions
                provideDeconflict-> [ targetLoc ]
            input-first
    JTAC
    CAOC
    AWACS
```

FIGURE 18.8 DEVSML specification for UAV.

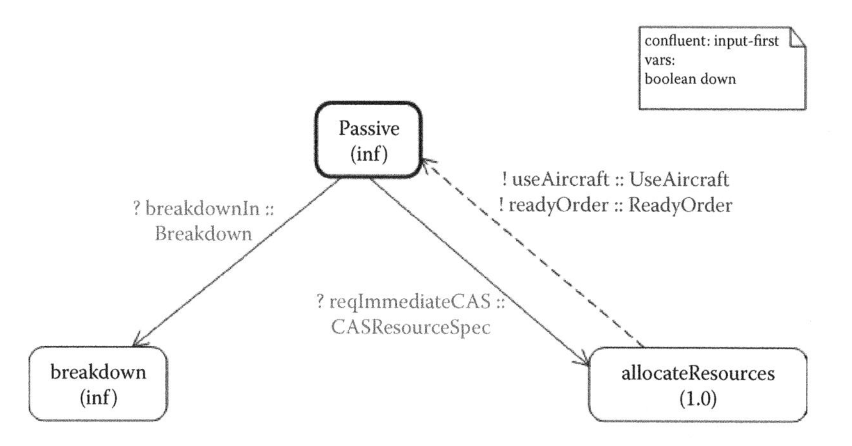

FIGURE 18.9 DEVS state machine for CAOC.

```
*jcas.fds
216
217
218
219   atomic CAOC{
220       vars{
221           boolean down
222       }
223       interfaceIO{
224           input CASResourceSpec reqImmidiateCas
225           input Breakdown breakdownIn
226           output UseAircraft useAircraft
227           output ReadyOrder readyOrder
228       }
229       state-time-advance{
230           passive infinity
231           breakdown infinity
232           allocateResources 1.0
233       }
234       state-machine{
235           start in  passive
236           deltext ( S: passive , X: [ breakdownIn]) => S"? breakdown
237           deltext ( S: passive , X: [ reqImmidiateCas]) => S"? allocateResources
238           deltint ( S: allocateResources ) => S"? passive
239           outfn ( S: allocateResources ) => Y: [ useAircraft readyOrder]
240           confluent input-first
241       }
242   }
243
244
245
```

```
Outline
jcasEnt
jcas
    jcasEnt.*
    JCASSystem
    USMCAircraft
    UAV
    JTAC
    CAOC
        Variables
            boolean down
        Inputs
            CASResourceSpec :: reqImmidiateCas
            Breakdown :: breakdownIn
        Outputs
            UseAircraft :: useAircraft
            ReadyOrder :: readyOrder
        State-Time-Advances
            passive : Inf
            breakdown : Inf
            allocateResources : 1.0
        DEVS State Machine
            start in : passive
            Internal-Transitions
                allocateResources -> passive
            External-Transitions
                passive [ breakdownIn ] -> breakdown
                passive [ reqImmidiateCas ] -> allocateResources
            Output-Functions
                allocateResources-> [ useAircraft readyOrder ]
            input-first
    AWACS
```

FIGURE 18.10 DEVSML specification for CAOC.

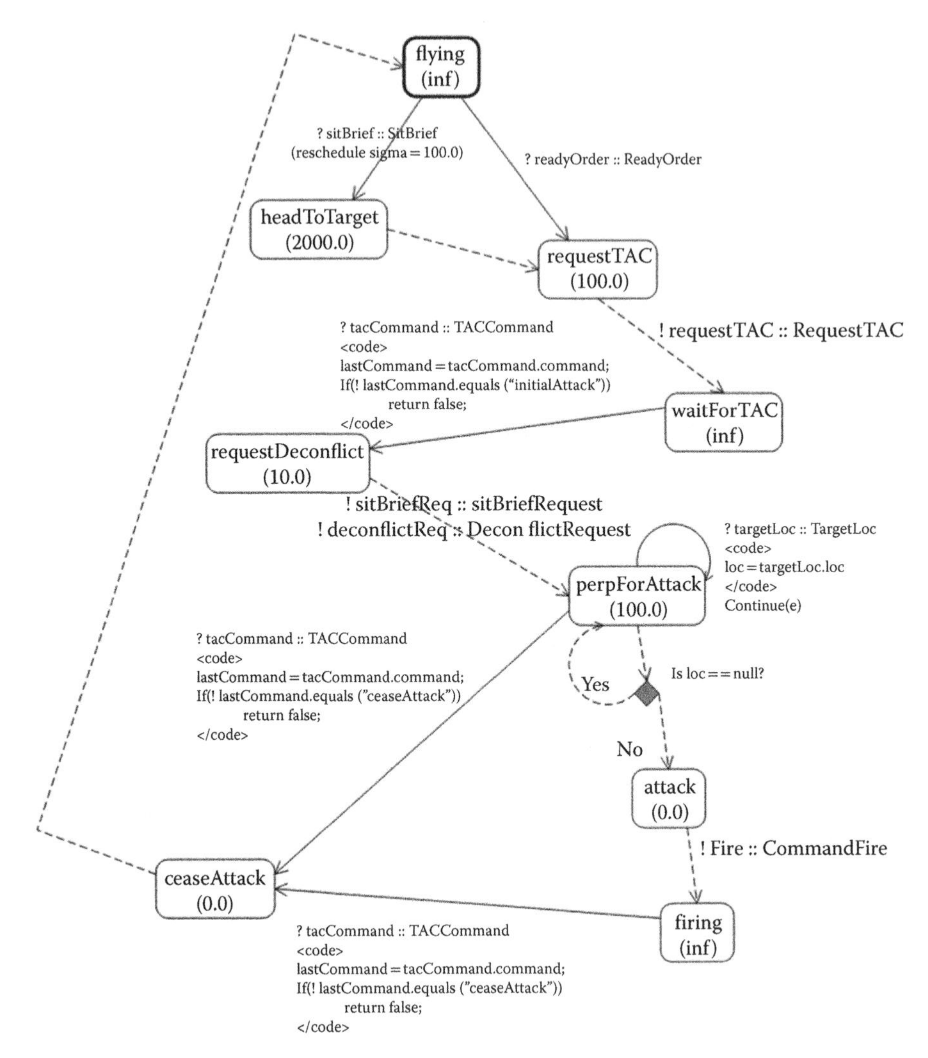

FIGURE 18.11 DEVS state machine for the USMC aircraft.

18.2.6 JCAS System Coupled Model

Having described the atomic components, we next define the couplings between the components so that the communication, as described in JCAS JMT Operational Scenario #1, can be realized. The coupled model in DEVSML is shown in Figure 18.14 and the automated code generation through the DEVSML editor leads it to the DEVSJAVA viewer (Figure 18.15). Please look at the SVN for the autogenerated DEVSJAVA code.

```
*jcas.fds
73
74  atomic USMCAircraft{
75      vars{
76          Location loc
77          String lastCommand
78      }
79      interfaceIO{
80          input SitBrief sitBrief          input ReadyOrder readyOrder          input TargetLocation targetLoc          input TACCommand tacCommand
81          output ReqTAC reqTAC             output SitBriefRequest sitBriefReq   output DeconflictRequest deconflictReq  output CommandFire fire
82      }
83      state-time-advance{
84          headToTarget 2000.0
85          requestTAC 100.0
86          waitForTAC infinity
87          requestDeconflict 10.0
88          attack 0.0
89          prepForAttack 100.0
90          flying infinity
91          ceaseAttack 0.0
92          firing infinity
93      }
94      state-machine{
95          start in  flying
96          deltint ( S: headToTarget ) => S"? requestTAC
97          deltint ( S: requestTAC) => S"? waitForTAC
98          deltint ( S: requestDeconflict) => S"? prepForAttack
99          deltint ( S: prepForAttack ) => S"? attack : prepForAttack{
100             "   if(loc == null) return false;   "}
101         deltint ( S: attack ) => S"? firing
102         deltint ( S: ceaseAttack )  => S"? flying
103
104         deltext ( S: flying , X: [readyOrder]) => S"? requestTAC
105         deltext ( S: flying , X: [ sitBrief ]) => S"? headToTarget reschedule ( sigma =  100.0)
106         deltext ( S: waitForTAC , X: [ tacCommand]) => S"? requestDeconflict{
107             "   lastCommand = tacCommand.get(0).command;    if(!lastCommand.equals(\"initialAttack\"))  return false;"  }
108         deltext ( S: prepForAttack , X: [ targetLoc]) => S"? prepForAttack  continue {
109             "loc = targetLoc.get(0).loc;
110             if (loc == null){
111                 loc = new Location(); //only for testing the state machine
112             }"}
113         deltext ( S: prepForAttack , X: [tacCommand ]) => S"? ceaseAttack {
114             " lastCommand = tacCommand.get(0).command;  if(!lastCommand.equals(\"ceaseAttack\")) return false;" }
115         deltext ( S: firing , X: [ tacCommand ]) => S"?ceaseAttack {
116             " lastCommand = tacCommand.get(0).command;
117             if(!lastCommand.equals(\"ceaseAttack\")) return false;  "}
118         outfn ( S: requestTAC) => Y: [ reqTAC ]
119         outfn ( S: requestDeconflict) => Y: [ sitBriefReq deconflictReq ]
120         outfn ( S: attack ) => Y: [ fire ]
121         confluent input-first
122     }
123 }
```

FIGURE 18.12 DEVSML specifications for the USMC aircraft.

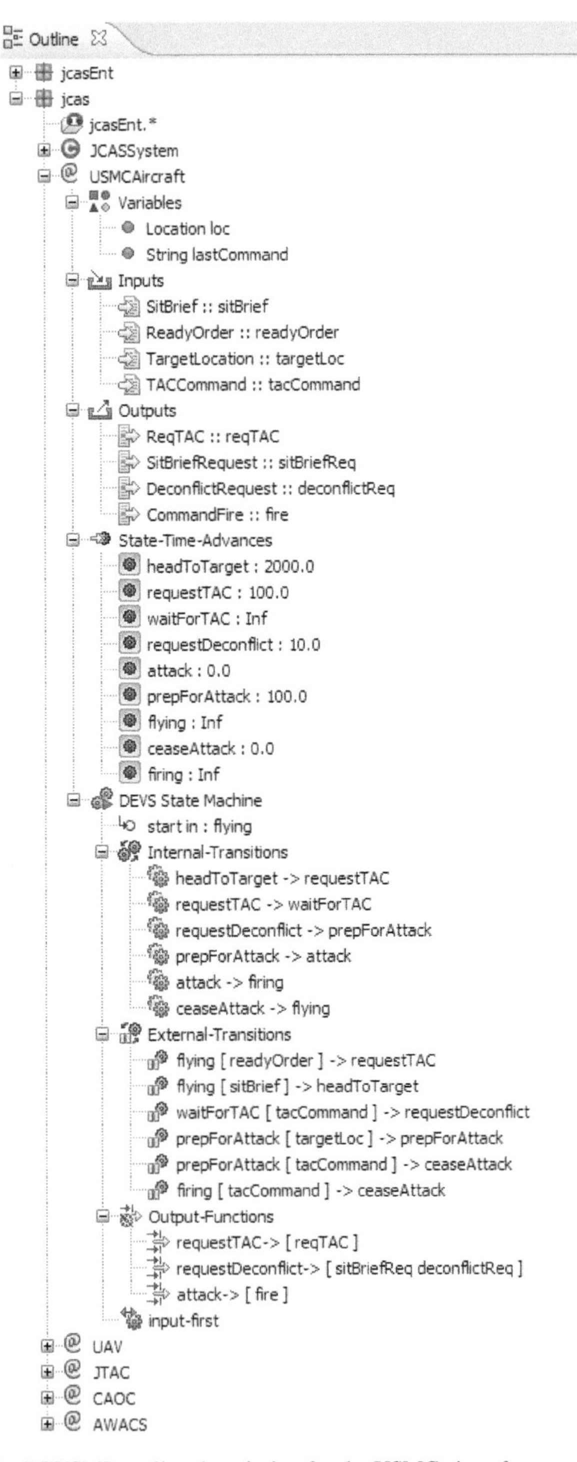

FIGURE 18.13 DEVSML outline description for the USMC aircraft.

```
package jcas{
    import jcasEnt.*

    coupled JCASSystem{
        models {
            atomic AWACS awacs
            atomic CAOC caoc
            atomic USMCAircraft usmcAir
            atomic JTAC jtac
            atomic UAV uav
        }
        interfaceIO{
            output CommandFire fire
        }
        couplings{
            ic usmcAir : reqTAC -> jtac : reqTAC
            ic usmcAir : sitBriefReq -> awacs : sitBriefReq
            ic jtac : casResourceSpec -> awacs : immideateCas
            ic awacs : casRecSpec -> caoc : reqImmidiateCas
            ic awacs : sitBrief -> usmcAir : sitBrief
            ic usmcAir : deconflictReq -> uav : deconflictReq
            ic caoc : useAircraft -> jtac : useAircraft
            ic caoc : readyOrder -> usmcAir : readyOrder
            ic uav : targetLoc -> usmcAir : targetLoc
            ic jtac : tacCommand -> usmcAir : tacCommand
            eoc usmcAir : fire -> this : fire
        }
    }
```

```
Outline
jcasEnt
jcas
  jcasEnt.*
  JCASSystem
    Components
      AWACS :: awacs
      CAOC :: caoc
      USMCAircraft :: usmcAir
      JTAC :: jtac
      UAV :: uav
    Inputs
    Outputs
      CommandFire :: fire
    Couplings
      EIC
      IC
        usmcAir.reqTAC -> jtac.reqTAC
        usmcAir.sitBriefReq -> awacs.sitBriefReq
        jtac.casResourceSpec -> awacs.immideateCas
        awacs.casRecSpec -> caoc.reqImmidiateCas
        awacs.sitBrief -> usmcAir.sitBrief
        usmcAir.deconflictReq -> uav.deconflictReq
        caoc.useAircraft -> jtac.useAircraft
        caoc.readyOrder -> usmcAir.readyOrder
        uav.targetLoc -> usmcAir.targetLoc
        jtac.tacCommand -> usmcAir.tacCommand
      EOC
        usmcAir.fire -> this.fire
  USMCAircraft
  UAV
  JTAC
  CAOC
  AWACS
```

FIGURE 18.14 DEVSML description for the JCAS system coupled model.

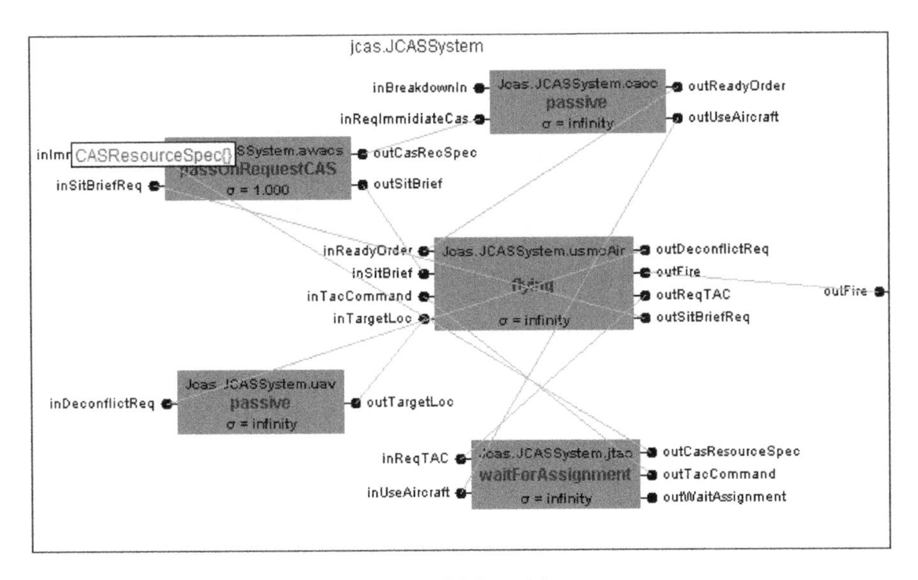

FIGURE 18.15 Coupled scenario for the JCAS model.

18.2.7 Execution of JCAS

The execution of the coupled model resulted in the simulation output (on console) of the successful message passing and scenario execution, as shown in Listing 18.1.

LISTING 18.1 SIMULATION LOG FOR JCAS SYSTEM MODEL

```
0.0    Jcas.JCASSystem.awacs      STATE:passive,SIGMA:
                                   infinity
0.0    Jcas.JCASSystem.uav        STATE:passive,SIGMA:
                                   infinity
0.0    Jcas.JCASSystem.jtac       STATE:reqImmediateCAS,
                                   SIGMA:0.000
0.0    Jcas.JCASSystem.usmcAir    STATE:flying,SIGMA:
                                   infinity
0.0    Jcas.JCASSystem.caoc       STATE:passive,SIGMA:
                                   infinity
0.0    Jcas.JCASSystem.jtac       OUTPUT:CASResourceSpec
0.0    Jcas.JCASSystem.awacs      INPUT:CASResourceSpec
0.0    Jcas.JCASSystem.awacs      STATE:passOnRequestCAS,
                                   SIGMA:1.000
0.0    Jcas.JCASSystem.jtac       STATE:waitForAssignment,
                                   SIGMA:infinity
1.0    Jcas.JCASSystem.awacs      OUTPUT:CASResourceSpec
1.0    Jcas.JCASSystem.awacs      STATE:doSurveillance,
                                   SIGMA:infinity
1.0    Jcas.JCASSystem.caoc       INPUT:CASResourceSpec
1.0    Jcas.JCASSystem.caoc       STATE:allocateResources,
                                   SIGMA:1.000
2.0    Jcas.JCASSystem.caoc       OUTPUT:UseAircraft
2.0    Jcas.JCASSystem.caoc       OUTPUT:ReadyOrder
2.0    Jcas.JCASSystem.jtac       INPUT:UseAircraft
2.0    Jcas.JCASSystem.jtac       STATE:waitForTACReq,
                                   SIGMA:infinity
2.0    Jcas.JCASSystem.usmcAir    INPUT:ReadyOrder
2.0    Jcas.JCASSystem.usmcAir    STATE:requestTAC,
                                   SIGMA:100.000
2.0    Jcas.JCASSystem.caoc       STATE:passive,
                                   SIGMA:infinity
102.0  Jcas.JCASSystem.usmcAir    OUTPUT:ReqTAC
102.0  Jcas.JCASSystem.jtac       INPUT:ReqTAC
102.0  Jcas.JCASSystem.jtac       STATE:provideTAC,
                                   SIGMA:10.000
102.0  Jcas.JCASSystem.usmcAir    STATE:waitForTAC,SIGMA:
                                   infinity
112.0  Jcas.JCASSystem.jtac       OUTPUT:TACCommand
112.0  Jcas.JCASSystem.jtac       STATE:continueExec,
                                   SIGMA:1000.000
112.0  Jcas.JCASSystem.usmcAir    INPUT:TACCommand
112.0  Jcas.JCASSystem.usmcAir    STATE:requestDeconflict,
                                   SIGMA:10.000
122.0  Jcas.JCASSystem.usmcAir    OUTPUT:SitBriefRequest
122.0  Jcas.JCASSystem.usmcAir    OUTPUT:DeconflictRequest
122.0  Jcas.JCASSystem.awacs      INPUT:SitBriefRequest
```

```
122.0 Jcas.JCASSystem.awacs       STATE:provideSitBrief,
                                   SIGMA:10.000
122.0 Jcas.JCASSystem.uav         INPUT:DeconflictRequest
122.0 Jcas.JCASSystem.uav         STATE:provideDeconflict,
                                   SIGMA:1.000
122.0 Jcas.JCASSystem.usmcAir     STATE:prepForAttack,
                                   SIGMA:100.000
123.0 Jcas.JCASSystem.uav         OUTPUT:TargetLocation
123.0 Jcas.JCASSystem.uav         STATE:passive,
                                   SIGMA:infinity
123.0 Jcas.JCASSystem.usmcAir     INPUT:TargetLocation
123.0 Jcas.JCASSystem.usmcAir     STATE:prepForAttack,
                                   SIGMA:98.000
132.0 Jcas.JCASSystem.awacs       OUTPUT:SitBrief
132.0 Jcas.JCASSystem.awacs       STATE:doSurveill`ance,
                                   SIGMA:infinity
132.0 Jcas.JCASSystem.usmcAir     STATE:prepForAttack,
                                   SIGMA:89.000
221.0 Jcas.JCASSystem.usmcAir     STATE:attack,SIGMA:0.000
221.0 Jcas.JCASSystem.usmcAir     OUTPUT:CommandFire
221.0 Jcas.JCASSystem.usmcAir     STATE:firing,
                                   SIGMA:infinity
1112.0 Jcas.JCASSystem.jtac       OUTPUT:TACCommand
1112 . 0 Jcas.JCASSystem.jtac     STATE:passive,SIGMA:infinity
1112.0 Jcas.JCASSystem.usmcAir    INPUT:TACCommand
1112.0 Jcas.JCASSystem.usmcAir    STATE:ceaseAttack,
                                   SIGMA:0.000
1112.0 Jcas.JCASSystem.usmcAir    STATE:flying,
                                   SIGMA:infinity
Terminated Normally before ITERATION 13, time: 1112.0
Terminated Normally before ITERATION 1, time: 1112.0
```

18.3 GENERATING THE TEST OBSERVER AGENTS FOR JCAS SYSTEM MODEL WITH DEVSML

We have seen in Chapters 5 and 6 the theory behind creating the DEVS observer agents from DEVS models. In the sections below, we will take one of the atomic components from the JCAS component set, namely, JTAC, and describe how to create an observer JTAC DEVS agent. Figure 18.16 shows two charts on a timeline. The top part shows the state transitions based on deltext and deltint specifications. The dotted line shows that if the transition had not happened, the state would continue indefinitely. Alternatively, a time-advance of infinity on the event of external transition makes a state change. The bottom part of Figure 18.16 shows the input and output messages corresponding to the input and output events exhibited by the JTAC component. As we have characterized them as events, they are correspondingly associated with a state change. Similarly, we can analyze other components, namely, CAOC, UAV, AWACS, and USMC aircraft.

Moving along further, now inverting the output events of JTAC to input events for the JTAC Observer, we get the input behavior of the JTAC Observer agent. The top part of Figure 18.17 shows the JTAC I/O behavior, and the bottom part shows the input behavior of the JTAC Observer agent. The unspecified transition times in the JTAC component have now been transformed to X, Y, and Z so that a meaningful observation can be made. Later in the sections ahead, we will relate these X, Y, and Z time advances to the verification and validation of the JTAC component. X, Y, and Z relate to the expected behavior of JTAC and may ultimately relate to the mission objectives or performance of JTAC.

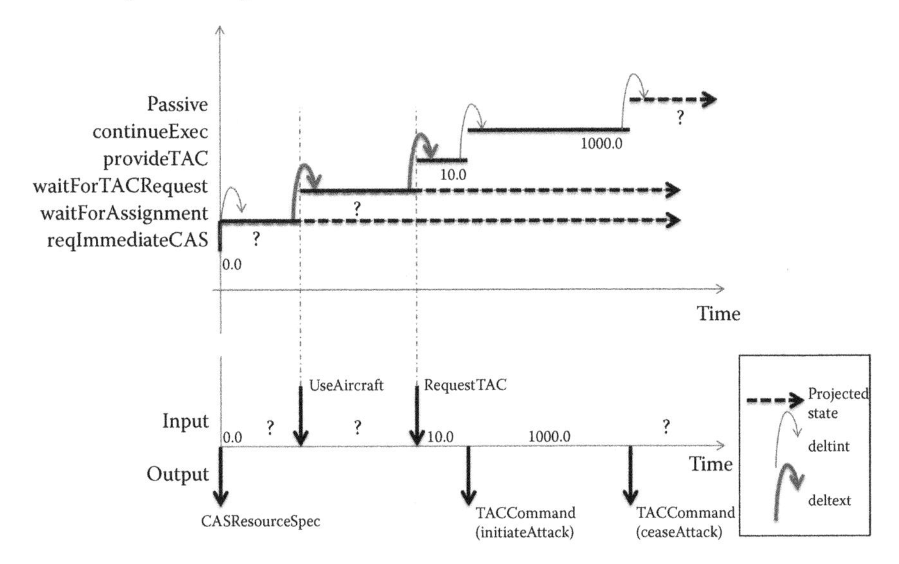

FIGURE 18.16 JTAC state-time-advance chart (top) correlated with I/O messages (bottom).

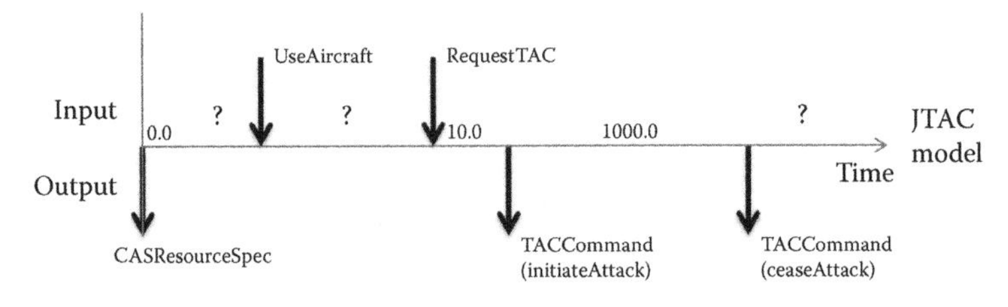

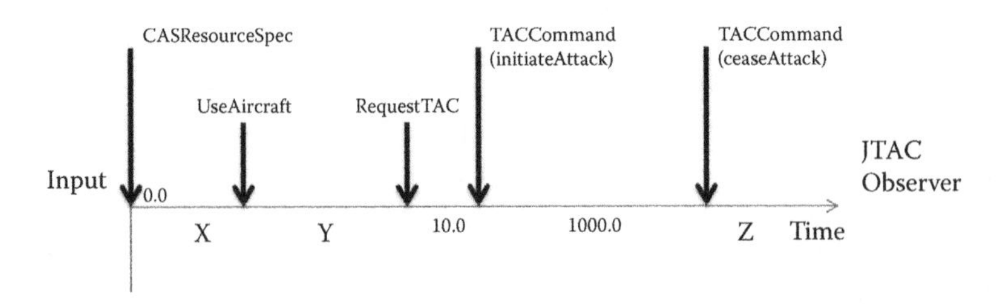

FIGURE 18.17 Inversion of JTAC output events to JTAC Observer input events.

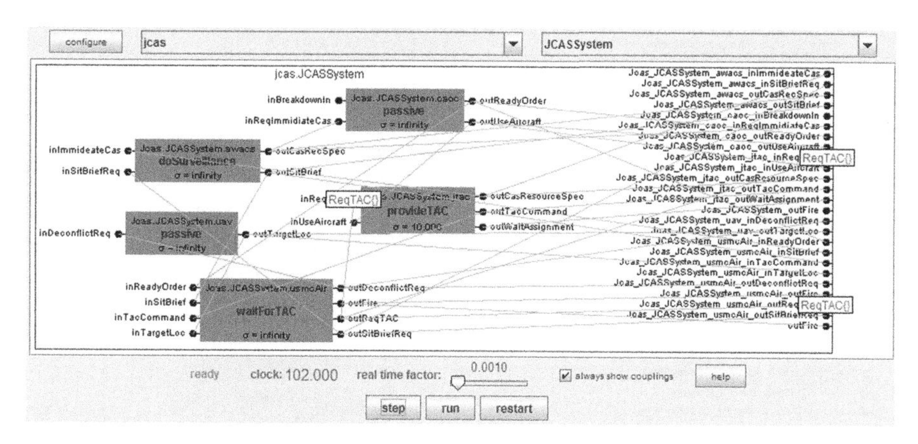

FIGURE 18.18 Outport autogeneration for the JCAS system model.

The next step is to ensure that the original model of JTAC sends both the received input and the generated output to the JTAC Observer agent. Consequently, new couplings have to be specified and autogenerated that take in the JTAC I/O couplings and transform them to the input couplings of the observer component. Figure 18.18 shows the final state of the JCAS system model after autogeneration of output ports that will couple with the respective observers at specific ports. For the JTAC observer model, following outports are added to the JCAS system model:

1. Jcas_JCASSystem_jtac_inCASResourceSpec
2. Jcas_JCASSystem_jtac_outInUserAircraft
3. Jcas_JCASSystem_jtac_outInRequestTAC
4. Jcas_JCASSystem_jtac_inTACCommand
5. Jcas_JCASSystem_jtac_inTACCommand

The port name is fully traceable down to the package name of the JTAC component so that the hierarchy of components with same message type is preserved. The autogeneration mechanism yield outport of the format *package_(subpackage*_) inMsg* for the JTAC input event and *package_(subpackage*_)outInMsg* for the JTAC output event. All the observer agents are generated in a separate subpackage. For example, the JCAS system model generated Java files in the *jcas* package. The observers are generated in the *jcas.observers* package.

Similarly, all the other observer agents are autogenerated for their DEVSML specifications. Figure 18.19 shows the observers for CAOC, AWACS, and so on. They are contained in a separate coupled model named as *jcas.JCASSystem.Observer.* By design, the JCAS system model is separated from the observer test suite so that it can be partitioned on a simulation platform for performance gains.

The input couplings of the observer coupled model are autogenerated in the same manner as laid out above. Figure 18.20 shows the JCAS system model coupled to the JCAS Observer Suite, exchanging messages. The coupling between these two coupled models is again autogenerated. The coupled models in Figure 18.20 are reduced to a black box hiding the internal components for efficient visualization. The federated model is generated in a separate package, *jcas.federations*. Summarizing, a single

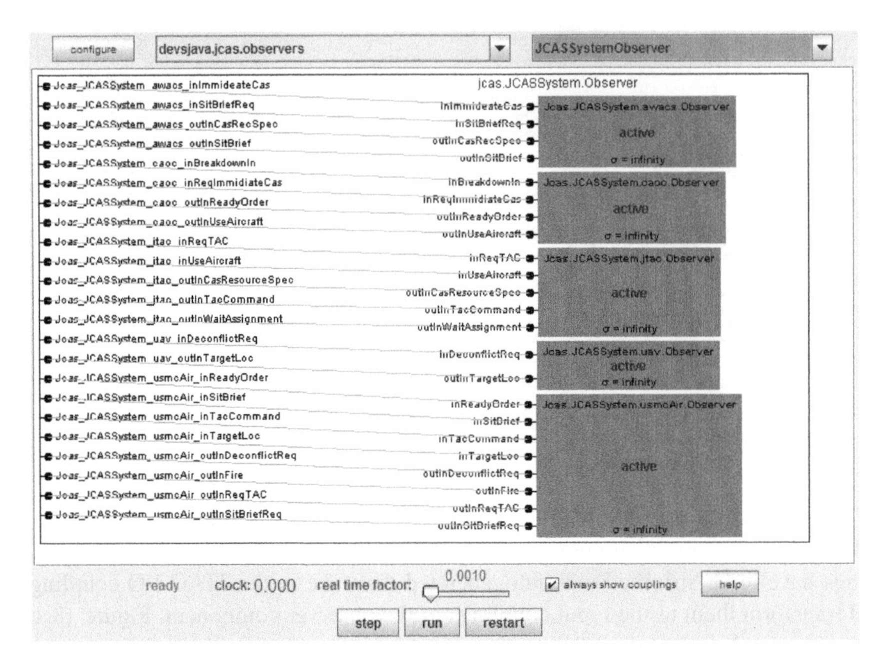

FIGURE 18.19 JCAS system observer test coupled model.

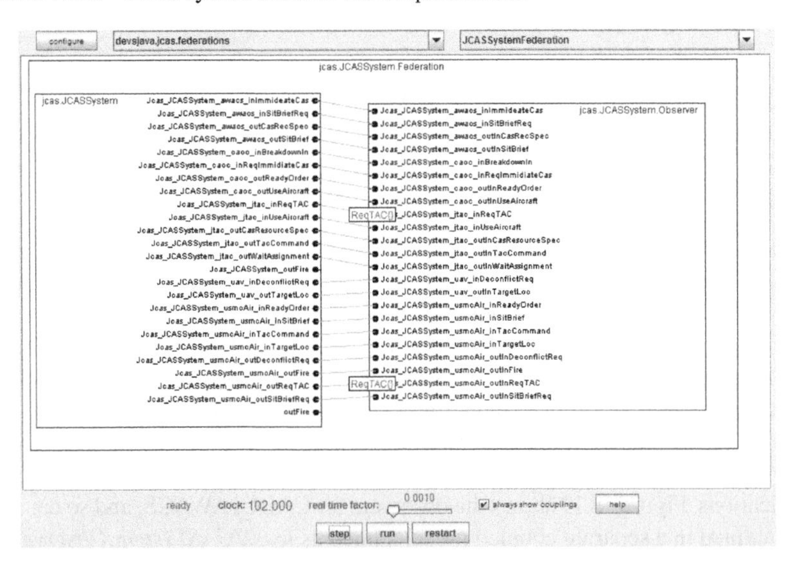

FIGURE 18.20 JCAS system federation containing the JCAS system model and the observer suite.

JCAS system DEVSML model gave us the DEVSJAVA JCAS system model and a JCAS Observer Suite that are simulation ready. Figure 18.21 shows the entire generated DEVSJAVA codebase on the left and the JCAS federation on the right.

The simulation log of *jcas.JCASSystem.Federation* that contains both the system model and the Observer Suite is shown in Listing 18.2. As can be seen, the simulation log yields the same result with no specification change. Each of the observers can now be independently analyzed and integrated with the experimental frames for further validation studies. Figure 18.22 shows the specific event log of the JTAC observer component as available in the DEVSJAVA viewer.

FIGURE 18.21 Generated DEVSJAVA files from the JCAS DEVSML specification. Code on the right shows the JCAS federation.

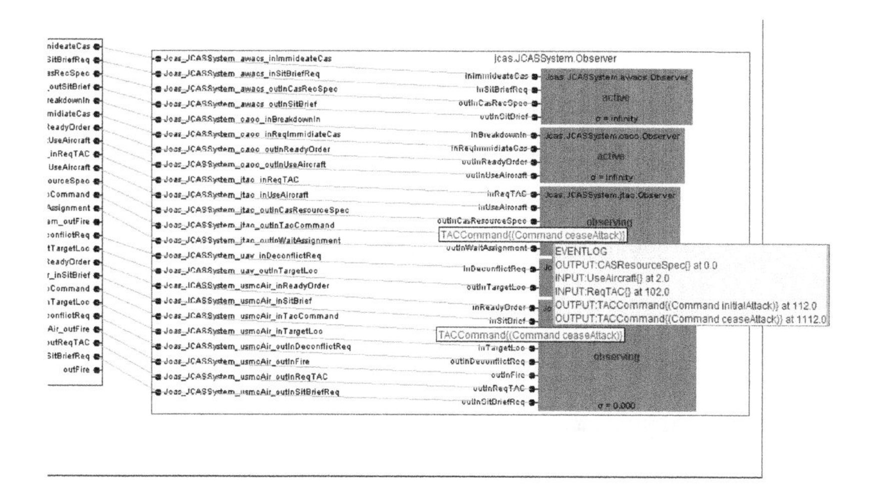

FIGURE 18.22 Event log for JTAC observer in an integrated simulation run.

LISTING 18.2 SIMULATION LOG FOR JCAS FEDERATION WITH SYSTEM MODEL AND TEST SUITE

```
0.0  Jcas.JCASSystem.usmcAir.Observer  STATE:active,
                                         SIGMA:infinity
0.0  Jcas.JCASSystem.jtac.Observer     STATE:active,
                                         SIGMA:infinity
0.0  Jcas.JCASSystem.uav.Observer      STATE:active,
                                         SIGMA:infinity
0.0  Jcas.JCASSystem.caoc.Observer     STATE:active,
                                         SIGMA:infinity
0.0  Jcas.JCASSystem.awacs.Observer    STATE:active,
                                         SIGMA:infinity
0.0  Jcas.JCASSystem.caoc              STATE:passive,
                                         SIGMA:infinity
0.0  Jcas.JCASSystem.uav               STATE:passive,
                                         SIGMA:infinity
0.0  Jcas.JCASSystem.jtac              STATE:
                                         reqImmediateCAS,
                                         SIGMA:0.000
0.0  Jcas.JCASSystem.usmcAir           STATE:flying,
                                         SIGMA:infinity
0.0  Jcas.JCASSystem.awacs             STATE:passive,
                                         SIGMA:infinity
0.0  Jcas.JCASSystem.jtac              OUTPUT:
                                         CASResourceSpec
0.0  Jcas.JCASSystem.jtac.Observer     INPUT:
                                         CASResourceSpec
0.0  Jcas.JCASSystem.awacs.Observer    INPUT:
                                         CASResourceSpec
0.0  Jcas.JCASSystem.jtac              STATE:
                                        waitForAssignment,
                                         SIGMA:infinity
0.0  Jcas.JCASSystem.awacs             INPUT:
                                         CASResourceSpec
0.0  Jcas.JCASSystem.awacs             STATE:
                                         passOnRequestCAS,
                                         SIGMA:1.000
0.0  Jcas.JCASSystem.jtac.Observer     STATE:active,
                                         SIGMA:infinity
0.0  Jcas.JCASSystem.awacs.Observer    STATE:active,
                                         SIGMA:infinity
1.0  Jcas.JCASSystem.awacs             OUTPUT:
                                         CASResourceSpec
1.0  Jcas.JCASSystem.caoc.Observer     INPUT:
                                         CASResourceSpec
1.0  Jcas.JCASSystem.awacs.Observer    INPUT:
                                         CASResourceSpec
1.0  Jcas.JCASSystem.caoc           INPUT:CASResourceSpec
```

```
1.0  Jcas.JCASSystem.caoc              STATE:allocateResources,
                                         SIGMA:1.000
1.0  Jcas.JCASSystem.awacs             STATE:doSurveillance,
                                         SIGMA:infinity
1.0  Jcas.JCASSystem.caoc.             STATE:active,
       Observer                          SIGMA:infinity
1.0  Jcas.JCASSystem.awacs.            STATE:active,
       Observer                          SIGMA:infinity
2.0  Jcas.JCASSystem.caoc              OUTPUT:UseAircraft
2.0  Jcas.JCASSystem.caoc              OUTPUT:ReadyOrder
2.0  Jcas.JCASSystem.usmcAir.          INPUT:ReadyOrder
       Observer
2.0  Jcas.JCASSystem.jtac.Observer INPUT:UseAircraft
2.0  Jcas.JCASSystem.caoc.Observer INPUT:UseAircraft
2.0  Jcas.JCASSystem.caoc.Observer INPUT:ReadyOrder
2.0  Jcas.JCASSystem.caoc              STATE:passive,
                                         SIGMA:infinity
2.0  Jcas.JCASSystem.jtac              INPUT:UseAircraft
2.0  Jcas.JCASSystem.jtac              STATE:waitForTACReq,
                                         SIGMA:infinity
2.0  Jcas.JCASSystem.usmcAir           INPUT:ReadyOrder
2.0  Jcas.JCASSystem.usmcAir           STATE:requestTAC,
                                         SIGMA:100.000
2.0  Jcas.JCASSystem.usmcAir.          STATE:active,
       Observer                          SIGMA:infinity
2.0  Jcas.JCASSystem.jtac.             STATE:active,
       Observer                          SIGMA:infinity
2.0  Jcas.JCASSystem.caoc.             STATE:active,
       Observer                          SIGMA:infinity
102.0  Jcas.JCASSystem.usmcAir         OUTPUT:ReqTAC
102.0  Jcas.JCASSystem.usmcAir.        INPUT:ReqTAC
         Observer
102.0  Jcas.JCASSystem.jtac.           INPUT:ReqTAC
         Observer
102.0  Jcas.JCASSystem.jtac            INPUT:ReqTAC
102.0  Jcas.JCASSystem.jtac            STATE:provideTAC,
                                         SIGMA:10.000
102.0  Jcas.JCASSystem.usmcAir         STATE:waitForTAC,
                                         SIGMA:infinity
102.0  Jcas.JCASSystem.usmcAir.        STATE:active,
         Observer                        SIGMA:infinity
102.0  Jcas.JCASSystem.jtac.           STATE:active,
         Observer                        SIGMA:infinity
112.0  Jcas.JCASSystem.jtac            OUTPUT:TACCommand
112.0  Jcas.JCASSystem.usmcAir.        INPUT:TACCommand
         Observer
112.0  Jcas.JCASSystem.jtac.           INPUT:TACCommand
         Observer
```

```
112.0  Jcas.JCASSystem.jtac          STATE:continueExec,
                                       SIGMA:1000.000
112.0  Jcas.JCASSystem.usmcAir       INPUT:TACCommand
112.0  Jcas.JCASSystem.usmcAir       STATE:requestDeconflict,
                                       SIGMA:10.000
112.0  Jcas.JCASSystem.usmcAir.      STATE:active,
         Observer                      SIGMA:infinity
112.0  Jcas.JCASSystem.jtac.         STATE:active,
         Observer                      SIGMA:infinity
122.0  Jcas.JCASSystem.usmcAir       OUTPUT:SitBriefRequest
122.0  Jcas.JCASSystem.usmcAir       OUTPUT:DeconflictRequest
122.0  Jcas.JCASSystem.usmcAir.      INPUT:SitBriefRequest
         Observer
122.0  Jcas.JCASSystem.usmcAir.      INPUT:DeconflictRequest
         Observer
122.0  Jcas.JCASSystem.uav.          INPUT:DeconflictRequest
         Observer
122.0  Jcas.JCASSystem.awacs.        INPUT:SitBriefRequest
         Observer
122.0  Jcas.JCASSystem.uav           INPUT:DeconflictRequest
122.0  Jcas.JCASSystem.uav           STATE:provideDeconflict,
                                       SIGMA:1.000
122.0  Jcas.JCASSystem.usmcAir       STATE:prepForAttack,
                                       SIGMA:100.000
122.0  Jcas.JCASSystem.awacs         INPUT:SitBriefRequest
122.0  Jcas.JCASSystem.awacs         STATE:provideSitBrief,
                                       SIGMA:10.000
122.0  Jcas.JCASSystem.usmcAir.      STATE:active,
         Observer                      SIGMA:infinity
122.0  Jcas.JCASSystem.uav.          STATE:active,
         Observer                      SIGMA:infinity
122.0  Jcas.JCASSystem.awacs.        STATE:active,
         Observer                      SIGMA:infinity
123.0  Jcas.JCASSystem.uav           OUTPUT:TargetLocation
123.0  Jcas.JCASSystem.usmcAir.      INPUT:TargetLocation
         Observer
123.0  Jcas.JCASSystem.uav.          INPUT:TargetLocation
         Observer
123.0  Jcas.JCASSystem. uav          STATE:passive,
                                       SIGMA:infinity
123.0  Jcas.JCASSystem.usmcAir       INPUT:TargetLocation
123.0  Jcas.JCASSystem.usmcAir       STATE:prepForAttack,
                                       SIGMA:98.000
123.0  Jcas.JCASSystem.usmcAir.      STATE:active,
         Observer                      SIGMA:infinity
123.0  Jcas.JCASSystem.uav.          STATE:active,
         Observer                      SIGMA:infinity
132.0  Jcas.JCASSystem.awacs         OUTPUT:SitBrief
```

```
132.0  Jcas.JCASSystem.usmcAir.      INPUT:SitBrief
          Observer
132.0  Jcas.JCASSystem.awacs.        INPUT:SitBrief
          Observer
132.0  Jcas.JCASSystem.usmcAir       STATE:prepForAttack,
                                       SIGMA:89.000
132.0  Jcas.JCASSystem.awacs         STATE:doSurveillance,
                                       SIGMA:infinity
132.0  Jcas.JCASSystem.usmcAir.      STATE:active,
          Observer                     SIGMA:infinity
132.0  Jcas.JCASSystem.awacs.        STATE:active,
          Observer                     SIGMA:infinity
221.0  Jcas.JCASSystem.usmcAir       STATE:attack,SIGMA:0.000
221.0  Jcas.JCASSystem.usmcAir       OUTPUT:CommandFire
221.0  Jcas.JCASSystem.usmcAir.      INPUT:CommandFire
          Observer
221.0  Jcas.JCASSystem.              STATE:firing,
          usmcAir                      SIGMA:infinity
221.0  Jcas.JCASSystem.usmcAir.      STATE:active,
          Observer                     SIGMA:infinity
1112.0  Jcas.JCASSystem.jtac         OUTPUT:TACCommand
1112.0  Jcas.JCASSystem.usmcAir.     INPUT:TACCommand
          Observer
1112.0  Jcas.JCASSystem.jtac.        INPUT:TACCommand
          Observer
1112.0  Jcas.JCASSystem.jtac         STATE:passive,
                                       SIGMA:infinity
1112.0  Jcas.JCASSystem.usmcAir      INPUT:TACCommand
1112.0  Jcas.JCASSystem.usmcAir      STATE:ceaseAttack,
                                       SIGMA:0.000
1112.0  Jcas.JCASSystem.usmcAir.     STATE:active,
          Observer                     SIGMA:infinity
1112.0  Jcas.JCASSystem.jtac.        STATE:active,
          Observer                     SIGMA:infinity
1112.0  Jcas.JCASSystem.usmcAir      STATE:flying,
                                       SIGMA:infinity
Terminated Normally before ITERATION 22,time: 1112.0
Terminated Normally before ITERATION 1,time: 1112.0
```

Exercise 18.6

Modify the UAV component as a Web Service and the associated DEVS component as a Web Service agent. Run the system locally while communicating with the UAV service. Describe the resulting architecture.

Exercise 18.7

Run the Experimental Frame developed in Exercise 18.4 with the system developed in Exercise 18.6.

Exercise 18.8

Run the system model in Exercise 18.7 in the DEVS netcentric environment. Describe the resulting architecture.

Exercise 18.9

Run the JCAS model federation in the DEVS netcentric environment with the model and the Test Suite on two different servers. Describe the architecture.

Exercise 18.10

Refine the requirements and run sample scenarios through the Experimental Frame and validate the requirements with system's effectiveness.

18.4 SUMMARY

This chapter has discussed JCAS and JMT sample scenarios with DEVS systems engineering principles. The process highlighted gaps in informal requirement specifications and encouraged formal requirements elicitation with the customer. We have shown how requirements can be refined and translated to a DEVS executable model along with the system test suite. The reader is encouraged to attempt the exercises given in this chapter.

REFERENCES

6212.01E, C. (2008). *Interoperability and Supportability of Information Technology and National Security Systems*. Chairman of the Joint Chiefs of Staff Instruction.

Behre, C. (2009). Applied Joint Mission Threads. *DoD Enterprise Architecture Conference*. St. Louis, MO.

Mittal, S. (2007). *DEVS Unified Process for Integrated Development and Testing on Service Oriented Architectures* (PhD dissertation). Tucson, AZ: University of Arizona.

Zeigler, B. (2007). *Standards Conformance Testing and M&S Web Service*. Arizona Center for Integrative Modeling and Simulation. Retrieved from http://acims.arizona.edu/PUBLICATIONS/Presentations/StandardsConformance.ppt.

19 DEVS Simulation Framework for Multiple Unmanned Aerial Vehicles in Realistic Scenarios

19.1 INTRODUCTION

This chapter presents an M&S framework for multiple Unmanned Aerial Vehicles (UAVs) based on DEVS framework for realistic military scenarios. To configure these scenarios, the M&S framework includes models for UAVs and their trajectories, terrain, radars, missiles, and prohibited flying zones. The results obtained by an evolutionary path planner act as a source to this framework. To provide better accuracy, the M&S framework utilizes more complex models than the path planner for all the aforementioned elements. The M&S environment works offline and online, by invoking the path planner in real time to recalculate parts of the original path, to avoid unexpected risks while the UAVs are flying. Since all the models demand intensive computing power, the model was parallelized using DEVS/SOA, a netcentric M&S framework. As described in previous chapters, DEVS is a formalism based on concepts derived from dynamic systems theory. DEVS/SOA is a cross-platform M&S framework that automatically deploys DEVS models using web services through an SOA. We performed both serial and distributed experiments in four different scenarios, which demonstrate that this framework can be used for real-flight missions.

UAVs are unpiloted aircrafts that can be controlled remotely or fly autonomously based on preprogrammed flight routes. They are used in a wide variety of fields, both civil and military, such as surveillance, reconnaissance, geophysical survey, environmental and meteorological monitoring, aerial photography, and search-and-rescue tasks (Stevens & Lewis, 2003).

In military missions, they work in dangerous environments, where it is vital to fly along routes that keep the UAVs away from any type of threat and prohibited zone. The threats to avoid are Air Defense Units (ADUs), which consist of detection radars to discover the UAVs. ADUs are sets of tracking radars that follow UAV trajectories and are capable of deploying a set of missiles to destroy them. The prohibited zones, also known as Non-Flying Zones (NFZs), are certain regions that the UAVs cannot visit due to mission restrictions.

The best routes for the UAVs are those that minimize the risk of destruction of each UAV and optimize some planning criteria (such as flying time and path length) while fulfilling all the physical constraints of the UAVs and their environment, plus the restrictions imposed by the selected mission (such as forcing the UAVs to visit some points of the map). Besada-Portas et al. (2010) presented a novel path planner for these kinds of problems based on evolutionary computation. This planner obtains an optimal 3D route for each UAV, codifying it as the natural cubic spline curve defined by the list of 3D points, called Way Points (WPs).

To evaluate the quality of the planner before using it in real missions, we validated the routes (and so the lists of returned WPs) in multiple experiments against a simulator that contains models for all the system requirements.

In this system, those elements are the lists of WPs, the UAVs, the radars, and the missiles, as well as the terrain, the online planners, and the controllers coupled with the UAVs that are responsible for translating the WPs in maneuverability instructions. The models of the radars and missiles are nondeterministic, incorporating stochastic behaviors related with the probability of detection and destruction of the UAVs.

To manage stochasticity, as well as the one inherent in the evolutionary planner in the experiments presented in this chapter, the seeds of the random generators were fixed at the beginning of each simulation.

Figure 19.1 depicts one solution obtained by the path planner for one given mission. The path planner assumed that all the ADU positions are known. In Figure 19.1, the initial and final positions of each UAV are represented as initial and final crosses. The intermediate crosses represent points that the UAVs are forced to visit. The ADUs,

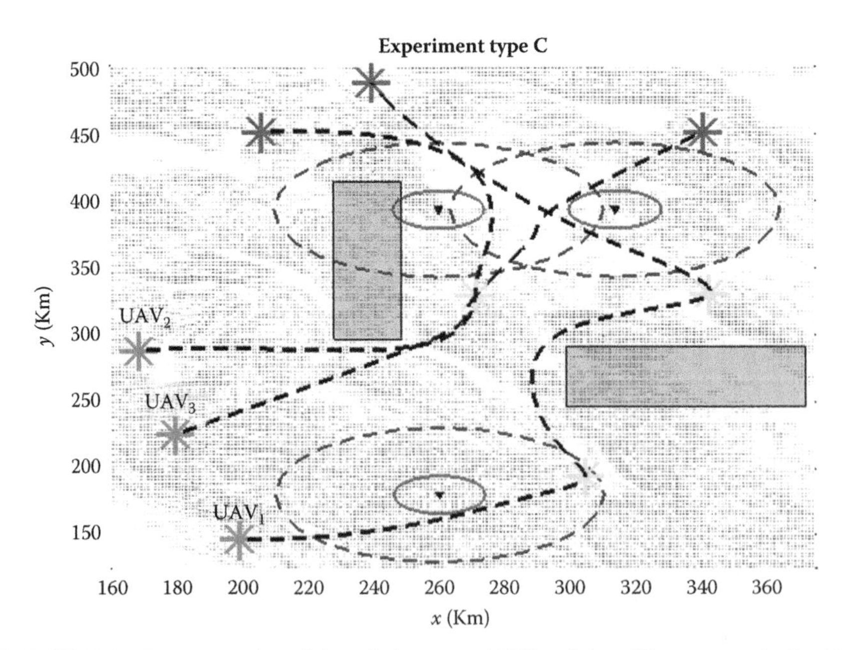

FIGURE 19.1 Representation of the mission named "C" and the offline routes obtained by the path planner. It is supposed that the ADU positions are known.

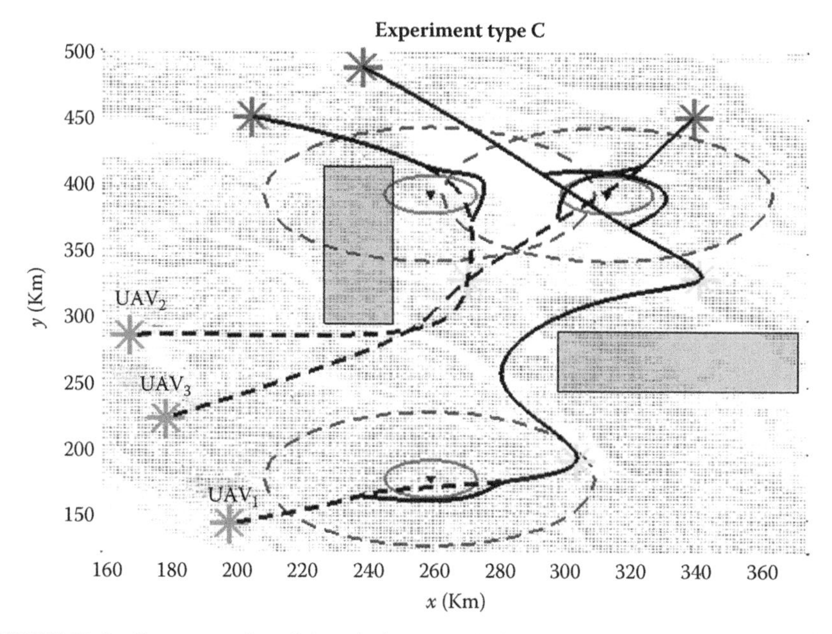

FIGURE 19.2 Representation of the mission named "C" and the offline routes (dashed path) obtained by the path planner. It is supposed that the ADU positions are unknown, and the online routes are obtained online by the simulator when a UAV detects an ADU.

which are initially known, are represented by the big external dashed circles (which show the maximum distance of detection of their radars) and by the small solid circles (which enclose the zones where the probability of destroying an UAV can be greater than 0). The NFZs are represented with the rectangular areas.

The original path obtained offline by the planner is not always valid in dynamic environments where the positions of all the threats are not known beforehand. Therefore, the planner is designed to also work online once the UAVs are flying in order to propose an alternative path during the mission of an UAV when a pop-up (unknown ADU) appears and is detected by the embedded radar in the UAV.

Figure 19.2 illustrates this point. Initially, the path planner does not know the ADU positions. Thus, it obtains shortest paths avoiding just NFZs. During the simulation run, when a UAV detects an ADU, it calls the path planner again to obtain new routes (solid paths), thereby avoiding new risks.

In the following sections, we first describe our DEVS model. Next, we describe some experiments, map the original DEVS model into a DEVS/SOA distributed architecture, and evaluate the performance of the framework.

19.2 DEVS MODEL

For this problem, each example is constructed upon multiple DEVS atomic components that exemplify UAV dynamics and behavior, together with multiple DEVS coupled components that characterize the line of action of an ADU (see Figure 19.3).

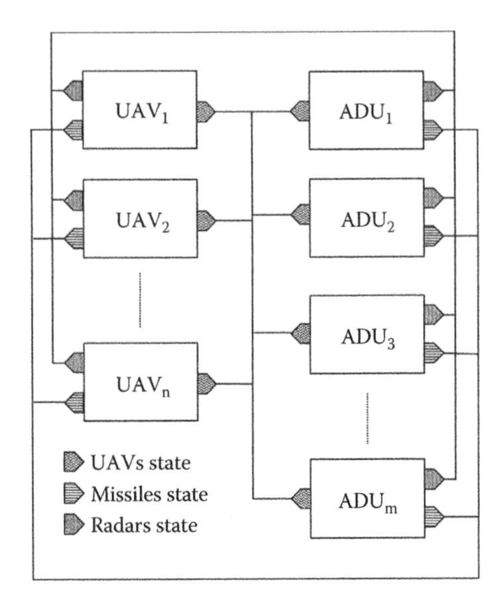

FIGURE 19.3 Root coupled model formed by *n* UAVs and *m* ADUs.

Each ADU is composed of various heterogeneous DEVS atomic components: detection radar, several tracking radars, and a certain number of missiles. Wiring rules are depicted in Listing 19.1.

LISTING 19.1 COUPLING PSEUDO CODE

```
For each uav in UAVs
  For each adu in ADUs
    Coupling (UAV State) from uav out(0) to adu in(0);
    Coupling (Radar State) from adu out(0) to uav in(0);
    Coupling (Missile State) from adu out(1) to uav in(1);
```

We begin our description with the coupling type, the structure, the behavior, and the coupled layout of each model.

19.2.1 Couplings

Couplings are directed edges that link output ports to input ports, defined by a type that bounds the data exchanged between models. The DEVS model described in this document uses up to four different coupling types (see Listing 19.2).

1. *UAV state*: UAV data type that encapsulates the necessary information to know *who*, *when*, *where*, and *how*
2. *Missile state*: Carries the missile information as an UAV, current missile *phase*, and a target identifier

3. *Radar tracking state*: Covers the tracking radar identifier, the ADU where it belongs, and the data of the UAV to track
4. *Lost target*: The identifier of the UAV lost during the tracking process

LISTING 19.2 COUPLING TYPES

```
UavState {
  String id;                      // identifier
  Time t;                         // Computed in time
  Double X,Y,Z;                   // X,Y and Z coordinates
  Double Vx,Vy,Vz;                // X,Y and Z velocities
  Double theta,phi,psi;           // roll, pitch, yaw
  Double Vtheta, Vphi, Vpsi;      // roll, pitch, yaw
                                     velocities
}

MissileState {
  UavState uav;                   // unmanned aerial vehicle
  String phase;                   // phase
  String target;                  // target identifier
}

RadarTrackingState {
  String id;                      // identifier
  String adu;                     // air defense unit
  UavState uav;                   // Unmanned aerial vehicle
}

String lostTarget;                // identifier of lost target
```

19.2.2 UAVs

UAVs are represented as models assembled with two input ports that receive states of tracking radars and missiles of each ADU correspondingly. One output port on gathering the computed time, identifier, position, orientation, and velocities sends its state out. Additionally, a UAV keeps an internal state variable with an array of these UAV states reflecting the UAV dynamics. Whenever a UAV realizes that it has been detected by an ADU, if possible, it starts an evasive maneuver to escape from the ADU firepower range and prevent being shot down.

Basically, the UAV model works in the following way. Every time the internal time event function is triggered (simulation time is equal to sigma), the next necessary collection of states is computed, unless this collection is not empty or the UAV has reached the end of the trajectory. These states store intermediate values of position, orientation, and velocity that describe the UAV's movement across the current coordinate to the next trajectory point. Then, sigma (time of next internal time event) is updated to the next computed time or set to infinity only if the UAV reached the end of the assign path. Whenever the external transition is executed (received an input), the UAV verifies if any radar is tracking its path and whether the distance from a missile aimed at overthrowing it is less than the established minimum. If the former case is positive, the UAV attempts to escape through an intersecting trajectory to flee away from the corresponding ADU and afterward updates sigma. If the latter is positive and according to a certain probability of destruction, the UAV is destroyed and sigma is set to infinity. On every occasion that the output function is activated, the current UAV state is sent thought the output port. Figure 19.4 and Table 19.1 depict this behavior.

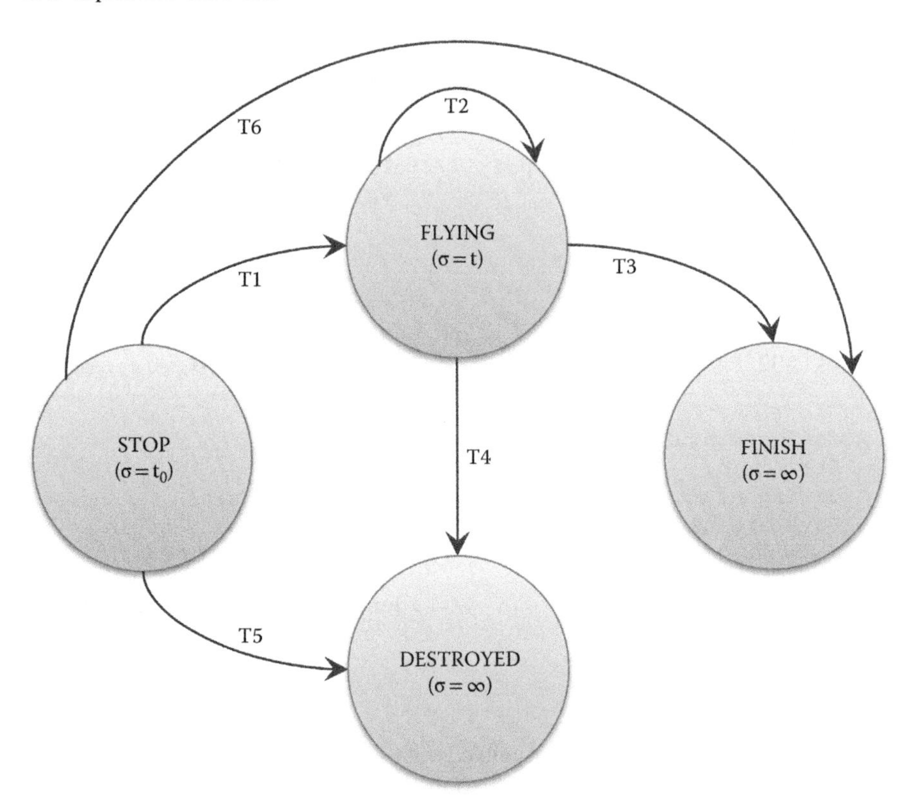

FIGURE 19.4 UAV state diagram.

TABLE 19.1
UAV States Transition

	S_k	S_{k+1}	δ_{int}	δ_{ext}
T1	STOP	FLYING	States is NOT empty.	
T2	FLYING	FLYING	States is NOT empty.	Dist. To missile is > MIN
T3	FLYING	FINISH	States is empty.	
T4	FLYING	DESTROYED		Dist. To missile is ≤ MIN
T5	STOP	DESTROYED		
T6	STOP	FINISH	States is empty.	

Listing 19.3 shows the source code of the two transition functions, as well as the output function.

LISTING 19.3 UAV TRANSITIONS AND OUTPUT FUNCTION

```
@Override
public void deltint() {
    boolean finish = false;
    if (super.phaseIs(phases[FINISH])) {
        finish = true;
    }
    if (uavStates.isEmpty() && !finish) {
      finish = uavModel.update(uavStates, refUavState.t,
                                  dt, refUavState);
    }

    if (!uavStates.isEmpty()) {
        double t0 = refUavState.t;
        refUavState = uavStates.remove();
        if (finish) {
            super.holdIn(phases[FINISH],
                              refUavState.t - t0);
        } else {
            super.holdIn(phases[FLYING],
                              refUavState.t - t0);
        }
    } else {
        super.holdIn(phases[FINISH], Constants.INFINITY);
    }

}

@Override
public void deltext(double e) {
    if (super.phaseIs(phases[DESTROYED])) {
        return;
    }

    if (!inTrackingRadars.isEmpty()) {
        Collection<RadarTrackingState> trackingRadarsState =
                                  inTrackingRadars.getValues();
        for (RadarTrackingState trackingRadarState :
                                      trackingRadarsState) {
              if (trackingRadarState.uavState.id.equals(name)
                  &&
              !escapes.contains(trackingRadarState.aduName)) {
                  if(planner.generateEscape(refUavState,
                                trackingRadarState.aduName))
                      uavStates.clear();
```

```
                    escapes.add(trackingRadarState.aduName);
                }
            }
        }
        if (!inMissile.isEmpty()) {
            Collection<MissileState> missilesState =
                                        inMissile.getValues();
            for (MissileState missileState : missilesState) {
                if (missileState.idTarget.equals(name)) {
                    double r =
                           this.distToMissile(missileState, e);
                    if (r <= Missile.MIN_DIST_TO_TARGET) {
                        super.holdIn(phases[DESTROYED],
                                    Constants.INFINITY);
                        return;
                    }
                }
            }
        }
    }

    @Override
    public void lambda() {
        out.addValue(refUavState);
    }
```

19.2.3 ADUs

ADUs are coupled models that are composed of one detection radar, multiple tracking radars, and multiple missiles. Figure 19.5 depicts the DEVS-based ADU model structure, and Listing 19.4 illustrates the pseudo code that clarifies how the wiring

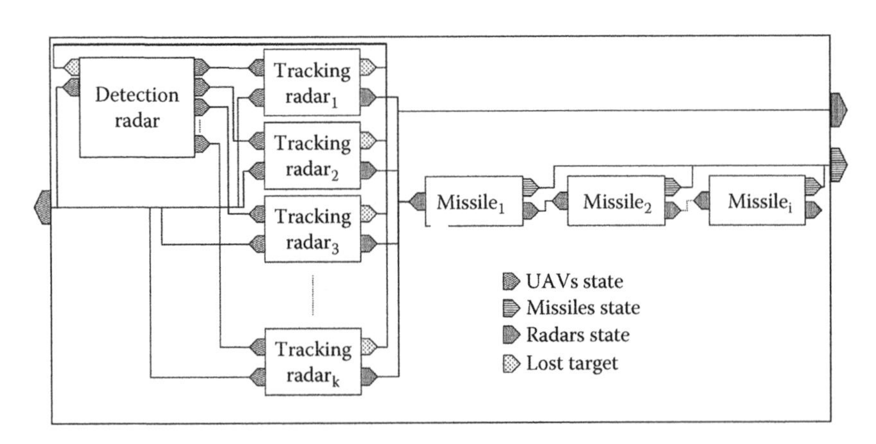

FIGURE 19.5 ADU DEVS model.

is set given the atomic models of all mentioned elements. Detection radars scan the skies seeking UAVs. If a detection radar detects a UAV in its proximity, it then looks for task-free tracking radars. It assigns the spotted target UAV to an unoccupied tracking radar. The tracking radar attempts to detect the target UAV, and if successful, alerts the first missile model in the row with the corresponding target. If the missile state is already fired, it hands the target to the next missile in the row. In essence, on every occasion an ADU detects an UAV, after a specified period, it shoots a missile to attempt to knock down the UAV. In the subsequent sections, each component is described in more detail.

LISTING 19.4 ADU COUPLINGS PSEUDO CODE

```
p=1
For each tr in Tracking Radars {
  Coupling (UAV State) from Detection Radar out(p)
                       to tr in(0);
  Coupling (Lost Target) from tr out(0)
                       to Detection Radar in(0);
  Coupling (Radar State) from tr out(1) to Missile1 in(0);
  Coupling (Radar State) from tr out(1) to ADU out(0);
  p++
}
j=1;
For each m in Missiles {
  Coupling (Missile state) from m out(0) to ADU out(1);
  If(j<i)
     Coupling (Radar State) from m out(1) to Missilej+1 in(0);
  j++;
}
```

19.2.3.1 Detection Radar

The radar detection component consists of one input port that receives the UAV's states, another input port that alerts if any tracking radar has lost its target, and one output port per each tracking radar model to transmit the UAV state to follow.

The detection radar stores as an internal state variable a collection that maps the assignment between incoming UAVs to tracking radars. When the detection radar receives one or more UAVs' states, it first attempts to discover the UAVs in its visibility field. If they are within its range and according to a certain probability of detection, it further checks if UAVs have not been already assigned to any tracking radar, and then, in that case, it searches for a task-free tracking radar to send them to track the UAV, and sets sigma to a certain response time. Otherwise, when it receives the notification of a lost target, it removes the mapping relationship from memory.

As with UAVs, Listing 19.5 shows the source code for the two transitions and the output function.

LISTING 19.5 DETECTION RADAR

```
@Override
public void deltint() {
  detectedTargets.clear();
  super.holdIn("active", tRef);
}

@Override
public void deltext(double e) {
  if (!inLostTargets.isEmpty()) {
    Collection<String> lostTargets =
                      inLostTargets.getValues();
    for (String lostTarget : lostTargets) {
      Integer portNumber = targetPort.get(lostTarget);
      if (portNumber != null) {
        targetPort.put(lostTarget, null);
        unAssignedPorts.add(portNumber);
      }
    }
  }
  if (!inTargets.isEmpty()) {
    Collection<UavState> targets = inTargets.getValues();
    for (UavState target : targets) {
      if (super.isDetected(target)) {
        detectedTargets.add(target);
        if (!targetPort.containsKey(target.id) &&
            !unAssignedPorts.isEmpty()) {
          targetPort.put(target.id,
                         unAssignedPorts.remove());
        }
      }
    }
  }
}

@Override
public void lambda() {
  if (detectedTargets.isEmpty()) {
    return;
  }
  for (UavState target : detectedTargets) {
    Integer portNumber = targetPort.get(target.id);
    if (portNumber != null) {
      outDetectedTargets.get(portNumber).
      addValue(target);
    }
  }
}
```

19.2.3.2 Tracking Radar

The DEVS-based tracking radar model is designed to operate as follows: Through the input port linked to the detection radar of the corresponding ADU, it obtains the state of a UAV to track, stores its value as an internal state variable, and waits for the reception of the same UAV state from the coupling wired to the UAV's models. It then verifies whether the UAV is within its detection field. If it fails and the elapsed time does not exceed the defined maximum, it estimates the UAV's position, orientation, and velocities and finally sends the UAV's state to the first model of the series of missiles. Otherwise, it reports to the detection radar that the target has been lost. Listing 19.6 shows the source code for the two transitions and the output function.

LISTING 19.6 TRACKING RADAR: TRANSITIONS AND OUTPUT FUNCTION

```
@Override
public void deltint() {
  if (super.phaseIs("LOST")) {
    assignedTarget = null;
  }
  super.passivate();
}

@Override
public void deltext(double e) {
  if (!inDetectedTarget.isEmpty()) {
    if (assignedTarget == null || timeEstimating > 0) {
      assignedTarget =
        inDetectedTarget.getSingleValue().clone();
      timeEstimating = 0.0;
    }
  }
  if (assignedTarget == null) {
    return;
  }
  if (!inTargets.isEmpty()) {
    Collection<UavState> targets = inTargets.getValues();
    for (UavState target : targets) {
      if (!assignedTarget.id.equals(target.id)) {
        continue;
      }
      if (!super.isDetected(target)) {
        if (updateTarget(e)) {
          super.holdIn("UPDATED", 0.0);
        } else {
          super.holdIn("LOST", 0.0);
        }
      } else {
        assignedTarget = target.clone();
```

```
                timeEstimating = 0.0;
                super.holdIn("DETECTED", 0.0);
            }
        }
    }
}

@Override
public void lambda() {
    if (!super.phaseIs("LOST")) {
        outRadarState.addValue(new
         RadarTrackingState(super.getName(), aduName,
                                   assignedTarget));
    } else {
        outLostTarget.addValue(assignedTarget.id);
    }
}
```

19.2.3.3 Missile

Missile models are composed of one input port that accepts states of UAVs intended to be blown down, one output port to give over the UAV's state to the next missile only if their status is fired, and another output port to communicate its state to the UAVs so that they can check whether they are destroyed or not. Like UAVs, they also keep an internal state variable with an array of missile states (same as UAVs) reflecting changes in time of the missile dynamics.

Essentially, as seen in Figure 19.6 and Table 19.2, missiles wait for an external command from any tracking radar model of its corresponding ADU to shift from the initial state stop to be fired. Then, sigma is updated from infinity to the next immediate state time. Afterward, every time the internal time event function is triggered, it jumps to the next computed state. Sigma is updated to the next computed time, unless the array of states is empty and the missile has reached its goal or exceeded its limits, or the distance sigma is set to ∞. This behavior is shown in Figure 19.6 and Table 19.2. The implementation is shown in Listing 19.7

19.3 EXPERIMENTS

We performed 16 different experiments on three UAVs and different number of ADUs and NFZs. The experiments differed in the initial and final positions of the UAVs, in the position of the ADUs and NFZs, in the number of initially known ADUs, and if the UAVs have their embedded radars enabled or disabled. According to those differences, we can classify UAVs into the following:

1. Groups: Experiments where all the ADUs are known initially (first group), where all the ADUs are disabled pop-ups (second group), where all the ADUs are enabled pop-ups and all the UAVs have their embedded radars disabled

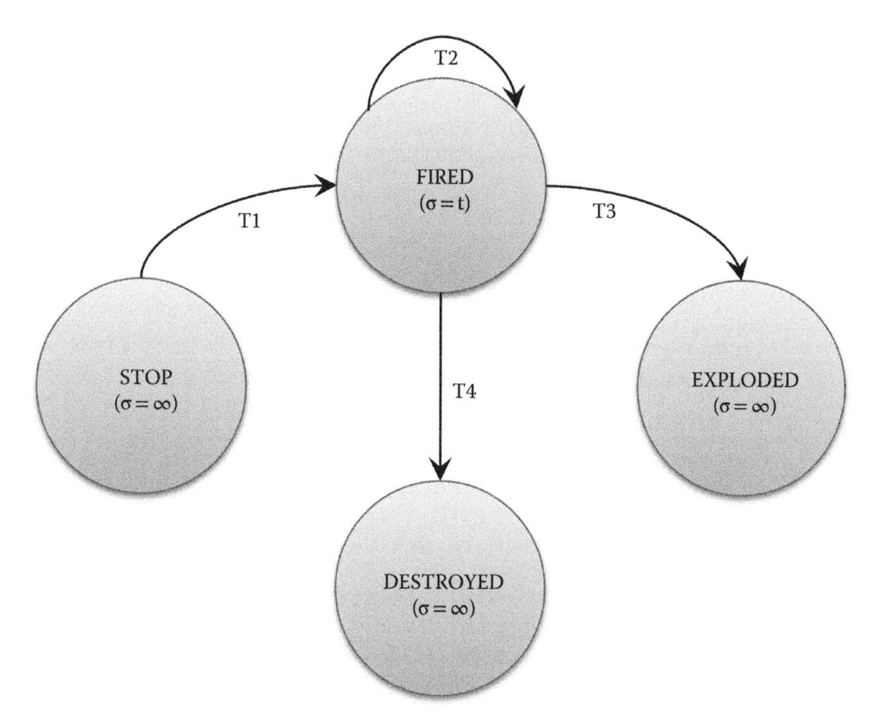

FIGURE 19.6 Missile state diagram.

TABLE 19.2
Missile State Transitions

	S_k	S_{k+1}	δ_{int}	δ_{ext}
T1	STOP	FIRED		NO assigned target
T2	FIRED	FIRED	States is not empty.	Assigned target
T3	FIRED	EXPLODED		Reached goal
T4	FIRED	DESTROYED		Exceeds limits or distance

LISTING 19.7 MISSILE: TRANSITIONS AND OUTPUT FUNCTION

```
@Override
public void deltint() {
  unAssignedTargets.clear();
  if (!missileStates.isEmpty()) {
    double t0 = refMissileState.t;
    refMissileState = missileStates.remove();
```

```
      super.holdIn(phases[FIRED], refMissileState.t - t0);
    } else {
      super.holdIn(refMissileState.phase,
                  Constants.INFINITY);
    }
  }

  @Override
  public void deltext(double e) {
    boolean update = false;
    if (!inRadarState.isEmpty()) {
      Collection<RadarTrackingState> radarStates =
                         inRadarState.getValues();
      for (RadarTrackingState radarState : radarStates) {
        UavState target = radarState.uavState;
        if (assignedTarget == null) {
          double dist =
                 Math.sqrt(Math.pow(refMissileState.xEast -
                 target.xEast, 2) +
                 Math.pow(refMissileState.yNorth -
                 target.yNorth, 2) +
                 Math.pow(refMissileState.h - target.h, 2));
          if (dist > Missile.RMEZ_MAX) {
            unAssignedTargets.add(radarState);
            continue;
          } else {
            assignedTarget = target;
            refMissileState.phase = phases[FIRED];
            refMissileState.idTarget = assignedTarget.id;
            update = true;
          }
        } else if (assignedTarget.id.equals(target.id)) {
          assignedTarget = target;
          update = true;
        } else {
          unAssignedTargets.add(radarState);
        }
      }
    }

    if (update) {
      if (!super.phaseIs(phases[STOP])) {
        updateTrajectory(e);
      }
    }
    super.holdIn(refMissileState.phase, 0.0);
  }

  @Override
  public void lambda() {
    if (super.phaseIs(phases[FIRED])) {
```

```
            outMissileState.addValue(refMissileState);
            outX.addValue(refMissileState.xEast);
            outY.addValue(refMissileState.yNorth);
        }
        if (!unAssignedTargets.isEmpty()) {
            outRadarState.addValues(unAssignedTargets);
        }
    }
```

(third group), and where all the ADUs are enabled pop-ups and all the UAVs have their embedded radars enabled (fourth group). On one hand, in the first group, all the ADUs are known from the beginning and the offline path planner uses their positions to find routes that avoid the threats. On the other hand, in the second, third, and fourth groups, the offline path planner considers no ADUs initially and obtains routes that are only restricted by the NFZs, terrain, and UAVs' maneuverability. Besides, in the second group, the pop-up ADUs are disabled and the initial routes are safe, whereas in the third and fourth groups, the pop-up ADUs are enabled and the initial routes can pass through ADUs' risky areas. Furthermore, the third group has the UAV-embedded radars disabled, the UAVs cannot detect the pop-ups, and the online planner does not recalculate an alternative route when the UAVs are attacked. Finally, the UAVs in the fourth group can detect the ADUs and use the online planner, which gives the UAVs chances to avoid the pop-ups in certain situations.

2. Types: These are experiments with the same initial and final positions for the UAVs, and locations of the ADUs and NFZs. There are four different types (A, B, C, and D), schematized in Figure 19.7. Experiment type C includes intermediate points that the UAVs are forced to visit. The ADUs, which are initially known in Figure 19.7 and pop-ups in Figure 19.2, are represented by big dashed circles (which show the maximum distance of detection of their radars) and by small solid circles (which enclose the zones where the probability of destroying an UAV can be greater than 0). The NFZs are represented with the rectangular shaded areas.

Figure 19.7 also represents with dashed lines, labeled as UAV1, UAV2, and UAV3, the initial offline trajectories of the UAVs for the experiments with known ADUs (first group). Similarly, Figure 19.2 represents with dashed black lines the initial offline paths of the UAVs for the experiments with pop-up ADUs (second, third, and fourth groups, type C). Figure 19.2 also includes, as solid black lines, the alternative routes calculated by the online planner during the simulations where all the ADUs are enabled pop-ups and all the UAVs have their embedded radars enabled (fourth group). The end of the initial route is usually covered by the end of the alternative route. The lack of connection between the beginning of the alternative route (solid) and the original (dashed) that occurs in some of the cases is because the alternative route starts at the point where the UAV should be after running the online planner and follow an escaping maneuver.

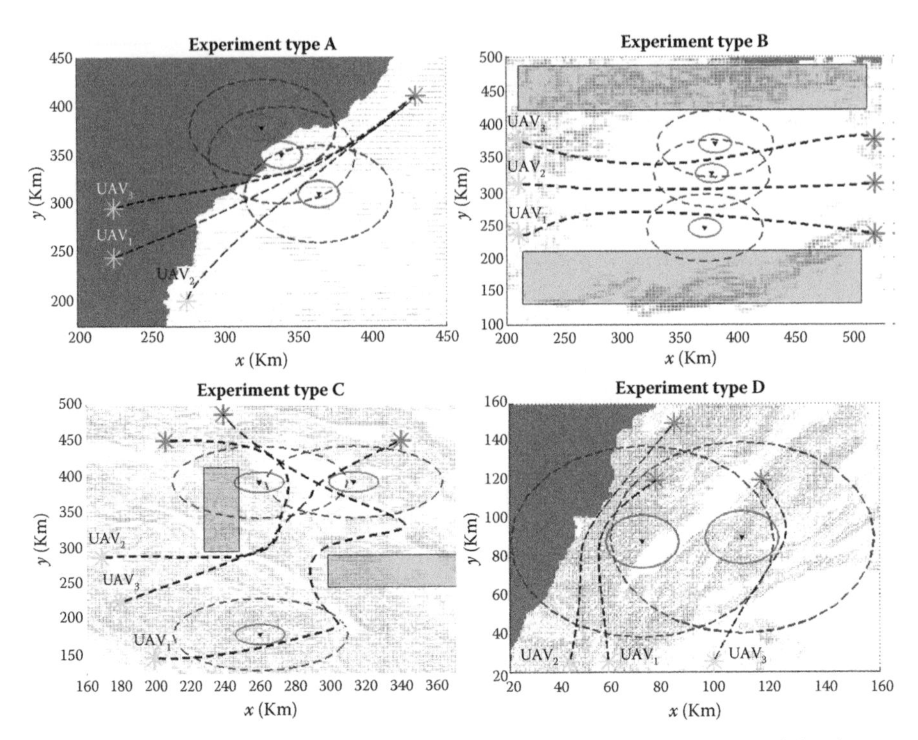

FIGURE 19.7 Representation of the four experiment types (A, B, C, and D) and the planner routes obtained for the first group.

Taking into account the offline routes of Figures 19.2 and 19.7 and the characteristics of each group, we expect that all the UAVs of the four experiments in the first group successfully reach their final destination because their offline routes fall outside the probable destroying zones (red circles). The behavior of the four experiments in the second group is similar, because although some of the offline routes fall inside the probable destroying zones, the pop-up ADUs are disabled. Finally, the behavior of the UAVs in the experiments of the third and fourth groups is more difficult to predict before running the simulations. However, we expect that the UAV's chances to survive in the four experiments of the third group decrease the longer the route is inside the probable destroying zone. Besides, the chance to survive for the same UAV and experiment type (A, B, C, and D) should be higher in the fourth group than in the third group, as in the former, the UAVs detect pop-ups, recalculate trajectories, and try to escape, whereas in the last, they maintain the original offline plan.

19.4 RESULTS

We analyzed behavior of three simulation scenarios (baseline, local, and distributed) over 16 experiments of the described system. For each of the 48 cases, we carried out 30 simulations, where the seeds of the random generators are fixed. On the basis of their results, we measure the success of the UAVs and the performance of the simulation scenarios.

The success is characterized by the number of UAVs that reached the target points in each of the 48 cases during the 30 simulations. The performance is measured according to the average simulation time needed in each of the 48 cases and 30 simulations.

Regarding simulation repeatability and reproducibility, the results of the comparison show that the simulations in DEV/SOA (local and distributed scenarios) are equivalent to the ones in centralized DEVS (baseline), because in each of the experiments, the three simulation scenarios obtain identical successful rates and final trajectories. Additionally, when we are measuring the simulation time, DEVS/SOA with their simulation services distributed in multiple computers (distributed scenario) outperform the other two scenarios, in spite of the high communication necessities of the problem under study.

19.4.1 UAVs' Success

Table 19.3 shows the results of measuring the successful arrivals of each UAV in each of the 16 experiments. These results can be organized according to the experimental group they belong to:

- First and second group: As Table 19.3 shows, all the UAVs arrive successfully at their final destinations, which coincides with the expected behavior, because the initial routes for the four possible experiments (A, B, C, and D) of the first group are safe, while the pop-up ADUs of the second are disabled.

TABLE 19.3
UAVs Success (+ Stands for 100% and – for 0%) for Each Example Type (A, B, C, and D) and Group (First, Second, Third, and Fourth)

	First Group			Second Group		
	UAV_1	UAV_2	UAV_3	UAV_1	UAV_2	UAV_3
A	+	+	+	+	+	+
B	+	+	+	+	+	+
C	+	+	+	+	+	+
D	+	+	+	+	+	+

	Third Group			Fourth Group		
	UAV_1	UAV_2	UAV_3	UAV_1	UAV_2	UAV_3
A	+	–	–	+	–	+
B	–	+	–	+	+	+
C	–	–	+	+	+	+
D	+	+	+	+	+	+

- Third and fourth groups: Their results, presented in Table 19.3, depend on the UAV, experiment type (A, B, C, or D), and group (third and fourth). For instance, if we focus on experiment B of the third group, UAV_2 survives because its initial trajectory is safe, and UAV_1 and UAV_3 are destroyed because their initial trajectory stays too long in the nonsafe zone. Similar explanations apply to the rest of the UAVs in the third group. Within an experiment type, the successful UAVs of the third group keep on surviving in the fourth, while the majority of the destroyed UAVs survive because the possibility of replanning lets them escape from the ADU attack. UAV_2 in experiment A is the only exception because the replanned route gets so close to the ADU that its tracking radars and missiles are able to estimate it and destroy this UAV.

Note that, regarding experiment of type D, all the three UAVs always arrive successfully at their final destination. The reason is that all the ADUs are located in mountains, very close to the UAVs' trajectory (the same altitude). Thus, the initial routes of UAV_1, UAV_2, and UAV_3 hide behind mountains and are always outside radars' coverage.

The results of all the simulations carried out in DEVS/SOA (local and distributed) and DEVS (baseline) are the same. This makes our framework suitable for the distributed simulation of any problem modeled following the DEVS formalism, after only generating the models in any of the currently supported centralized implementations of DEVS.

19.4.2 Performance

The analysis of the performance of DEVS/SOA for each of the 16 experiments performed in each scenario (baseline, local, and distributed) is presented in this section. It is based on the mean value of the execution times of the 30 simulations belonging to each of the 46 cases, which is used to calculate the speedups of the local and distributed scenarios with respect the baseline one.

The speedup ratios, presented in Figure 19.8, are calculated as the quotient of the mean execution time of the baseline scenario to the mean execution time of each scenario. Therefore, ratio values greater than 1 mean speedup improvements with respect to the baseline scenario. The speedup percentages, presented in Table 19.4, are measured as the difference between the mean execution time of the baseline scenario and the mean execution time of the local or distributed one, divided by the mean execution time of the baseline scenario. Therefore, positive percentages also mean speedup improvements with respect to the baseline scenario.

For each of the 16 experiments, we can see that the simulation carried out with DEVS/SOA in a single machine (local) is always the slowest: it has a speedup ratio smaller than 1 and so a negative speedup percentage. This behavior is justified because many messages are exchanged between the UAVs and the ADUs. The amount of messages plus the unnecessary web communication protocol overloads the local simulations with extra computational effort that is not used in the baseline ones that only need the time for simulating the environment and send messages within DEVS.

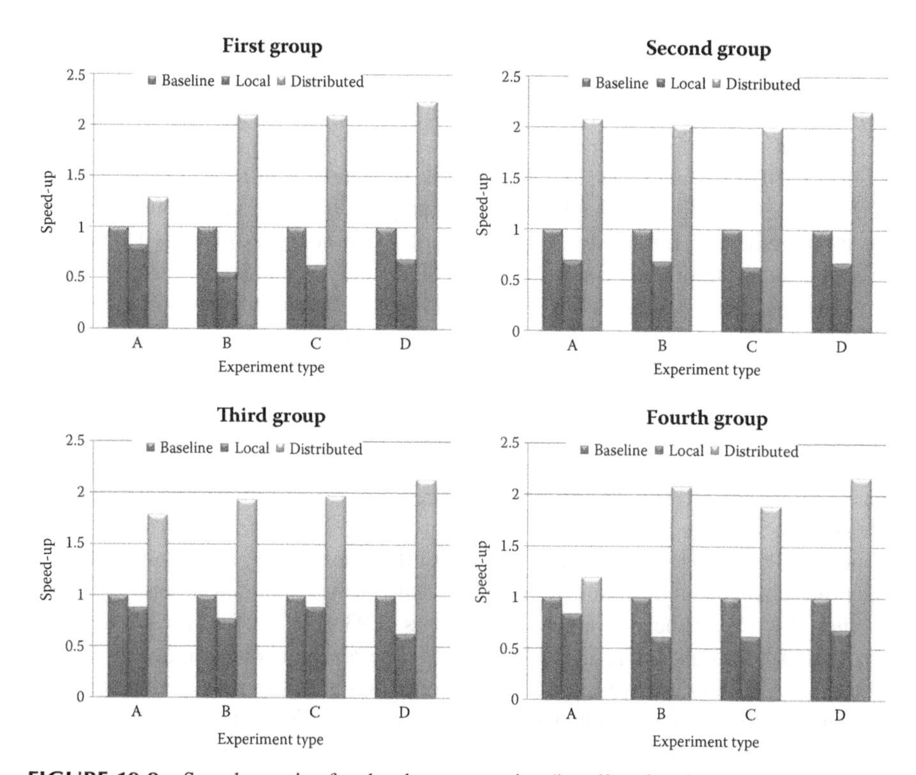

FIGURE 19.8 Speedup ratios for the three scenarios (baseline, local, and distributed) and each example type (A, B, C, and D) and group (first, second, third, and fourth).

TABLE 19.4
Speedup Percentages of the Distributed (Dist) and Local (Loc) Scenarios with Respect to the Baseline (Bas) One for Each Example Type (A, B, C, and D) and Group (First, Second, Third, and Fourth)

	First Group		Second Group	
	Loc vs. Bas	**Dist vs. Bas**	**Loc vs. Bas**	**Dist vs. Bas**
A	−20.98%	22.58%	−43.83%	51.86%
B	−79.66%	52.46%	−46.12%	50.58%
C	−58.65%	52.44%	−57.55%	50.13%
D	−44.47%	55.27%	−47.37%	53.69%

	Third Group		Fourth Group	
	Loc vs. Bas	**Dist vs. Bas**	**Loc vs. Bas**	**Dist vs. Bas**
A	−13.28%	44.18%	−19.38%	16.36%
B	−29.36%	48.34%	−62.96%	51.99%
C	−12.24%	49.25%	−59.73%	47.17%
D	−59.36%	53.12%	−46.00%	54.00%

The opposite happens with the simulations carried out with DEVS/SOA distributed in multiple machines (distributed), which is always the quickest: it has a speedup ratio greater than 1 and so a positive speedup percentage. This is because the time that is used to simulate the behaviors of the UAVs and ADUs is distributed among the different machines, and so the total distributed simulation time is significantly smaller than the one needed for simulating everything in the same machine. The improvement is limited by the number of machines, the time needed by the message transformation operation of DEVS/SOA, and the number of messages that are exchanged during the simulation of the elements. The main contribution to final speedup is the number of missiles in flight. When a missile is fired, its dynamics is integrated every 5×10^{-3} seconds. As a result, a huge amount of messages is sent from missiles to UAV, trying to compute if the target UAV has been destroyed or not. We may verify this hypothesis examining the speedups of the distributed vs. baseline scenarios in Figure 19.8. In the first, third, and fourth groups, we obtain an average speedup of 1.93, 1.96, and 1.84, respectively. On the contrary, we reach the best average speedup of 2.07 in the second group. It is because in this group there are no missiles in flight (ADUs are disabled in this case). Thus, the number of messages interchanged is minimal. Finally, the third group obtained a better average speedup (1.96) than the first and fourth groups (1.93 and 1.86), whereas there are more missiles in flight (more UAVs destroyed). The reason is that commonly two of the three UAVs are destroyed in the third group, and relatively early in the simulation. Thus, in this case, we found just one UAV flying the 75% of the simulation time, which improves the speedup considerably because, again, the number of messages is reduced.

To conclude, the average speedup of the 16 experiments of the distributed scenario is 1.95. Thus, our experiments show that the DEVS/SOA simulator, when distributed among several machines, can obtain quicker results than the centralized DEVS one. The speedup improvement depends on the number of machines used in the distribution; the way the root system is divided among the different machines; and the proportion of time used in the centralized simulation to simulate the atomic behaviors and to exchange the messages, which, in the distributed simulation, will go through the DEVS/SOA translation module and the web.

REFERENCES

Besada-Portas, E., de la Torre, L., de la Cruz, J. M., & de Andrés-Toro, B. (2010). Evolutionary trajectory planner for multiple UAVs in realistic scenarios. *Robotics, IEEE Transactions on*, *26*(4), 619–634.

Stevens, B. L., & Lewis, F. L. (2003). Aircraft Control and Simulation. (p. 664). Hoboken, NJ: John Wiley & Sons.

20 Generic Network Systems Capable of Planned Expansion

From Monolithic to Netcentric Systems

20.1 OVERVIEW

The Systems Capable of Planned Expansion (SCOPE) command is a highly automated, high-frequency global communication system (HFGCS) that links U.S. Air Force (USAF) command and control (C2) functions with globally deployed strategic and tactical airborne platforms. The SCOPE command replaces the existing USAF high-power, high-frequency (HF) stations with a communication system featuring operational ease of use, dependability, and seamless end-to-end connectivity comparable to commercial telephone services. The network consists of 15 worldwide HF stations (Figure 20.1) interconnected through various military and commercial telecommunications media (Figure 20.2). It increases overall operational and mission capabilities while reducing operation and maintenance costs.

HFGCS provide communications support (Hurd, 2007) to

- Foreign dignitaries
- State Department
- White House
- Joint Chief of Staff (JCS)
- Defense Information Systems Agency (DISA)
- Air Mobility Command (AMC)
- Air Combat Command (ACC)
- Air Force Space Command (AFSPC)
- U.S. Air Force Europe (USAFE)
- Pacific Air Forces (PACAF)
- Air Weather Service (AWS)
- United States Navy
- North Atlantic Treaty Organization (NATO)
- Civil Air Patrol
- Department of Homeland Defense

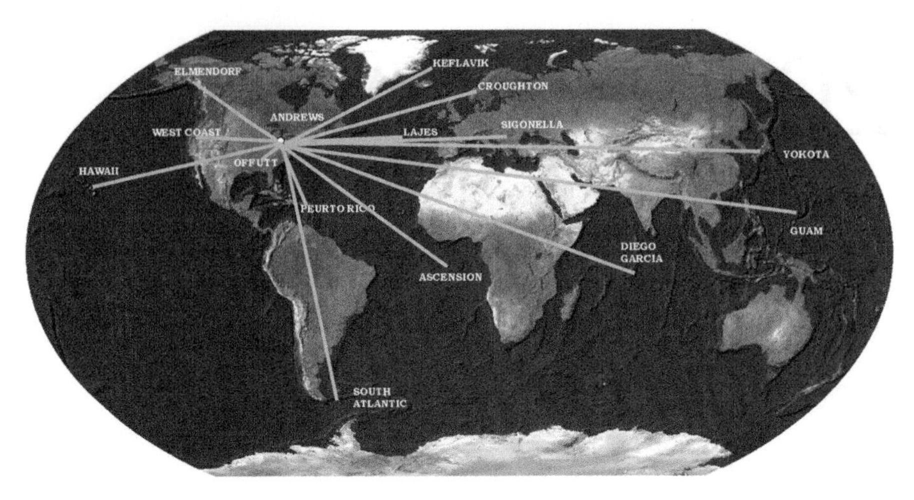

FIGURE 20.1 Geographic locations of fixed stations.

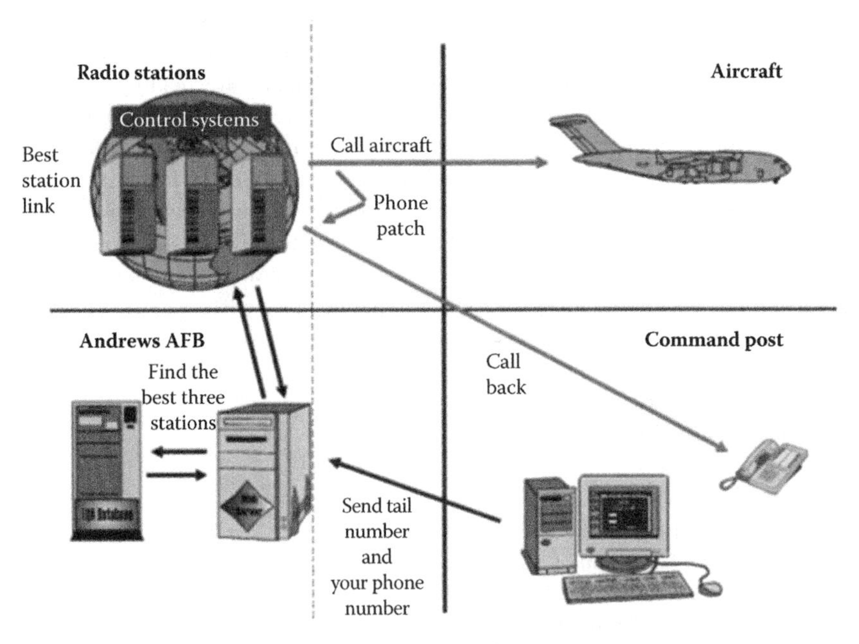

FIGURE 20.2 Communication flow diagram for SCOPE command.

The HF radio equipment includes the Collin's Spectrum DSP Receiver/Exciter, Model RT-2200. The radios feature automatic link establishment (ALE) (MIL-STD-188-141B, 1998) and link quality analysis (LQA) capabilities and are adaptable to future ECCM FSK waveforms (MIL-STD-188-110B; STANAG5066). The transmit subsystem includes 4-kW solid-state power amplifiers, a high-power transmit matrix, and a combination of receiver/multicoupler antenna matrix. A typical SCOPE command station includes operator consoles (HFNC), circuit switching equipment (digital electronic switch [DES], DSN, and LCO), HF radios (ALEs), RF matrixes (RTs), and antennas (RXs and TXs). A non-blocking DES connects the station to the local military or commercial telecommunication services. The switch features unlimited conferencing, modular sizing, a digital switch network, a precedence function, and capacity for up to 2016 user lines.

The SCOPE command uses a modular, open-system design to automatically manage and control all network operations, including those at split-site stations. To achieve maximum flexibility, the system uses commercially available standards-based software and a multitasking operating system. This approach permits 14 out of 15 network stations to operate "lights out" (unmanned) and to be economically controlled from a central location. The control system also includes local area network (LAN) software, servers, and routers to support unlimited LAN/wide area network (WAN).

The program includes a systems integration lab (SIL) and test bed facility located in Rockwell Collins's Texas facility. The SIL is used to predict the impact and risk that any changes or upgrades will have on system performance, integrity, or costs before actual implementation begins. The SIL includes a fully functional SCOPE command station for performing baseline design verification and interface compatibility and functional verification tests.

Joint Interoperability Test Command (JITC) is the only government agency that is assigned the task to validate and authorize information technology (IT) systems for military operations. The HF SCOPE command system has also been evaluated by JITC. In collaboration with Dr. Eric Johnson, a simulator called NetSim-SC was developed in the C language around 1997 that was validated and eventually used by both the government and the industry to conduct experiments and run scenarios. The simulator was an exhaustive and comprehensive effort with respect to the details it implemented and served its purpose well. However, in today's circumstances, the same simulator stands obsolete due to the heterogeneous nature of today's network traffic, in which e-mail occupies a considerable percentage of traffic. These demands stem from the possibility of expansion of the current infrastructure of the SCOPE command for both the USAF and U.S. Navy. Questions arise such as how many stations need be added to service a required workload. Also needing to be investigated are trade-offs such as whether it is more economical to add more stations or increase the number of internal radio levels within stations to meet the anticipated demands. The simulator was upgraded at the Arizona Center for Integrative Modeling and Simulation (ACIMS) lab in 2006–2007 to make it more useful for current demands. An architecture and simulation model called the Generic Network Model for Systems Capable of Planned Expansion (GENETSCOPE) was architected by Mittal in the capacity of the lead developer at ACIMS, at the University of Arizona in collaboration with Northrop Grumman and RTSync.

The GENETSCOPE perspective of the world is based on date, time, sunspot number (SSN), equipment location (latitude and longitude), and equipment specifics (power, antenna, bandwidth, signal-to-noise ratio [SNR], etc.). The HF communication is made possible by calculating the SNR of HF Skywave propagation through the atmosphere. In the developed version of GENETSCOPE, propagation was available for any location in the world. For Skywave propagation, this is valid whether the target locations are 150 or 1500 mi apart. Once the first phase of the GENETSCOPE was completed for USAF, more requirements poured in from the U.S. Navy to extend the existing architecture toward the incorporation of a littoral model, ground waves, and manpack mobiles that carry radios on their backpack.

Given a requirement to use GENETSCOPE in littoral operations (0–300 mi.), it was decided to add the ground wave calculations to the available Skywave predictions. Ground wave can in some cases provide a better signal than Skywave propagation. The navy has used surface wave to communicate between ships for years. These operations, normally 50–150 mi., provide a reasonably robust environment without the variations of Skywave. The ocean as the path between targets is both a drawback and one of the pluses in surface wave environments. The ocean provides a good medium for surface wave communications. The disadvantages are the frequencies used and the attenuation of any landmass between the targets. Given a typical mission, the ship (fixed location) sends a marine team ashore. A requirement is to maintain communications to that team as it moves inland. The user of the model could change antenna types, frequencies used, or a number of other options to determine what may be needed to communicate. Additionally, the other functionality of GENETSCOPE is to include aircraft or other fixed stations. So a scenario could be developed with a ship, manpack, and aircraft to answer the question of whether the manpack can communicate with both the ship and the aircraft on the same antenna or frequencies.

20.1.1 GENETSCOPE Feature Set

In summary, GENETSCOPE is a complex system that has a hardware radio whose operational performance was initially specified for the SCOPE command but due to the advancements in technology and scope of mission operations, a state-of-the-art model was needed that could perform both in AMC and in littoral scenarios. The component-based design of the GENETSCOPE allowed such extensions and incorporations of the new requirements through the DEVS experimental frame design. The overall functional feature set of GENETSCOPE 2.0 is as follows:

- *Mission presets*: Each station (level/mobile) can be configured according to various preset settings. Each preset consists of the radio mode, scan list used by that preset, and many other settings.
- *Non-ALE traffic*: The station can choose to send traffic using either the ALE radio linkage mechanism or the non-ALE radio.
- Dynamic Ionospheric Communications Enhanced Profile Analysis and Circuit Prediction Program (ICEPAC) and Voice of America Coverage Analysis Program (VOACAP) SNR values based on a user-entered SSN, month, and year of simulation configuration.
- Traffic generation based on mission preset and user-provided parameters.
- Comprehensive logs output as a result of simulation execution.
- Extensive locations database that can aid flight path planning for trips worldwide.
- Capability to add new presets to the existing repository of presets used.

- Capability to reload existing simulation configuration and update it toward a new simulation configuration through the graphical user interface (GUI).
- Graph reporting for top-level analysis of simulation execution.
- Animation capability to display call links of mobile stations on a worldwide map during post-simulation analysis.
- *Reporting*: Detail, Select, and Summary levels. The Detail and the Select reporting levels are shown in the various logs and graphs. The Summary file contains various statistics like the following:
  - Preset-based analysis
  - Channel utilization in minutes on a per station basis for both sender and receiver
  - Average linking time
  - Average Link Quality Assessment (LQA) score
  - Number of failed ALE links during ALE establishment procedure
  - Number of voice and data calls generated and complete on preset basis
  - Number of total fixed and mobile stations participating as sender and receiver contributing to channel occupation
  - Total number of propagation calls made during simulation run
- The GENETSCOPE model uses a set of the default antennas that are provided with the ITSHF propagation package. The propagation program uses the antenna data to determine the antenna gain for a path. The GENETSCOPE model has a number of default antennas such as LTO, HTO, RLP, Rosette, Diploe, Probe, and so on.
- Littoral and ground wave calculations have been added to support short-range communications requirements. The user has the ability to describe the littoral environment, radio details, and traffic to be used in the scenarios.

20.2 METHODOLOGY TO COMPONENTIZE A LEGACY APPLICATION

Air traffic has increased manifold since 1997, along with the computing technology. Consequently, the transition effects need to be monitored more closely, and the overall system response time* needs to be documented. The significant parameters that have the most impact on system performance have to be identified. To more easily address such questions, an effort was made to modularize Johnson's 15K lines of code into a component-based structure. Johnson's code was written in procedural C programming language and had a customized discrete event scheduler, where events were dynamically inserted in a time-ordered queue. There was a lack of hierarchical structure, data encapsulation, and other benefits that an object-oriented paradigm brings with it. Since DEVS is implemented as an object-oriented technology, migration of the original code to DEVS

* Response time of a system is defined as the time taken by the system to display significant effect caused by any update in the configuration parameters.

state machines (atomic components) was not possible. To add to the complexity, the code had a very limited documentation. Since there was an event scheduler that linked each defined structure, it was very problematic to understand the message flow between the entities through the C function calls. The notion of an event structure was heavily overloaded with redundant information being processed at the scheduler level. Consequently, an exhaustive study of Johnson's codebase was done to understand the modular system and its behavior. The study was done in a twofold manner: first, with the help of subject matter experts from JITC, and second, with the help of code-visualization tools. The process is depicted in Figure 20.3.

As we developed an increased understanding of the system structure and behavior, we migrated about 80% of the code directly to Java using automated code authoring scripts with Perl and Extensible Markup Language (XML). The rest of the code was either not required or was newly written to accommodate the modular system structure and behavior. After a grueling exercise of 8 months, the modularization succeeded in the system entity structure (SES), as shown in Figure 20.4. Once "componentized," the components were made DEVS compliant resulting in a DEVS-based simulation package to support the systems engineering needs of the SCOPE command.

To study the effect of changes/upgrades introduced to the existing SCOPE Command system, the effort built the experimental frame, based on DEVS principles for the modular DEVS-NetSim simulation model, named GENETSCOPE (2007).

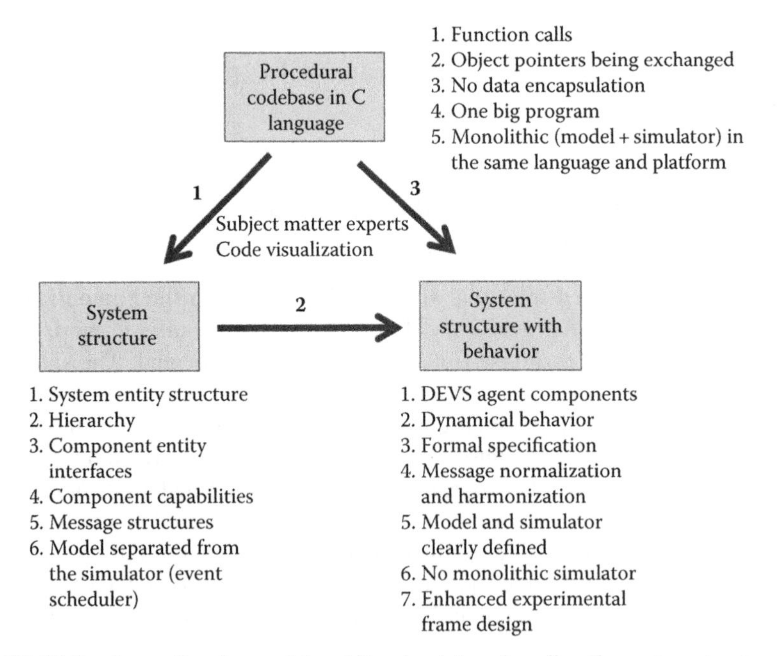

FIGURE 20.3 Separating the model and the simulator; decoding the system structure and behavior.

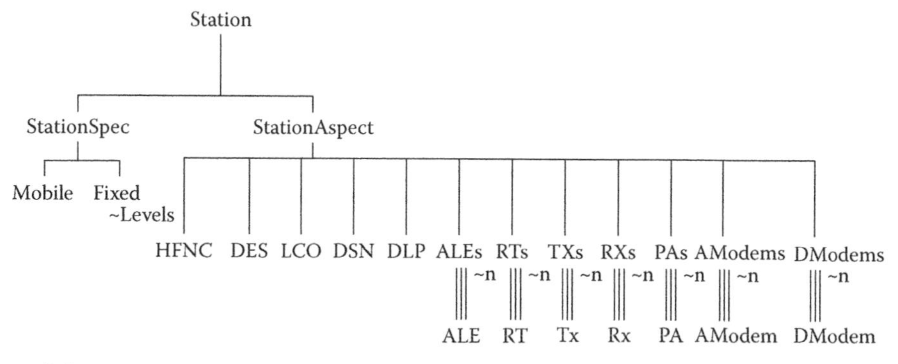

Rules:
1: if selecting Mobile from StationSpec then n = 1 else if selecting Fixed then (n = Levels) specified by number of Levels.

FIGURE 20.4 System entity structure for SCOPE command system showing the fixed and the mobile (aircraft) stations.

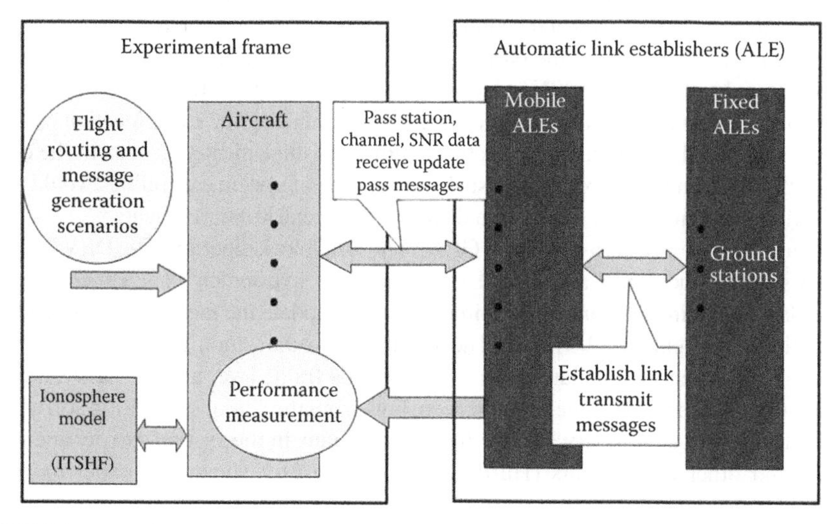

FIGURE 20.5 Conceptual simulation architecture for modularized SCOPE Command.

Figure 20.5 shows the block architecture of the simulation model. The right-hand box is the system phenomenon that contains the ALE and STANAG 5066 protocols used for establishing links and exchanging data messages between mobile stations and fixed stations. The left-hand box is the experimental frame that generates various scenarios and parameters under study. The scenarios and parameters are fed into the model and performance characteristics are obtained from it, which are then visualized and analyzed in real time as per the extended Model Simulator View Controller (MSVC) architecture.

20.3 ARCHITECTURE IMPLEMENTATION USING ENHANCED MSVC

Figure 20.6 shows the detailed simulation architecture for GENETSCOPE using the concepts laid out in Chapter 7, especially the enhanced MSVC paradigm that allows intervention at the simulation layer and provides functionalities such as pause, resume, modify, and resume. The ionosphere model used in the architecture is ICEPAC_DATA. It is worth stressing that the initial NetSim model written in C language has ICEPAC database (text files) tightly coupled with the model. In our present implementation, it was made modular so that it can be replaced by any other database that could provide the channel propagation values through the ionosphere, for example, VOACAP. In the current implementation, there is no ICEPAC database included but the complete ICEPAC software that is executed at run-time. This is one of the biggest advantages in separating ICEPAC from the model itself. The ICEPAC software is configured through the experimental frame parameters and is made available for real-time execution as an independent thread for different stations that are active in the running DEVS model. The real-time execution of ICEPAC software involves creation of a dynamic ICEPAC configuration file that contains information about the two stations, their geographical locations in latitude and longitude, the SSN, and the time of year, month, and day. This implementation allows us to get the ionospheric SNR values for any location at any time of the year (for SSN) unlike the earlier implementation (NetSim-SC) where we were limited to only a handful of SSN values (10, 70, 100, and 130) with locations specified in five-degree increments. This has the added benefit of using the exact location of any mobile station rather than using projections within the implemented grid as in the earlier NetSim-SC. In future extensions, such partitioning of system capabilities would aid in making the entire model deploy in a distributed netcentric environment.

Figure 20.6 also shows a high-level GENETSCOPE block diagram. The DEVS layer comprises both the model and the DEVS simulation environment. The experimental frame layer contains the controls required to modify/update the model and a simulator controller as per enhanced MSVC. The simulation visualization is modular in construction and reflects the updates in the experimental frame layer and the DEVS layer.

Next, we will present some of the screenshots from the actual simulation software to emphasize the complexity and configurable options in the system's experimental frame and other components (Hurd, 2007). Figure 20.7 shows the experimental

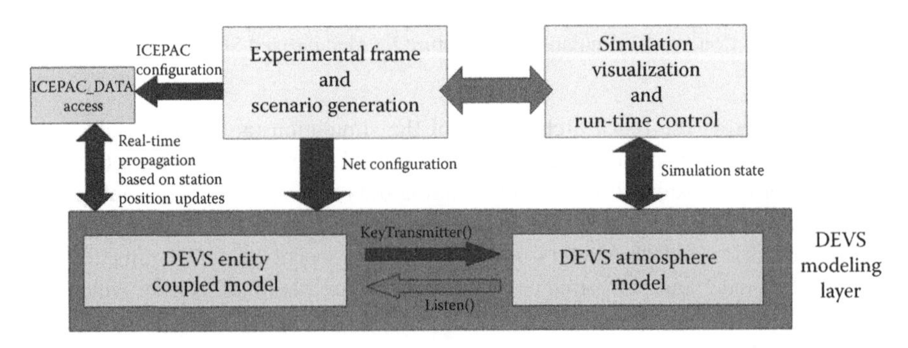

FIGURE 20.6 Simulation architecture for GENETSCOPE.

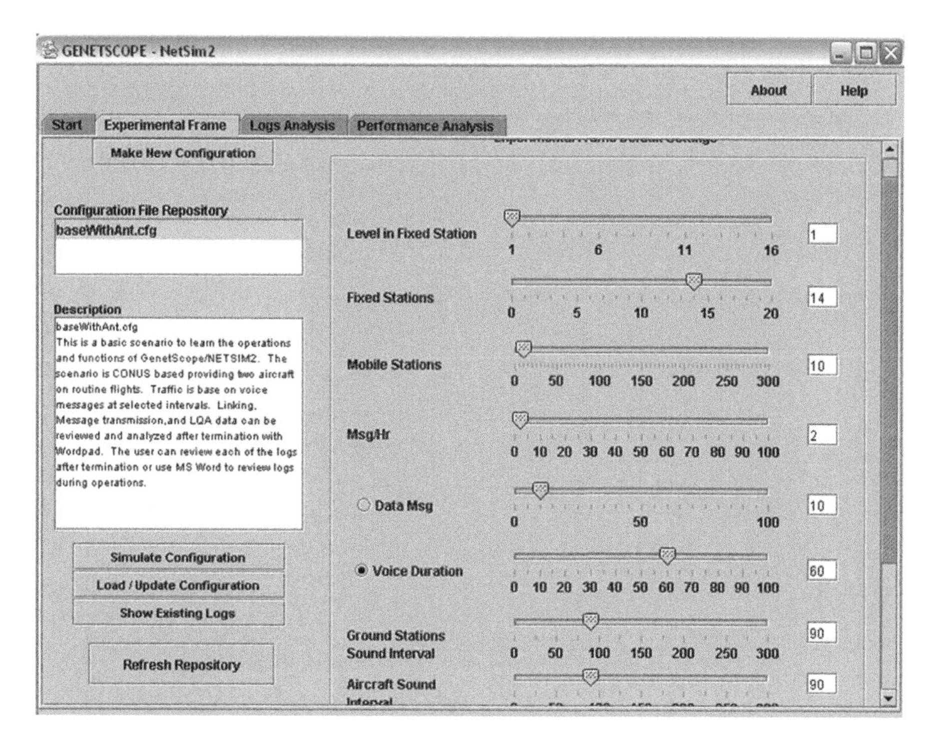

FIGURE 20.7 Experimental frame for GENETSCOPE.

frame and various parameters (along with their default values) used in scenario configuration.

Once the experimental frame parameters are configured, these parameters are channeled down to the individual components. The top-level design parameters then bind the other internal component parametric settings. The other important aspect of this process is that during simulation run-time, if the experimental frame parameters are changed to study any particular parameter, that change is channeled across the whole system model configuration using "interrupts," thereby exploiting the discrete event simulation methodology. The update of any experimental frame parameter is taken by the simulation model as an "external" event.

Figure 20.8 shows a typical configuration of the ground station at San Francisco port in a littoral scenario. The left column in Figure 20.8 shows two fixed stations, and the individual details about each station can be seen by pressing the Lookup button. Other internal details of station configuration can be seen in the GENETSCOPE software user's manual (Mittal, et al., 2007).

In Figure 20.8, the ship at San Francisco port is considered a fixed station in this scenario. The Fixed Station tab also contains various other configuration options such as Preset, Radio Levels, Ale Parameters, Scan List, and Gnd (ground) Infrastructure.

Similarly, a mobile station configuration panel is shown in Figure 20.9. The user can select any specific mobile aircrafts bounded by the number of mobile stations specified in the experimental frame (Figure 20.7). Figure 20.10 basically lets the user enter *call signs* to these mobile stations and invites the user to enter aircraft-specific

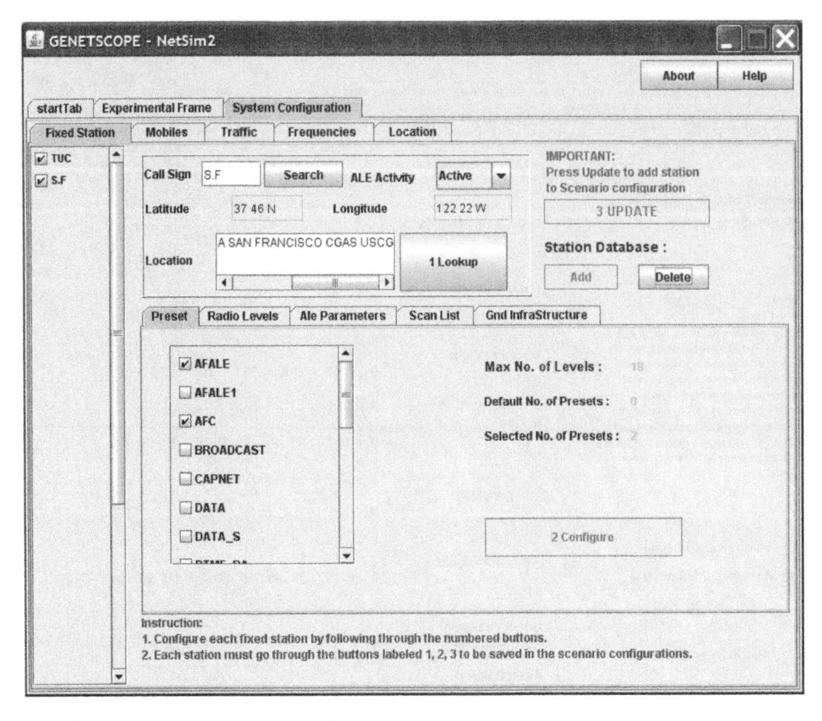

FIGURE 20.8 Ground station configuration screen.

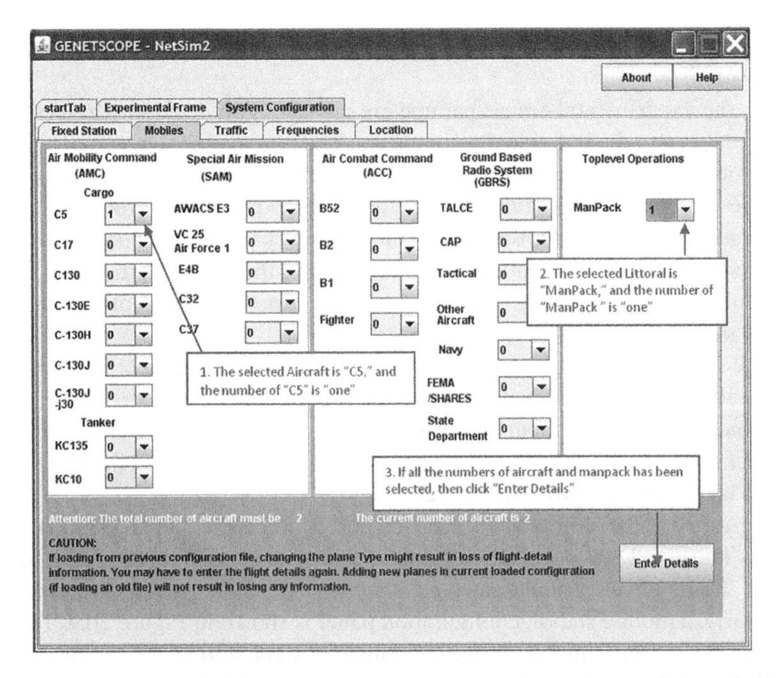

FIGURE 20.9 Mobile station configuration screen where the total count is bounded by the experimental frame.

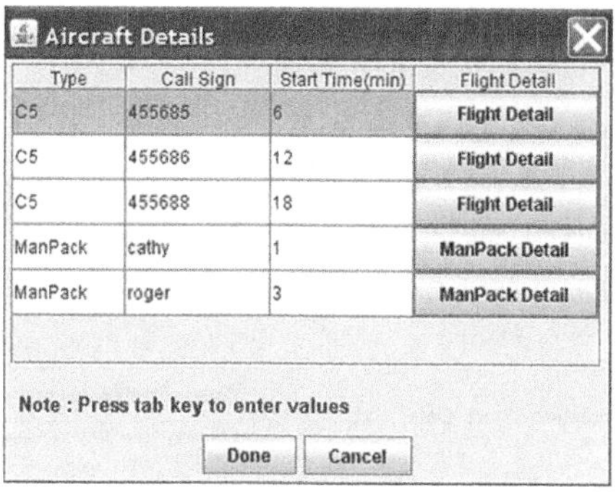

FIGURE 20.10 Call sign entry for a mobile station.

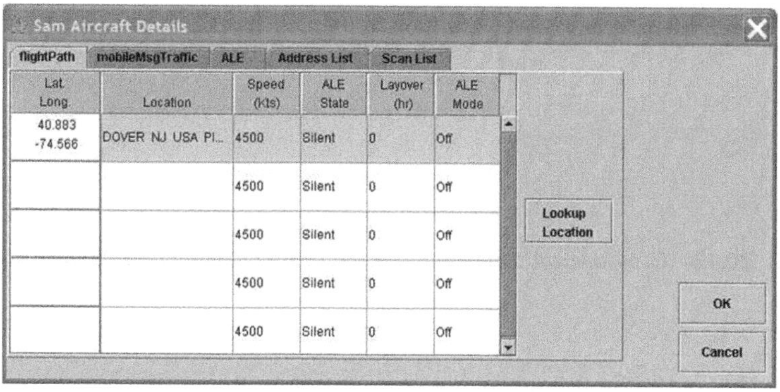

FIGURE 20.11 Flight path of mobile aircraft and other details.

details like message traffic, flight path (see Figure 20.11), radio parameters, and channel frequencies being used.

The last piece of information being fed through the experimental frame is the ICEPAC setting, based on the SSN. Once the system model is configured through the experimental frame settings, the user is directed toward the simulation setup. Figure 20.12 shows the final setup screen after which the user moves on to the run-time simulation screen (see Figure 20.13) to execute the simulation. When the user clicks the *Write Files* button in Figure 20.12, it results in writing up of the detailed configuration file for repository purposes.

Figure 20.13 shows the simulation clock as it happens in real time and the obtained statistics. The snapshots (Figures 20.8–20.13) complete the architectural components specified in Figure 20.6. Figure 20.13 has the functionalities that are described earlier in Chapter 7, for example, run-time configuration update and simulation control. It has three buttons at the top of the screen, which are as follows:

1. *Run/Resume Model* (using detailed parametric settings)
2. *Pause* (to interrupt the simulation)
3. *Terminate* (to end the simulation)

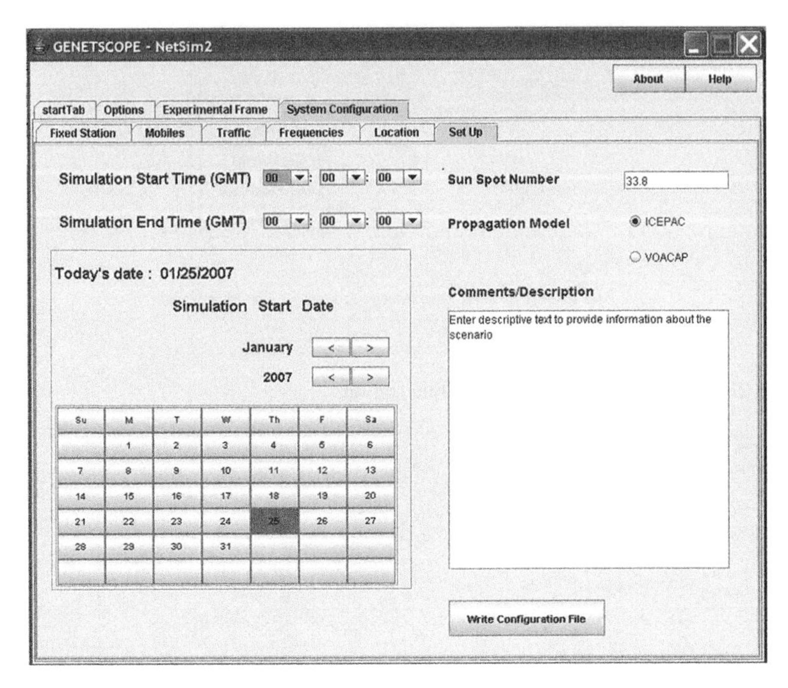

FIGURE 20.12 Experimental frame and ICEPAC data configuration through selection of SSN.

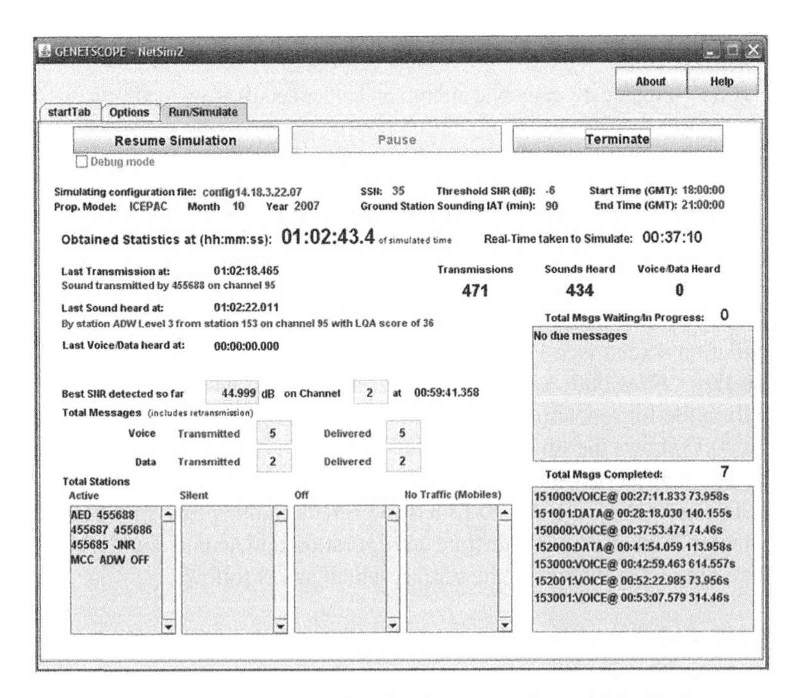

FIGURE 20.13 Run-time simulation visualization screen for rapid feedback.

The *Pause* button is of special interest here as the user can interrupt the running simulation (manual reactive control described in Chapter 7) and change the experimental frame or system configuration settings while the simulation is in action. Once the parameters have been updated, the user can resume the simulation and can see the impact of that update on the above "active" simulation visualization screen. One such example may be the two obtained values of total transmissions and total *sounds* heard. If the number of *sounds* heard is not up to the mark (with respect to a validated real-world scenario), the user may change sound interval time or any other parameter that would impact this number or may conclude that the model is not "performing" correctly. The rapid impact of any such parameter can be studied by *pausing* the simulation, changing it, and then observing the effects in the simulation pane.

The DEVS layer in Figure 20.6 is implemented in the following manner. The simulation engine running behind uses the following code:

```
NetsimSC net = new NetsimSC(createdConfigFile, debugOption);
tCoord = new TunableCoordinator(net);
tCoord.initialize();
tCoord.setTimeScale(0.0001);
tCoord.simulate(Integer.MAX_VALUE);
```

The model configuration is written into a configuration file that is used to create the DEVS digraph model, with automated coupling using the SES shown in Figure 20.4. The DEVS model is then passed on to the *TunableCoordinator* derived from DEVS *RTcoordinator* class. The *TunableCoordinator* is initialized and is then directed to simulate for a maximum number of iterations, which means that simulation will proceed indefinitely (in logical sense). The *Pause* button executes the following line:

```
tCoord.interrupt();
```

After the simulation is paused and updates are made, the simulation is restarted by simply calling the coordinator to "simulate."

```
tCoord.simulate(Integer.MAX_VALUE);
```

The simulation core functionality provided by the DEVS simulation protocol facilitates interrupting the coordinator and makes real-time parametric and component structures at run-time.

Figure 20.13 contains a very limited set of aggregated information. However, run-time graphs and projections are very well aligned with this visualization to see result patterns and the direction in which the simulation is proceeding. Logs are generated for each simulation run. This visualization pane shows the important information of the experimental frame and the run-time information from the system model, which, needless to say, is according to the enhanced MVC (through the development of appropriate interfaces between these layers). The View layer (see Chapter 7) in the current example shows only the model and the Experimental Frame control visualization.

The Experimental Frame control is *controller B* in Figure 7.1, that is, parameters that "control" the model. The lowest layer, that is, *controller A* in the enhanced MVC process, was not the focus of the GENETSCOPE project and consequently not illustrated here. Its implementation is illustrated by Nutaro and Hammonds (2004).

20.4 MOE, MOP, AND NR-KPP

Having laid out the framework to conduct and design the experiments, the next item on the agenda is to identify the *measures of effectiveness* (MoEs) that eventually will be considered in making recommendations for any update or modification needed in the current SCOPE command infrastructure. Since the SCOPE command is a deployed system, we were given various statistical reports by JITC to determine these MoEs. The point of this exercise is to provide sufficient analysis through simulation of the modeled system so that the impact of any particular infrastructural change intended in the system can be observed on these MoEs. Some of the MoEs that were identified are as follows:

- Longest time taken by any e-mail on HF network
- Number of e-mails sent and number of e-mails actually delivered
- Average message transmission time at any station per hour
- Messages attempted versus messages received per hour
- Bandwidth usage at central network command station (CNCS*)
- Number of planes in "good" SNR range per hour

The parameters that are to be set to recommend any upgrades in the current infrastructure can be listed as follows:

- Average number of daily flights
- Minimum number of messages attempted by any station
- Number of fixed stations participating in any mission scenario
- Number of active levels within a fixed station
- Minimum and maximum message size in KB
- Minimum and maximum duration of a phone call (voice message)
- Minimum data rate by any ALE radio modem

As can been seen clearly, there is no one-to-one mapping between MoEs and experimentation parameters. The MoEs tell us about the effectiveness of any mission that would be executed. They are holistic measures that tell about the fitness, capacities, and limitations of the system. Modeling and simulation (M&S) is the preferred means for assessing the impact of parameters on MoEs, with the goal of determining the most significant parameters. A simulation execution environment can help this investigation through a rapid feedback cycle where the analyst can change parameter

* CNCS is the gateway for any land-based network (Secret Internet Protocol Router Network (SIPRNET) or Non-classified Internet Protocol Router Network (NIPRNET)) to be connected to the SCOPE command HF network. All e-mails are routed through CNCS.

values on the fly and quickly assess their impact on holistic measures. These MoEs impact evaluations very well and become part of the result set, while the parameters identified become part of the experimental frame layer, as shown in Figure 20.13.

Similarly, for any Department of Defense Architectural Framework (DoDAF), the MoEs are also specialized for that particular architecture. Considering the breadth of the SCOPE command system, some of the MoEs mentioned in the preceding discussion also apply to any netcentric architecture. Within the DoD, JITC has the sole responsibility of certifying the IT and National Security Systems (NSS) for interoperability purposes (Buchheister, 2004). The major testing and evaluation (T&E) problem identified today by JITC is how to verify that a solution provided by any architecture is syntactically integrated and netcentric in operation. The traditional T&E approaches are optimized to verify performance and effectiveness of point solutions, but new criteria are needed to reflect the realities of systems operating within networked systems. According to Buchheister, such criteria are just beginning to emerge and are not yet matured for immediate and widespread use of T&E.

The net ready key performance parameter (NR-KPP) assesses net-readiness information assurance (IA) requirements and end-to-end operational effectiveness of that exchange with respect to the communities of interest (COIs) mentioned in Chapter 12.

Description of key interface profile (KIP) with relation to this scenario is beyond the scope of this chapter and reader is encouraged to review the work by Mittal, et al. (2006). The major objective underlying NR-KPPs is to identify verifiable performance parameters and associated metrics required to evaluate timely, accurate, and complete exchange and use of information to satisfy the information needs for a given capability (Buchheister, 2004).

Section 20.5 describes a suite of simulation logs that get generated after a simulation run and various performance logs that help address the MoEs for the GENETSCOPE project.

20.5 SIMULATION EXECUTION AND LOGS

A simulation run generates a total of 11 log files with each entry time-stamped. Out of the 11 log files, 8 are meant for the end user while the remaining 4 are for debugging and analysis purposes. The list is as follows:

- *ALE log*: This log shows the state of ALE radios and their operational behavior on a particular frequency. The parameters logged are location, time stamp, channel, destination, SNR, link quality and ALE state.
- *Channel log*: This log shows channel occupancy and transmissions by a particular station (or ALE radio).
- *LQA log*: This log reports the LQA scores by all the stations (fixed and mobile). It shows which station is able to hear the other stations at what SNR.
- *Linking log*: This log reports the ALE link established between two ALE radios during the course of simulation and the time it took to establish the link through atmospheric propagation.
- *Message log*: Once the link is established between any two ALE radios, this log reports if any message (voice or data) has been transmitted over the

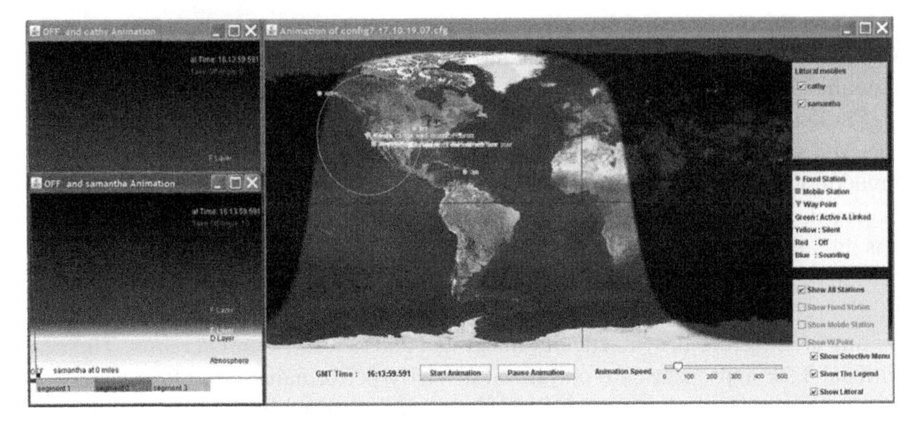

FIGURE 20.14 Animation screen with a link highlighted. (From www.rtsync.com.)

link. It shows the message transmission duration, source, destination, and message ID.
- *Mobiles log*: This log reports the position of mobile ALE stations as they move in a scenario through a series of waypoints.
- *GWave NVIS log*: This log is specifically for littoral operations and correlates Ground wave (GWave) and near vertical incident Skywave (NVIS) SNR values with the mobile's distance traveled from the fixed station.
- *Summary log*: This log provides a summary of the simulation run and correlates information from all the logs above.
- *Propagation log*: This log reports the execution of ICEPAC SNR values when ICEPAC is invoked by the ALE radio to evaluate if the link is of good quality based on its current location, time, and other factors. Any time a link quality is evaluated, the result from ICEPAC software is reported here. This log is used for debugging and analysis.
- *Error log*: This log is used to report any simulation errors.
- *Event log*: This log is used for debugging purposes and it is huge as each event is logged.

After simulation of a specific configuration, the entire simulation can be played as an animation. The animation panel contains initial fixed stations, mobile stations, and waypoints (Figure 20.14).

20.6 SIMULATION PERFORMANCE

JITC conducted various benchmarking tests on the performance of GENETSCOPE simulation software. Table 20.1 shows two of the results from the work by Hurd, 2007.

As can be easily seen, for a realistic 1-day scenario of 24 hours, the simulation takes more than 3 days to complete and provide the results. It is a good thing that the software did not crash and is robust, but running the software on a desktop still has its limitations. A 3-day return window for a 1-day scenario may not be acceptable to

TABLE 20.1
Simulation Performance of GENETSCOPE

Scenario Duration	ALE Activities (Sound, Listen, Call)	Transmissions (Sound, Link)	Calls to Propagation Program	ALE Links	Time Taken for Simulation Run
Small scenario: 6 stations, 2 aircrafts, 9 channels					
1 hour	217	54	723	2	2 hours
Large scenario: 14 stations, 100+ aircraft, 9 channels					
24 hours	10,000+	5,000	100,000	500	72+ hours

the decision makers. As the software is modular, we now look at a potential methodology to make it executable in a distributed netcentric platform. While netcentricity was not implemented in any version of GENETSCOPE, we would like to go through an exercise of making a legacy application netcentric.

20.7 MAKING GENETSCOPE NETCENTRIC

We begin our analysis with the bottleneck identification as making an application execute on a distributed platform should alleviate the existing problem, thereby having a positive cumulative effect on the performance. Making an application execute on a distributed parallel infrastructure does not guarantee that the application is netcentric. A netcentric application certainly is distributed by nature. What separates a distributed architecture from netcentric architecture is the inclusion of standards. These standards define the protocols and modularity of the interacting components. A netcentric application built on web services has standards like Web Service Description Language (WSDL) and Simple Object Access Protocol (SOAP) over various transport protocols like Hypertext Transfer Protocol (HTTP), Simple Mail Transfer Protocol (SMTP), and so on. In Section 20.2 (Figure 20.3), we described the modularization process. The next step to achieve netcentricity is the standardization process, as shown in Figure 20.15. Step 4 in Figure 20.15 standardizes the system components' interfaces and the messaging backbone's protocols. Step 5 identifies the service components' interfaces and the message backbone's protocols.

In the case of GENETSCOPE, consider that two bottlenecks exist at

- *Log writing*: This consumes the precious CPU time. It is generally known that disk input and output (I/O) operations are most time consuming.
- *Propagation calls*: This indicates an unusually high number of calls to the ICEPAC/VOACAP software.

To solve the first issue, we introduce a messaging backbone that interfaces with the ALE engine, the channel model, and the propagation engine. As these engines go about their business of execution, they publish to an enterprise service bus (ESB) and the file I/O operations are handled at a remote location. This saves the CPU

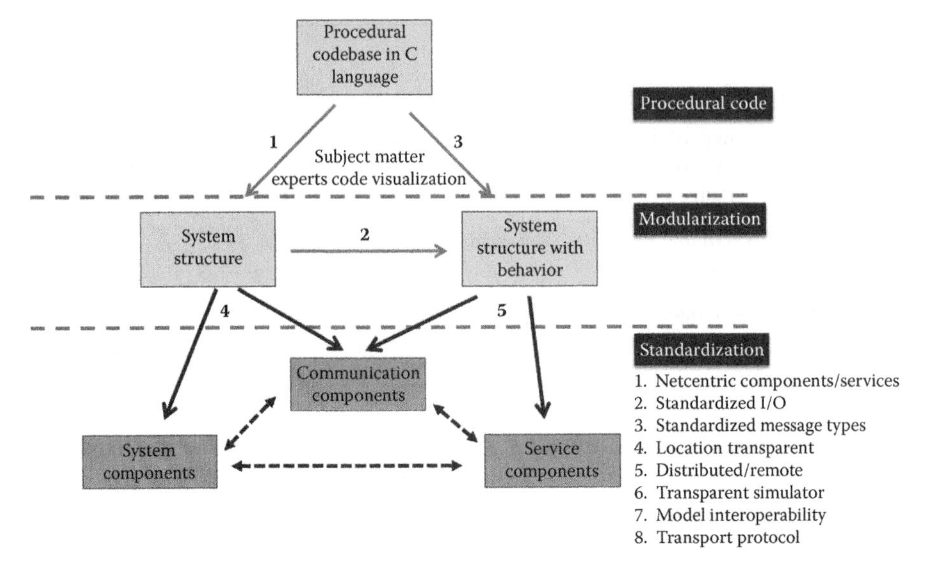

FIGURE 20.15 Standardization of system components, communication, and service interfaces.

cycles at the simulation nodes. In the current architecture, the logging objects are serialized from bytecode to the text file. The standardization aspect of this exercise would serialize the bytecode into XML and push the XML messages onto the ESB, where they are routed to respective log handlers. The ESB have an inbuilt buffer mechanism, so the messages are queued and do not slow down the engines during the file I/O blocking calls.

To solve the second issue, the ICEPAC/VOACAP software is put behind a web service interface. The software is called using Java Runtime class and parameters are passed through a function call. Considering a large scenario of 10,000 mobiles, equivalent to 10,000 ALE radios, the odds of blocking on the function call are quite high. A typical ICEPAC software call takes about few hundred milliseconds. After doing some numbers, assuming we arrive at a number wherein we have a thread pool to respond to 10 requests simultaneously, we can develop a system wherein no call to ICEPAC software is blocked. The thread pool can be implemented by a load balancer that can dynamically create additional ICEPAC software sites behind the web service interface.

The resulting architecture incorporating the two solutions is shown in Figure 20.16.

20.8 SUMMARY

This chapter has described a simulation software architecture named GENETSCOPE. The name was coined during the development efforts as it helped specify an architecture that is generic and network-oriented and geared toward SCOPE. The old version of this simulator was a monolithic system with no extensibility features. In the new version, new features were added and placeholders were provided that allow addition

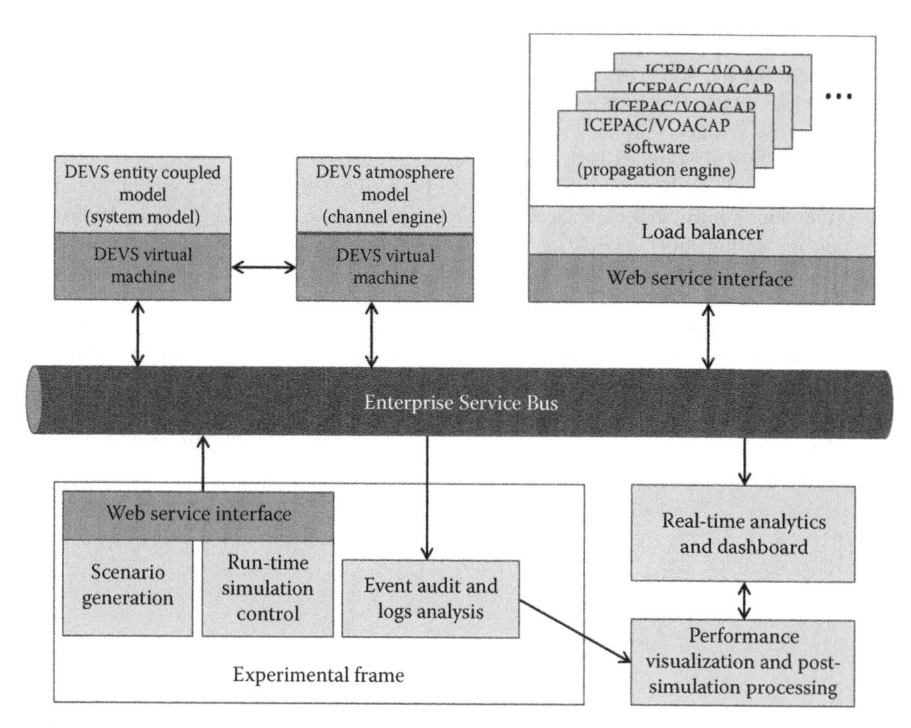

FIGURE 20.16 Netcentric architecture for GENETSCOPE.

of new capabilities, mobility models, terrains, hardware models, and scenario development. The architecture was made component-based and modular, and the behavior of system components was separated from the system structure. The structure was specified using the SES theory. The current system is currently in use at USAF and U.S. Navy. We also described the simulation architecture of GENETSCOPE and how component-based architecture lends itself to a focused and localized simulation capability. Such component-oriented and MoE-based requirement specifications guide the data collection efforts and help address many of the validation issues. As the entire architecture is modular, netcentric architecture of GENETSCOPE is described by standardizing the modular component interfaces and attending to the performance bottlenecks.

REFERENCES

Buchheister, J. (2004). Net-centric test & evaluation. *Command and Control Research and Technology Symposium: The Power of Information Age Concepts and Technologies* San Diego, CA.

GENETSCOPE. (2006), DISA, JITC Networks & Transport Division (JTE), High Frequency Test Facility, accessible at: http://jitc.fhu.disa.mil/it/gscope.html, last accessed October 2012.

Hurd, D. (2007, July 19). *High Frequency Test Facility and HF Network Modeling Brief.* Retrieved from http://www.hfindustry.com/meetings_presentations/presentation_materials/2007_feb_hfia/presentations/13_hfia_briefing_1feb07v3.pdf.

MIL-STD-188-110B, D. (n.d.). *DoD Interface Standard.* Retrieved from http://hflink.com/standards/MIL_STD_188_110B.pdf.

MIL-STD-188-141B, D. (1998). *Interoperability and Performance Standards for Medium and High Frequency Radio Systems.* Retrieved from http://militarycomms.tripod.com/-mil-std-188-141-b/index.htm. Last accessed: June 2012.

Mittal, S., Mak, E., & Nutaro, J. (2006). DEVS-based dynamic model reconfiguration and simulation control in the enhanced doDAF design process. *Journal of Defense Modeling and Simulation, 3*(4), 239–267.

Mittal, S., Seo, C., Kim, T., & Cheon, S. (2007). *GENETSCOPE (Beta Version). Software User's Manual.* Tucson, AZ: ACIMS, University of Arizona.

Nutaro, J., & Hammonds, P. (2004). Combining the model/view/control design pattern with the DEVS formalism to achieve rigor and resuability in distributed simulation. *Journal of Defense Modeling and Simulation, 1*(1), 19–28.

STANAG5066. (n.d.). *NATO Standardization Agreement: Profile for Maritime High Frequency (HF) Radio Data Communication.* Retrieved from zmailer.org: http://ham.zmailer.org/oh2mqk/HF-data/stanag5066.pdf. Last accessed: June 2012.

21 Executable UML

21.1 OVERVIEW OF UML

It is difficult to develop a simulation model in the early phase of system development since it requires a high-level knowledge of the following three aspects: modeling techniques, the system domain, and the model execution paradigms. The model development platform may be completely different from the model execution platform, but most of the time both are treated as one. The developed model is so much customized to the problem at hand that it impedes extensibility on the said aspects. It becomes necessary to answer questions about properties (most notably behavior) of the whole system. It requires intensive cooperation among domain experts and modeling experts as both demand totally different expertise. The entire systems M&S process starts with eliciting system requirements and ultimately translating them into an executable modeling code. The simple act of executing the model is termed as *simulation*. In addition, the Department of Defense (DoD) has strongly recommended applying M&S techniques to validate the requirements during the system development (Department of Defense, 2006). We need a practical and efficient way of applying M&S to the development of systems under design in early phases of system development. Most importantly, we must separate the art of modeling with the model platform so that the subject experts could focus on the model abstraction rather than the modeling platform. In other words, a platform-independent model (PIM) is the preferred way that would aid the subject-matter expert (SME) to participate in the modeling process directly.

Unified Modeling Language (UML) is one of the preferred means of communication between the domain experts and the modeling experts. UML is very powerful in terms of its graphical representation but diminishes in quality when it comes to execution of the UML model. Executable UML (Mellor & Balcer, 2002) is a working draft, and various groups are trying to make it operational in its current state today. In this chapter, we address the issue of executable UML with a set of transformations to generate an executable simulation model from a UML graphic specification using DEVS that provides a system theoretic foundation.

Many different paradigms, such as System Entity Structure (SES) and DEVS hierarchical modeling, can very well be used to interface with UML structure diagrams. However, the problem comes at the level of atomic components that contain finite state machines as their behavior models. Although the UML specification contains statecharts, their mapping with the DEVS state machine results in augmentation of UML statecharts with new added information for which there is no UML specification present, for example, timeouts for each state (i.e., *time advance* in DEVS). This problem has been highlighted during our work on transforming DoDAF 1.0 to DEVS in Chapter 11, and an argument is presented that DEVS is more rigorous when it comes to modeling a state-based system. DEVS is more known in the academic community, while UML is widely practiced in the industry. The aim of this chapter is to specify the graphic language so that systems engineering modelers may learn how to apply and use UML to build DEVS models, both structure and behavior. At the structure design level, we utilize UML component, package, and class diagrams. At the behavior design level, we use UML use case, sequence, timing, and state machine diagrams. To provide widespread adoption of the proposed executable UML, we coin the acronym *eUDEVS*, which stands for *executable* UML based on DEVS. While the basic conceptual mapping was done as a part of the DoDAF 1.0 study (Mittal, 2006), this chapter provides a detailed implementation.

The proposed UML-based M&S method takes three steps. First, we synthesize the static structure defined by a DEVS, SES, or UML model. Second, we specify its behavior using the DEVS Modeling Language, as described in Chapter 5. At this stage, the models are totally platform independent. Lastly, we take these PIMs and autogenerate the platform-specific model (PSM) using various transformations that depend on the DEVS simulation engine syntactical requirements.

There are many UML computer aided software engineering (UML CASE) tools, such as IBM Rational Rose and Poseidon, and all of them provide simulation functionalities, tracing states change or signal invocation. All these tools have proprietary simulation engines. Further, most of the simulation engines are not extensible toward performance-related requirements. For example, a typical OPNET model takes days to complete an execution. A parallel simulation engine would be needed, but because of the proprietary nature, such extensions are not possible. For our purpose, we need an open-source specialized simulation engine that can take over the details of the simulation process (event management, simulation time management, etc.), provide extensibility, and be an implementation of the DEVS formalism. We are henceforth using DEVSJAVA version 3.0 (ACIMS, n.d.) and Microsim/JAVA to develop our case.

After demonstrating the transformation of UML models into a DEVS component-based system, we will go a step further to make these components fall under an overarching DEVS Unified Process (DUNIP). It is a logical extension that allows UML-based models, once made executable, to form a component in a unified process much like the IBM Rational Unified Process (RUP).

To provide an overview, this chapter has the following objectives:

1. To unify the UML community with the DEVS community
2. To facilitate the execution of UML models, especially behavior models, using the DEVS formalism
3. To demonstrate that a behavior can be represented using an XML-based DEVS formalism with some caveats and limitations
4. To illustrate that UML and DEVS can be cross transformed
5. To make UML models as a component in an overarching systems engineering-based DUNIP

There are other approaches to represent DEVS models, such as the Scalable Entity Structure Modeler (SESM) from Arizona State University. It is suitable for developing component-based hierarchical models. It offers a basis for the modeling behavioral aspect of atomic models by providing the structural specification and storage of the model using XML, but this approach is very close to the simulation expert instead of the domain expert. Furthermore, there are no means to develop the atomic state machine or behavioral model explicitly. This is largely a structure tool and needs further work to represent atomic models using XML.

In the UML-based M&S domain, several authors have approached this subject from various perspectives. Choi et al. (2006) utilize UML sequence diagrams to define the system behavior. Hong and Kim (2004) introduce eight steps to make DEVS models using UML, but in these cases they need many human decisions to transform the model. A formal mapping from DEVS to UML is presented by Zinoviev (2005). Within this mapping, input and output ports are mapped to UML events. DEVS state variables that are non-continuous are mapped to UML states, and DEVS state variables that are continuous are mapped to attributes of UML states. Zinoviev also employs a combination of a special timeout event and use *after* events for handling internal transitions. The mapping presented is elegant. However, his UML representation is not intended to provide a unified representation on top of a modeling formalism for the purpose of composition, but as a replacement for the original DEVS specification. Huang and Sarjoughian (2004) present a mapping for coupled models into UML-RT structure diagrams, but the use of the UML profile for schedulability, performance, and time specification (defined by the Object Management Group in 2012) is unnecessary when mapping from DEVS to UML. They conclude that UML-RT is not suitable for a simulation environment, and they assert that the design of software and simulation is inherently distinct, as we address in this chapter. Borland and Vangheluwe (2003) develop a methodology to transform hierarchical statecharts into DEVS. The formal transformation of timed input/output automata into a DEVS simulation model is presented by Giambasi (2003). However, the timed automata are hard to communicate and develop the simulation model. We earlier proposed a UML-based M&S method, making use of UML state machine diagrams to define the system behavior, where by using the State Chart eXtensible Markup Language (SCXML) (W3C, 2012), time events are clearly defined in order to generate executable atomic models (Risco-Martín et al., 2007). Finally, an informal mapping from DEVS to an equivalent STATEMATE statechart is done by Schulz et al. (2000), who note that DEVS has greater expressive capabilities than statecharts and that any DEVS model can be represented via STATEMATE activity charts along with an appropriate naming convention for events.

Extending the domain to the model-driven architecture (MDA) paradigm, Tolk and Muguira (2004) show how the complementary ideas and methods of the High Level Architecture (HLA) and DEVS can be merged into a well-defined M&S application domain within the MDA framework. HLA is a distributed simulation architecture regardless of computing platforms. It provides an interface that each constituting simulation engine must conform to in order to participate in a distributed simulation exercise. While it is widely used in the defense industry, its adoption into the industry has been prohibited by its lack of expressive power.

As it is clear, none of this work has attempted to provide a unifying framework to bridge the UML-DEVS gap and executable UML in particular. We analyze not only the specification of both the structure and the behavior of DEVS models in UML, but also a general-purpose modeling procedure for DEVS models, supporting the specification, analysis, design, verification, and validation of a broad range of DEVS-based systems.

21.2 METAMODELS

In the field of software engineering, UML is a standardized visual specification language for object modeling (Hopcroft, et al., 2006). UML is a general-purpose modeling language that includes a graphical notation used to create an abstract model of a system, referred to as a UML model.

UML is officially defined by The Object Management Group (2012) by the UML metamodel. The UML metamodel and UML models may be serialized in XML. UML was designed to specify, visualize, construct, and document software-intensive systems. However, UML is not restricted to modeling software. UML is also used for business process modeling, systems engineering modeling, and representing organizational structures. UML has been a catalyst for the evolution of model-driven technologies. By establishing an industry consensus on a graphic notation to represent common concepts like classes, components, generalization, aggregation, and behaviors, UML has allowed software developers to concentrate more on design and architecture.

UML models may be automatically transformed to other representations (e.g., Java) by means of XSLT or QVT-like transformation languages, supported by the Object Management Group. In addition, UML is extensible, offering the following mechanisms for customization: profiles and stereotype. The semantics of extension by profiles have been improved with the UML 2.0 major revision.

UML 2.0 has 13 types of diagrams, which can be categorized hierarchically, as shown in Figure 21.1. Structure diagrams emphasize what things must be in the system being modeled and include class diagram, component diagram, composite structure diagram, deployment diagram, object diagram, and package diagram. Behavior diagrams emphasize what must happen in the system being modeled. They include activity diagram, state machine diagram, and use case diagram. UML also includes interaction diagrams, a subset of behavior diagrams used to define the flow of control and data among the entities in the system being modeled. Interaction diagrams are communication diagram, interaction overview diagram, sequence diagram, and timing diagram.

The UML metamodel shown in Figure 21.1 is put together within the SES ontology (Chapter 10). UML diagram is made of two different perspectives, namely,

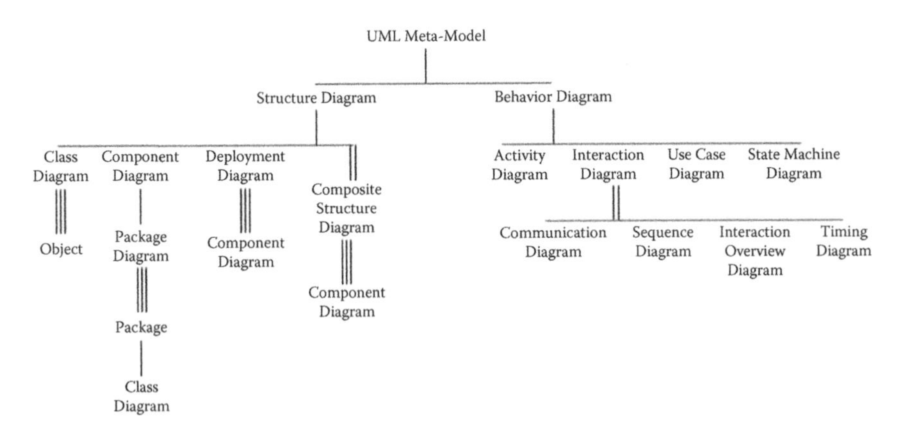

FIGURE 21.1 UML metamodel.

structure and behavior diagrams. For an effective UML model, in fact for any effective system model, both the diagrams are necessary. Similarly, behavior diagram is made of activity diagram, interaction diagram, use case diagram, and state machine diagram. However, there can be many ways in which interaction diagrams can be implemented. Depending on the project requirements, an interaction diagram can be a communication diagram, sequence diagram, interaction overview diagram, or timing diagram. The structure diagram consists of class diagram, component diagram, and deployment diagram. In addition to this decomposition perspective of the structure diagram, it is specialized as a composite structure diagram that is built using many entities of the type component diagram. A component diagram is made of a package diagram that contains many class diagram entities.

21.2.1 SES Representation of DEVSML

Chapter 5 described the DEVS Modeling Language (DEVSML) in the Xtext EBNF grammar. As stated earlier, Xtext is based on the Eclipse Modeling Framework, which has an underlying ecore meta-metamodel. This gives a hierarchical structure to any Xtext-based language that may be expressed in structured formalisms like XML. Similarly, SES is also a formal structured language that can be expressed in EBNF and XML. Further, the SES axioms position SES to be at the meta-metamodeling level, where it can be used to generate various other formal representations with a specific hierarchical structure. Figure 21.2 shows how SES relates to the model-driven engineering's classic levels and its relationship with the DEVS DSL or DEVSML.

Now, having described DEVSML in EBNF, we will now show the SES representation of DEVSML. Figure 21.3 shows the structure of organizing packages, imports, and types. There is a concept of an Abstract Element that can be a package, an import, or a type. A package is decomposed into a qualified name and an abstract element. This structure allows hierarchical and nested packages. Figure 21.4 shows the top-level basic types, namely, entity, atomic, and coupled.

Figure 21.5 shows the entity details. An entity is made up of its supertype and a set of variables. Each variable has a VarType, which may be simple or complex. A complex VarType references an entity by its qualified name. A message is a

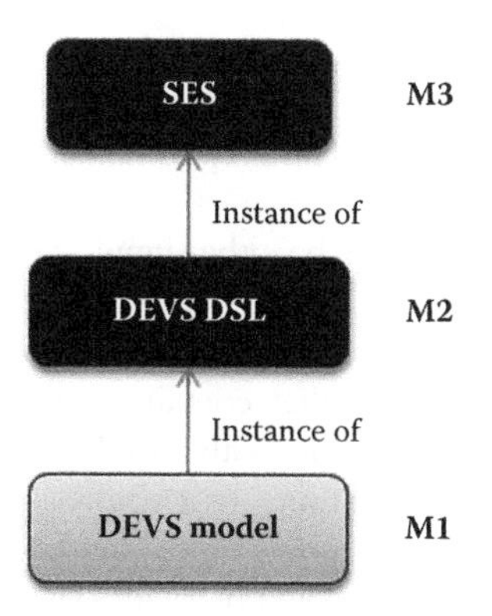

FIGURE 21.2 Model-driven engineering's description of SES and DEVSML.

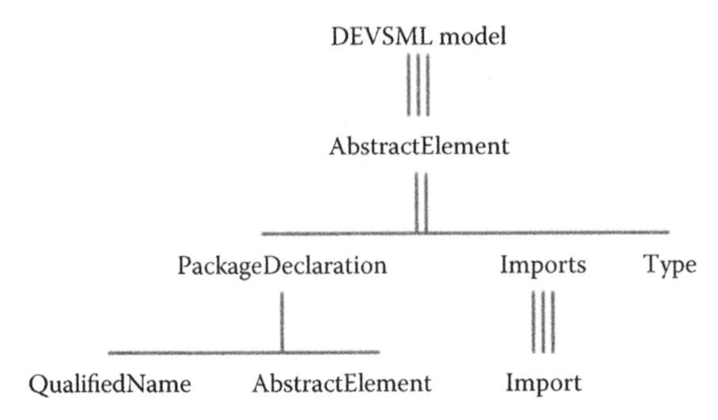

FIGURE 21.3 Hierarchical package structure and namespace organization in DEVSML.

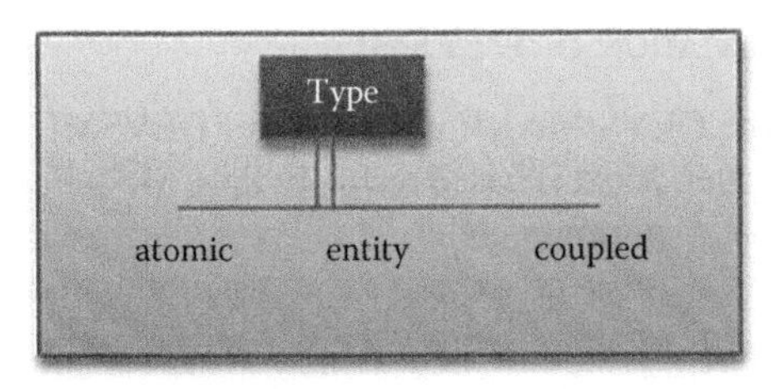

FIGURE 21.4 Top-level SES types in DEVSML.

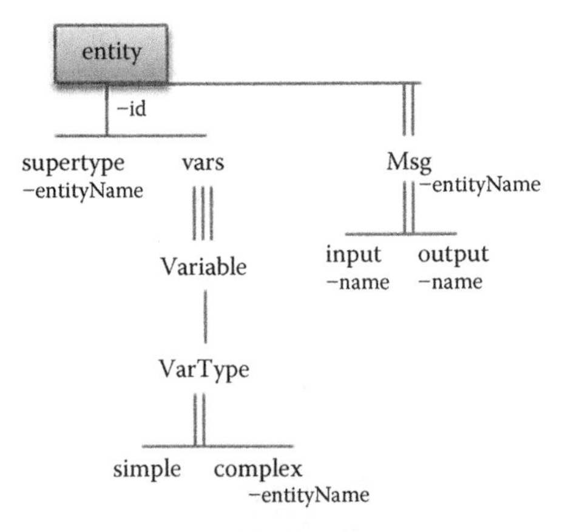

FIGURE 21.5 DEVSML entity type in SES representation.

specialization of an entity, and it can be either input or output. A message references an entity with a qualified name.

Figure 21.6 shows the atomic details. An atomic is made up of a set of variables described as above, an interfaceIO that contains a set of messages, and a set of state-time-advances that associate a state with a well-defined time-advance value. The time-advance value can be of float, infinity, or a locally scoped variable. In addition, an atomic is also decomposed into a state-machine. The state-machine has an initial state (defined in the state-time-advance set) and behavior. The behavior is decomposed into Deltint, Deltext, confluent, and Outfn. The remaining figure is self-explanatory.

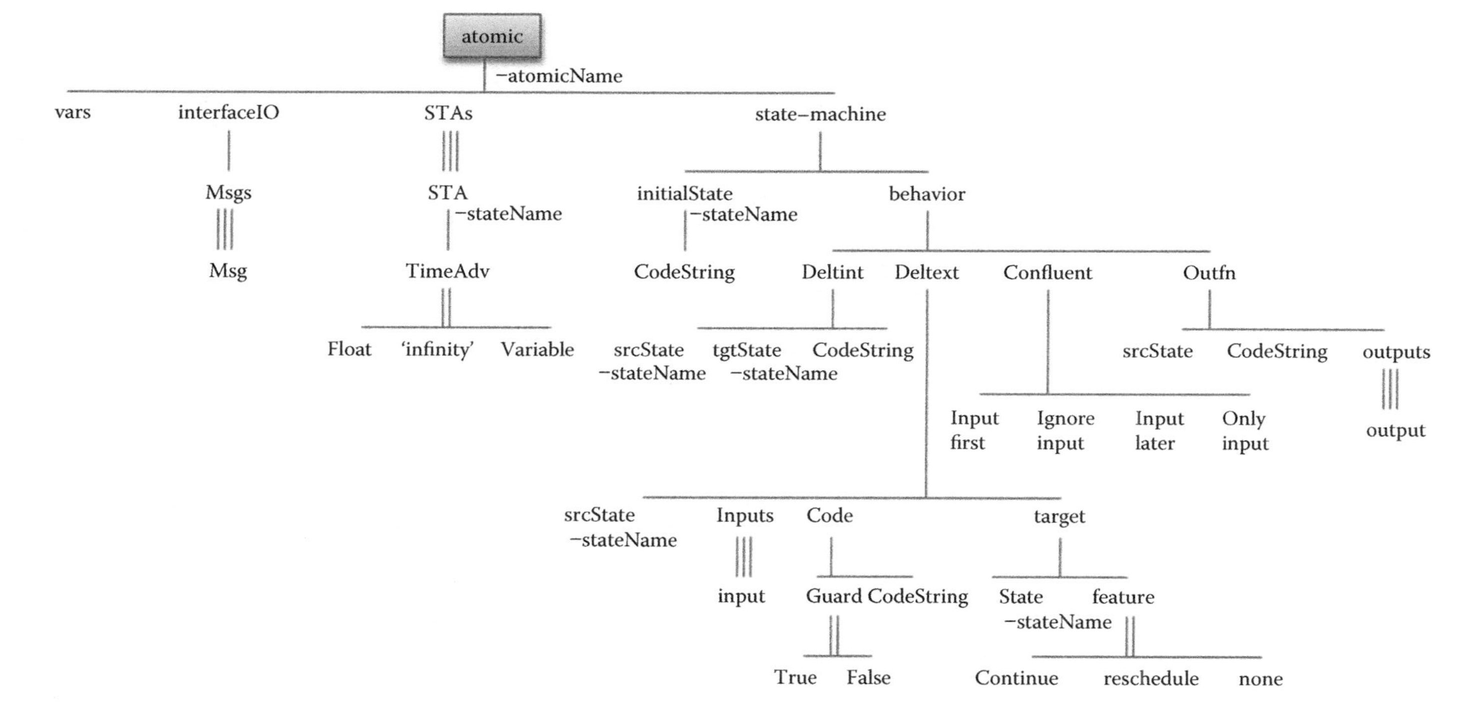

FIGURE 21.6 DEVSML atomic type in SES representation.

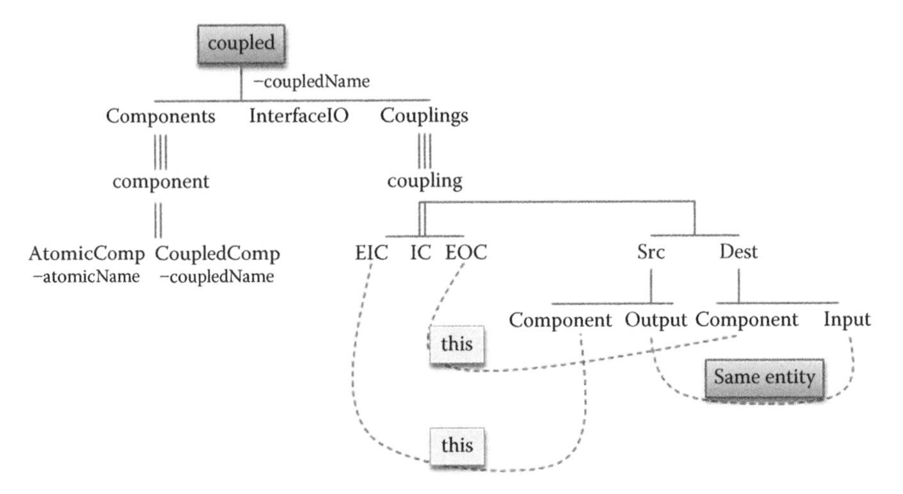

FIGURE 21.7 DEVSML coupled entity type in SES representation.

Figure 21.7 shows the coupled entity details. A coupled component is made up of components that may of qualified atomic or coupled reference, an InterfaceIO, and a set of couplings. A coupling can be EIC, IC, and EOC and so it goes. The dotted lines show constraints that are enforced through the SES meta-metamodeling framework and that are specific to the DEVSML. For example, in EOC, the destination component must be the "self" component referred as "this," and both the input and the output messages have to be of the same entity type.

21.3 MAPPING UML TO DEVS IN eUDEVS

21.3.1 Overview

Modeling is an art of abstraction of real systems to generate the behavior by specifying a set of instructions, rules, and equations to represent the structure and the behavior of the system. The structural elements of a system include the components of the system and their interactions (inputs, outputs, connections, etc.). The behavioral elements include the sequence of interactions, the timing constraints of the interactions, and the operations of each component. Modeling provides the means of specifying the structure of a system, behavior of a system over time, and the mechanism for executing the instructions, rules, or equations.

We provide a new approach for both the structural and the behavioral graphic descriptions. Currently, it is a common practice for systems engineers to use a wide range of modeling languages, tools, and techniques on large systems projects. Since UML unified the modeling languages used in the software industry, our approach is intended to unify the diverse modeling languages currently used by DEVS-based systems engineers. It will improve communication among the various stakeholders who participate in the systems development process and promote interoperability among DEVS-based M&S tools.

Our DEVS UML diagram taxonomy is shown in Figure 21.8. This diagram has its origin in Figure 21.1, but focuses on only those UML elements that contribute toward

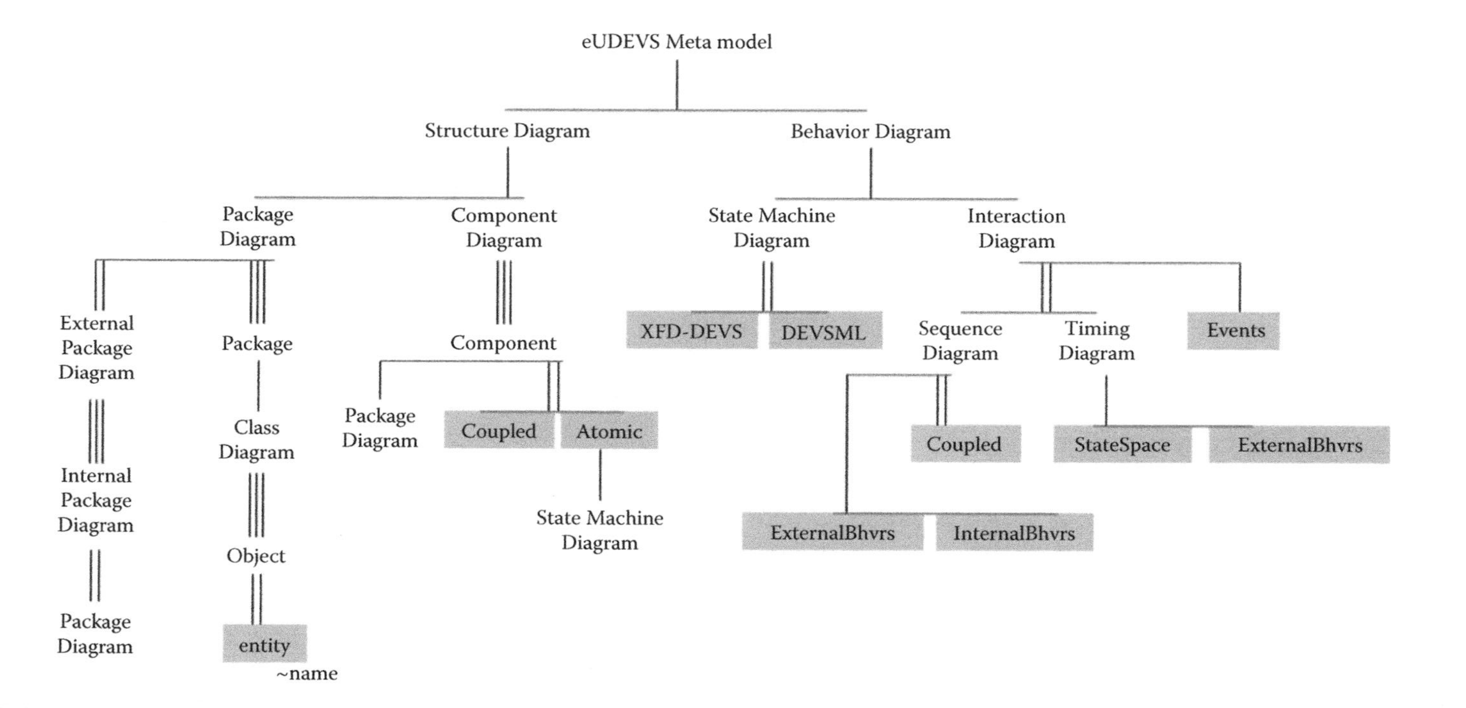

FIGURE 21.8 Executable UML using DEVS: eUDEVS metamodel.

a DEVS-based system model. A DEVS model, like any other system specification, is defined by means of structure and behavior diagrams. To define the structure, we make use of UML component diagrams, package diagrams, and/or class diagrams. To define the behavior, we utilize UML use case diagrams, state machine diagrams, sequence diagrams, and/or timing diagrams. To develop an interface between DEVS and any other modeling framework, such as UML here, we should first enumerate the information that is needed to develop a DEVS model:

1. Entities as objects and their hierarchical organization
2. Finite state machines of atomic models
3. Timeouts for each of the phases (states) in atomic models
4. Entity interfaces as input and output ports
5. External incoming messages at an entity's interface at a specified duration in a specific state
6. External outgoing messages at an entity's interface at a specified duration in a specific state
7. Coupling information derived from hierarchical organization and interface specifications
8. Experimental frame specifications

Having known the information needed to develop DEVS, the following subsections will show how this information is extracted from UML elements.

21.3.2 DEVS UML Structure Diagrams in eUDEVS

This section defines the static and structural constructs used to describe DEVS UML structure diagrams, including the component diagram, package diagram, and class diagram.

21.3.2.1 Component Diagrams

UML component diagrams have evolved substantially from version 1.x of the language to the current version 2.0. In the latest version of the language, deployment, object, and component diagrams have been merged into a single class of deployment-object-component diagrams. These new diagrams have a rich language suitable for elaborated models, but for the purpose of this chapter, we need to mention only components, delegation connectors, interfaces, and ports.

A UML component diagram is a set of components, ports, connections, and interfaces. UML components may contain subcomponents in a hierarchical fashion similar to DEVS coupled models containing models. Each UML component has a set of externally visible ports. UML components may be connected to one another and attached via ports similar to DEVS. UML ports can be unidirectional (input or output) or bidirectional. The direction of the port is defined by the types of its interfaces. Required interfaces ("antennas") define output ports, and provided interfaces ("lollipops") define input ports. In DEVS, connections between ports are unidirectional. Therefore, in a UML component diagram, all ports should be unidirectional; that is, a port may either provide an interface or require an interface, but not both since this implies bidirectionality. Components may be connected in two different ways. First, we can connect provided and required ports by a so-called assembly connector. In contrast, the delegation connector connects two ports of the same type between a component and one of its subcomponents; thereby, hierarchical components become possible.

When we use parallel DEVS as the target formalism, the mapping of component connections to model couplings and their association with model ports becomes relatively easy. Each component connection maps to a set of couplings $c \in \{\text{EIC} \cup \text{EOC} \cup \text{IC}\}$. Assembly connectors map to internal couplings, and delegation connectors map to external input and output couplings.

In UML, ports may have a multiplicity greater than one; this is not the case in DEVS (and hence not allowed in our UML component diagrams). In UML, ports may be unnamed; they must be named in DEVS and thus in our UML component diagrams. In UML, connectors need not attach to components (more correctly parts) via ports; this is not an option in DEVS and hence not an option in UML. If ports are specified to provide or require an interface, there should be only one such interface specified in the component diagram.

Finally, a UML component represents either another UML component diagram or a UML state machine diagram. We will consider that a DEVS coupled model may only include components and that a DEVS atomic model may only contain state machine diagrams.

A simple UML component diagram is shown in Figure 21.9, which corresponds to the DEVS coupled model. Figure 21.9 depicts two components, M1 and M2. These components are in turn subcomponents of component M. Component M has one output port, "out," with interface Event1 and one input port, "in," with the same interface. M1 has two input ports, in1 and in2, with interface Event1, and one output port, "out," with interface Event2. The M2 has one input port, in1, with interface Event2, and two output ports, out1 and out2, with interface Event1. Ports connected must send/receive compatible interfaces.

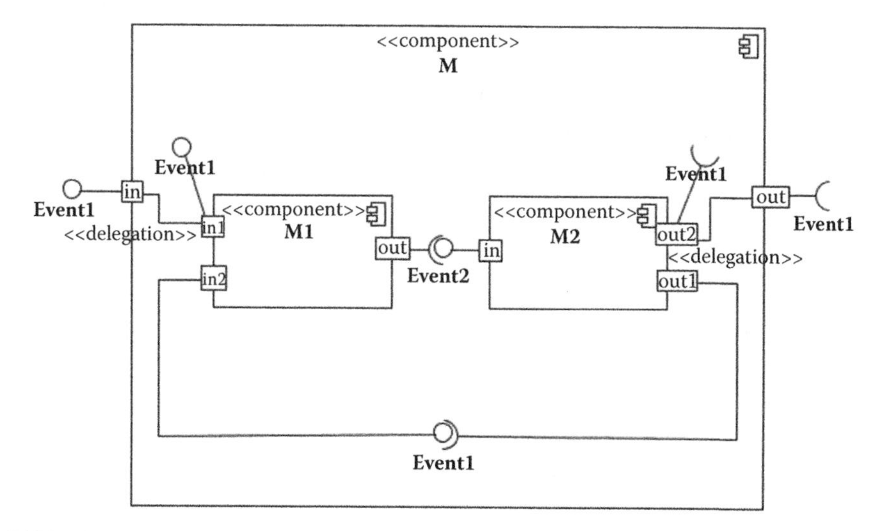

FIGURE 21.9 A DEVS UML component diagram.

Interoperability between different simulation platforms is possible at this level of message abstraction using XML. The message structure with a port-value pair is very well the foundation of interoperable and cross-platform system integration. The latest development of the DEVS/SOA platform is an evidence of such XML-based message passing between different DEVS components. In Figure 21.9, M1 could be implemented using one simulation engine and M2 by another one. The only constraint is that messages (or events) sent by M1::out must satisfy the same interface that those received by M2::in. If this constraint is not satisfied, another component, M12, should be defined allowing transformations to make interfaces compatible, as shown in Figure 21.10. This issue has been adequately dealt within the DEVSML in Chapter 5 with entity validations for input and output couplings in the coupled model creation. Table 21.1 shows the relation between the DEVS structural formalism and UML component diagrams.

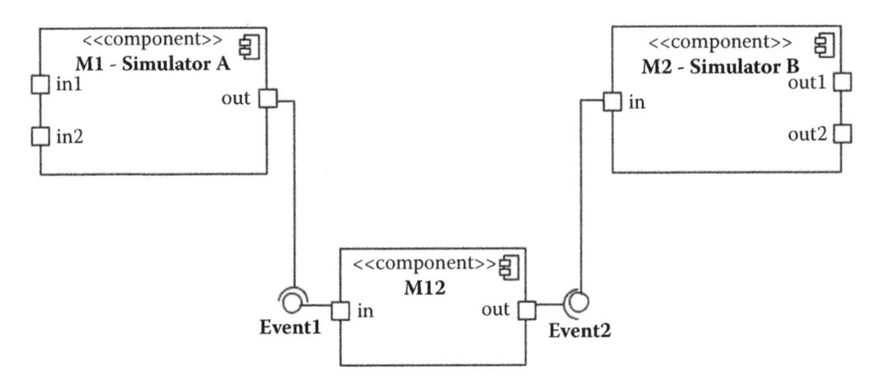

FIGURE 21.10 Interoperability between components.

TABLE 21.1
Relation between the DEVS Structural Formalism and UML Component Diagram

DEVS	UML Component Diagram
Atomic Model (Component Containing UML State Machine)	
IP	Ports with provided interface
OP	Ports with required interface
Coupled Model (Component Containing Components)	
IP	Ports with provided interface
OP	Ports with required interface
D	Components
EIC	Input delegation connectors
EOC	Output delegation connectors
IC	Interfaces connected

21.3.2.2 Package Diagrams

In UML, a package diagram depicts how a system is split up into logical groups by showing the dependencies between them. As a package is typically thought of as a directory, package diagrams provide a logical hierarchical decomposition of a system. Packages are usually organized to maximize internal coherence within each package and to minimize external association among packages. With these guidelines in place, the packages are good management elements. Each package can be assigned to an individual or team, and the dependencies among them indicate the required development order.

To define the structure of an M&S-based system, we define two kinds of representations in terms of package diagrams: external and internal (Figure 21.8). Packages in external representation will contain only packages, that is, the internal representation. Packages in internal representation will contain various class diagrams or models. The external representation encapsulates three parts of an M&S-based system: (1) simulation engines required to implement the system, (2) components utilized in the components diagram, and (3) supporting classes that are not executed directly by the simulation engine, such as port interfaces and extra data types. Internal representation defines the external representation structure in more detail. At this layer, we require one package per component in the components diagram. To this end, we have defined several stereotypes—engine, model, coupled, atomic, and support—that correspond to designing packages for simulation engines, the entire DEVS model, coupled and atomic components, and supporting classes.

Figure 21.11 shows the external representation (left side) and the internal representation (right side) of the component diagram. There are three packages, M, M1, and M2, in the internal representation that correspond to each component in the model. DEVSJAVA is the selected simulation engine to implement the model, and we need some port interfaces as supporting classes.

In contrast to component diagrams, package diagrams do not define the structure of a DEVS system. They need the support of class diagrams and sequence diagrams to define both the structure and the behavior.

21.3.2.3 Class Diagrams

UML 2 class diagrams are the mainstay of object-oriented analysis and design. UML 2 class diagrams show the classes of the system, their interrelationships (inheritance, aggregation, and association), and the operations and attributes of the classes. Class diagrams are used for a wide variety of purposes, including both conceptual/domain modeling and detailed design modeling.

By using the information contained in the component diagram and/or package diagrams, we are able to distinguish classes that are directly used by the simulation engine from supporting classes that are not.

Figure 21.12 depicts an illustrative example of two object diagrams. Figure 21.12a shows the M package (as per the package design in the last subsection, the package name is same as the component name). The class CoupledM represents a DEVS coupled model. CoupledM contains two atomic models, AtomicM1 and AtomicM2,

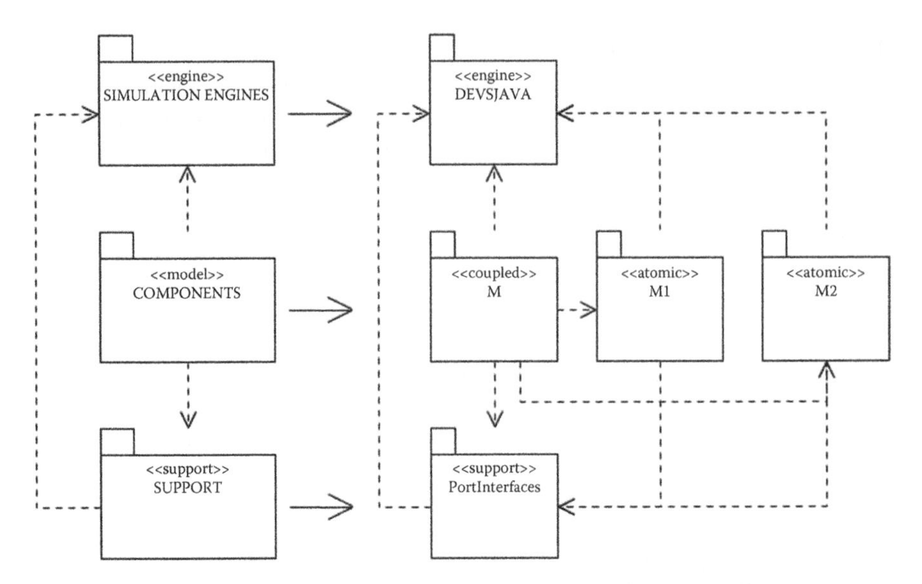

FIGURE 21.11 External and internal structure representation in package diagrams.

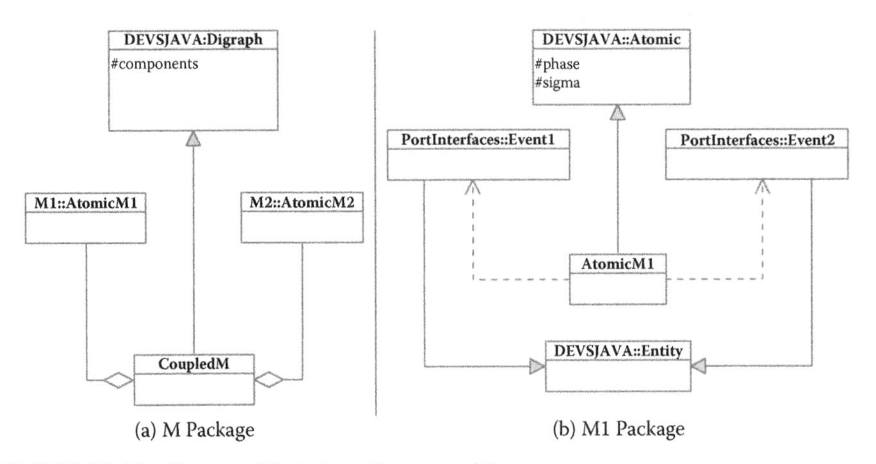

FIGURE 21.12 Two simplified class diagrams of the system.

both coming from component packages M1 and M2, respectively. Finally, since we are using the DEVSJAVA simulation engine, CoupledM extends the functionality of digraph, which is the base class for coupled models in DEVSJAVA. Figure 21.12b shows the M1 package. The AtomicM1 class is an atomic model, so it inherits atomic, which is the base class used in DEVSJAVA to implement atomic models. AtomicM1 sends and receives messages composed by elements that implement the interfaces Event1 and Event2. Both interfaces must extend the entity DEVSJAVA class as shown in Figure 21.12b, since it is the base class used to send messages between components. Please note that, for example, the DEVS component M, which corresponds to the M package, could be formed by more than one class. The example provided is quite simple; in this case, every DEVS component is implemented using only one class.

We have defined three diagrams that allow defining the structure of a DEVS-based system. Component diagrams are able to describe a DEVS model structure by themselves. However, package and class diagrams provide a software development point of view, which is important to design the modeling environment. Combining the information of these diagrams, we may generate the skeleton of such models, in terms of classes, attributes, empty member functions, and so on. This generation of software artifacts is platform dependent though. Again, we must stress that at this stage, we have only an executable structure model that is devoid of any behavior. The next section will address this issue.

21.3.3 DEVS UML Behavior Diagrams in eUDEVS

This section specifies the behavioral definition of a DEVS model in terms of use case diagrams, sequence diagrams, timing diagrams, and state machine diagrams.

21.3.3.1 Use Case Diagrams

Use case diagrams describe behavior in terms of the high-level functionality and usage of a system, by the stakeholders, and other members such as developers who build the system. The use case diagram describes the usage of a system (subject) by its actors (environment) to achieve a goal, which is realized by the subject providing a set of services to selected actors. The use case can also be viewed as functionality and/or capabilities that are accomplished through the interaction between the subject and its actors. Use case diagrams include the use case and actors and the associated communication between them. Actors represent classifier roles that are external to the system that may correspond to users, systems, and/or other environmental entities. They may interact either directly or indirectly with the system.

The use case relationships are "communication," "include," "extend," and "generalization." Actors are connected to use cases via communication paths, which are represented by an association relationship. The "include" relationship provides a mechanism for factoring out common functionality that is shared among multiple use cases, and is always performed as part of the base use case. The "extend" relationship provides optional functionality that extends the base use case at defined extension points under specified conditions. The "generalization" relationship provides a mechanism to specify variants of the base use case.

The use cases are often organized into packages with the corresponding dependencies between them.

Figure 21.13 depicts how use cases help delineate a specific kind of goals associated with driving and parking a vehicle. In Figure 21.13, the "extends" relationship specifies that the behavior of a use case may be extended by the behavior of another (usually supplementary) use case. The "Start the Vehicle" use case is modeled as an extension of "Drive the Vehicle." This means that there are conditions that may exist that require the execution of an instance of "Start the Vehicle" before an instance of "Drive the Vehicle" is executed.

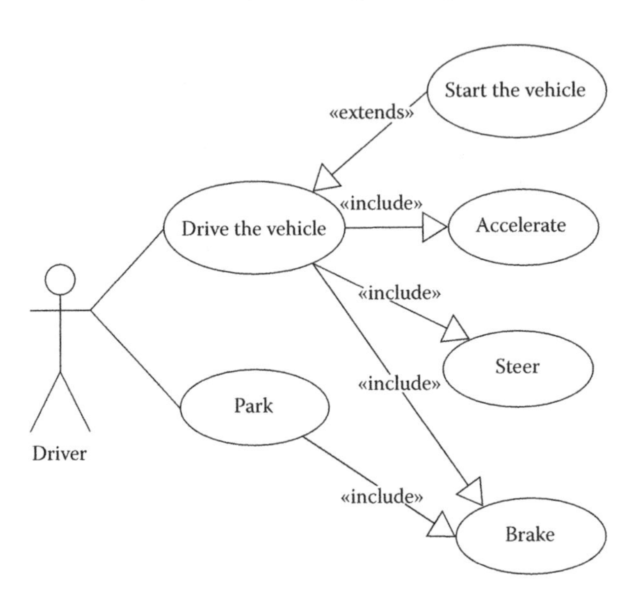

FIGURE 21.13 Use case diagram.

The use cases "Accelerate," "Steer," and "Brake" are modeled using the include relationship. Include is a directed relationship between two use cases, implying that the behavior of the included use case is inserted into the behavior of the including use case. The including use case may only depend on the result (value) of the included use case. This value is obtained as a result of the execution of the included use case. This means that "Accelerate," "Steer," and "Brake" are all part of the normal process of executing an instance of "Drive the vehicle."

In many situations, the use of the include and extend relationships is subjective and may be reversed, based on the approach of an individual modeler.

With respect to DEVS-based M&S, use case diagrams are used to establish the system context—defining system boundaries and multilevel resolutional capabilities at an appropriate hierarchical level. They should be used like a starting point in the model development. However, there are no special rules to design the best use cases. The resolution depends mainly on the model context under development.

21.3.3.2 Sequence Diagrams

The sequence diagram describes the flow of control between actors and systems or between parts of a system. This diagram represents the sending and receiving of messages between the interacting entities called lifelines, where time is represented along the vertical axis. The sequence diagrams can represent highly complex interactions with special constructs to represent various types of control logic, reference interactions on other sequence diagrams, and decomposition of lifelines into their constituent parts.

We make use of state diagrams defined in Choi's work (2006), which uses three operators (seq, alt, and loop), and adds the DEVS sigma information to explicitly specify the timing constraint among components of a system. We define both the phase and sigma of the model by means of constraints in the sequence diagram. Sigma specifies a point in time and the event occurs at the specified time; the phase provides information about the model's global state. For example, the constraint "active, 5s" means that the event occurs after 5 seconds, while the global state of the model is "active." All the sending and receiving events should have sigma and phase, except those lifelines that represent coupled models. If sigma is infinity, it is denoted as "inf," which implies that the object waits for the incoming event indefinitely until any message arrives. Finally, if sigma is not defined explicitly, it is supposed that sigma is updated using the elapsed time of the coming event, that is,

$$\sigma = \sigma - \varepsilon$$

where ε is the elapsed time.

Figure 21.14 depicts an illustrative example. From M1 point of view, there is an initialization message that initializes the phase to active and sigma to 5 seconds. If no external transition happens, the next event (message) is sent after 5 seconds followed by an internal transition, which sets the phase to passive and sigma to infinity. After 9 seconds, M1 receives a message from M2 and an external transition happens. It sets the phase of M1 to passive and sigma to infinity. Note that in the case of M2, its sigma is updated to 4 seconds after the M1 internal transition (9 seconds minus elapsed time, 5 seconds) implicitly.

Although sequence diagrams are presented in the behavior section, they are able to represent the DEVS structure. Table 21.2 summarizes the relations between the DEVS formalism and UML sequence diagrams.

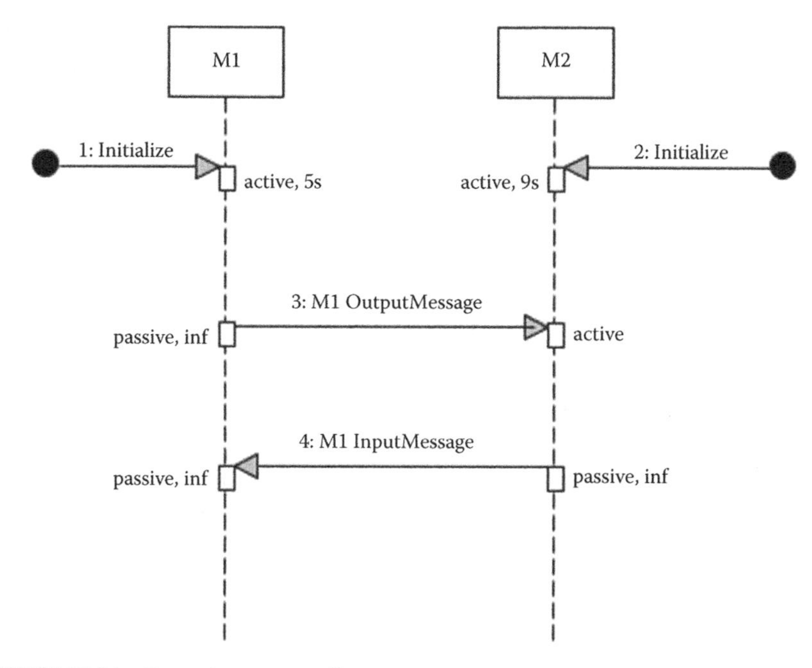

FIGURE 21.14 Example sequence diagram.

TABLE 21.2
Relation between DEVS and UML Sequence Diagram

DEVS	UML Sequence Diagram
Atomic Model (Lifeline with AtomicStereotype)	
IP	Incoming messages' names
OP	Outgoing messages' names
X	Incoming messages
S	Constraints (phase, sigma)
Y	Outgoing messages
Coupled Model (Lifeline with CoupledStereotype)	
IP	Incoming messages' names
OP	Outgoing messages' names
X	Incoming messages
Y	Outgoing messages
D	Other lifelines
EIC	External incoming messages
EOC	External outgoing messages
IC	Messages between lifelines

21.3.3.3 Timing Diagrams

Timing diagrams are one of the new artifacts added to UML 2. They are used to explore the behaviors of one or more objects throughout a given period. There are two basic flavors of timing diagrams: the concise notation and the robust notation.

Figure 21.15 depicts an example of both concise and robust notations of timing diagrams. M1 starts in active state for 5 seconds. After that, M1 sends a timeout message called M1OutputMessage and makes an internal transition changing its state to passive. The message changes the state of M2 from "active, 9s" to "active" (sigma = 4s implicitly) through an external transition in M2. Four seconds later, M2 sends a timeout message and makes an internal transition, changing its state from active to passive. The message is called M1InputMessage, which executes an external transition in M1 without effects.

Timing diagrams are valid only for DEVS atomic models. Therefore, they are able to define the models' behavior. Table 21.3 summarizes the relation between the DEVS formalism and timing diagrams.

21.3.3.4 State Machine Diagrams

UML state machine diagrams define a set of concepts that can be used for modeling a discrete behavior through finite state transition systems. The state machine represents a behavior as the state history of an object in terms of its transitions and states. The activities that are invoked during the transition of the states are specified along with the associated event and guard conditions following the format "event [guard]/activity."

In our previous work, we established a set of procedures to define a DEVS state machine by using UML state machine diagrams using SCXML, used later in this chapter (W3C, 2012). We used the IBM Rational Software Architect to export UML

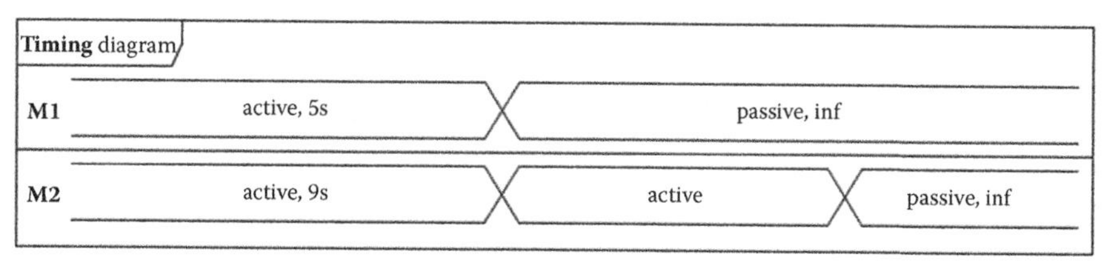

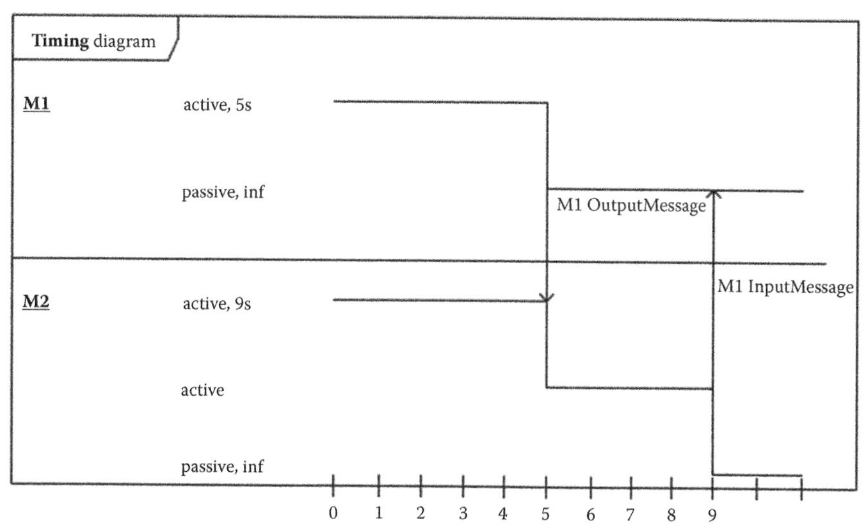

FIGURE 21.15 Timing diagram.

TABLE 21.3
Relation between Atomic DEVS and UML Timing Diagram

DEVS	UML Timing Diagram
IP	Incoming time messages' names
OP	Outgoing time messages' names
X	Incoming time messages
S	State/condition (phase, sigma)
Y	Outgoing time messages

state machine diagrams into SCXML and then used the XSLT mechanism to export it to XFD-DEVS. In its completeness, we showed how to transform UML state machine diagrams into a DEVS executable code. In the present approach, which is guided by the XFD-DEVS formalism, we follow a slightly different notation, where the state stores a list of two parameters: DEVS phase and sigma (timeout for that state). Output messages are defined by activities in the transition with the keyword "deltint" as the name of the event. It means that an internal transition happens, and just before it, an output message is sent. Input messages are specified using events in

TABLE 21.4
Relation between Atomic DEVS and UML State Machine Diagram

DEVS	UML State Machine Diagram
X	Events
S	State (phase/sigma)
Y	Activities

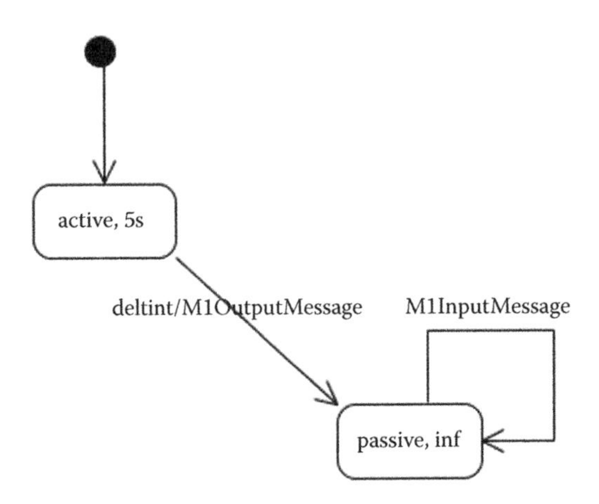

FIGURE 21.16 M1 state machine diagram.

the transition, which denotes a DEVS external transition. Optionally, we utilize the guard condition to define the interface sent or received, as we show in the example section. Table 21.4 summarizes the relation between the DEVS formalism and state machine diagrams.

Figure 21.16 shows the state machine diagram for M1. M1 starts in the state "active" for 5 seconds. After that, the message "M1OutputMessage" is sent, and an internal transition happens, changing the state to "passive, inf." If M1 receives the message "M1InputMessage," an external transition is executed, without changes in the state of M1.

21.4 TRANSFORMATIONS

In the previous section, we described how to define a DEVS model (both structure and behavior) by means of seven different UML diagrams. Some of them are utilized to define the DEVS model at a high level of abstraction (external package diagrams to represent the structure and use case diagrams to represent the behavior). However, it is important to emphasize that it is sufficient to implement a UML component diagram and a UML state machine diagram to define a DEVS executable model, but in this case, the development process is closer to modeling experts than to domain experts.

Modeling a DEVS model through the UML diagrams as described in earlier sections may follow a certain order. First, the structure must be defined in terms of component diagrams, package diagrams, and class diagrams. This UML information can be very easily represented by an SES diagram as well, which is entirely XML-based or DSL-based in Xtext in its latest implementation. Second, the behavior is defined by means of use case diagrams and sequence diagrams or timing diagrams or state machine diagrams that are augmented by more information as per DEVS DSL requirements.

At this stage, we have information coming from UML, SES, and DEVSML models. Our task is to remove redundancy and take the intersection of this information set guided by the minimalist information that is needed to create a DEVS M&S-based system. The information extraction process is largely attributed to various XML-based technologies, such as XSLT, XPATH, XSD, DOM, and JAXB, in case of XFD-DEVS and to graph transformations using Xtext in case of DEVSML. As laid out in Section 3 and bounded by Figure 21.8, we define our transformations leading to a DEVS PSM.

Figure 21.17 depicts the set of possible transformations between formats (transformation between DEVSJAVA and DEVSML are described in Chapter 5). The transformations from XML files to Java files are implemented using XSLT. Transformations from DEVSJAVA files to XML files are implemented using Java XML libraries such as JavaML. Use cases are out of the transformations because they do not provide relevant information for the DEVS executable model.

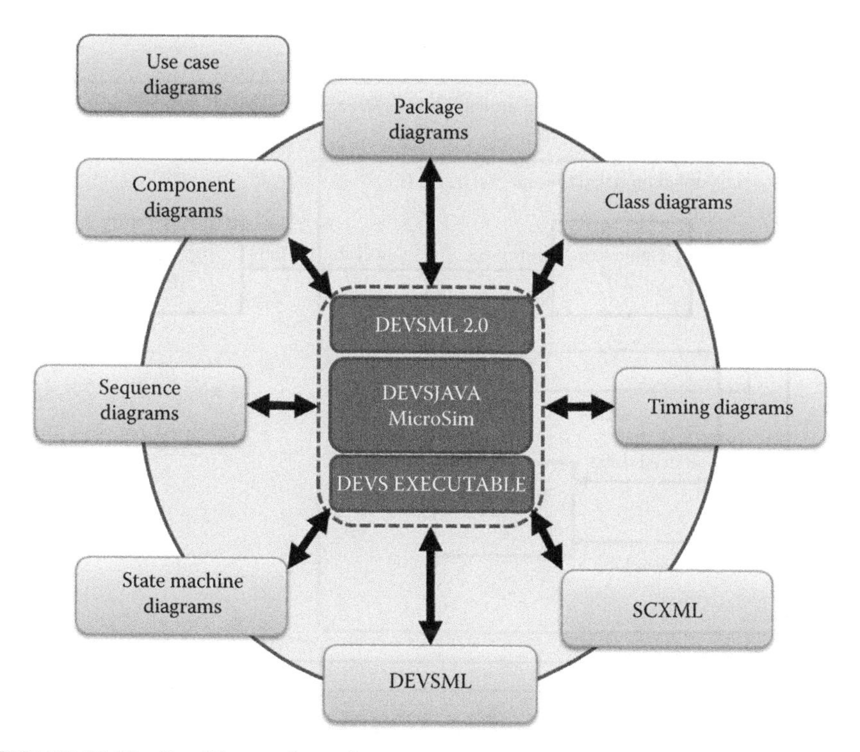

FIGURE 21.17 Possible transformations.

21.5 CASE STUDY: XFD-DEVS AND UML TOGETHER

In this section, we show some of the possible cross transformations described in Section 21.4 between eUDEVS and XFD-DEVS, a precursor to DEVSML. We have chosen these two paradigms because these are the most complicated and recent ones. Representing a state machine entirely in DSL as a PIM has been the foundation of DEVS automation, netcentric execution, model interoperability, and code generation. We implemented such transformations using Java (DOM libraries) plus a set of XSLT documents. We developed a graphical user interface called Transformed UML and DEVS (TUDEVS), by means of which a user is able to select the UML model as well as an XFD-DEVS model to execute the transformation in the chosen direction. The current version of TUDEVS has some limitations. The process of generating a platform-specific code or PSM is handled by the XFD-DEVS framework; that is, the executable DEVS code is generated by XFD-DEVS, so we only need to transform UML into XFD-DEVS. Transformations to package and class diagrams are not implemented. In addition, component coordinates are not generated, so the UML diagram generated needs some manual editions.

21.5.1 EF-P Model

The EF-P model is a simple coupled model consisting of three atomic models (Figure 21.18).

The Generator atomic model generates job messages at fixed time intervals and sends them via the out port. The Transducer atomic model accepts job messages from the

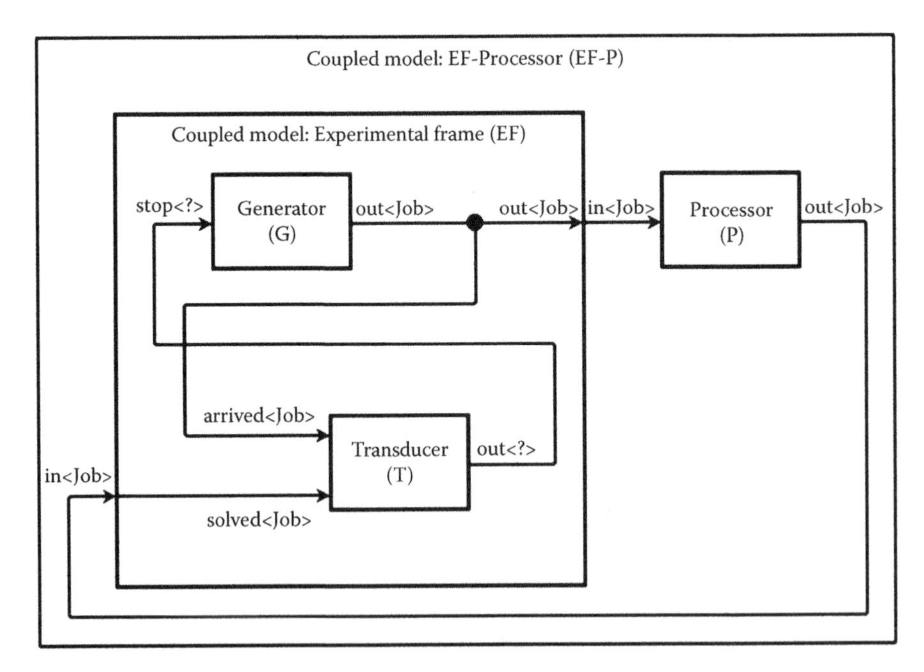

FIGURE 21.18 Experimental frame (EF) processor (P) model; boxes: models; arrows: couplings; arrow labels: input/output port names.

Generator at its arrived port and remembers their arrival time instances. It also accepts job messages at the solved port. When a message arrives at the solved port, the Transducer matches this job with the previous job that had arrived on the arrived port earlier and calculates their time difference. Together, these two atomic models form an Experimental Frame coupled model. The Experimental Frame sends the Generator's job messages on the out port and forwards the messages received on its in port to the Transducer's solved port. The Transducer observes the response (in this case the turnaround time) of messages that are injected into an observed system. The observed system in this case is the Processor atomic model. Processor accepts jobs at its in port and sends them via the out port again after some finite, but non-zero time period. If the Processor is busy when a new job arrives, it discards the job. Finally, the Transducer stops the generation of jobs by sending any event from its out port to the stop port at the Generator.

21.5.2 EF-P UML Model

TUDEVS is designed to take both kinds of inputs: XFD-DEVS and UML models. It is a transformer that transforms in either direction. We will start with the UML design of the EF-P example.

Figures 21.19 and 21.20 show the EF-P model. The component diagram (Figure 21.19a) depicts the structure of the EF-P model in terms of components, ports, interfaces, and delegation connectors. Since all the atomic models generate job objects, the interface defined to create the connections between ports is precisely

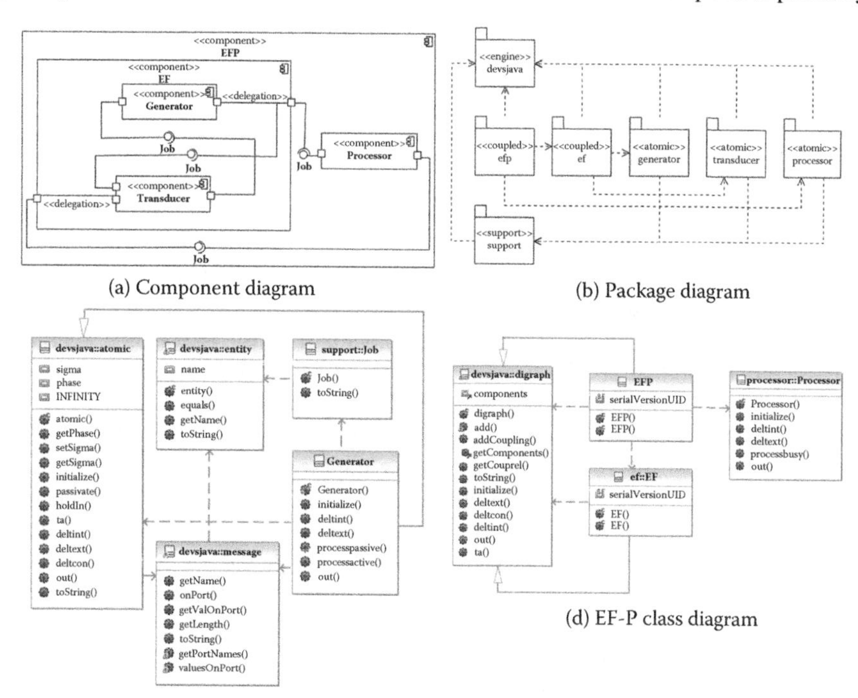

(a) Component diagram (b) Package diagram

(c) Generator class diagram (d) EF-P class diagram

FIGURE 21.19 Some diagrams of the EF-P UML structure.

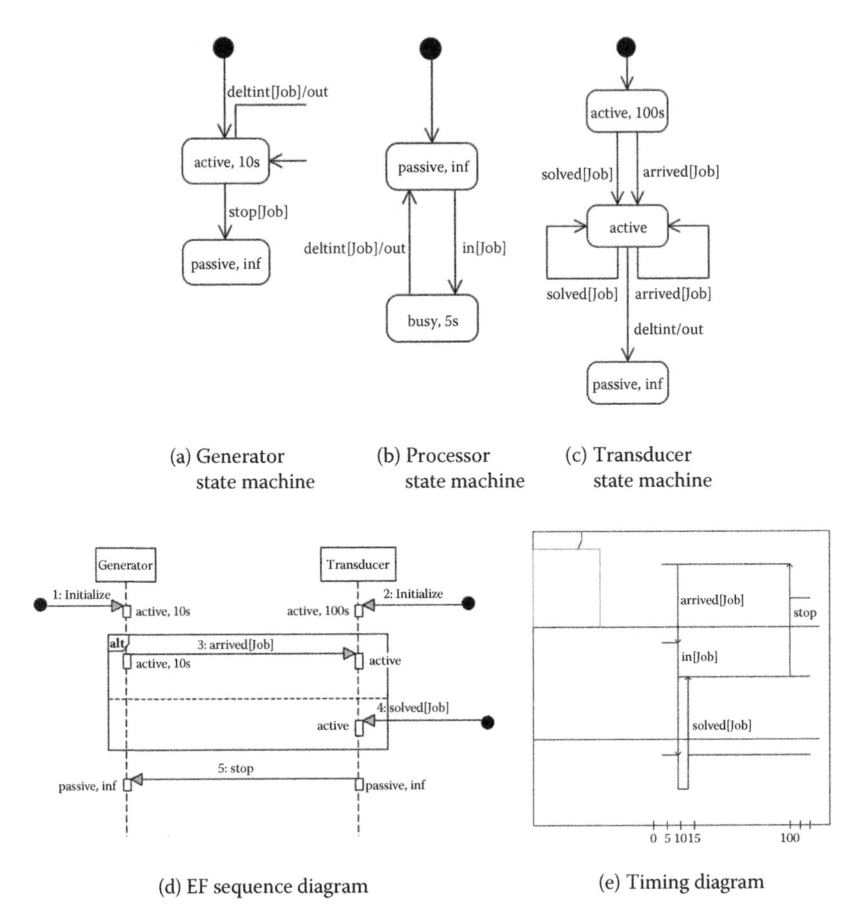

FIGURE 21.20 Some diagrams of the EF-P UML behavior.

job. The internal package diagram (Figure 21.19b) shows the software design using the simulation engine selected, supporting classes, and components that are present in the model, such as EFP as the root coupled model, EF as the Experimental Frame coupled model, and three atomic models: Generator, Transducer, and Processor. Finally, class diagrams (Figure 21.19c and d) show the final implementations of the EF-P model. Some dependencies of DEVSJAVA have been removed for clarity reasons. In Figure 21.19c, the Generator class inherits devsjava::atomic, which is the base class for atomic models in DEVSJAVA, whereas the EFP and EF classes (Figure 21.19d) inherit devsjava::digraph as the base class for coupled models. Figure 21.20 depicts the behavior diagrams. We did not define use cases since such diagrams are dedicated to a high-level description of the system. Figure 21.20a, b, and c shows the state machine diagrams of all the atomic models included in the EF-P. The left state machine is the Generator atomic model; at the middle, the Processor; and, finally, the Transducer at the right side. Figure 21.20d shows the sequence diagrams of the EF coupled model. Finally, Figure 21.20e shows the timing diagrams of all the atomic models.

In the following, we present the XMI specification of the EF-P model. Listing 21.1 presents the EF-P component diagram (just the root-coupled model), Listing 21.2 presents the EF component diagram (generator and transducer), and, finally, Listing 21.3 presents the generator atomic model. As can be seen, the behavior of the generator atomic model has been defined in terms of SCXML.

LISTING 21.1 EFP.UML

```
<?xml version="1.0" encoding="UTF-8"?>
<uml:Package xmi:id="devsjava" xmi:version="2.1"
  xmlns:uml="http://www.eclipse.org/uml2/3.0.0/UML"
  xmlns:xmi="http://schema.omg.org/spec/XMI/2.1">
  <packagedElement name="ExpFrameProc"
                 xmi:id="ExpFrameProc"
                    xmi:type="uml:Component">
    <packagedElement name="ExpFrame" xmi:id="ExpFrame"
                      xmi:type="uml:Component"/>
    <packagedElement name="Processor" xmi:id="Processor"
                      xmi:type="uml:Component"/>
    <packagedElement memberEnd="ExpFrame:out
        Processor:in"
                      xmi:id="1"
                      xmi:type="uml:CommunicationPath"/>
    <packagedElement memberEnd="Processor:out
        ExpFrame:in"
                      xmi:id="2"
                      xmi:type="uml:CommunicationPath"/>
  </packagedElement>
</uml:Package>
```

LISTING 21.2 EF.UML

```
<?xml version="1.0" encoding="UTF-8"?>
<uml:Package xmi:id="devsjava" xmi:version="2.1"
  xmlns:uml="http://www.eclipse.org/uml2/3.0.0/UML"
  xmlns:xmi="http://schema.omg.org/spec/XMI/2.1">
  <packagedElement name="ExpFrame" xmi:id="ExpFrame"
                    xmi:type="uml:Component">
    <packagedElement name="Generator"
                   xmi:id="Generator"
                      xmi:type="uml:Component"/>
```

```
    <packagedElement name="Transducer"
                     xmi:id="Transducer"
                     xmi:type="uml:Component"/>
    <ownedAttribute name="in" xmi:id="ExpFrame:in"
                    xmi:type="uml:Port"/>
    <ownedAttribute name="out" xmi:id="ExpFrame:out"
                    xmi:type="uml:Port"/>
    <packagedElement memberEnd="Generator:out
                                Transducer:arrived"
                    xmi:id="1"
                    xmi:type="uml:CommunicationPath"/>
    </packagedElement>
</uml:Package>
```

LISTING 21.3 GENERATOR.UML

```
<?xml version="1.0" encoding="UTF-8"?>
<uml:Package xmi:id="devsjava" xmi:version="2.1"
      xmlns:uml="http://www.eclipse.org/uml2/3.0.0/UML"
      xmlns:xmi="http://schema.omg.org/spec/XMI/2.1">
  <packagedElement name="Generator" xmi:id="Generator"
                   xmi:type="uml:Component">
    <ownedAttribute name="stop" xmi:id="Generator:stop"
                    xmi:type="uml:Port"/>
    <ownedAttribute name="out" xmi:id="Generator:out"
                    xmi:type="uml:Port"/>
  </packagedElement>
  <scxml initialstate="active">
    <state id="active">
      <onentry>
        <send delay="10.0" event="outJob"/>
      </onentry>
      <transition event="Job" target="passive"/>
      <transition event="*" target="active"/>
    </state>
    <state id="passive">
      <onentry/>
      <transition event="Start" target="active"/>
      <transition event="*" target="passive"/>
    </state>
  </scxml>
</uml:Package>
```

21.5.3 From UML to XFD-DEVS

Figure 21.21 shows how TUDEVS generates an XFD-DEVS model from UML component and state machine diagrams for the structure and behavior, respectively.

First, we select the UML component diagram that will generate the corresponding XFD-DEVS structure, that is, a hierarchical coupled DEVS model. It includes all the XML files, input and output ports, and connections for atomic or coupled models. Second, we select the UML XMI file. Third, we select the target directory, where the generated files will be placed. Finally, we update the current XFD-DEVS model (which is initially empty) generating the aforementioned structure.

Listing 21.4 shows an XML document, which is the EF.xml file generated:

Next, as Figure 21.22 depicts, we generate the behavior of each atomic model selecting the corresponding UML state machine diagram. In this case, the generator XML file is updated with the behavior (states, transitions, and outputs).

The following XML (Listing 21.5) shows the part of the behavior section of the generator in the XFD-DEVS model updated. The structure (input and out ports) has been omitted for brevity.

Finally, the following code (Listing 21.6) shows the XSLT used in TUDEVS to implement UML component and SCXML model definition transformations to XFD-DEVS.

The complete source code of the EF-P XFD-DEVS model is available online (Mittal, 2011). Recall that this XML representation of a finite state machine is completely executable using the DEVSJAVA or DEVS.net or MicroSim simulation framework. In the next subsection, we will see how a reverse transformation from XFD-DEVS to UML can be attempted.

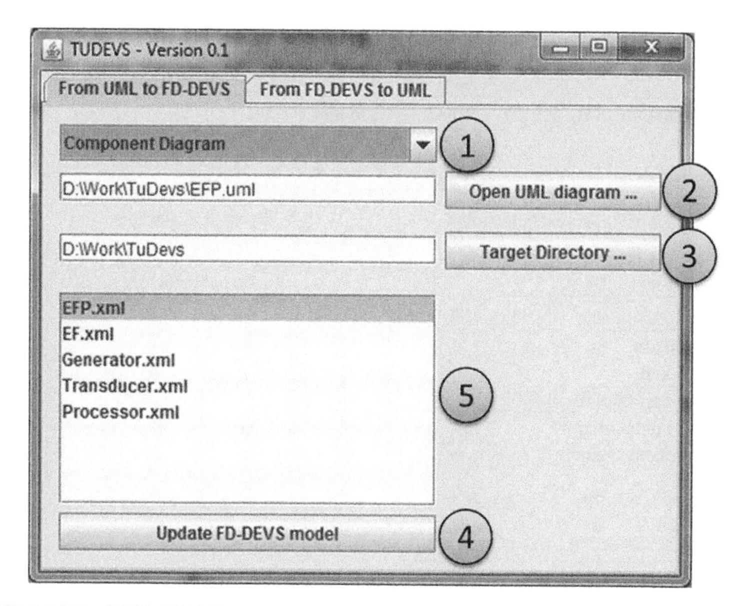

FIGURE 21.21 XFD-DEVS structure generation from a UML component diagram.

LISTING 21.4 EF.XML

```
<digraphType name="ExpFrame">
  <inports>
    <inport>in</inport>
  </inports>
  <outports>
    <outport>out</outport>
  </outports>
  <Models>
    <Model>
      <devs>Generator</devs>
    </Model>
    <Model>
      <devs>Transducer</devs>
    </Model>
  </Models>
  <Couplings>
    <Coupling>
      <SrcModel>Generator</SrcModel>
      <outport>out</outport>
      <DestModel>Transducer</DestModel>
      <inport>arrived</inport>
    </Coupling>
  </Couplings>
</digraphType>
```

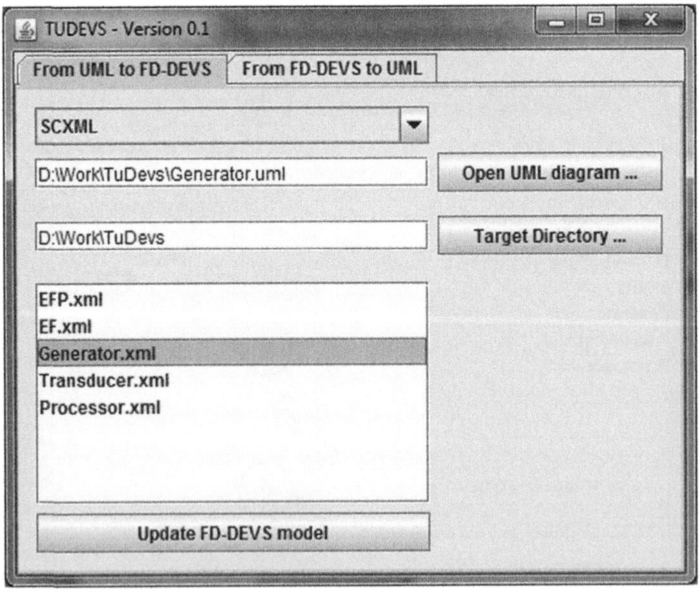

TUDEVS - Version 0.1

From UML to FD-DEVS | From FD-DEVS to UML

SCXML

D:\Work\TuDevs\Generator.uml | Open UML diagram ...

D:\Work\TuDevs | Target Directory ...

EFP.xml
EF.xml
Generator.xml
Transducer.xml
Processor.xml

Update FD-DEVS model

FIGURE 21.22 XFD-DEVS behavior generation from a statechart XML.

LISTING 21.5 GENERATOR.XML

```
<atomicType name="Generator">
  <states>
    <state>active</state>
    <state>passive</state>
  </states>
  <TimeAdvance>
    <ta>
      <state>passive</state>
      <Timeout>INF</Timeout>
    </ta>
    <ta>
      <state>active</state>
      <Timeout>10.0</Timeout>
    </ta>
  </TimeAdvance>
  <LamdaSet>
    <lamda>
      <state>active</state>
      <outport>outJob</outport>
    </lamda>
  </LamdaSet>
  // …
  <deltint>
    <InternalTransition intTransitionID="2">
      <transition>
        <StartState>active</StartState>
        <NextState>active</NextState>
      </transition>
    </InternalTransition>
    <InternalTransition intTransitionID="1">
       <transition>
        <StartState>passive</StartState>
        <NextState>passive</NextState>
      </transition>
    </InternalTransition>
  </deltint>
  <deltext>
    <ExternalTransition extTransitionID="2">
      <IncomingMessage>Job</IncomingMessage>
      <transition>
        <StartState>active</StartState>
        <NextState>passive</NextState>
      </transition>
      <scheduleIndicator>true</scheduleIndicator>
    </ExternalTransition>
    <ExternalTransition extTransitionID="1">
      <IncomingMessage>Start</IncomingMessage>
```

```
      <transition>
        <StartState>passive</StartState>
        <NextState>active</NextState>
      </transition>
      <scheduleIndicator>true</scheduleIndicator>
    </ExternalTransition>
  </deltext>
</atomicType>
```

LISTING 21.6 UMLTOFDDEVS.XSL

```
<?xml version="1.0" encoding="UTF-8"?>
<xsl:stylesheet version="1.0"
      xmlns:xsl="http://www.w3.org/1999/XSL/Transform"
      xmlns:xmi="http://schema.omg.org/spec/XMI/2.1">
  <xsl:output method="xml" indent="yes"/>
  <xsl:template match="/">
    <devsModel>
      <xsl:attribute name="name">
        <xsl:value-of select="*/packagedElement/@name"/>
      </xsl:attribute>
      <ports>
        <xsl:apply-templates
         select="//ownedAttribute[@xmi:type =
           'uml:Port']"/>
      </ports>
      <xsl:if test="count(*/packagedElement
                    /packagedElement[@xmi:type =
                    'uml:Component'])>0">
        <Models>
          <xsl:apply-templates
           select="*/packagedElement/packagedElement
                   [@xmi:type = 'uml:Component']"/>
        </Models>
      </xsl:if>
      <xsl:if test="count(*/packagedElement
                    /packagedElement[@xmi:type =
                     'uml:CommunicationPath'])>0">
        <Couplings>
          <xsl:apply-templates select="*/packagedElement
                     /packagedElement[@xmi:type =
                     'uml:CommunicationPath']"/>
        </Couplings>
      </xsl:if>
      <xsl:if test="count(*/scxml)>0">
```

```
      <xsl:apply-templates select="*/scxml"/>
    </xsl:if>
  </devsModel>
</xsl:template>

<xsl:template match="ownedAttribute">
  <port>
    <xsl:value-of select="@name"/>
  </port>
</xsl:template>

<xsl:template match="packagedElement">
  <xsl:if test="@xmi:type = 'uml:Component'">
    <Model>
      <devs>
        <xsl:value-of select="@name"/>
      </devs>
    </Model>
  </xsl:if>
  <xsl:if test="@xmi:type = 'uml:CommunicationPath'">
    <Coupling>
      <SrcModel>
        <xsl:value-of select="substring-before
          (@memberEnd, ':')"/>
      </SrcModel>
      <outport>
        <xsl:value-of select="substring-before(
                substring-after(@memberEnd,':'),' ')"/>
      </outport>
      <DestModel>
        <xsl:value-of select="substring-before(
               substring-after(@memberEnd,' '),':')"/>
      </DestModel>
      <inport>
        <xsl:value-of select="substring-after(
           substring-after(@memberEnd,':'),':')"/>
      </inport>
    </Coupling>
  </xsl:if>
</xsl:template>

<!-- ///////////////////////////////////////////// -->
<xsl:template match="scxml">
  <states>
    <xsl:for-each select="state">
      <state><xsl:value-of select="@id"/>
        </state>
    </xsl:for-each>
```

```
  </states>
  <TimeAdvance>
    <xsl:for-each select="state">
      <ta>
        <state><xsl:value-of select="@id"/></state>
        <xsl:if test="count(onentry/send)>0">
          <Timeout>
            <xsl:value-of select="onentry/send/
              @delay"/>
          </Timeout>
        </xsl:if>
        <xsl:if test="count(onentry/send)=0">
          <Timeout>INF</Timeout>
        </xsl:if>
      </ta>
    </xsl:for-each>
  </TimeAdvance>
  <LamdaSet>
    <xsl:for-each select="state">
      <xsl:if test="count(onentry/send)>0">
        <lamda>
          <state>
            <xsl:value-of select="@id"/>
          </state>
          <outport>
            <xsl:value-of select="onentry/send/
              @event"/>
          </outport>
        </lamda>
      </xsl:if>
    </xsl:for-each>
  </LamdaSet>
  <deltint>
    <xsl:for-each select="state">
      <xsl:if test="count(transition[@event='*'])>0">
        <InternalTransition>
          <xsl:attribute name="intTransitionID">
            <xsl:value-of select="position()"/>
          </xsl:attribute>
          <transition>
            <StartState>
              <xsl:value-of select="@id"/>
            </StartState>
            <NextState>
              <xsl:value-of select="transition[@
                event='*']
                                            [1]/@target"/>
```

```
          </NextState>
        </transition>
      </InternalTransition>
    </xsl:if>
  </xsl:for-each>
</deltint>
<deltext>
  <xsl:for-each select="state">
   <xsl:for-each select="transition[@event!='*']">
     <ExternalTransition>
       <xsl:attribute name="extTransitionID">
        <xsl:value-of select="position()"/>
       </xsl:attribute>
       <IncomingMessage>
         <xsl:value-of select="@event"/>
        </IncomingMessage>
        <transition>
          <StartState>
            <xsl:value-of select="../@id"/>
          </StartState>
          <NextState>
            <xsl:value-of select="@target"/>
          </NextState>
        </transition>
        <scheduleIndicator>true</scheduleIndicator>
      </ExternalTransition>
    </xsl:for-each>
   </xsl:for-each>
  </deltext>
 </xsl:template>
</xsl:stylesheet>
```

Exercise 21.1

Develop an XSLT to transform UML to DEVSML.

21.5.4 From XFD-DEVS to UML

The procedure to generate UML diagrams from XFD-DEVS models goes through the same kind of steps as in the reverse manner. Figure 21.23 shows how we generate the component diagram. First, we select the directory where the XFD-DEVS model is located. Second, we select the XFD-DEVS file needed to generate the desired UML diagram. For example, if we want to generate a pure UML component diagram, we must select the source coupled model (the root coupled model in Figure 21.23). Next, we select the type of the diagram we want to generate. Finally, we generate the UML diagram into a new.uml file.

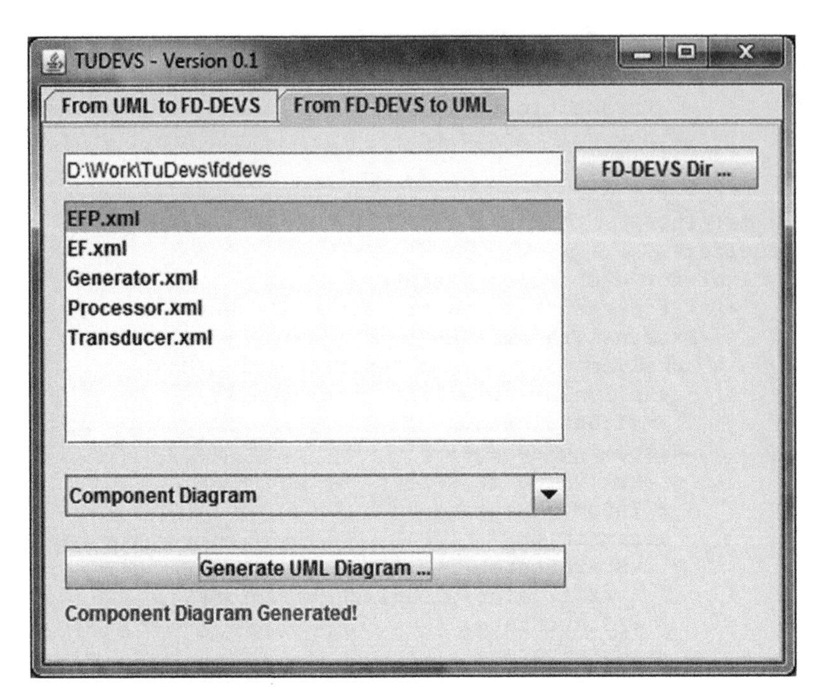

FIGURE 21.23 UML component diaagram generation from the XFD-DEVS model.

LISTING 21.7 FDDEVSTOUML.XSL

```
<?xml version="1.0" encoding="UTF-8"?>
<xsl:stylesheet version="1.0"
     xmlns:xsl="http://www.w3.org/1999/XSL/Transform">
  <xsl:output method="xml" indent="yes"/>
  <xsl:template match="/">
    <uml:Package xmi:version="2.1"
         xmlns:xmi="http://schema.omg.org/spec/XMI/2.1"
         xmlns:uml="http://www.eclipse.org/uml2/3.0.0/UML"
         xmi:id="devsjava">
      <packagedElement>
        <xsl:attribute name="xmi:type">
          uml:Component
        </xsl:attribute>
        <xsl:attribute name="xmi:id">
          <xsl:value-of select="*/@name"/>
        </xsl:attribute>
        <xsl:attribute name="name">
          <xsl:value-of select="*/@name"/>
        </xsl:attribute>
        <xsl:apply-templates select="//Model/devs"/>
```

```
      <xsl:apply-templates select="//inports/inport"/>
      <xsl:apply-templates select="//outports/outport"/>
      <xsl:apply-templates select="//Couplings/
        Coupling"/>
    </packagedElement>
    <xsl:apply-templates select="*/states"/>
  </uml:Package>
</xsl:template>
<xsl:template match="devs">
  <packagedElement>
    <xsl:attribute name="xmi:type">
      uml:Component
    </xsl:attribute>
    <xsl:attribute name="xmi:id">
      <xsl:value-of select="."/>
    </xsl:attribute>
    <xsl:attribute name="name">
      <xsl:value-of select="."/>
    </xsl:attribute>
  </packagedElement>
</xsl:template>
<xsl:template match="inport">
  <ownedAttribute>
    <xsl:attribute name="xmi:type">
      uml:Port
    </xsl:attribute>
    <xsl:attribute name="xmi:id">
      <xsl:value-of select="../../@name"/>
      :<xsl:value-of select="."/>
    </xsl:attribute>
    <xsl:attribute name="name">
      <xsl:value-of select="."/>
    </xsl:attribute>
  </ownedAttribute>
</xsl:template>
<xsl:template match="outport">
  <ownedAttribute>
    <xsl:attribute name="xmi:type">uml:Port
      </xsl:attribute>
    <xsl:attribute name="xmi:id">
      <xsl:value-of select="../../@name"/>
      :<xsl:value-of select="."/>
    </xsl:attribute>
    <xsl:attribute name="name">
      <xsl:value-of select="."/>
    </xsl:attribute>
  </ownedAttribute>
</xsl:template>
```

```
<xsl:template match="Coupling">
  <packagedElement>
    <xsl:attribute name="xmi:type">
      uml:CommunicationPath
    </xsl:attribute>
    <xsl:attribute name="xmi:id">
      <xsl:value-of select="position()"/>
    </xsl:attribute>
    <xsl:attribute name="memberEnd">
      <xsl:value-of select="SrcModel/."/>
      :<xsl:value-of select="outport/."/>
      <xsl:text> </xsl:text>
      <xsl:value-of select="DestModel/."/>
      :<xsl:value-of select="inport/."/>
    </xsl:attribute>
  </packagedElement>
</xsl:template>
<xsl:template match="states">
  <scxml>
    <xsl:attribute name="initialstate">
      <xsl:value-of select="state[1]/."/>
        </xsl:attribute>
    <xsl:for-each select="state">
      <state>
        <xsl:attribute name="id">
          <xsl:value-of select="."/></xsl:attribute>
          <xsl:call-template name="EventList">
            <xsl:with-param name="stateName"
              select="."/>
          </xsl:call-template>
      </state>
    </xsl:for-each>
  </scxml>
</xsl:template>

<xsl:template name="EventList">
  <xsl:param name="stateName"/>
  <onentry>
    <xsl:for-each select="../../LamdaSet/
                            lamda[state/. = $stateName]">
      <send>
        <xsl:attribute name="event">
          <xsl:value-of select="outport/."/>
        </xsl:attribute>
        <xsl:attribute name="delay">
          <xsl:value-of select="//ta[state/. =
            $stateName]/Timeout/."/></xsl:attribute>
      </send>
```

```
        </xsl:for-each>
      </onentry>
      <transition>
        <xsl:attribute name="event">
          <xsl:value-of select="//ExternalTransition
                                /transition[StartState/. =
                                $stateName]/../
                                  IncomingMessage
                                /."/>
        </xsl:attribute>
        <xsl:attribute name="target">
          <xsl:value-of select="//ExternalTransition
                                /transition[StartState/. =
                                $stateName]/NextState/."/>
        </xsl:attribute>
      </transition>
      <transition>
        <xsl:attribute name="event">
          <xsl:text>*</xsl:text>
        </xsl:attribute>
        <xsl:attribute name="target">
          <xsl:value-of select="//InternalTransition
                                /transition[StartState/. =
                                $stateName]/NextState/."/>
          </xsl:attribute>
      </transition>
    </xsl:template>
  </xsl:stylesheet>
```

The principles are exactly the same as those in the previous subsection. In the following, we show the source code of the XSLT (Listing 21.7) used to transform XFD-DEVS models to UML Component and SCXML definitions:

Exercise 21.2

Develop an XSLT to transform DEVSML to UML.

We have shown how TUDEVS can take in a UML model and can generate an XFD-DEVS executable simulation model, and how an XFD-DEVS model can deliver a UML model. We are working toward removing the deficiencies in the generated UML models from TUDEVS so that they can be viewed by IDEs that have the capabilities to import/export XMI files. This example has demonstrated the proof of concept and the validation of the underlying eUDEVS metamodel that made these transformations easier.

21.6 SYNOPSIS

UML has become the *defacto* standard for modeling in the industry, and many vendors provide rich graphical tools conforming to the UML 2.0 standard. There are many open-source tools, such as Eclipse, that propel the development of extensions related to UML additions and research. UML profiles have been defined for various application areas, such as DoDAF, systems engineering (i.e., SysML), and others. However, UML lacks a system theoretic foundation. DEVS, on the other hand, is founded on systems theoretic principles and has advocated the component-based engineering since its inception. With the advancement of UML in recent years, the DEVS community has made advances in mapping the DEVS elements with those of the UML elements. There is a consensus that DEVS is more rigorous than UML but lacks the expressive power for the nonengineer who is proficient with the graphical notations of UML. All the DEVS groups, most notably, Sarjoughian, Vangheluwe, and Zinoviev, have tried to address the problem of mapping DEVS and UML, but without an underlying metamodel or ontology. We have proposed eUDEVS using an SES ontology that binds both UML and DEVS in a metamodel framework. This has many advantages. Having such a foundation allows the eUDEVS framework to develop cross-transformations, tools, and editors based on the eUDEVS metamodel. The other major advantage of such a framework is inclusion of UML and its usage in a much larger systems engineering-based DUNIP that allows creation of a Test-suite, along with the simulation model.

Modeling and simulation are two independent areas. The art of modeling involves dealing with the problem domain, whereas simulation involves simply the execution of the model using state-of-the-art technologies and multiple platforms. Historically speaking, modeling and simulation have been integrated by the term "simulation model", which would address any specific problem at hand. This brings many problems in extending the model as well as performance of simulation itself, when both are inseparable. In the first place, a framework is warranted that would allow such separation.

The impact of M&S cannot be underestimated and is currently in the mainstream with technologies and tools such as UML, Statemate, and MATLAB®. In this chapter, we are dealing only with the software-modeling domain, where UML is the preferred means and a standard. It is a graphical language that allows designers to develop their architecture model using various graphical elements. UML lies strictly in the modeling domain as per say and simulation of the UML model, though is not a standard. There have been attempts to develop executable UML.

DEVS has a rich history and it categorically separates the model and the simulator. However, it needs improvement at the graphical end, which accounts for its non-acceptance in the commercial non-engineering domain. The DEVS formalism exists in multiple platform-specific implementations, such as DEVSJAVA, DEVS/C++, and DEVS/.net. Recently, subsets of DEVS, namely, XFD-DEVS and DEVSML, are made available as platform-independent implementation that can lead to any of the platform-specific code executions. DEVS simulators have been made executable on P2P, RMI, CORBA, HLA, and SOA, which allows the same DEVS models to be simulatable on various distributed platforms as well.

In this chapter, we have attempted to bridge the gap between the UML graphical modeling elements and the DEVS formalism by means of XML and a subset of DEVS known as XFD-DEVS, which gives us the needed edge when interfacing with a powerful formalism such as DEVS. The reader is encouraged to work on XSLT transformations that transform DEVSML to UML and vice versa. Although, this is not the first attempt, it has been the most comprehensive in terms of mapping various implementations and addressing the problem in a platform-independent manner using XFD-DEVS.

This chapter has shown the following:

1. UML can be used within a systems theoretical framework such as DEVS.
2. Metamodel for an executable M&S framework using ontology such as SES can be used as a foundation to develop transformations from UML to DEVS and vice versa.
3. UML models are executable using the DEVS formalism.

Despite being graphically rich, UML suffers from lack of systems theory in the background. Extensions like SysML exist, but again, they are extensions, not the needed foundation. We have developed metamodels of DEVS, UML, and the proposed executable UML-Based DEVS, eUDEVS.

These metamodels are built using the SES ontology. This is by far the most important contribution where UML now is backed by the systems theoretical framework.

We have described the essential mappings that need to be done to extract information from UML, or augment UML toward a DEVS model. We incorporated only the required UML elements that we thought would suffice to develop a DEVS model. These UML elements are shown as mapping to DEVS in Figure 21.10, which is also the metamodel for eUDEVS. We have placed eUDEVS in a much bigger framework of DUNIP. This opens UML to the entire suite of integrated systems development using the DEVS theory.

We have also demonstrated the entire life cycle of doing the cross-transformation between UML and XFD-DEVS by an example that has a hierarchical component structure. We have shown how the transformations could be done with the developed tools like XFD-DEVS workbench and TUDEVS.

Finally, we have established that we have a means to develop mapping between practical frameworks (UML) with an engineering framework (DEVS) using XML transformations and SES, which lead to a platform-independent code. This chapter closes the gap between UML and executable UML using the DEVS modeling formalism and the underlying DEVS simulation protocol.

REFERENCES

ACIMS—Software site. (n.d.). Retrieved May, 10, 2012, from www.acims.arizona.edu/SOFTWARE/ software.shtml.

Borland, S., & Vangheluwe, H. (2003). Transforming statecharts to DEVS. *Summer Computer Simulation Conference* (pp. 154–159) Montreal, Canada.

Choi, K., Jung, S., Kim, H., Bae, D.-H., & Lee, D. (2006). UML-based modeling and simulation method for mission-critical real-time embedded system development. *IASTED Conference on Software Engineering* (pp. 160–165).

Department of Defense. (2006). *DoD Acquisition GuideBook, V2006*. Retrieved May, 10, 2012 from https://akss.dau.mil/dag.

Giambiasi, N., Paillet, J.-L., & Chane, F. (2003). From timed automata to DEVS models. *Proceedings of the 2003 International Conference on Machine Learning and Cybernetics* (pp. 923–931). IEEE. doi:10.1109/WSC.2003.1261512

Hong, S., & Kim, T. (2004). Embedding UML subset into object-oriented DEVS modeling process. *Proceedings of the Summer Computer Simulation Conference* (pp. 161–166). San Jose, CA.

Hopcroft, J. E., Motwani, R., & Ullman, J. D. (2006). *Introduction to Automata Theory, Languages, and Computation, Third Edition*. (p. 535). Addison-Wesley Boston, Massachusetts, USA.

Huang, D., & Sarjoughian, H. (2004). Software and simulation modeling for real-time software-intensive systems. *Proceedings of the 8th IEEE International Symposium on Distributed Simulation and Real-Time Applications* (pp. 196–203).

Mellor, S. J., & Balcer, M. J. (2002). *Executable UML: A Foundation for Model-Driven Architecture*. (p. 416). Addison-Wesley Professional Indianapolis, IN, USA.

Mittal, S. (2006). Extending DoDAF to allow integrated DEVS-based modeling and simulation. *The Journal of Defense Modeling and Simulation Applications Methodology Technology*, *3*(2), 95–123.

Mittal, S. (2011). XFD-DEVS, an example: efp. Retrieved November 02, 2012, from http://duniptechnologies.com/research/efxfd.php.

Schulz, S., Ewing, T. C., & Rozenblit, J. W. (2000). Discrete event system specification (DEVS) and StateMate StateCharts equivalence for embedded systems modeling. *Proceedings Seventh IEEE International Conference and Workshop on the Engineering of Computer Based Systems (ECBS 2000)* (pp. 308–316). IEEE Comput. Soc. doi:10.1109/ECBS.2000.839890

The Object Management Group. (2012). The Unified Modeling Language (UML). Retrieved May 10, 2012 from www.uml.org.

Risco-Martín, J. L., Mittal, S., Zeigler, B. P., & Cruz, J. M. (2007). From UML State Charts to DEVS State Machines Using XML. *MPM'07: Proceedings of the Workshop on Multi-Paradigm Modeling: Concepts and Tools at the 10th International Conference on Model-Driven Engineering Languages and Systems* (pp. 35–48). Nashville, TN.

Tolk, A., & Muguira, J. A. (2004). M&S within the Model Driven Architecture. *Interservice/Industry Training, Simulation, and Education Conference* (pp. 1–13).

W3C. (2012). State Chart XML (SCXML): State machine notation for control abstraction. Retrieved November 02, 2012 from www.w3.org/TR/scxml/.

Zinoviev, D. (2005). Mapping DEVS models onto UML models. *Arxiv preprint cs0508128*. Retrieved November 02, 2012, from http://arxiv.org/abs/cs/0508128.

22 BPMN to DEVS
Application of MDD4MS Framework in Discrete Event Simulation

22.1 INTRODUCTION

An organization needs to improve process efficiency and quality to stay competitive and operate effectively by adapting its strategy, structure, management, and operations to the changes in its business environment. Management consultants and business analysts provide expertise and recommendations to improve the business performance of their clients. A good understanding of the business process is essential to support organizational decisions.

Business process modeling (BPM) and business process simulation (BPS) help to analyze and improve business processes. There are numerous methods and tools available for BPM and some of the tools provide ways to simulate models. Although most of the BPM tools provide easy model building capabilities and BPS tools provide good animation features, many of them lack formal semantics and capabilities for formal model verification (Van Nuffel et al., 2009; van der Aalst et al., 2003). In order to draw useful and correct conclusions from the modeling and simulation results, the BPM and BPS models need to be valid. Establishing the correctness of models is easier when the models have been built using formal methods.

This chapter presents a formal method for business process modeling and simulation by applying model-driven engineering principles. The model-driven development for modeling and simulation (MDD4MS) framework (Çetinkaya et al., 2011) is used as the underlying conceptual framework. The remainder of the chapter is organized as follows: Section 22.2 provides background information about BPM. Section 22.3 presents the application of the MDD4MS framework to Discrete Event Systems Specification (DEVS)-based simulation of Business Process Modeling and Notation (BPMN) models. An overview of the MDD4MS prototype is given in Section 22.4, and an implementation of a case example with the prototype is explained in Section 22.5. Finally, a model transformation from the MDD4MS DEVS metamodel to DEVSML 2.0 (DEVS Modeling Language) is presented in Section 22.6.

22.2 BUSINESS PROCESS MODELING

BPM is the activity of defining a graphical representation of either the current or the future processes of an organization in order to analyze and improve their efficiency and quality (Recker et al., 2009). A business process model is a visual representation of a set of related activities. BPS tools are used for the evaluation of the dynamic behavior of business processes. BPM is also known as static modeling while BPS is also known as dynamic modeling (Bosilj-Vuksic et al., 2007).

Different domain-specific languages have been used to develop business process models (Recker et al., 2009). Integrated DEFinition method (IDEF) and BPMN are the most common BPM techniques. In this chapter, BPMN (OMG, 2011) was selected because it is an industry-wide standard for creating business process models. BPMN is a specification for the high-level representation of business processes. It provides a notation that is readily understandable by business analysts and technical developers. BPMN follows the tradition of flowcharting notations for readability and flexibility. In addition, the BPMN execution semantics is fully formalized. There are five basic categories of elements in BPMN. These are flow objects, data elements, connecting objects, swimlanes, and artifacts.

Flow objects are the main graphical elements that define the behavior of a business process. There are three types of flow objects: events, activities, and gateways. Data elements are represented by the four following types: data object, data input, data output, and data store. There are four ways of connecting the flow objects to each other or to other elements: sequence flows, message flows, associations, and data associations. There are two ways of grouping the primary modeling elements through swimlanes: pools and lanes. Artifacts are used to provide additional information about the process. BPMN also defines the visual representation of each element.

22.3 DISCRETE EVENT SIMULATION OF BUSINESS PROCESS MODELS

Discrete event simulation (DES) is an effective technique for analyzing and designing complex systems. In a discrete event system, the system's state changes at discrete points in time upon the occurrence of an event. DEVS is a well-known mathematical formalism based on system theoretic principles (Zeigler et al., 2000). Models that are expressed in the basic formalism are called atomic models. Hierarchical DEVS is the extended version of the basic formalism that defines the means for coupling the DEVS models. The composite models are called coupled models. DEVS is introduced with the input and output ports to provide an elegant way of building composite models.

22.3.1 From BPMN Elements to DEVS Components

The DEVS-based simulation of BPMN models requires a transformation process. The modeling elements of BPMN have to be mapped to DEVS components in order to be able to simulate their behavior in a DEVS simulation environment. In our case, composed elements such as pools and lanes are mapped to coupled

DEVS, whereas other basic elements such as events and gateways are mapped to atomic DEVS. Flow connections are expressed by coupling relations in coupled models. Input and output ports are defined for each component. For every atomic model, a state diagram is defined. A supplementary component called Resource Manager is designed to support the simulation functionality for resource allocation and waiting queues.

The Distributed Simulation Object Library (DSOL) is selected to provide the simulation and execution functionalities (Jacobs et al., 2002). DSOL is an open source multi-formalism simulation suite that is full featured and very effective as a generic purpose simulation tool. Various simulation projects have been conducted using DSOL in a broad range of application areas, including flight scheduling, emulation, and web-based supply chain simulation. DSOL also supports the execution of simulation models based on the DEVS formalism through the DEVSDSOL library (Seck & Verbraeck, 2009). DEVSDSOL provides an object-oriented conceptualization of DEVS language constructs and implements a DEVS compliant modeling and simulation environment using the event-scheduling worldview.

DSOL and DEVSDSOL are both written in the Java programming language. DEVSDSOL defines AtomicModel and CoupledModel abstract classes. Building an atomic model is done by extending the abstract class AtomicModel, instantiating input and output ports, creating state variables and phases, and overriding the abstract methods specifying the DEVS functions (deltaExternal, deltaInternal, lambda, and timeAdvance functions). Similarly, building a coupled model is done by extending the abstract CoupledModel class, adding input and output ports, adding model components, and defining the connections within the coupled model. InputPort and OutputPort are defined as member classes of the AtomicModel and the CoupledModel.

Because of the underlying event-scheduling simulator, the simulation of DEVS models in DSOL is managed differently from the simulation in classical DEVS. Each subcomponent uses its container's simulator, which recursively leads to a unique event-scheduling simulator. However, the low-level details of hierarchical simulation management are not visible to the modeler. Thus, the separation of concerns between modeling and simulation is guaranteed from the modelers' perspective.

22.3.2 Applying the MDD4MS Framework

The MDD4MS framework, which is explained in Chapter 9, prescribes the separation of platform-independent and platform-specific details for a simulation model. In the example of this chapter:

- BPMN is used as the conceptual modeling language to define conceptual models
- DEVS is used as the formal specification language to define platform-independent simulation models
- Java with the DEVSDSOL library, in particular, is used as the underlying simulation model programming language to define platform-specific simulation models

Thus, following the MDD4MS method from Chapter 9, the following metamodels and model transformations are required during the application of the MDD4MS framework:

- BPMN metamodel as the simulation conceptual modeling metamodel (CM metamodel)
- DEVS metamodel as the simulation model specification metamodel (PISM metamodel)
- DEVSDSOL metamodel as the simulation model implementation metamodel (PSSM metamodel)
- BPMN-to-DEVS transformation as the CM_to_PISM transformation
- DEVS-to-DEVSDSOL transformation as the PISM_to_PSSM transformation
- DEVSDSOL-to-Java transformation as the PSSM_to_Code transformation

The BPMN-to-DEVS transformation produces atomic and coupled models with ports, couplings, and templates for the system dynamics. The BPMN metamodel, The DEVS metamodel and the BPMN to DEVS transformation are adopted from (Çetinkaya et al., 2012). The DEVS-to-DEVSDSOL transformation produces valid DEVSDSOL models with information for Java classes. The DEVSDSOL-to-Java transformation generates Java code. Figure 22.1 illustrates the application of MDD4MS for BPMN simulation. The implementation of the concepts with the MDD4MS prototype is explained in Section 22.5. Before that, the following section provides a brief overview of the MDD4MS prototype.

22.4 MDD4MS PROTOTYPE

The MDD4MS framework defines the methodology for model-driven development of simulation models through model transformations between well-defined metamodels. The MDD4MS prototype is an Eclipse-based application of the tool architecture, as defined in the framework (Çetinkaya et al., 2011).

The prototype is based on the Eclipse Modeling Project and the Eclipse Modeling Framework (EMF) (Eclipse, 2008, 2009). All of the tools that have been used are subprojects of the top-level Eclipse Modeling Project. EMF is a modeling framework and code generation facility for building tools and other applications based on a structured data model. From a model specification described in XML Metadata Interchange (XMI), EMF provides tools and run-time support to produce a set of Java classes for the model, along with a set of adapter classes that enable viewing and command-based editing of the model, and a basic editor. The core EMF framework includes a meta-metamodel (Ecore) for describing models and run-time support for the models, including change notification, persistence support with default XMI serialization, and a very efficient reflective API for manipulating EMF objects generically. The remainder of this section provides information about the used tools and projects for different activities in the MDD4MS framework.

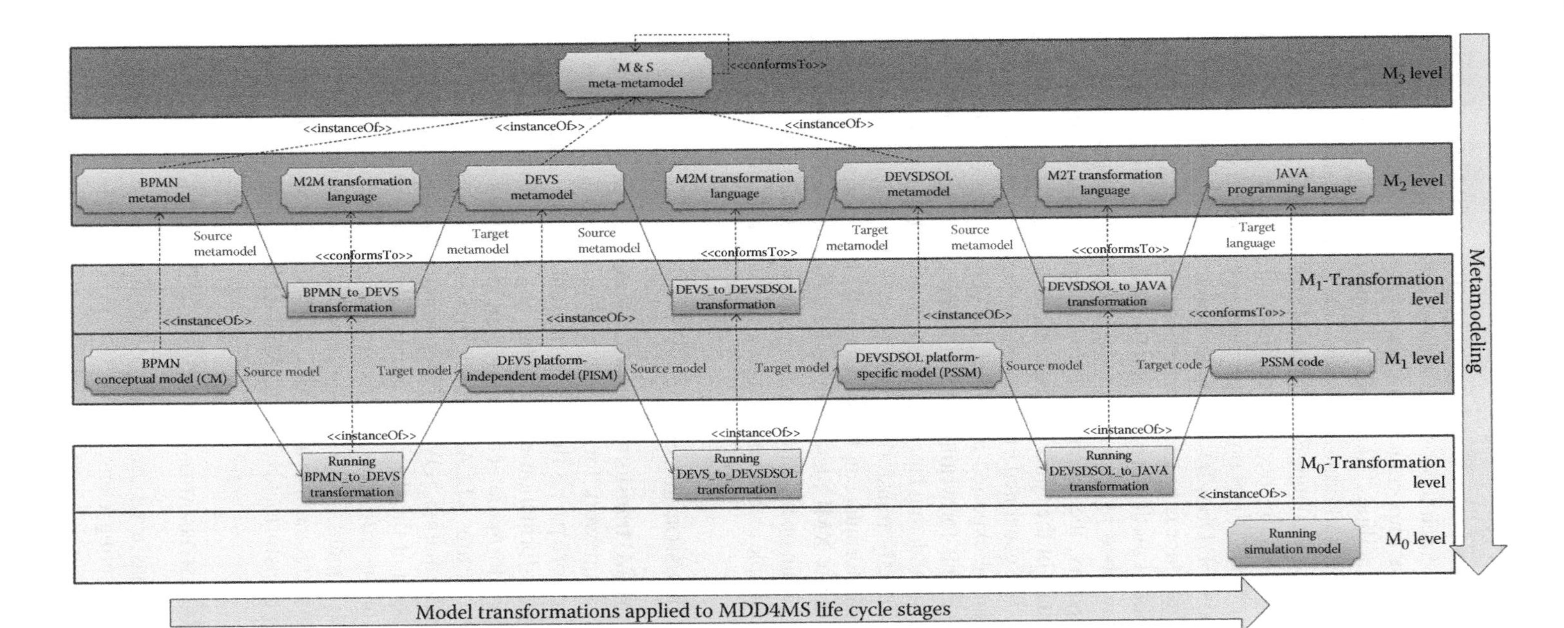

FIGURE 22.1 Application of MDD4MS framework.

22.4.1 Metamodeling with the GEMS Project

The Eclipse-based Generative Modeling Technologies (GMT) project provides a set of prototypes and research tools in the area of model-driven development (MDD). Historically, the most important operation was model transformation, but other model management facilities, like model composition, are also part of the GMT project (Eclipse, 2006). Different subprojects are proposed in the GMT project.

The Generic Eclipse Modeling System (GEMS) in the GMT project is a configurable tool kit for creating domain-specific modeling and program synthesis environments for Eclipse (2008). The GEMS project provides a visual metamodeling environment based on EMF and GEF/Draw2D. It includes a code generation framework in which a graphical modeling editor is generated automatically from a visual metamodel specification. The graphical modeling editor can be used for editing instances of the modeling language described by the metamodel. Although the generated editor uses a default concrete syntax, the graphical representation of the modeling elements and the visual appearance of the editor can be customized by changing a Cascading Style Sheets (CSS) file. The generated graphical modeling tool is based on EMF, GEF, and Draw2D and it can be exported as an Eclipse plug-in. Besides, it can be extended through extension points for writing model interpreters, triggers, or constraints.

The built-in metamodeling language is based on the UML class diagram notation. Metamodels are directly transformed into Ecore meta-metamodel. Metamodels in other Ecore readable formats can be used as well. The models defined by the generated tools are saved as XMI files. Metamodel constraints can be specified in Java. The GEMS project is an open source project and the MDD4MS prototype is based on GEMS, with some simple extensions for increasing the graphical modeling capabilities.

22.4.2 M2M Transformations with ATL

Model-to-model (M2M) transformation is a key aspect of MDD. The Eclipse M2M project presents a framework for defining and using model-to-model transformation languages. The core part is the transformation infrastructure. Transformations are executed by transformation engines that are plugged into the infrastructure. There are three transformation engines that are developed in the scope of Eclipse M2M project, which are ATLAS Transformation Language (ATL), Procedural Query/View/Transformation (QVT) (Operational), and Declarative QVT (Core and Relational). Each of the three represents a different category, which validates the functionality of the infrastructure from multiple contexts. The ATL integrated development environment (IDE) aims to ease the development and execution of ATL transformations (Jouault et al., 2008). The MDD4MS prototype uses the ATL IDE for the M2M transformations in the MDD4MS framework.

22.4.3 M2T Transformations with Visitor-Based Model Interpreters

The GEMS project has a visitor-based model interpretation mechanism (Eclipse, 2008). For each modeling element specified in the metamodel, a corresponding *visit<modeling_element_name>(...)* method is generated in the interface. A Java project with a class that implements these methods can be developed as a model interpreter. And then, each element in a particular model can be visited separately and the required code can be generated. The interpreters are registered to the editor by adding an extension point to the MANIFEST.MF file in the META-INF directory. The current MDD4MS prototype makes use of this model interpretation mechanism. However, any other M2T (model-to-text) tool that accepts the Ecore metamodels and models serialized with XMI can be used as well.

22.5 IMPLEMENTATION EXAMPLE WITH THE MDD4MS PROTOTYPE

This section presents an implementation of the BPMN-to-DEVS transformation with the MDD4MS prototype and the Eclipse-based tools that were described above. The tool architecture used during the implementation process is given in Figure 22.2. The following sections explain the specified metamodels and model transformations.

22.5.1 BPMN Metamodel

The BPMN metamodel defined in Ecore is given in Figure 22.3. BPMNDiagram represents the business process model. The main graphical element of the diagram is BPMNFlowObject. BPMNEvent, BPMNActivity, and BPMNGateway inherit from the flow object. An event is something that happens during the course of a process.

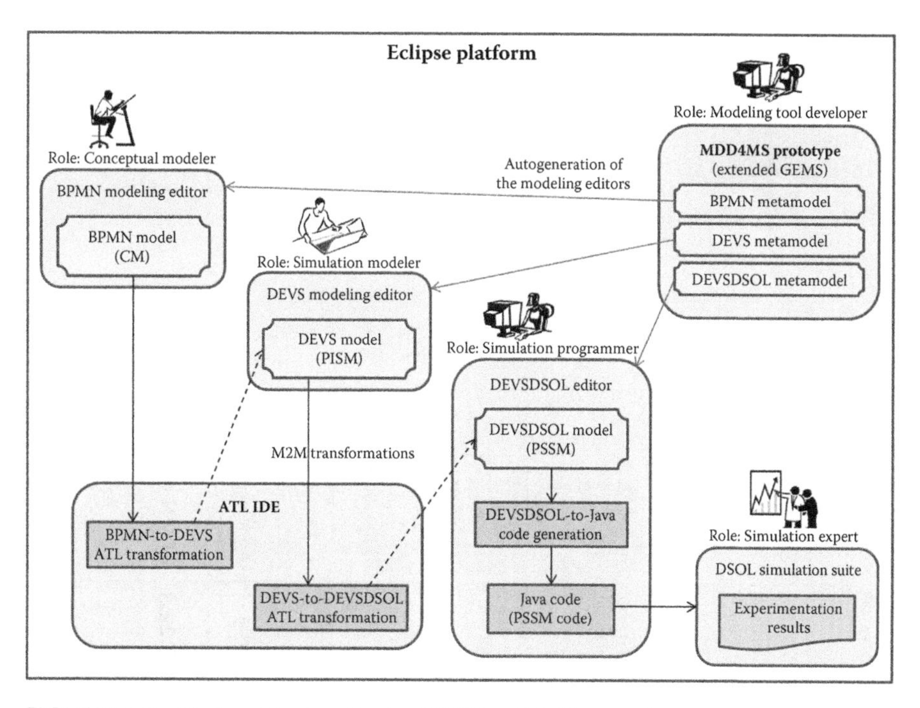

FIGURE 22.2 Tool architecture for the BPMN-to-DEVS example.

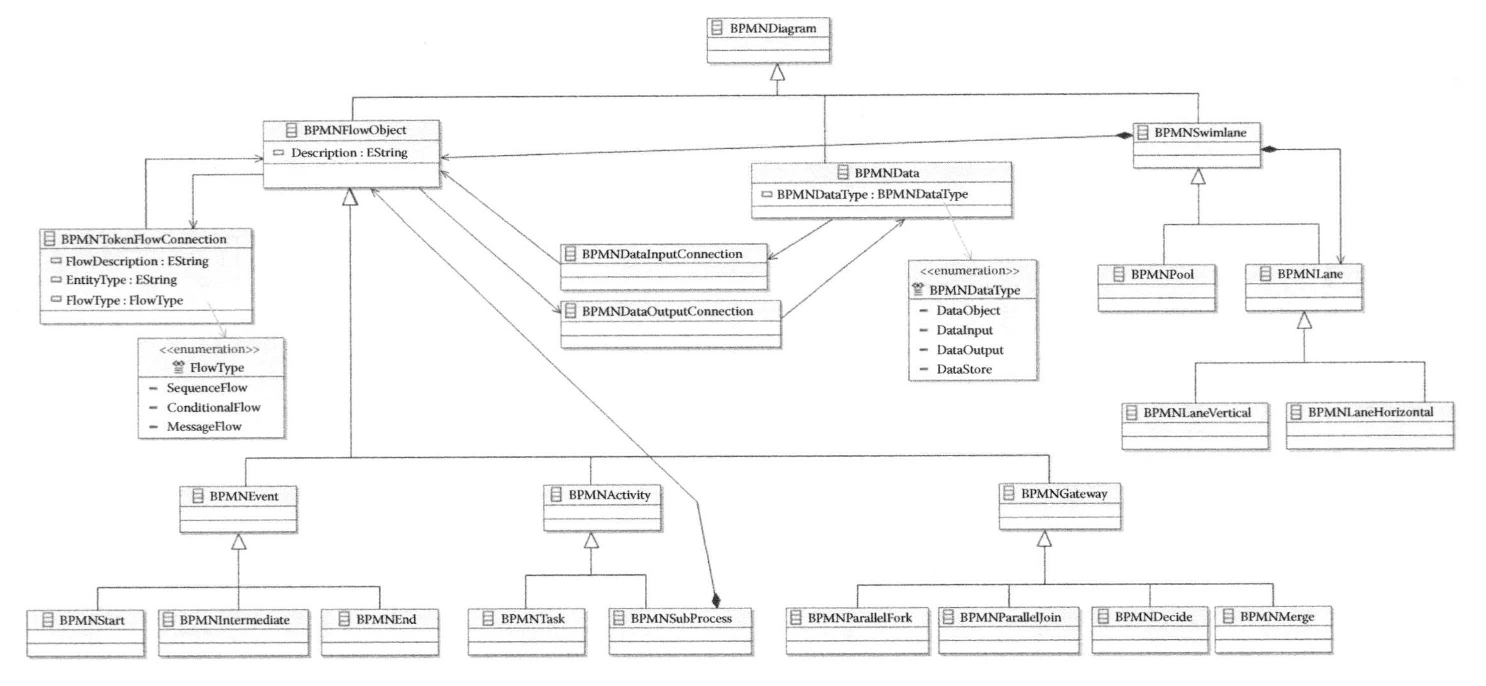

FIGURE 22.3 BPMN metamodel.

There are three types of events, based on when they affect the flow: start event, intermediate event, and end event. An activity is a task performed in a process. An activity can be an atomic task or a compound subprocess. A gateway is used to control the divergence and convergence of the sequence flows in a process. Thus, it determines the branching, forking, merging, and joining of paths. Internal markers indicate the type of behavior control. Each type of control affects both the incoming and outgoing flow. BPMNParallelFork and BPMNParallelJoin are defined for parallel forking and joining; and BPMNDecide and BPMNMerge are defined for exclusive decision and merging.

BPMNTokenFlowConnection represents the flow between the flow objects. The type of flow is determined by the FlowType attribute. The sequence flow is used to control the order of the activities in a process.

Swimlanes are used to group activities. A pool is used as a graphical container for partitioning a set of activities. A lane is a subpartition within a process, sometimes within a pool, and will extend the entire length of the process, either vertically or horizontally. Lanes are used to organize and categorize activities. The metamodel in Figure 22.3 represents the basic modeling elements of BPMN. A visual BPMN modeling editor that works on Eclipse platform is automatically generated from this metamodel in the MDD4MS prototype.

22.5.2 DEVS Metamodel

The DEVS metamodel for the Hierarchical DEVS formalism with ports (Zeigler et al., 2000) has been defined in Ecore and is shown in Figure 22.4. The underlying atomic DEVS model is defined as a 7-tuple (Zeigler et al., 2000) classic DEVS:

$$M = < X, Y, S, \delta_{int}, \delta_{ext}, \lambda, ta>$$

where

$X = \{(p,v) \mid p \in \text{InPorts}, v \in X_P\}$ is the set of input ports and values
$Y = \{(p,v) \mid p \in \text{OutPorts}, v \in Y_P\}$ is the set of output ports and values
S is the set of states
$\delta_{int}: S \rightarrow S$ is the internal transition function that defines how the state of the system changes internally (when the elapsed time reaches to the lifetime of the state)
$\delta_{ext}: Q \times X \rightarrow S$ is the external transition function that defines how an input event changes the state of the system, where
$Q = \{(s,t_e) \mid s \in S, 0 \leq e \leq ta(s)\}$ and t_e is the time that has elapsed since the last transition
$\lambda: S \rightarrow Y$ is the output function that defines the output (when the elapsed time reaches to the lifetime of the state) and only available just before δ_{int}
ta: $S \rightarrow R_{0,\infty}$ is the time advance function that is used to determine the life span of a state

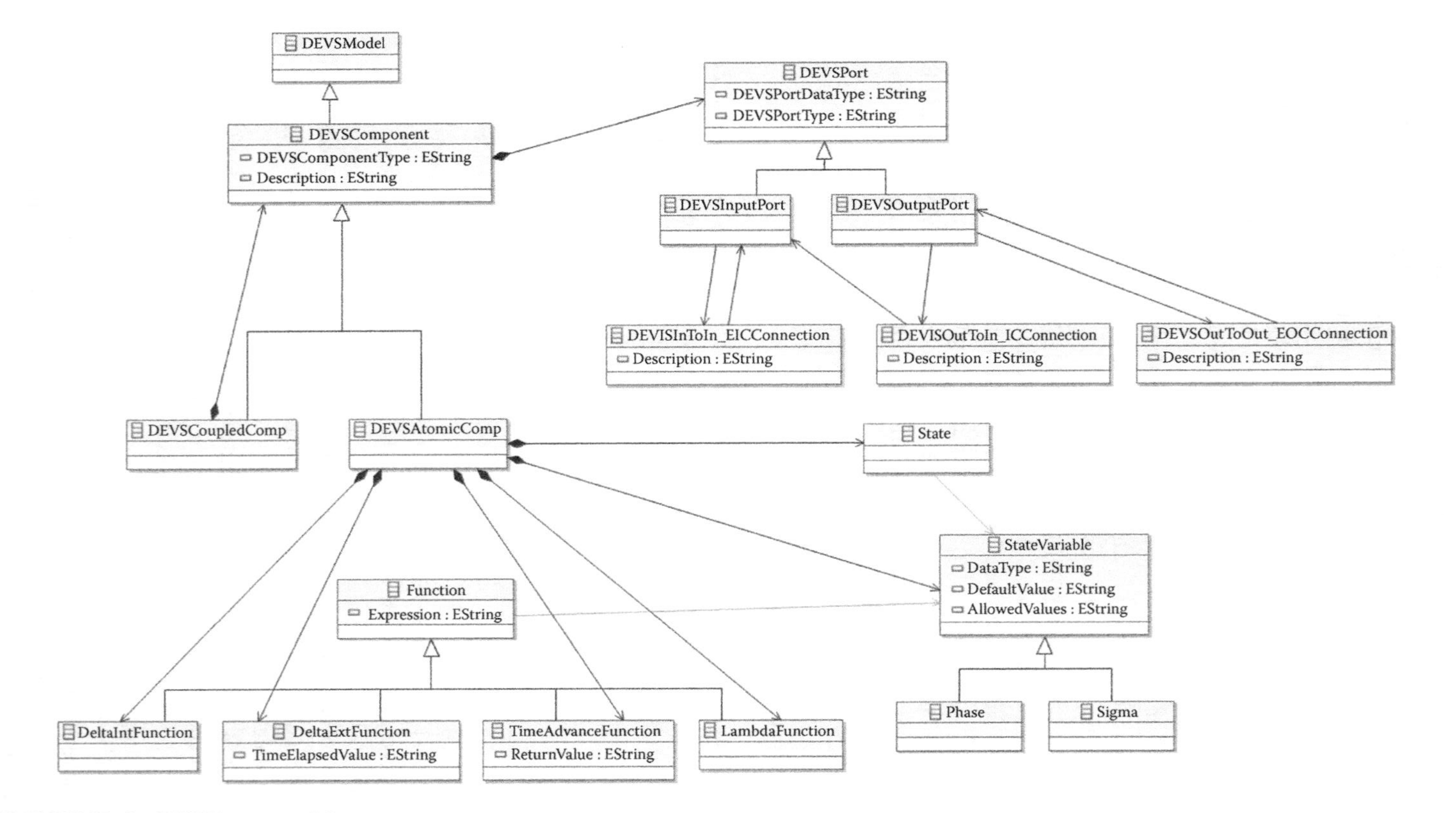

FIGURE 22.4 DEVS metamodel.

DEVSModel represents the platform-independent simulation model. The main graphical elements of the metamodel are DEVSCoupledComp and DEVSAtomicComp, which inherit from DEVSComponent. Each DEVS component has input and output ports. Coupled models are defined hierarchically and couplings are represented via connecting the ports. Atomic models have state variables, states, and functions. Phase and Sigma are already defined as state variables. Functions of the atomic components are: DeltaExtFunction, DeltaIntFunction, TimeAdvanceFunction, and LambdaFunction.

The behavior of the functions is represented via a pseudo-code metamodel that is linked by the use of an Expression attribute. Any expression can be a function call, a conditional block, or an assignment. A conditional block can be a while loop block, an if-block, or an if-else block. Each block contains other expressions. A visual DEVS modeling editor is automatically generated from this metamodel.

22.5.3 DEVSDSOL Metamodel

The DEVSDSOL metamodel defined in Ecore is shown in Figure 22.5. DEVSDSOLModel represents the platform-specific simulation model. The abstract DEVSDSOLComponent represents the Java classes, which extend from either AtomicModel or CoupledModel abstract classes in the DEVSDSOL library. Hence, each component has import definitions for the required Java packages and port definitions. Each DEVSDSOLPort has a PortType, which can be either an input port or an output port, and a PortDataType, which defines the data type.

Atomic and coupled components inherit from DEVSDSOLComponent and both of them have a constructor. The coupled constructor has connection definitions and subcomponent definitions. The atomic constructor has an initialization code. DEVSDSOLAtomicComp has four other functions—namely, lambda, timeAdvance, deltaInternal, and deltaExternal. Besides, it has state variables and specific states. States are defined with a number of conditions, based on the state variables.

The metamodel includes both a structural and a behavioral abstraction of the DEVSDSOL implementation. Although a visual modeling editor is automatically generated from this metamodel, it is not used for modeling purposes. The editor is utilized for direct code generation from DEVSDSOL models.

22.5.4 Model Transformation from BPMN to DEVS

Once the source and target metamodels are available, model transformation rules from the source models to the target models can be specified. A model-to-model transformation from BPMN to DEVS is defined by using the BPMN metamodel and DEVS metamodel. The transformation is written in ATL, as proposed in the MDD4MS prototype. The transformation has two steps. In the first step, all BPMN model elements are transformed into specific DEVS model elements; and all connections are transformed into internal couplings from an output port in the source component to an input port in the target component. Ports are also generated. Table 22.1 shows the specific DEVS modeling elements that are mapped to BPMN modeling elements in step 1.

If the connections in the source model connect only elements of the same level, then the output of the first step becomes a valid DEVS model and the second step is

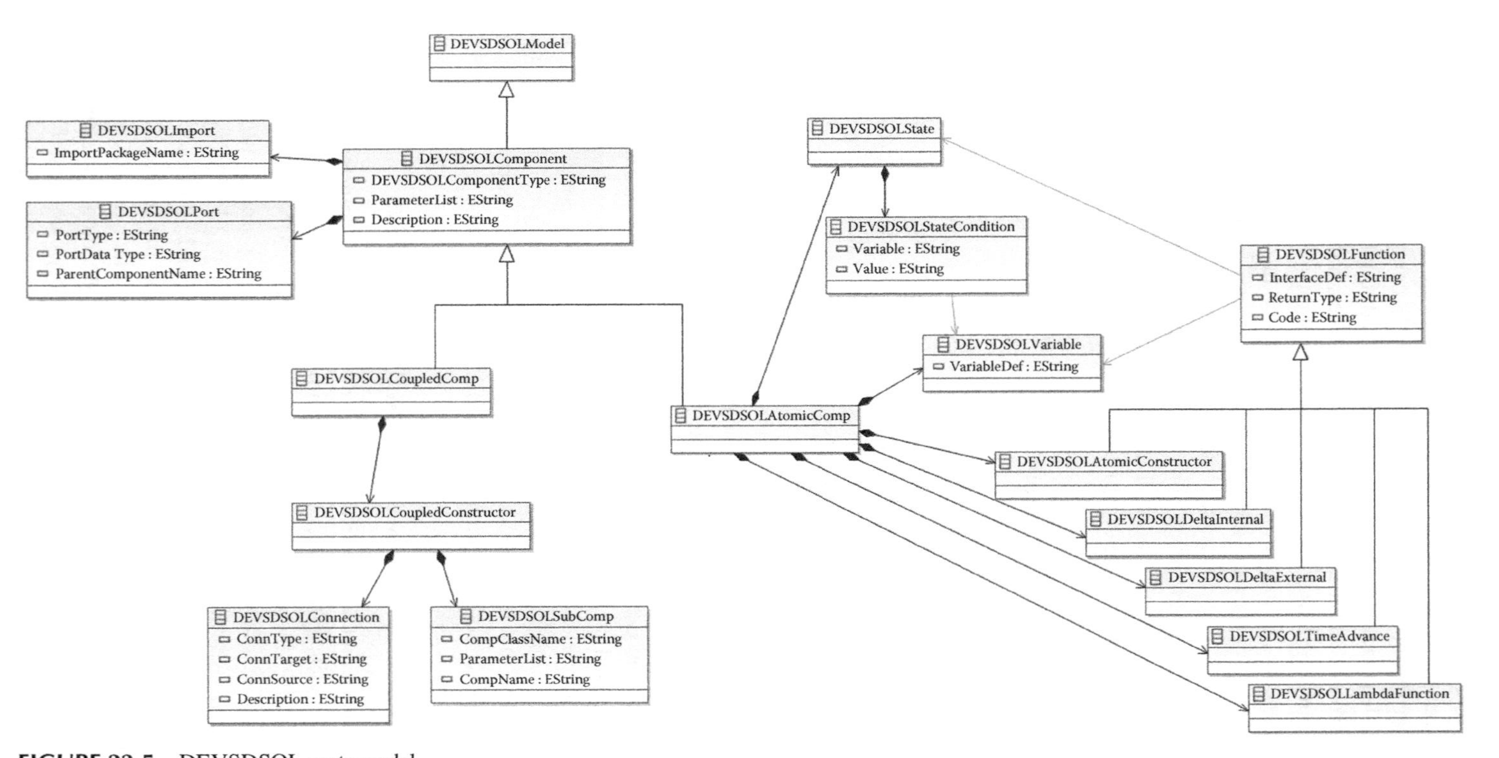

FIGURE 22.5 DEVSDSOL metamodel.

TABLE 22.1
BPMN-to-DEVS Transformation Pattern

BPMN Metamodel	DEVS Metamodel
BPMNDiagram	DEVSModel
BPMNPool	DEVSCoupledComp
BPMNLane (BPMNLaneVertical, BPMNLaneHorizontal)	DEVSCoupledComp
BPMNSubProcess	DEVSCoupledComp
BPMNTask	DEVSAtomicComp
BPMNEvent (BPMNStart, BPMNEnd, BPMNIntermediate)	DEVSAtomicComp
BPMNGateway (BPMNDecide, BPMNMerge, BPMNParallelFork, BPMNParallelJoin)	DEVSAtomicComp
BPMNTokenFlowConnection	DEVSOutputPort, DEVSOutToIn_ICConnection, DEVSInputPort
BPMNData	DEVSAtomicComp
BPMNDataInputConnection	DEVSInputPort, DEVSOutputPort, DEVSOutToIn_ICConnection
BPMNDataOutputConnection	DEVSOutputPort, DEVSOutToIn_ICConnection, DEVSInputPort

skipped. However, this does not apply in most cases. BPMN models generally have connections which cross more than one modeling elements, that is, which connect the modeling elements of different levels. Therefore, the internal couplings generated in the first step need to be refined. So, the output of the first step is only a temporary model.

In the second step, the external input couplings (EICs) and external output couplings (EOCs) are defined for the nested components. In this way, the DEVS compatibility of the target models is guaranteed. The required number of EIC and EOC is determined with the number of the nested components that a flow crosses. Figure 22.6 illustrates the all-inclusive case when the source component is nested *n* levels and the target component is nested *m* levels. In this case, *n* times DEVSOutputPort and *m* times DEVSInputPort are generated. Besides, *n* times DEVSOutToOut_EOCConnection and *m* times DEVSInToIn_EICConnection are defined.

Figure 22.7 shows a sample transformation for the experimental frame processor (EFP) hierarchical model, which is illustrated in Chapter 5, Section 5.3.2.

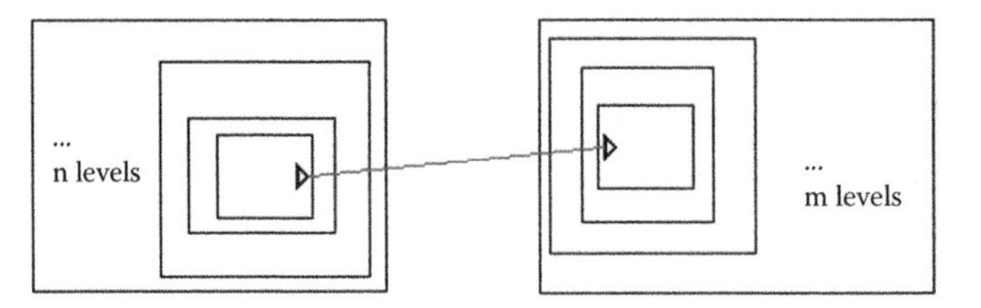

FIGURE 22.6 Transformation of the hierarchical models.

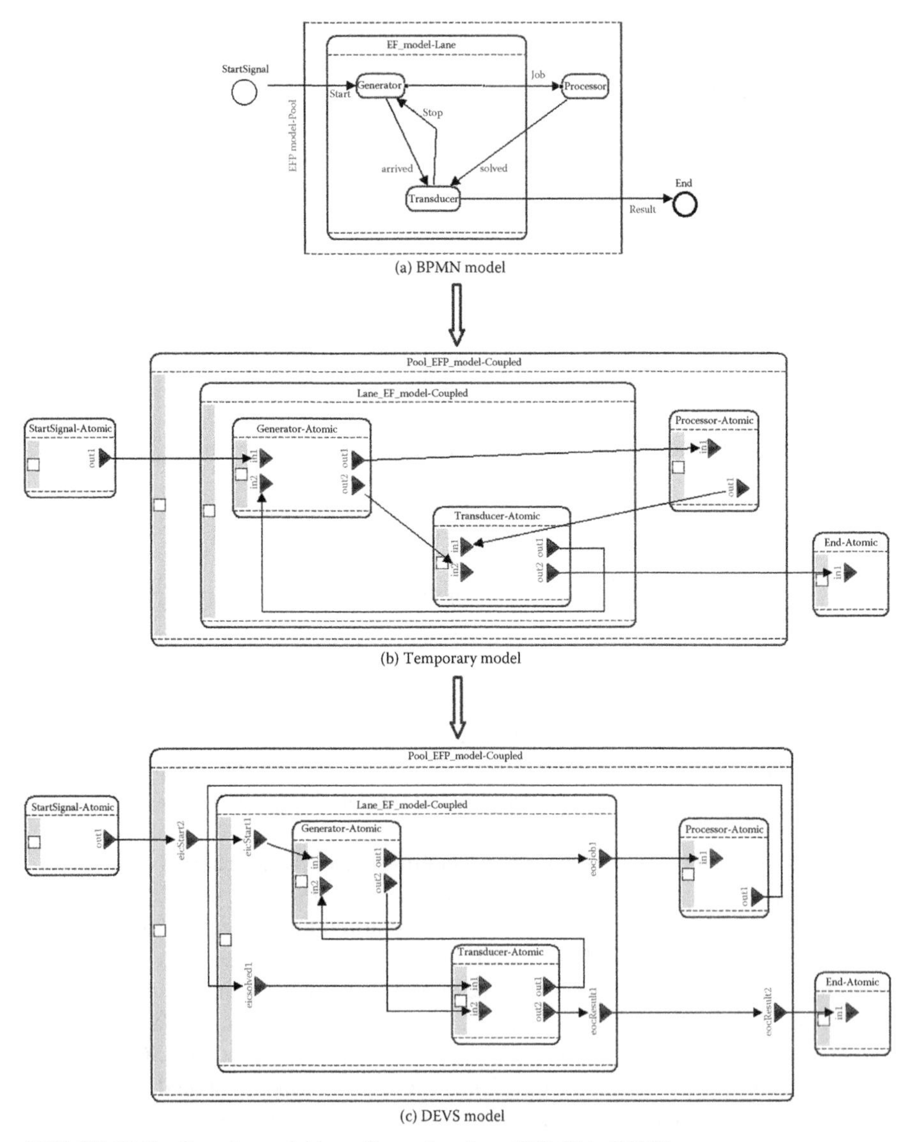

FIGURE 22.7 Sample model transformation from BPMN to DEVS.

The EFP model is first conceptualized with BPMN. The BPMN model for the EFP example contains a start event, an EFP_model pool, and an end event. The EFP_model pool contains an EF_model lane and a processor task (Note: the EF_model could be a subprocess as well; this is a choice of the conceptual modeler). The EF_model internally contains a generator task and a transducer task. The generator task generates and sends jobs at a periodic rate. The processor receives the generated job and gets busy for a constant operation time. On completion, it reports to the transducer task that keeps a count of generated as well as processed jobs. The generator starts with a start signal. The transducer has an observation time, after which it reports the result to the end event. First, the BPMN source model (a) is transformed into the temporary model (b); and then the target DEVS model (c) is generated.

Sample rules for the BPMN-to-DEVS transformation are given in Listing 22.1. The source pattern starts with the keyword *from* and declare which element type of the source model has to be transformed. The target pattern starts with the keyword *to* and declares into which element(s) of the target model the source pattern has to be transformed. It may contain one or several target pattern elements. Each target pattern element consists of a variable declaration and a sequence of bindings (assignments). These bindings consist mainly of left arrow constructs. An attribute of the target model *t* (on the left side of the arrow) receives a return value of an Object Constraint Language (OCL) expression (on the right side of the arrow) that is based on the source model *s*. In this sense, the right side of the arrow may consist of an attribute of *s* or a call to a helper function (which is an OCL expression to define global variables and functions).

LISTING 22.1 SAMPLE BPMN-TO-DEVS TRANSFORMATION RULES IN ATL

```
module BPMN_To_DEVS;
create OUT: SM_Metamodel from IN: CM_Metamodel;

--Transform Root Object
rule TransformRoot {
  from
    s: CM_Metamodel!Root (true)
  to
    t: SM_Metamodel!Root (
      RealRoot <- s.RealRoot
    )
}

--Transform Pool to a Coupled Model
rule BPMNPoolToDEVSCoupled_inRoot {
  from
    s: CM_Metamodel!BPMNPool (true)
  to
```

```
    t: SM_Metamodel!DEVSCoupledComp (
        SMParentModel <- s.CMParentModel,
        DEVSComponents <- s.BPMNActivities,
        DEVSComponents <- s.BPMNLanes,
        Id <- s.Id + 1,
        Name <- 'Pool_' + s.Name,
        X <- s.X,
        Y <- s.Y,
        Width <- s.Width,
        Height <- s.Height,
        ExpandedWidth <- s.ExpandedWidth,
        ExpandedHeight <- s.ExpandedHeight,
        Expanded <- true
    )
}
```

22.5.5 Model Transformation from DEVS to DEVSDSOL

A model-to-model transformation from DEVS to DEVSDSOL is defined by using the DEVS metamodel and DEVSDSOL metamodel. The transformation is written in ATL and has two steps.

In the first step, DEVSComponent instances are transformed into DEVSDSOLComponent instances. For each DEVSAtomicComp model, a DEVSDSOLAtomicComp, a DEVSDSOLSubComp (subcomponent definition), and a DEVSDSOLAtomicConstructor are generated. In the same way, for each DEVSCoupledComp model, a DEVSDSOLCoupledComp, a DEVSDSOLSubComp, and a DEVSDSOLCoupledConstructor are generated. Input ports and output ports are transformed into a DEVSDSOLPort by retaining the PortType (either input or output). All couplings are transformed into a DEVSDSOLConnection by retaining the ConnType (as EIC, IC, or EOC). In the second step, subcomponents and connections are replaced by the related coupled components. The generated DEVSDSOL models include all the required information to generate Java classes in a graphical model.

22.5.6 Code Generation from DEVSDSOL to Java

In the last step, a code generator for DEVSDSOL models is used to generate the Java code automatically. The code generator is a visitor-based model interpreter and has been written in Java. Java files for each DEVS component are generated separately. Coupled component files include the package imports, class definition, port definitions, constructor definition, contained component definitions, and couplings. Coupled component files are fully transformed and they are ready for compiling.

Atomic component files include imports, class definition, port definitions, and constructor definition. Also, deltaExternal(double e, Object inp), deltaInternal(), lambda(), and timeAdvance() functions are generated, which need to be checked for only the user-defined expressions.

A part of the code is given in Listing 22.2. The visitDEVSDSOLCoupledComp function calls the visitPorts and visitCoupledConstructor functions. The VisitPorts

LISTING 22.2 FUNCTIONS FOR THE CODE GENERATION

```
package devsdsolcode.devsdsoldiagram;

import java.io.File; …

public class DEVSDSOL2JAVA_Interpreter
   .extends org.eclipse.gmt.gems.model.actions
       AbstractInterpreter
   implements devsdsolcode.devsdsoldiagram
       .DEVSDSOLDiagramVisitor
  {

  private static final long serialVersionUID = 1L;
  String projectPath = "../";
  FileWriter outCurrent = null;

  //constructor
  public DEVSDSOL2JAVA_Interpreter() {
    super("devsdsolcode.devsdsoldiagram
       .DEVSDSOL2JAVA_Interpreter");
  }

  @Override
  public ModelActionResult interpret(ModelObject root,
     String eid)
  {
    // TODO Auto-generated method stub
    DirectoryDialog chooser =
       new DirectoryDialog(Display.getDefault()
           .getActiveShell());
    chooser.setFilterPath(projectPath);
    String path = chooser.open();
    if (path != null) projectPath = path + "\\";
    root.accept(this);
    return null;
  }
```

```
@Override
public void visitDEVSDSOLCoupledComp(
     DEVSDSOLCoupledComp tovisit) {
  // TODO Auto-generated method stub
  String CCname = tovisit.getName();
  String filePath = projectPath + CCname + ".java";
  File a = new File(filePath);
  try {
    outCurrent = new FileWriter(a);
  } catch (IOException e1) {
    e1.printStackTrace();
  }
  //start writing
  write_file("package efp_test;\n", 0);
  //write imports
        .
        .
        .
  //write ports
  write_file("private static final long
     serialVersionUID = 1L;\n", 1);
  visitPorts(tovisit);
  write_file("\n", 0);

  //constructor
  visitDEVSDSOLCoupledConstructor(
     tovisit.getDEVSDSOLConstructor().get(0), CCname);
  //finish
  write_file("}", 0);
  try {
    outCurrent.close();
  } catch (IOException e1) {
    // TODO Auto-generated catch block
    e1.printStackTrace();
  }
}

public void visitPorts(Container cont) {
  for(Object o:cont.getChildrenOfType
     (DEVSDSOLPort.class)){
    ModelObject mo = (ModelObject)o;
    mo.accept(this);
  }
}
public void visitDEVSDSOLCoupledConstructor(
     DEVSDSOLCoupledConstructor tovisit,
       String n) {
```

```
        // TODO Auto-generated method stub
        write_file("public " + n + "(CoupledModel
          parentModel) {", 1);
        write_file("super(\"" + n + "\",
          parentModel );\n", 2);

        visitConstructorSubComponents (tovisit);
        write_file("", 0);
        visitConstructorConnections (tovisit);

        write_file("", 0);
        write_file("}", 1);
    }

    public void visitConstructorSubComponents(Container
        cont){
        for(Object o:cont.getChildrenOfType
          (DEVSDSOLSubComp.class)){
          ModelObject mo = (ModelObject)o;
          mo.accept(this);
        }
    }

    public void visitConstructorConnections
        (Container cont) {
        for(Object o:cont.getChildrenOfType(
           DEVSDSOLConnection.class)){
          ModelObject mo = (ModelObject)o;
          mo.accept(this);
        }
    }
```

function gets all ports of the component and calls the visitor function for each of them. visitCoupledConstructor calls two more functions, which will visit all the sub-components and couplings of the coupled component. The code generator is added as an extension to the DEVSDSOL modeling editor and it can be called for each model from a right click menu. As an example, the generated code for the EF coupled model is given in Figure 22.8 and the generated code for the generator atomic model is given in Figure 22.9.

The behavior of the atomic model is defined visually with the autogenerated DEVS modeling editor, as illustrated in Figure 22.10. After code generation, the generated code can be compiled and checked for any compilation errors. After fixing any compile errors, like try/catch blocks, missing imports and so on, the simulation model is ready to be run. A screenshot of the running model is given in Figure 22.11.

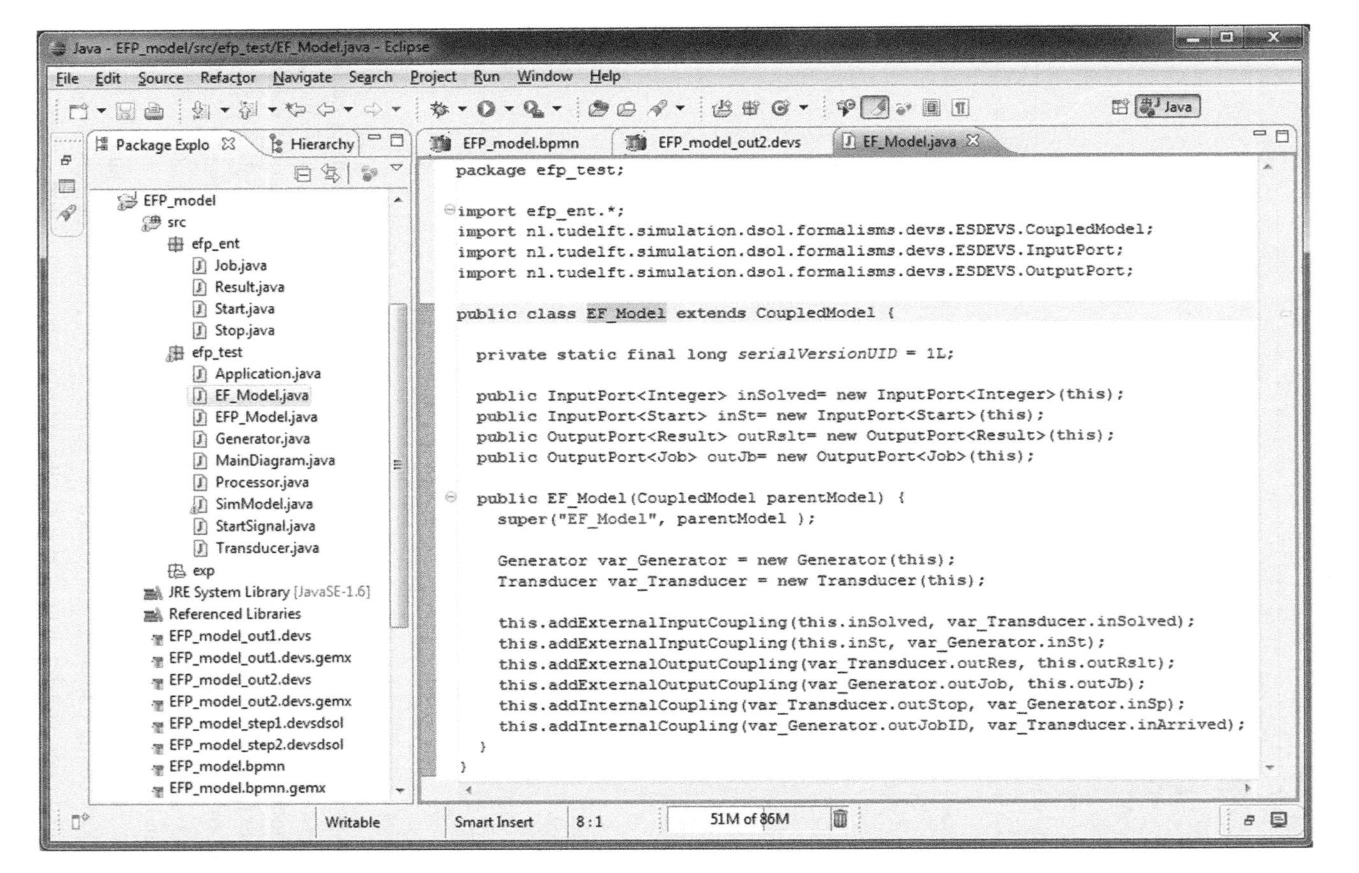

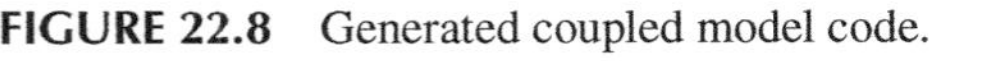
FIGURE 22.8 Generated coupled model code.

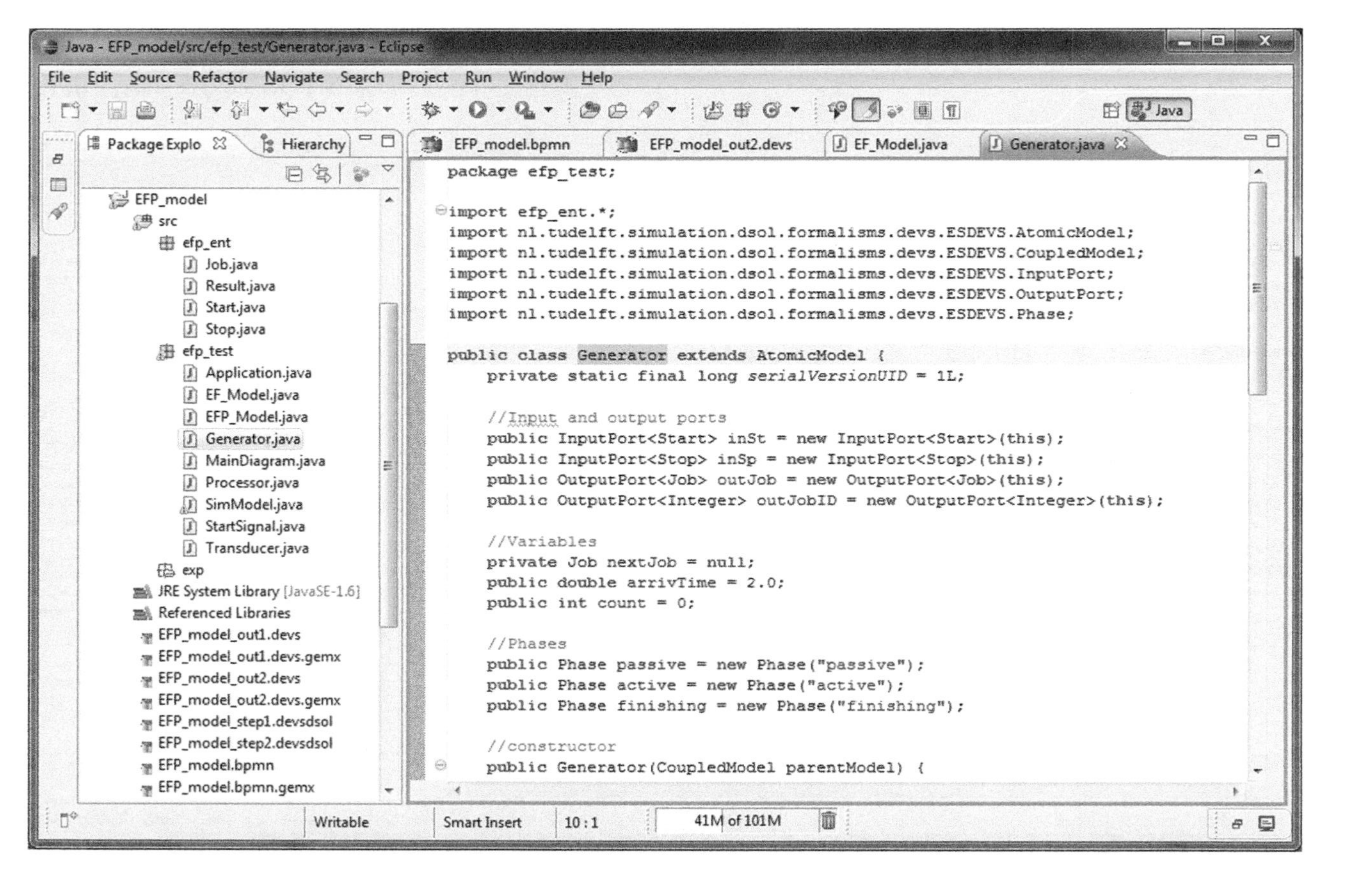

FIGURE 22.9 Generated atomic model code.

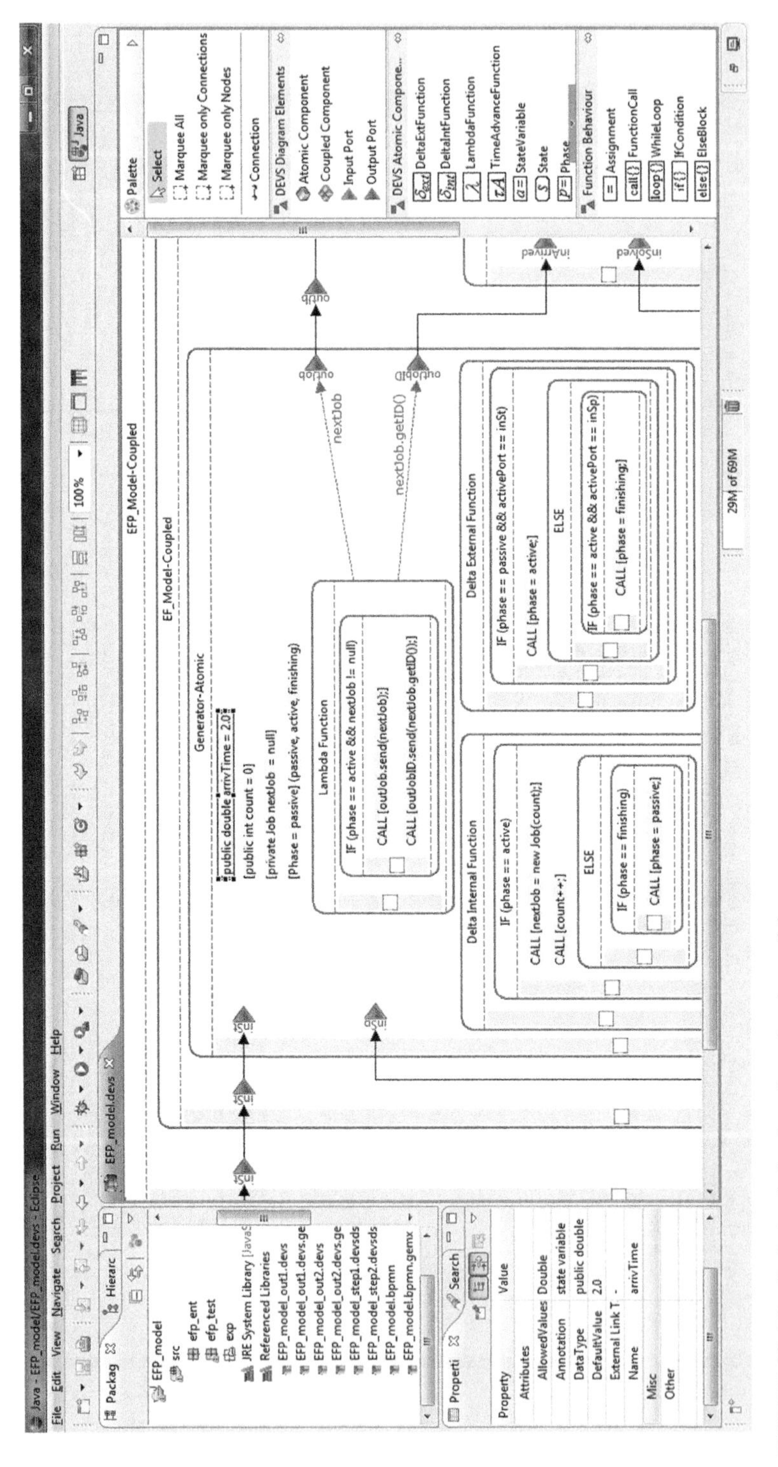

FIGURE 22.10 Defining behavior of an atomic model.

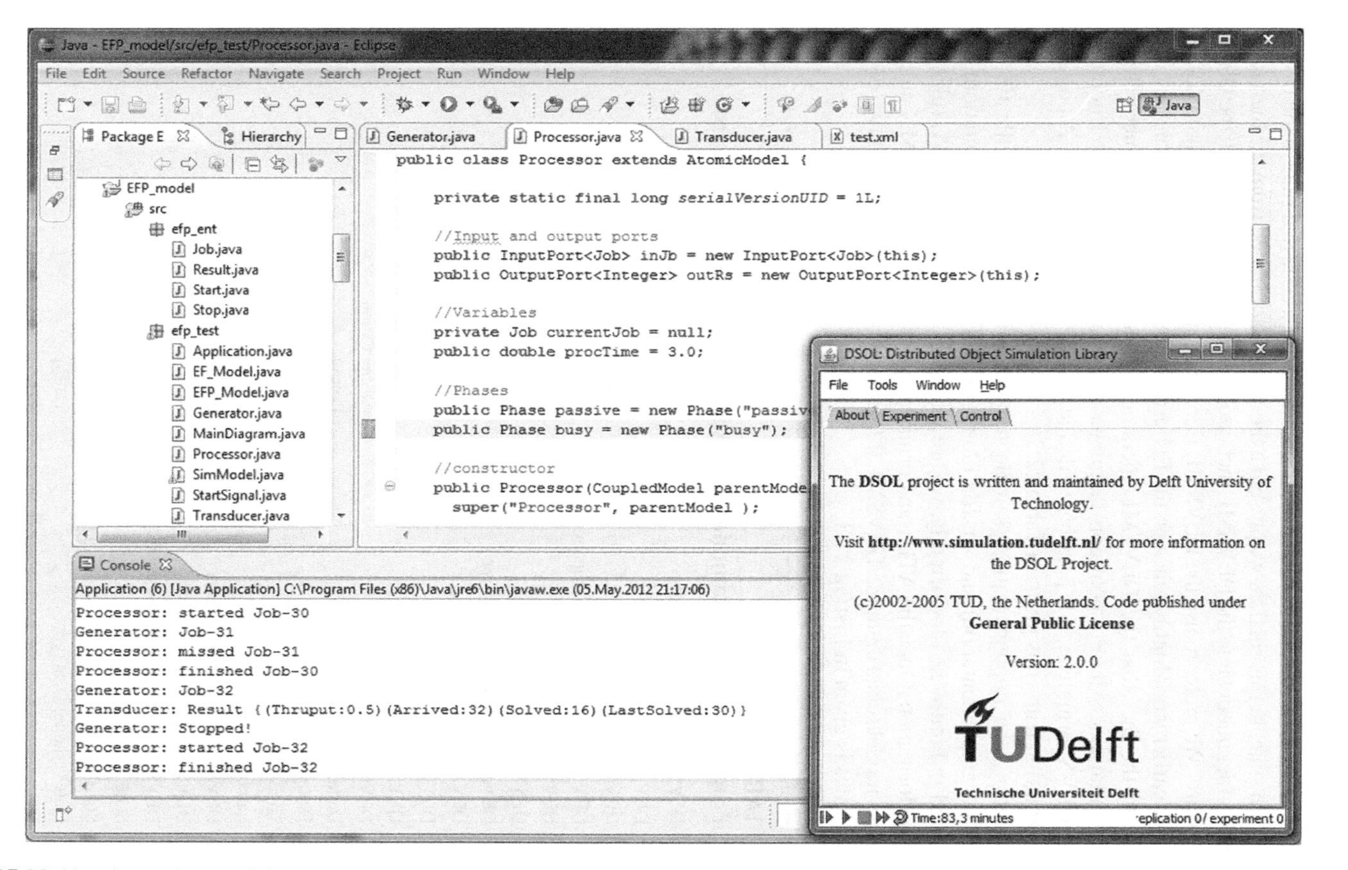

FIGURE 22.11 A running model.

22.6 INTERACTING WITH DEVS MIDDLEWARE THROUGH DEVSML

The example case presented in Section 22.5 generates Java code that runs in the DSOL simulation suite. In this section, we present a model transformation from the hierarchical DEVS metamodel into the DEVSML metamodel in order to interact with the DEVS middleware. We prefer to use the DEVSML Ecore metamodel, given in Figure 22.12, since our earlier work highly relies on the Eclipse platform and the Ecore metamodel.

Due to the fact that the DEVSML metamodel and the presented hierarchical DEVS metamodel share a common language and provide a higher level notation for the DEVS formalism, it is possible to define a transformation pattern. A possible transformation pattern is given in Table 22.2.

Regarding the implementation, sample mappings between a **.devs** file (in the MDD4MS implementation) and an **.fds** file (in the DEVSML implementation) are illustrated in Figure 22.13. Currently, we are working on drafting the ATL rules for this transformation. Once the ATL transformation is available, visual BPMN models and **.devs** models can run in the DEVSML 2.0 framework. An example of two of the written rules is given in Listing 22.3.

TABLE 22.2
Proposed DEVS Metamodel to DEVSML Transformation Pattern

DEVS Metamodel Concepts	DEVSML Concepts
DEVSModel	Model
DEVSComponent	Component
DEVSComponent: DEVSComponentType	Type (Coupled, Atomic, Entity)
DEVSPort	Msg
DEVSPort: DEVSPortType	Msg: type
DEVSPort: DEVSPortDataType	Entity
DEVSInputPort	Msg: type = input
DEVSOutputPort	Msg: type = output
DEVSCoupledComp	CoupledComp
DEVSAtomicComp	AtomicComp
DeltaIntFunction	Deltint
DeltaExtFunction	Deltext
TimeAdvanceFunction	TimeAdv
LambdaFunction	Outfn
DEVSInToIn_EICConnection	EIC
DEVSOutToIn_ICConnection	IC
DEVSOutToOut_EOCConnection	EOC
State	STA
StateVariable	Variable
StateVariable: DataType	VarType
Phase	STA
Sigma	Variable
Function: Expression	Code: str

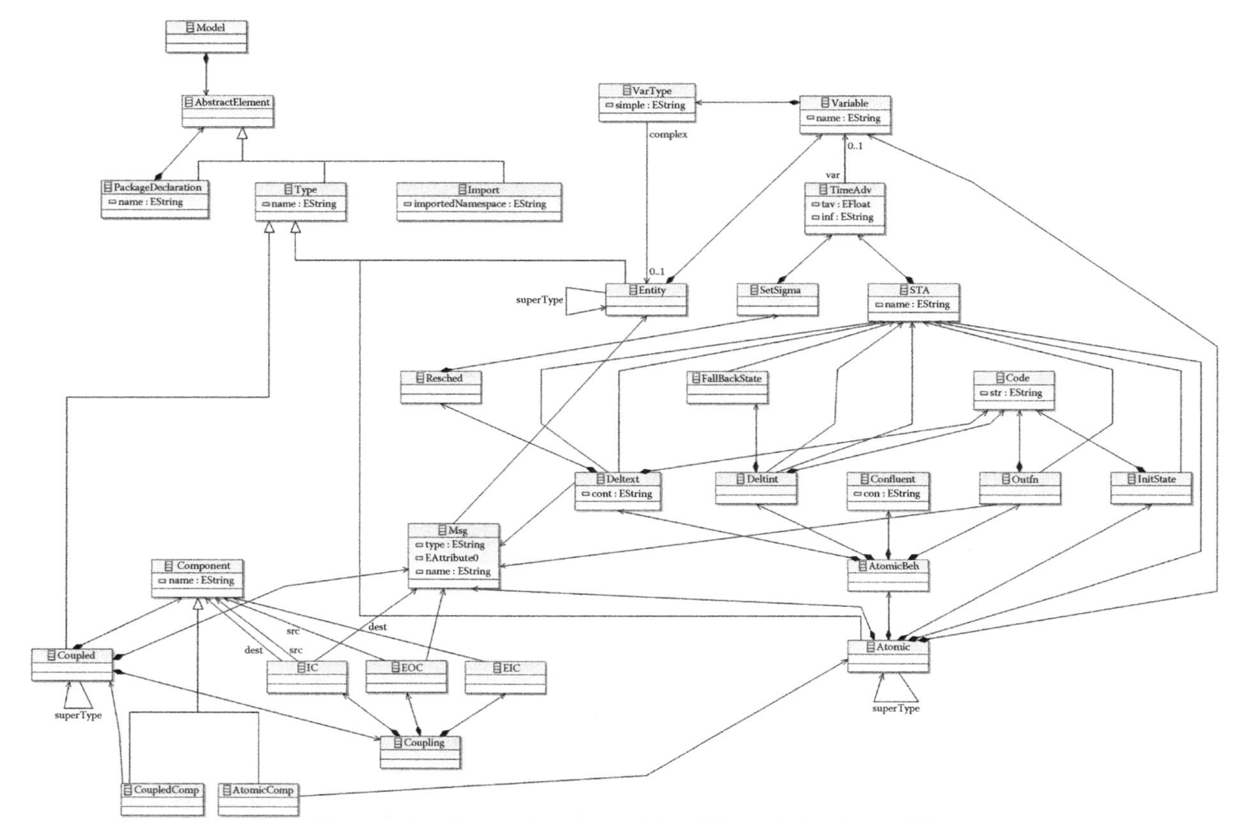

FIGURE 22.12 DEVSML metamodel in Ecore.

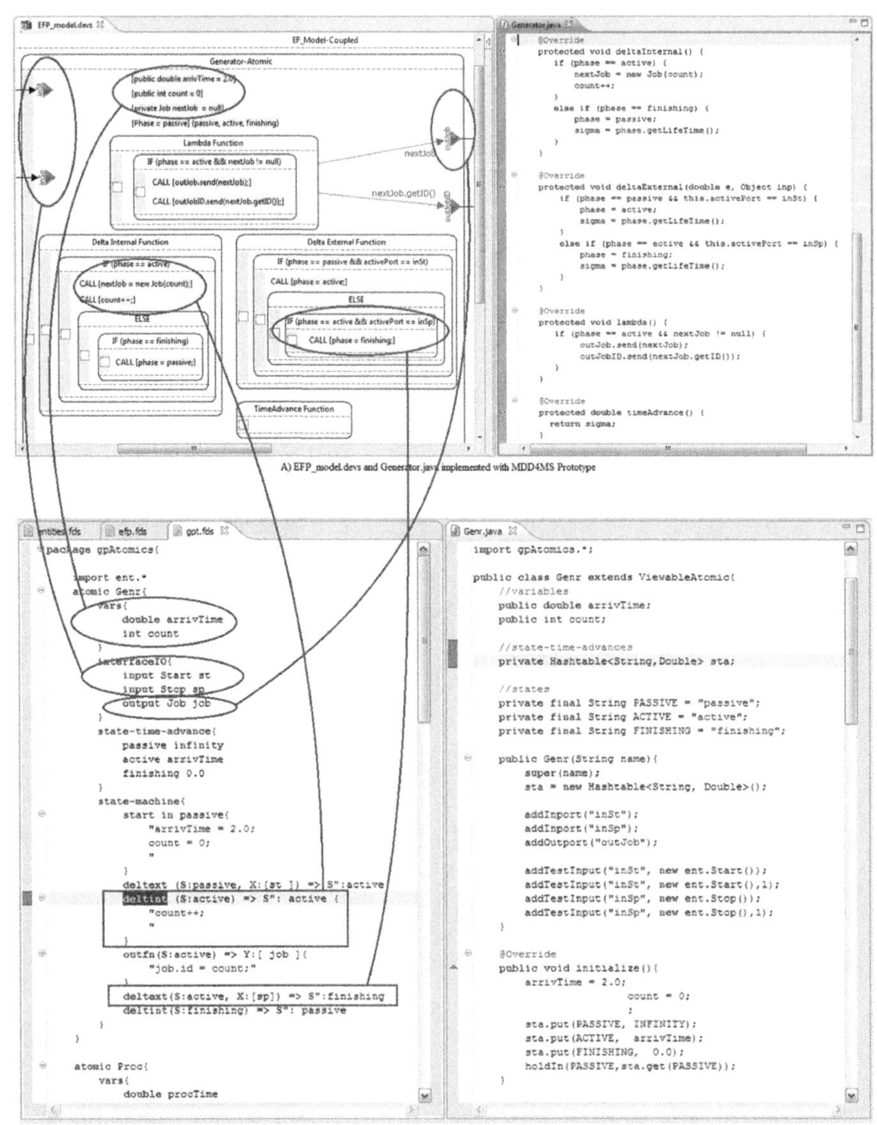

FIGURE 22.13 Sample mappings from hierarchical .devs to .fds.

LISTING 22.3 FROM HIERARCHICAL .DEVS TO .FDS

```
module DEVS_To_DEVSML;
create OUT: DEVSML_Metamodel from IN: DEVS_Metamodel;

rule DEVSModel_to_DEVSMLModel {
```

```
    from
      s: DEVS_Metamodel!DEVSModel (s.isMainDiagram())
    to
      t: DEVSML_Metamodel!Model (
        Elements <- s.DEVSComponents
      )
  }

  rule DEVSCoupled_to_DEVSMLCoupled {
    from
      s: DEVS_Metamodel!DEVSCoupledComp (true)
    to
      t: DEVSML_Metamodel!CoupledComp (
        CP <- cp_new
      ),
      cp_new: DEVSML_Metamodel!Coupled (
        Name <- s.Name,
        SuperType <- s.DEVSCompParent,
        Components <- s.DEVSComponents,
        Msgs <- s.DEVSPorts,
        Couplings <- s.getCouplings()
      )
  }
```

22.7 SUMMARY

This case study proposes a new BPM approach based on a proven theory. It provides an easy way of DEVS modeling and simulation for business analysts with a limited knowledge of DEVS. Applying MDD4MS to BPM projects and supporting the life cycle via graphical modeling tools provides a formal way to construct the simulation models in a faster, better, and more reliable way.

The transformation from hierarchical DEVS metamodel into DEVSML 2.0 enables the transformation of BPMN models to DEVSML 2.0. The BPMN-to-DEVSML transformation can be obtained easily with two successive transformations—first from BPMN to **.devs** and then, from **.devs** to **.fds** in DEVSML.

Our future work includes the integration of the DEVSDSOL library into DEVSML 2.0 stack for more powerful DEVS-based M&S environment integration.

REFERENCES

Bosilj-Vuksic, V., Ceric, V., & Hlupic, V. (2007). Criteria for the evaluation of business process simulation tools. *Interdisciplinary Journal of Information, Knowledge, and Management*, *2*, 73–88.

Çetinkaya, D., Verbraeck, A., & Seck, M. D. (2011). MDD4MS: A model driven development framework for modeling and simulation. *Proceedings of the Summer Computer Simulation Conference (SCSC)* (pp. 113–121). The Hague, Netherlands.

Çetinkaya, D., Verbraeck, A., & Seck. M. D. (2012). Model transformation from BPMN to DEVS in the MDD4MS Framework. *Proceedings of the Theory of Modeling and Simulation: DEVS Integrative M&S Symposium 2012* (pp. 304–309).

Eclipse. (2006). Eclipse Generative Modeling Technologies (GMT) Project. Retrieved April 24, 2012, from http://eclipse.org/gmt/.

Eclipse. (2008). Generic Eclipse Modeling System (GEMS). Retrieved April 24, 2012, from http://www.eclipse.org/gmt/gems.

Eclipse. (2009). Eclipse Modeling Framework (EMF) Project. The Eclipse Foundation. Retrieved April 24, 2012, from http://www.eclipse.org/modeling/emf/.

Jacobs, P. H. M., Lang, N. A., & Verbraeck, A. (2002). D-SOL: A distributed java based discrete event simulation architecture. In E. Yücesan, C.-H. Chen, J. L. Snowdon, & J. M. Charnes, *Proceedings of the 34th Winter Simulation Conference* (pp. 793–800). San Diego, CA: Winter Simulation Conference.

Jouault, F., Allilaire, F., Bézivin, J., & Kurtev, I. (2008). ATL: A model transformation tool. *Science of Computer Programming*, *72*(1–2), 31–39. doi:10.1016/j.scico.2007.08.002.

OMG. (2011). *Business Process Model and Notation (BPMN) Version 2.0*. Retrieved from http://www.omg.org/spec/BPMN/2.0/PDF/. Last accessed October 2012.

Recker, J., Indulska, M., Rosemann, M., & Green, P. (2009). Business process modeling—a comparative analysis. *Journal of the Association for Information Systems*, *10*(4), 333–363.

Seck, M. D., & Verbraeck, A. (2009). DEVS in DSOL: Adding DEVS operational semantics to a generic event-scheduling simulation environment. *Proceedings of the Summer Computer Simulation Conference* (pp. 261–266). Istanbul, Turkey.

Van der Aalst, W. M. P., ter Hofstede, A. H. M., & Weske, M. (2003). Business process management: a survey. In W. M. P. van der Aalst, A. H. M. ter Hofstede, M. Weske, G. Goos, J. Hartmanis, & J. van Leeuwen, *Business Process Management, Lecture Notes in Computer Science, 2678* (Vol. 2678, pp. 1–12). Berlin, Heidelberg: Springer-Verlag. doi:10.1007/3-540-44895-0_1.

Van Nuffel, D., Mulder, H., & Van Kervel, S. (2009). Enhancing the formal foundations of BPMN by enterprise ontology. In A. Albani, J. Barjis, J. L. G. Dietz, W. Aalst, J. Mylopoulos, M. Rosemann, M. J. Shaw, et al., *Advances in Enterprise Engineering III, Lecture Notes in Business Information Processing, 34* (Vol. 34, pp. 115–129). Berlin, Heidelberg: Springer-Verlag. doi:10.1007/978-3-642-01915-9_9.

Zeigler, B. P., Praehofer, H., & Kim, T. G. (2000). *Theory of modelling and simulation: integrating discrete event and continuous complex dynamic systems* (2nd ed.). Academic Press, San Diego, CA, USA.

Section V

Next Steps

23 Netcentric Complex Adaptive Systems

23.1 INTRODUCTION

So far, we have described systems that can be analyzed in an Object-oriented manner, systems that are designed, and systems that are artificial or human-made. We described various processes, methodologies, and frameworks that help define system structure and behavior. The DEVS formalism is one such formalism that lends itself to a formal study of system structure and behavior at various levels of abstractions through the notion of an event. A netcentric system is just another system where the component subsystems are heterogeneous in nature and may have different execution platforms. However, they are designed to solve two major issues: integration and interoperability. While integration is mostly a technical exercise with heavy emphasis on technological solutions, interoperability is more complex, as it happens at various levels of systems specifications, for example, pragmatic, semantic, and syntactic. Integration can largely solve syntactic interoperability, but for any two systems to interoperate or perform an operation together, they have to understand each other's semantics and context (pragmatics). A netcentric system brings forth a set of standards that help automate syntactic interoperability through Service Oriented Architecture (SOA), where each system is made modular and finally integrated. Much work is needed in the research area of semantic and pragmatic interoperability. This is an open-ended problem that needs to be further explored.

Humans have always sought inspiration from nature—from development of simple machines to complex aircrafts. A natural system is not a monolithic system, but a heterogeneous system made up of disparity and dissimilarity, devoid of any larger goal. The system just "is." Examples of such systems include ant colonies, the biosphere, the brain, the immune system, the biological cell, businesses, communities, social systems, and stock markets. These systems are classified as complex adaptive systems (CAS). These systems can be analyzed at different levels of abstraction much the same way as the communication network can be modeled at a packet level (e.g., Opnet) or as a fluid flow solution to determine the capacity of the network. Moreover, when the system becomes so large or acts as a component in a larger system of systems (SoS), various nonlinear effects come into the picture that give a unique emergent property to the SoS. Events happening in one system can impact other systems in unpredictable ways. Such systems are adaptable systems where emergence and self-organization are factors that aid evolution. According to Holland (1992, p. 1), "CAS are systems that have a large number of components, often called agents that interact and adapt or learn." In the context of the SoS, agents could be entire systems. Most artificially modeled systems that exhibit complex adaptive behavior are driven by multiresolution bindings and interconnectivity at every level of the system behavior. The need for modeling and simulation (M&S) to make progress in understanding CASs has been well acknowledged by Holland (2006). Ziegler's approach to CASs has been through the quantization of continuous phenomena and how quantization leads to abstraction. Any CAS must operate within the constraints imposed by space, time, and resources on its information processing (Pinker, 1997). Evidence from neuronal models and neuron-processing architectures and from fast and frugal heuristics provide further support to the centrality of discrete event abstraction in modeling CASs when the constraints of space, time, and energy are taken into account. Zeigler stated that discrete event models are the right abstraction for capturing CAS structure and behavior (Zeigler, 2004).

In this chapter, we look at various features of CASs and how a netcentric SoS resembles a CAS. Eventually, we would like to provide a framework wherein such a netcentric CAS can be modeled and simulated.

23.2 CHARACTERISTICS OF COMPLEX ADAPTIVE SYSTEMS

CASs are occasionally modeled by means of agent-based models and complex network-based models. Multiagent systems (MASs) compose the area of research that deals with such study. However, CASs are fundamentally different from MASs in portraying features such as self-similarity (scale-free), complexity, emergence, and self-organization that are at a level above the interacting agents. A CAS is a complex, scale-free collectivity of interacting adaptive agents, characterized by a high degree of adaptive capacity, giving them resilience in the face of perturbation. Indeed, designing an artificial CAS requires formal attention to these specific features. Complexity is a phenomenon that is multivariable and multidimensional in a space–time continuum. Complexity must not be confused with "complicated." Complicated worlds are reducible to individual parts, whereas complex ones are not. There is an inherent notion of emergent behavior that emerges in a complex system and each component does its bit. A complex system is like an orchestra where music emerges.

Let us now look at some of the fundamental characteristics of CASs. We will attempt to describe the inherent meaning of complexity and adaptive nature. While we know intuitively what "complex" is, we take a hard look at what makes a system complex. We also are aware of the term "adaptation" and we will look at it from the perspective of a complex system. What makes a complex system adaptive? We address this question from three perspectives:

1. Network topology
2. Agent/system actions
3. Adaptation through self-organization and emergence in a contingent environment

23.2.1 Network Topology

Complex networks are the backbone of complex systems, and each complex system is a network of interactions among numerous network elements. Some networks are geometric or regular in 2D or 3D space, and some have long-range connections that are not spatial at all. Network topology or anatomy is important to characterize because structure affects function and vice versa. The dynamic nature of a network is one of the keys to understanding complexity. Each network comes with a peculiar set of properties, and the manifested behavior of the components is bounded by the constraints the network imposes on them (Barabási, 2003). CASs are the most complex systems that display adaptive behavior through growth and preferential attachment and are abundant in nature (Newman, 2001). The abundant of CASs at every level of natural phenomena is due to their scale-free nature. Some of the characteristics of scale-free networks as described in the works of Newman and Barabasi are as follows:

1. *Incremental growth*: A scale-free network is a natural product of nodes incrementally linking to other nodes or situating themselves in a persistent environment. A persistent environment is an environment that has a spacio-temporal character to it.
2. *Self-organization*: Nodes may organize themselves to achieve a common objective. Particle swarms and robotic MASs are examples of this category.
3. *Critical state transition*: At an appropriate time during network growth when agents display a fundamentally different behavior from their earlier manifestation, they change their roles and some nodes manifest new behavior. A hierarchy either emerges or is broken at these critical transitions.
4. *Emergence of hubs and clusters*: A cluster is a dense network of nodes connected together. Hubs are nodes that have a high degree of connectivity within the cluster and weak, long-range connectivity outside the cluster displaying a small-world effect (Barabási, 2003; Watts, 1999). Hubs tend to break small system (local network) boundaries and connect to other clusters.
5. *Display of power law behavior*: Some nodes assume the role of the network enabler structurally and reduce their functional contribution. They start acting as a hub forming long, weak ties across clusters. A hub redefines network linkages and ultimately topology, bringing in new affordances and/or constraints.
6. *Nonlinear interactions*: Agents or constituent systems are connected through these hubs, introducing nonlinear behavior in a larger SoS, wherein a change in the agent's state may impact other clusters in unpredictable ways. It is difficult to define a boundary in an SoS. CASs are mostly *open* systems.
7. *Preferential attachment*: Nodes are independent systems with inherent properties and display affinity or aversion in connecting with the environment or other agents. Based on the intensity of these two behaviors, the rate at which links are acquired or dropped, it redefines the topology and the behavior of the node, making it a hub (a structural enabler) or a performer node.
8. *Vulnerability to attacks if targeted at hubs*: Any attack on hubs has a cascaded effect on multiple clusters, thereby breaking the scale-free nature and weak ties, and reducing the network to independent clusters. However, multiple attacks on nodes other than hubs do not cause much damage. The network may also align itself toward self-organized criticality, where

it disintegrates to individual nodes when it reaches a critical state. For example, sand piles and avalanches display such behavior.

9. *Threshold levels*: Each node, including hubs, has a threshold model that specifies its degree of preferential attachment. A low threshold at the hub implies that the communication travels far without much censoring. This distribution of critical threshold levels within the network nodes is a property of the network that explains phenomena such as virus spread, fire spread, adoption of innovative ideas, and so forth.
10. *Concurrency and multitasking*: Nodes and constituent systems are modular; that is, they have defined interfaces and capabilities. Each node brings value to the network, and depending on its topological significance, it impacts the entire network accordingly.

To understand the modeling aspect of CASs, we ask the following:

I. How are the rules inside these agents triggered when "critical state transition" occurs?
II. Does the agent already have an inherent capacity, and does the presence of a conducive environment enforce the agent to manifest a totally different behavior?
III. Does the agent learn new rules by active observation in an environment?
IV. Do inherent capabilities make the agent vulnerable to external influence?
V. What makes a node or an agent evolve into a hub?
VI. What dynamics happen within a node that urges it to extend a link?
VII. What transpires through when a link is formed between two nodes?

We will address these questions within the context of dynamic structure modeling in sections ahead.

23.2.2 Agent/System Actions

Agents or systems in an SoS are multi-component modular systems. In the real world, each agent node and the environment are persistent; that is, they have memory. If agents simply respond to a persistent environment and have a fixed set of rules devoid of any complex learning and/or memory apparatus, then agents are reactive. However, for an agent to transform from a node to a hub, advanced learning and memory apparatus must be present that can be triggered when critical state transition occurs. Both agents and systems are components that stand independently and have a defined behavior both internally and externally. Consequently, these components take actions both internally and externally. We classify these actions into two broad categories, intra-actions and inter-actions, as follows:

1. *Intra-actions*: These are the actions taken by the node internally; that is, these actions impact the node itself first and may impact other nodes through various interactions in which this node participates. These actions are initiated by the internal dynamics of the node in the absence of any external input.

Inter-actions: These are the actions taken by the node in an external environment; that is, these actions impact other nodes in the network. This impact to other nodes is communicated through various modes of communication as allowed by the environment or specified by the properties of the network. Such interactions have been further classified by Keil and Goldin (2003) as

1. *Direct interactions*: These interactions are point-to-point connections, and the actions of one node directly affect another node and *nobody* else.
2. *Indirect interactions*: These interactions follow a publish–subscribe phenomenon where a node publishes its actions (through specific messages) to a shared environment (acting as a persistent medium) and other nodes that are subscribed to this particular type of messages are affected through the medium of exchange. Keil and Goldin (2003) prove that the behavior of computational agents that make use of indirect interactions via the real world is richer than the behavior of agents that interact directly.

The capacity to detect and be affected by changes in the environment as a result of an action of another node in a spatiotemporal manner is guided and simultaneously constrained by both the properties of a network and the internal structure of the node. The capacity *to detect* is a property analogous to preferential attachment and affinity. The capacity *to be affected* is a property analogous to the threshold model. The capacity to form weak links with other individuals separated by space and time is made possible through the persistent environment and the persistent initial state of the agents.

23.2.3 Adaptation through Self-Organization and Emergence in a Contingent Environment

Self-organization as defined by Wolf and Holvoet (2005, p. 7) "is a dynamical and adaptive process where systems acquire and maintain structure themselves, without external control."

Note the keywords *dynamical*, *adaptive*, *acquire*, *maintain*, *structure*, and *without external control*. The context can easily be understood with respect to the scale-free network and its emerging topologies. Wolf and Holvoet (2005) acknowledge the identification of the "boundary" of the system that relates to the modularity principle in scale-free networks.

Self-organization is also defined as the interaction of a set of processes at a lower level of system to yield structures at a higher global level (Holland & Melhuish, 1999). This does not imply that self-organization evolves into a hierarchical organization. For a hierarchy, that is, a hub to appear, the self-organization must be coupled with other constructs such as the scale-free nature of the underlying network, coherent emergents, and upward as well as downward causations. Self-organization leading to hierarchy has been observed in most natural CASs. Similar work by Brooks (1991) acknowledges that the fundamental decomposition of intelligent systems is not in the identification of individual processing producers that must interface with each other, but in the interactions that these producers have with the world through perception and action.

Emergence has gained widespread attention in the last two decades partly due to the analysis capabilities afforded by massive computational power and partly due to widespread complex systems in everyday use such as the World Wide Web. Emergence has been a native of the land of complex systems, and we begin with a definition of emergence as given by Wolf and Holvoet (2005, p. 3):

> A system exhibits emergence when there are coherent emergents at the macro-level that dynamically arise from the interactions between the parts at the microlevel. Such emergents are novel with respect to the individual parts of the system.

Other definitions of emergence can be found in the work of Deguet et al., (2006). Banabeau and Dessalles (1997) in their definition give significant importance to the detection of "something." They also described how emergence is handled in a hierarchical complex system:

> When a detector becomes active in such a hierarchy, the active detectors from the lower level that are connected to it can be omitted from the description. [...] Emergence is thus a characteristic feature of detection hierarchies (Banabeau & Dessalles, 1997, p. 5).

We acknowledge that emergence is an observer phenomenon. This is also supported by Miller and Page (2007). We would add that the emergence happens only at levels above the interacting agents; that is, a hierarchy must be present to detect emergence. The observer is always at a higher level of perception to detect something emerging. In the classic ant colony example, an ant does not know that it is building a colony. It is the human observer or a higher animal or a detector agent that witnesses the existence of such structural functional affordances. Although these affordances may be present in the environment that would result in the manifested behavior, an observer must be present to label such an "affordance" in an artificial system that self-organizes at the collective behavior at the level above it. In an artificial system, the notion of observer takes a stronghold as various observers can be defined at various levels of a hierarchical system that can detect the occurrence and variations in key indicators at specific levels of resolution to categorize as emergence. Another way to look at it is through the identification of such key indicators. These key indicators are not a part of the system they are meant to observe. Sometimes these indicators can be derived from the lower level constructs, and sometimes they are beyond such deduction and are completely novel (Muller, 2003). This demarcation in detection capabilities was formalized by Banabeau and Dessalles (1997) as $L_{(n-1)}$ and L_n levels. $L_{(n-1)}$ is a level where individuals interact in a persistent environment. L_n is a level above $L_{(n-1)}$ that observes these global collective properties of $L_{(n-1)}$. Both $L_{(n-1)}$ and L_n have their own grammar and representation. Obviously, if a pattern in a lower level $L_{(n-1)}$ is not formalized as a detector at L_n, then such a pattern will fail to "emerge" at L_n. This correlates with deducible emergence (Baas, 1994) where two disjoint levels are linked by a computational process, that is, how $L_{(n-1)}$ and L_n interact computationally. Alternatively, we, as humans, see what we are entrained to see through our rich set of experiences. If we are unable to identify a pattern, it is highly unlikely that we will label it as such; that is, we do not understand the grammar at L_n!

When we design intelligent agents, we could also imbue them with the power to observe the collective whole. The agent actions are based on this persistence wherein the collectives have been realized in the environment in a spatiotemporal matrix. Now, having both a persistent environment and a sophisticated detector mechanism in this artificial agent will make the agent more competitive to other agents due to its handling of additional percepts. From a performer at a specific level of system, on detection of emerging patterns, agents can become an enabler, or a hub. This is the rise of hierarchy in a self-organized manner as has been described in Barabási's work and the coherent property (Wolf & Holvoet, 2005) mentioned in the first definition of emergence. This aspect of having the act of observation inside the agent is also congruent with ideas of strong and weak emergence by Muller (2003). Muller also added that in strong emergence, the observer has causal powers.

The system displays strong emergence when the emergent behavior is irreducible to either the agent or the environment, as both interact in a dynamic spatiotemporal manner. The emergent behavior also has a downward causation at lower levels (Chalmers, 2006), changing the very nature of the network beneath it. This marks the rise of hierarchy and the critical state transition. In the case of weak emergence, the emergence phenomenon is reducible to its constituent components and the effect of causality on lower levels is questionable. The "Game of Life" and connectionist networks are examples of weak emergence, where laws encoded in low-level rules result in high-level structures and patterns. The collective behavior is manifested in the persistent dynamic environment that the agent is part of. The agent detects such changes and acts over them.

While both emergence and self-organization are dynamic properties of complex systems, the network's and the agent's robust internal properties together will decide if both of them are simultaneously portrayed by the system. Three cases arise (Wolf & Holvoet, 2005):

1. *Self-organization without emergence*: MASs organize to perform a specific task. Loss of an entity does not impact the execution of the task. The behavior is reducible to encoded rules and the observed behavior is not novel at all.
2. *Emergence without self-organization*: Consider the property of a gas occupying volume. This property is unique as a function of its atoms and can exhibit chaos that emerges from interactions of these atoms but no self-organization, as there is no collective behavior.
3. *Emergence and self-organization together*: This occurs in most natural systems where a persistent environment and producing agents are present. In such a dynamical system, the agents can hierarchically self-organize and display adaptive emergent phenomena through their defined actions that result in dynamic scale-free topologies. Providing a structure to such an emergent CAS a priori is almost impossible as the system structure itself is based on the persistent nature of components, dynamic interactions, and the resulting topology. As has been evident in the evolution of scale-free networks that the hubs tend to reduce complexity (Barabási, 2003), similar results from Shalizi (2001, p. 118) are present in complexity research that state, "... self-organization increases statistical complexity, while emergence, generally speaking, reduces it."

Because these systems are intricately linked, they display a nonlinear behavior (Camazine et al., 2001; Heyligen, 2001) where a small perturbation can lead to a large effect due to amplification by positive or negative feedback loops. Consequently, in a self-organizing emergent system, the interplay of positive, negative feedback loops, amplifications, suppressions, taken all together as preferential attachment, thresholds and affinities, become a function of the environment and the agent taken together. In a complex adaptive self-organizing emergent system, the system continues to redefine the topology, displays the emergent properties, and refines the properties themselves that initially defined the interactions.

23.3 MODELING A CAS

Let us move now to the modeling aspect of CASs. We will now describe some recent extensions of DEVS and SES formalism that allow modeling of an adaptive behavior.

23.3.1 DEVS Variants for Adaptive Behavior Modeling

DEVS system is a hierarchical complex dynamical system *closed under coupling* (similar to closed under composition) with modularity at its core. In its default description, while DEVS can specify a structurally static system, the formal atomic and coupled models cannot sufficiently describe the network dynamics and the adaptive behavior as needed for natural and biological systems. However, in its current state, it can certainly describe weak emergence in isolation. To model hierarchical self-organization, it is imperative to have a variable structure capability (Uhrmacher & Zeigler, 1996) that can reconfigure the component system both structurally and functionally. The structural capability is manifested externally, outside the component boundary, whereas the functional capability is manifested internally, within a component. DEVS systems have a continuous time base, but their execution is event based. A variable-structure discrete event system adds a temporal nature to the structure of the system itself. The structure of a system can be dynamic at three levels:

1. *Component level*: Entire substructures are removed from or added to a live system.
2. *Connection level*: Interactions are reconfigured in a live system.
3. *Interface level*: Interface of the component itself is subject to reconfiguration.

The behavior of a system can be dynamic in four ways:

1. State space
2. Time advances of each state
3. Transition functions (e.g., δ_{int}, δ_{ext}, δ_{con}, λ)
4. Initial state

More technically, such dynamism must be traceable to the levels of system specification described in Chapter 2. Table 23.1 provides the mapping of how dynamism is introduced at various levels. It shows what an outcome would be of such a dynamic

TABLE 23.1
Introducing Dynamism at Various Levels of System Specifications

Level	Name	How Dynamism Is Introduced	Outcome	Impact in a Scale-Free Network
4	Coupled systems	1. System substructure 2. System couplings 3. Subsystem I/O interfaces 4. Subsystem active/dormant	1. Dynamic component structures 2. Dynamic interaction	II, IV, VII
3	I/O system	1. Addition/removal of states 2. Augmentation of transitions with constraints/guard conditions	1. Dynamic states 2. Dynamic transitions 3. Dynamic outputs	I, III, IV, VI, VII
2	I/O function	1. Initial state 2. Addition/removal of initial state 3. Addition/removal of I/O pairs	Dynamic initial state	IV
1	I/O behavior	1. Time scale between the I/O behavior 2. I/O mapping changing the behavior itself 3. Allowed behavior 4. Addition/removal of I/O pairs	Dynamic I/O behavior	I, II, III, V, VII
0	I/O frame	1. Allowed values 2. I/O to port mapping	Dynamic interfaces	III, VII

activity. Table 23.1 is reproduced from Chapter 7 with the addition of the last column, which relates it to the list of questions we encountered in understanding the nature of scale-free networks (Section 23.2.1) earlier.

The dynamic structure outcomes have been adequately dealt with in our earlier work (Hu et al., 2005), and formally by Barros (1995; 1997; 1998), Uhrmacher (2001), Uhrmacher and Priami (2005), and Uhrmacher et al. (2006). Here we will discuss the formal undertaking of the dynamic structure by Uhrmacher et al. (2007) and Uhrmacher et al. (2011), as the structural change is initiated from within the system components rather than a specialized component called Network Executive in Barros' DSDEVS. The first version is named as DynPDEVS. The underlying idea behind this DEVS extension is to interpret models as a set of models successively generating themselves by model transitions. Model and network transitions are introduced mapping the current state of a model into a set of models that the model belongs to. The formalism supports models that adapt their own interaction structure and their own behavior as a result of those interactions through a newly added transition function ρ_α The structural changes are induced bottom-up and are communicated through another newly defined transition function, ρ_λ. Specific types of input and output interfaces are introduced that communicate these structural changes to other models. This version refers to dynamic components and dynamic coupling in a live system.

The second more advanced type is built on DynPDEVS, and it introduced dynamic port interfaces. The ports X and Y are part of the incarnations of model M. This is the most critical of capabilities required for metamorphosis of the component allowing plasticity (Mittal et al., 2005), for example, in neuronal ensembles that add dendrites and axons to support the Hebbian hypothesis. A DEVS neuron with dynamic interfaces requires this capability of dynamic interfaces as it strengthens or weakens its connections with other neurons. The second version is named ρDEVS. Formally, it is described as a structure $< m_{\text{init}}, M, X_{\text{SC}}, Y_{\text{SC}} >$ with $m_{\text{init}} \in M$ being the initial model, and $m_i \in M$ being the ith incarnation, X_{SC} and Y_{SC}, and the ports to communicate structural changes, and M, the least set with the following structure:

$$M_{\rho\text{DEVS}} = \left\langle M_{\text{P-DEVS}}, s_0, \rho_\alpha, \rho_\lambda \right\rangle,$$

where

$M_{\text{P-DEVS}}$ is the parallel DEVS
$s_0 \in S$ is the initial state
$\rho_\alpha : S \times X_{\text{SC}} \rightarrow M$ is the model transition function
$\rho_\lambda : S \rightarrow Y_{\text{SC}}$ is implied structural network change

A reflective, higher order network, a ρNDEVS, is the structure ρNDEVS = $\langle n_{\text{init}}, N, X_{\text{SC}}, Y_{\text{SC}} \rangle$, where $n_{\text{init}} \in N$ is the start configuration; X_{SC} and Y_{SC}, the ports to communicate structural changes; and N, least set with the following structure:

$$N_{\rho\text{DEVS}} = \langle X, Y, C, \text{MC}, \rho_N, \rho_\lambda \rangle,$$

where

C is the set of components that are of type ρDEVS
MC is the set of multicouplings
$\rho_N : S^N \times X_{\text{SC}} \rightarrow N$ is the network transition function
$\rho_\lambda : S^N \rightarrow Y_{\text{SC}}$ is the structural output function

The value of ρ_N preserves the state and the structure of models that belong to the old and the new composition of the network. A multicoupling $\text{mc}_i \in \text{MC}$ in this formalism determines how the outputs are distributed from output to input ports. In regular DEVS (Chapter 3), if more than one input port is linked to an output port, the output values are cloned at all the inports. When the artifacts and messages are in real world and consumable physical objects, this may not be desirable. The standard strategy is useful when the information is to be broadcast. In natural systems, the capability warrants a function that selects the output port for consumable resources.

A random selection strategy may very well be used in the MC function. For rigorous mathematical analysis of this formalism, see Uhrmacher et al. (2006).

The dynamic structure capability thus far defined by ρDEVS is manifested externally in the topology. An atomic model can be reincarnated as a coupled model and hierarchy can emerge. However, the coupled component still acts as a container of other components without any state and behavior representation. Hubs cannot form without displaying a behavior. To alleviate this problem, state and transition functions are introduced at the coupled level in Multilevel-DEVS (ML-DEVS) (Uhrmacher et al., 2007). ML-DEVS is an extension of ρDEVS and consists of Micro-DEVS (atomic) and Macro-DEVS (coupled). Let us look at Macro-DEVS first. A Macro-DEVS has structured input, output, and state sets, X, Y, and S, respectively. An λ output function produces output for the output ports, and a set C of components is specified. A set of multicoupling functions MC allows specification of value couplings. The state transition function δ takes into account the current state, the components, and multicouplings to calculate the new state. A function p associates ports with each state. The structural change function sc defines the correlation between the set of components and multicouplings for the current state. The downward causation is enabled by v_{down} that couples Macro-DEVS' current state variables to the input ports of Micro-DEVS. The downward activation is done by λ_{down} function that allows synchronous activation of Micro-DEVS models in an event-based manner. The upward causation is enabled by the port transition function, as all the available ports at Micro-DEVS level are available at the Macro-DEVS level. The transition function δ at Macro-DEVS accounts for any change in ports at the Micro-DEVS in calculation of the next Macro-DEVS state.

A Macro-DEVS is defined as a structure:

$$N_{\mathrm{mlDEVS}} = \langle X, Y, S, s_{\mathrm{init}}, p, C, \mathrm{MC}, \delta, \lambda_{\mathrm{down}}, \mathrm{v}_{\mathrm{down}}, \mathrm{sc}, \mathrm{act}, \lambda, \mathrm{ta} \rangle,$$

where

p is the function that maps ports with each states
C is the set of submodels that are type Micro-DEVS or Macro-DEVS
MC is the set of multicouplings, $\{m | m : 2^P \to 2^P\}$
$\delta : X \times Q \times 2^{C \times P}\ S$ is the state transition function
$\lambda_{\mathrm{down}} : \mathrm{S} \to 2^{Y \times C \times P}$ is the downward output function
$\mathrm{v}_{\mathrm{down}} : V_s \to P$ is the value coupling downward
sc: $S \to 2^{\mathrm{C}} \times 2^{\mathrm{MC}}$ is the structural change function
$\mathrm{act}_{\mathrm{up}} : S \times 2^{C \times P} \to \{\mathrm{true, false}\}$ is the activation function

For more detailed mathematical analysis, see Uhrmacher et al. (2007). The application of ML-DEVS has been in the areas of computational chemistry and biology. As a result, the formalism was designed to satisfy the needs of these disciplines, where agents are essentially reactive. Micro-DEVS is a simplified version of *parallel* DEVS (P-DEVS) in which there is no δ_{int} and δ_{con}, but only δ_{ext} to account for external messages. This simplification is undesirable when the agent is proactive and adaptive with the learning behavior. The agent's internal state is equally important and is much needed. Consequently, Micro-DEVS is unsuitable for modeling CAS. We recommend using the atomic ρDEVS for CAS. Macro-DEVS, being a coupled model, holds components, but also has state and various transition functions that enable upward and downward causation. Other examples in literature that deal with the variable structure in MASs are agent-oriented DEVS (Uhrmacher & Zeigler, 1996). However, their atomic DEVS specification has to be integrated with Macro-DEVS to model the transformation of a node into a hierarchical node, that is, a hub.

Coming back to our discussion of CASs, let us now look at how the dynamic-structure DEVS lends itself to describe a scale-free CAS.

23.3.2 DEVS Application to CAS and the Needed Augmentations

The feature list presented in Table 23.2 lists just some of the features that we identified and that can help in modeling CASs with DEVS. Our analysis is based on scale-free topologies and co-occurrence of self-organization and emergence in an interconnected network of persistent agents and persistent environments. The last column in Table 23.2 shows the state of the art in modeling CASs using dynamic-structure DEVS extensions.

The current state of both ρDEVS and ML-DEVS together is fully equipped to specify a CAS with some necessary augmentation. Augmentation of the strong emergence capability, that is, embedding the observer functionality inside an agent model, and the augmentation of the clustering capability, that is, transformation of a node into a hub at both the structural and the behavioral levels, would specify CAS-DEVS (Figure 23.1).

TABLE 23.2
Features Required for Modeling Scale-Free CAS, Capable of Self-Organization and Emergence

ID	Property	What Is Answered?	Contributions by DEVS Engineering
A	Clustering	How does a node become a hub? How does the network handle hubs?	ρDEVS and Macro-DEVS formalism together. While the clustering can easily be implemented using value couplings, the transformation of a node into a hub and the dynamic behavior of such transformation need to be investigated.
B	Scale-free topology	How do the network structures in the presence of power law behave? How does the network connect nodes, clusters, and hubs in a scale-free topology?	P-DEVS formalism Co-occurrence of hubs and nodes with dynamic couplings and dynamic components

TABLE 23.2 (*Continued*)
Features Required for Modeling Scale-Free CAS, Capable of Self-Organization and Emergence

ID	Property	What Is Answered?	Contributions by DEVS Engineering
C	Preferential attachment	How does the new node in the network choose its neighbor based on affinity?	ML-DEVS formalism Value couplings allow development of contingency-based links that could reflect affinity and thresholds in a dynamic manner. ML-DEVS framework with contingency-based systems concepts (Chapter 10) can help specify the preferential attachments at various levels of system specifications.
D	Growth and decay	How do the network linkages increase or decrease for a node?	ρDEVS formalism Internal transition functions can direct inport and outport couplings along with dynamic component structures.
E	Threshold and affinity	How does the agent act upon various thresholds and how does it reconfigure its behavior?	P-DEVS formalism Transition functions can have threshold and affinity models.
F	Inter-connectivity	How is the dynamic nature of network specified?	ρDEVS formalism
G	Modularity	How does the external interface of an agent guide its role in network?	P-DEVS formalism It is the very foundation of DEVS systems.
H	Hierarchy	How do clusters and hubs reduce their connectivity and change their role from a performer to an enabler?	P-DEVS formalism DEVS complex systems are hierarchical by design.
I	Agent persistence	How does an agent handle persistent state? How is memory defined in an agent?	P-DEVS formalism Agents have state variables and are persistent. The state variables persist along the entire life cycle of the agent.
J	Environment persistence	How does an environment handle persistence? How do the affordances provided by the environment persist?	Loosely coupled, agent is modular and environment is external and unpredictable. Environment is available as an external activity through a netcentric infrastructure. The agents developed in P-DEVS as implemented using the DEVS/SOA framework are loosely coupled with external web services through modular interfaces.

(*Continued*)

TABLE 23.2 (*Continued*)
Features Required for Modeling Scale-Free CAS, Capable of Self-Organization and Emergence

ID	Property	What Is Answered?	Contributions by DEVS Engineering
K	Interactive transition systems	How does an agent or a system specify its transition functions in an interactive manner?	P-DEVS formalism The three transition functions (δ_{int}, δ_{ext}, and δ_{con}) are based on a notion of abstract event that triggers an internal transition or an external transition or both. A message exchange is an indication of an event at both the sender's and the receiver's end and is formally dealt with.
L	Self-organization	How does an agent system organize itself toward a global behavior? How does it reconfigure its behavior?	ρDEVS and ML-DEVS formalism ρDEVS handles structural dynamism, that is, components, behavior, and value couplings. ML-DEVS allows specification of constraints through value couplings that dictate coupling formation. Possible integration with the KCGS framework (Chapter 10) may allow constraints specification.
M	Weak emergence	How does a system display global behavior greater than the behavior of its constituents?	P-DEVS formalism Emergence is an outcome. Specific observer agents can be coupled to the system that detects emergent parameters and activity.
N	Strong emergence	How is an observer embedded in a persistent agent, such that it reconfigures its external behavior, moves to a higher level hierarchy to enable causal behavior at lower level?	P-DEVS formalism The observer is a DEVS agent that observes another DEVS agent or any external modular component. Such an observer can cause behavior change in the observed agent. A tightly coupled agent + observer coupled system becomes a composite agent with an embedded observer. A partial workable solution is thus provided.
O	Nonlinearity	How does an event cascade in a network resulting in cascaded effects?	Quantized-DEVS and ML-DEVS formalism Value couplings communicate messages at various levels of hierarchy resulting macro-micro effects.
P	Concurrency	Agent displays many parallel executing behaviors.	P-DEVS formalism The P-DEVS formalism is a complex dynamical system that allows concurrency.

TABLE 23.2 (*Continued*)
Features Required for Modeling Scale-Free CAS, Capable of Self-Organization and Emergence

ID	Property	What Is Answered?	Contributions by DEVS Engineering
Q	Upward causation	How do the nodes in a hierarchical environment communicate information to hubs, thereby eliciting reaction at a level above it?	ML-DEVS formalism The system components at lower levels of hierarchy when updated dynamically due to their adaptive nature can trigger macro-level behavior at higher levels of hierarchy.
R	Downward causation	How do the hubs cause changes at lower levels of hierarchy?	ML-DEVS formalism The macro-level decision at higher levels of hierarchy can be communicated instantly at multiple levels of lower hierarchy due to existence of ports. The macro-level coupled models have complete knowledge about the micro-level component interfaces.

Source: Adapted from Mittal, S., *Cognitive Systems Research*, retrieved from http://dx.doi.org/10.1016/j.cogsys.2012.06.003, 2012.

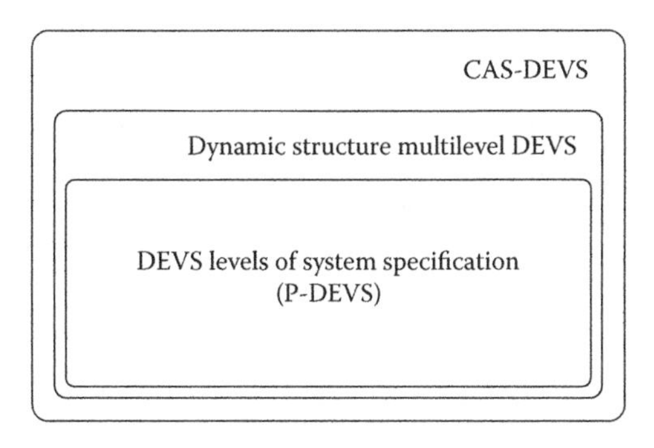

FIGURE 23.1 CAS-DEVS as extensions of dynamic-structure DEVS.

Since ML-DEVS is based on ρDEVS, the ML-DEVS extension should be augmented to

1. *Transform an atomic component to a cluster component*: This requires addition and augmentation of new transition functions in a live system such that it performs a macro-role rather than a micro-role. This is related to rise in abstraction at the DEVS atomic level. Augmentation should result in an algorithm that transforms a node into a hierarchical node with a Macro-DEVS behavior.
2. *Strong emergence*: This capability requires the agent to reconfigure its behavior based on its observation of the micro- and macro-patterns in a downward causal manner as designed by the designer of the artificial system.

Features listed in Table 23.2 operate at various levels of abstractions, and an implementation of these features at the appropriate level of abstraction yields the desired effect. The presence of the same feature at different levels of DEVS specification implies that the feature needs to be implemented at all those levels. For example, feature A should be specified at levels 1, 2, 3, and 4 simultaneously to get the clustering effect. We introduced a coupling abstraction level in the coupled system at level 4 to clearly mark the features that impact connectivity of atomic and coupled components. This may imply that there is an additional level of abstraction between the DEVS atomic and coupled components that formally specifies a dynamic coupling relation. The dynamic coupling relation has been described with reference to ρDEVS and ML-DEVS. As can be seen from Table 23.3, a coupled system at level 4 is mirroring the feature set of an atomic system at level 3 with the exception of features of hierarchy (including containment) and environment persistence. This also reaffirms our thesis that a coupled model specification needs to have a behavior of its own and not just act as a container. Further, the coupling abstraction may also cater to features such as clustering, topology, preferential attachment, growth and decay, interconnectivity, and self-organization. This implies the following:

- There may be a way to formally define a rich coupling specification that has above-mentioned aspects encoded.
- Some of the behaviors encoded in the nodes can become the behaviors of the networked system where the formal coupling specification manifests the properties of the complex network.

Another important aspect warranting discussion is the *closure under coupling* property of complex systems. The DEVS levels of system specification are closed under coupling; that is, the behavior of a coupled DEVS can be specified as an atomic DEVS. This property helps build hierarchical complex systems, and the current DEVS formalism is positioned to support weak emergence, whereby the emergent behavior can be reduced to a lower level behavior of the constituent components of the system. To display strong emergence, what is needed is an extension of the closure-under-coupling property of CAS such that the novel emergent behavior that

TABLE 23.3
Abstraction Levels of Scale-Free CAS Features Portraying Self-Organization and Emergence

Level	Name	Features of Scale-Free CAS
4	Coupled system	A, B, C, D, E, G, H, I, J, K, L, M, N, P, Q, R
	Couplings	A, B, C, D, F, L
3	I/O system	A, B, C, D, E, G, I, K, L, M, N, O, P, Q, R
2	I/O function	A, C, D, F, I, N, Q, R
1	I/O behavior	A, C, E, F, H, I, K, O, L, M, N, Q, R
0	I/O frame	F, G, K, N, L, O, Q, R

is irreducible to the constituent components can be accommodated. This implies that the new observed behaviors (or emergents) that are not part of the system (at level $L_{(n-1)}$) be made available as observers at a higher level of hierarchy at level L_n become acquired behaviors at level L_n. Such an acquired behavior should then reconfigure the Macro-DEVS behavior specification to incorporate the new abstraction and concepts as provided by the observers at L_n.

23.4 NETCENTRIC CAS

Having laid the foundation for DEVS as a CAS modeling platform, we turn our focus on applying netcentricity concepts in modeling CAS. CASs have been studied in various independent disciplines, such as economics, financial markets, cell biology, neurobiology, power grids, and language. Although these systems can be analyzed, as we have shown earlier in the chapter, on the basis of various identified features (Table 23.2), there is a fundamental requirement of formalizing the information flow within and without these systems. The closure-under-coupling property defines a clear boundary in DEVS complex hierarchical systems. However, CASs are *open* systems with new information added at various levels of the system. The two augmentations we suggested point to this issue:

1. When a node transforms into a hub, there is new information added. It can be argued that rules may be encoded earlier that lay dormant until a threshold crossing occurs, switching the node's behavior to a hub. However, in truly adaptive systems that learn, new information is added or old information is abstracted to a new higher level. This is the way nature has been managing complexity. The knowledge to abstract a totally new concept is external information. We, as humans, do it readily based on our rich set of experiences and developing analogies. Automating it in a computational agent is a challenge.
2. When a node acts as an active observer that has downward causation, the abstractions are translated to behaviors and actions at multiple levels of hierarchy. This capacity can also be pre-encoded. However, the question still remains: What are the mechanisms to formalize new behaviors, abstractions, in a live system?

To address this issue, we put various ideas presented in the book in the perspective of a netcentric SoS that turns out to be a netcentric CAS. These can very well be taken as requirements for a netcentric CAS:

- *Agents in their own domain-specific languages*: The designed agents contain rich domain knowledge. The DSLs are rich and standalone, and satisfy the requirements of a particular domain. These DSLs are analogous to various silos, or monolithic systems that satisfy a need.
- *Dynamic agents*: The agents may have a dynamic behavior and have a perceive-think-act or perceive-act cycle. Various communication and negotiation protocols are refined during the course of agent's lifecycle.

- *Situated agents*: The agents can be either reactive or proactive based on the environment they are in. Multilayered contingencies situate the agent in the environment by formally addressing the context. Agents have structural and functional affordances.
- *DSL interoperability*: The DSLs should be interoperable at all the three levels: pragmatic, semantic, and syntactic. Various agent grammars at various levels of hierarchy must align computationally for agents to communicate semantically. Semantic Web is a step in this direction.
- *Standardized communication*: The agents, through DSLs, communicate through a medium that enables interoperability. A netcentric system is integrated on XML, and work is underway to streamline semantic and pragmatic communication and the associated communication and negotiation protocols.
- *Usage of new knowledge*: Through the semantic interoperability, and integration with ontological frameworks or within the Semantic web, the agents can leverage the capabilities of other domain agents.
- *Service and capabilities oriented*: Agents declare their structural and functional affordances through services they provide and capabilities they need. These can be cross-domain and the resulting SoS may be required to achieve a common objective. Various resource requirements are specified at the pragmatic level and delegated to agents at the semantic level.
- *Abstraction hierarchy*: The agents, through the use of additional knowledge via semantic interoperability, are able to situate themselves better at the pragmatic level in the environment, leading to new structural and functional affordances.
- *Adaptive and learning behavior*: The agents have cognitive capacities and employ both supervised and unsupervised learning techniques that preserve the new abstractions and knowledge affordances. Metalevel learning is introduced that binds pragmatics to semantics. Artificial intelligence (AI) learning mechanisms (e.g., abduction, deduction, induction, and rule-based) and machine learning mechanisms (e.g., reinforcement learning) are applied.
- *Event based*: A heterogeneous system has communication between various components at various levels of hierarchy. The communication should be abstracted to events rather than messages. It is the granularity of the event where pragmatics, semantics, and syntactics are defined at a given instant of the dynamic system.
- *Overarching frameworks*: Frameworks such as DoDAF and SES provide a means to define the context where these cross-domain agents are brought together in a netcentric environment to interoperate toward a common objective.
- *Observer agents*: A netcentric CAS is a modular system. Observer agents can be injected at any level of hierarchy that can intercept various Event clouds and netcentric enterprise buses and sniff on various Service interfaces for instrumentation and control purposes. These observer agents may be pattern-recognizing agents performing complex event processing that can interoperate with other agents to keep the complex agent within the operating bounds of the overarching purpose.

- *Data analytics*: Through the help of observer agents, massive amount of data could be collected, mined, and abstracted toward domain ontology that feeds back into the live system.
- *Human in the loop*: The capability to incorporate humans as live operators who can redefine pragmatics in a live system is a necessity. A dynamic unfolding scenario in a military warfighter domain is a reality of today. The usage of Remotely Piloted Aircrafts (RPAs) is an example of man's role in a complex netcentric system.
- *Formal M&S infrastructure*: The capability to design a formal executable model that is integrated with live netcentric systems such as the Global Information Grid (GIG) and Netcentric Enterprise Services (NCES) provide the much-needed capability to realize the vision of model-driven development (MDD) and model-integrated computing (MIC) in an SoS.
- *Formal verification and validation*: CASs are open systems where defining boundaries is problematic. Utilizing CAS within the overarching frameworks helps define the context, the purpose, and the requirements for the constituents systems. Since there would be an underlying formal M&S framework, the verification and validation efforts can be aligned toward the design of experimental frames at the pragmatic levels.

The ideas presented above are summarized in Figure 23.2. There are multiple domains interacting to achieve a common objective. This is analogous to a human with a rich knowledge of multiple domains. Consider a human engaged in an activity

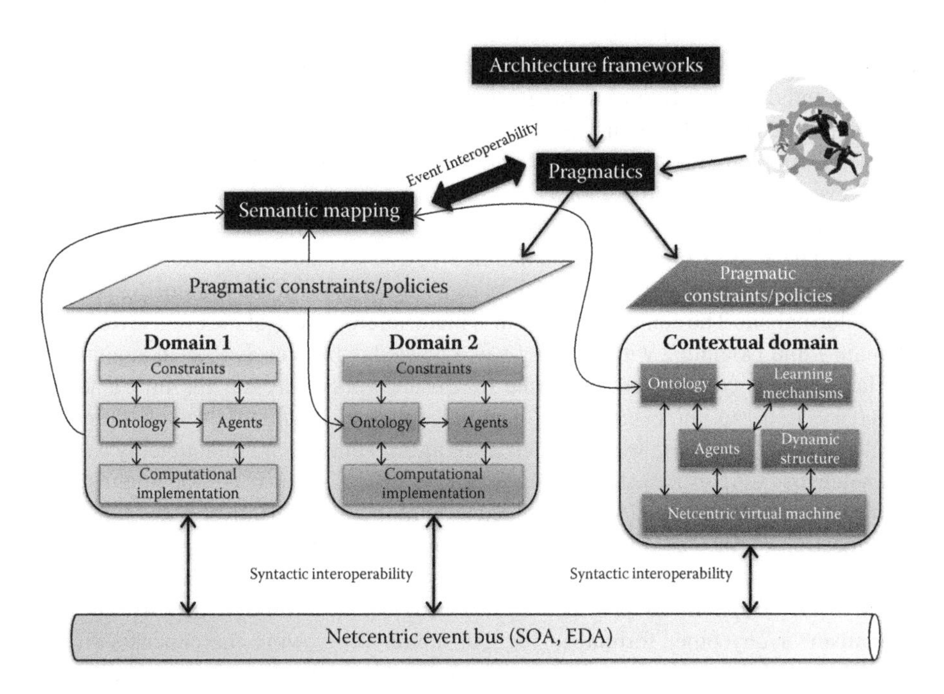

FIGURE 23.2 Netcentric complex adaptive system.

of driving home via a new neighborhood. He has knowledge of his car, his GPS, his destination, and so on. Information is called upon from a particular domain at a specific instant during the pragmatic course of his journey until the journey is completed. This is a simple scenario if taken in isolation. Now imagine the same scenario in a situation in which he hears news on the radio that multiple bank failures have occurred, and he has to run to an ATM on his way home to withdraw cash from his account. To adapt, he has to find the closest ATM and interact with a circumstance that is complex and adaptive where events are unfolding in a larger SoS. He has to bring in knowledge of other relevant domains into his current goal sequence and his actions generate events that interleave between these relevant domains at that particular instant. Similarly, on his way home, many other situations may arise where his interactions with other existing complex systems guide his imminent goal and his primary goal of reaching home.

In Figure 23.2, there are two existing domains (Domains 1 and 2). A domain is defined by its ontology, a set of constraints, an agent system that realizes the ontology and adds spaciotemporal character, and the computational representation of both the ontology and the agent framework. Domain 1 has a definite boundary. Domain 1 can interoperate at the syntactic level with Domain 2. This has been the founding idea behind the evolution of XML and integration of monolithic systems. In a larger context, the ontologies of both Domain 1 and Domain 2 are semantically mapped through the process of *harmonization* (Chapter 10), such that it may lead to a new derived ontology. To use this new ontology in context, pragmatics is needed that specify the creation of the *contextual domain*. A new set of constraints is overlaid on the derived ontology that reflects the application of pragmatics as derived from the architecture frameworks such as DoDAF mission threads. In addition, new components such as learning mechanisms and dynamic structure capabilities are introduced in the contextual domain that make both the agents and the resulting system structure adaptive and dynamic. Likewise, pragmatic constraints are also applied to Domain 1 and Domain 2. While Domain 1 and Domain 2 are available through various services in a netcentric domain, the contextual domain uses these services in its agent system to achieve pragmatic interoperability. Further, the presence of learning mechanisms refines the contextual domain ontology and the behavior of agents in the contextual domain. The logical topology is maintained by the dynamic structure component that provides advanced structural capabilities as highlighted in the dynamic structure DEVS formalism. The contextual domain is made interoperable at the event level with Domain 1 and Domain 2 with influx of new knowledge from each of the domains.

Having these concepts implemented in a framework addresses the information boundary issue to an extent. The new information is added through domain ontologies (e.g., Semantic web), created by humans. Humans are undoubtedly a part of a netcentric CAS as they define pragmatics. While a CAS just "is," the pragmatics define various ways the CAS is subjected to a specific objective. Incorporation and injection of various domain agents and interfacing them with learning agents allow containment of new knowledge in the existing agents imbued with learning mechanisms. Capabilities provided by agents described through the ML-DEVS formalism or any other formalism for that matter (borrowing the concepts from ML-DEVS) along with domain interoperability through events, in a netcentric environment, realize a netcentric CAS.

On another note, more inspiration is taken from social systems as the inherent features tend to produce complexity. Social agents must continually make choice either through direct cognition or through stored heuristics to make action, thereby continually weaving the connections and reassessing contingencies in their environment and with other agents (Miller & Page, 2007). Incorporation of a human element or a cognitive agent in a CAS is a step in the right direction. While the M&S community can help simulate these systems to a good degree of fidelity, the absence of the human operator model that is cognitively plausible presents results that are difficult to map in the real world. To address this critical component in larger DoD frameworks such as DoDAF or the Ministry of Defense Architecture Framework (MoDAF, UK) (MoD, 2008), a human view is proposed that addresses this critical need (Handley et al., 2009). These architecture frameworks produce common systems engineering approaches to development, presentation, and integration of current and future SoS. Newer architectures like DoDAF V2.0 address netcentric, SoSs, and system/services concepts.

Human view is to enable effective Human System Integration (HSI) processes within the design of these complex, large-scale, sociotechnical systems. The North Atlantic Treaty Organization (NATO) *Human View Handbook* facilitates design decisions by identifying relevant elements emphasizing the explicit need of merging seamlessly and efficiently with sound systems engineering practice. It establishes a logical and systematic framework for HSI studies and makes explicit human, crew, and team sociobehavioral processes as integral to total systems performance. Although HSI is a fundamental component of a total systems approach, the successful integration of HSI into systems engineering and acquisition life cycles continues to be a challenge (Phillips, 2010).

23.5 SUMMARY

Complexity is a multifaceted topic and each complex system has its own properties. However, some of the properties, such as high interconnectedness, large numbers of components, and adaptive behavior, are present in most natural complex systems. We looked at the mechanisms behind interconnectedness using network science that describes many natural systems in the light of power laws and self-similar scale-free topologies. Such scale-free topologies bring their own inherent properties to the complex system such that the entire system is subjected to the network's structural and functional affordances. We also discussed the elusive topics of self-organization and emergence in CAS and how DEVS may provide a mechanism to address the modeling aspects of such issues to model an adaptive behavior in a contingent environment.

We looked at the information boundaries in modeling CAS and how netcentric CAS are positioned to address the issues using the concepts of pragmatic, semantic, and syntactic interoperability. We listed some of the features of a netcentric CAS and that it also incorporates a human component at the pragmatic level.

The fundamental ideas presented in the book relate to the model-driven design paradigm and the model-continuity principles, where the model is the actual artifact in a deployed system. Netcentricity is a concept that pushes the MDD and MIC concept further ahead by executing a model in an environment where other capabilities are available for their reuse. A deployed model in a netcentric environment is akin to the *model-as-a-service* (MaaS) paradigm analogous to the model-continuity principle. With frameworks like DEVS Unified Process and DEVSML 2.0 stack, the MaaS paradigm is readily realized. Opening the model to a netcentric environment makes it a part of a larger CAS. The issues related to CAS are presented in this chapter to bring the problem into perspective and to effectively design models deployable in an open environment. The management, design, and reuse of models and services in a complex adaptive environment are facilitated with data-driven architecture frameworks such as the Department of Defense Architecture Framework 2.0. While DoDAF 2.0 is a step in the right direction, more work is needed at the semantic and pragmatic levels to achieve true interoperability. The human operator in a netcentric system is a challenge of its own. Incorporating human at the pragmatic level solves only a part of this challenge. The human operator model has received its due attention just recently, and netcentric CAS makes the problem even more complex. This book introduces various concepts to tackle the model-based integration and interoperability challenges. The journey has just begun.

REFERENCES

Baas, N. (1994). Emergence, hierarchies and hyperstructures. *Artificial Life, 3.*

Banabeau, E., & Dessalles, J. (1997). Detection and emergence. *Intellica, 2*(25).

Barabási, A.-L. (2003). *Linked: How Everything Is Connected to Everything Else and What It Means for Business, Science and Everyday Life.* New York, NY: Penguin Books.

Barros, F. (1995). Dynamic structure discrete event system specifications. *Proceedings of the 1995 Winter Simulation Conference.* Arlington, VA.

Barros, F. (1997). Modeling formalisms for dynamic structure systems. *ACM Transactions on Modeling and Computer Simulation, 7*(4), 501–515.

Barros, F. (1998). Abstract simulators for the DSDE formalism. *Proceedings of the 1998 Winter Simulation Conference.* Washington, DC.

Brooks, R. (1991). Intelligence without reason. *12th Int'l Joint Conference on Artificial Intelligence.* San Mateo, CA.

Camazine, S., Deneubourg, J., Franks, N., Sneyd, N., Theraulaz, G., & Bonabeau, E. (2001). *Self-Organization in Biological Systems.* Princeton, NJ: Princeton University Press.

Chalmers, D. J. (2006). Strong and weak emergence. In P. Clayton & P. Davies (Eds.), *The Re-Emergence of Emergence.* Oxford: Oxford University Press.

Deguet, J., Demazeau, Y., & Magnin, L. (2006). Elements about the emergence issue: a survey of emergence definitions. *Complexus, 3*, 24–31.

Handley, H. A., Smillie, R. J., & Knapp, B. Architecture frameworks and the human view. National Defense Industrial Association. Retrieved September 1, 2009, from www.ndia.org/Divisions/Divisions/SystemsEngineering/Documents/HSI%2.

Heyligen, F. (2001). The science of self-organization and adaptivity. In L. D. Kiel (ed.). Knowledge Management, Organizational Intelligence and Learning, and Complexity, in: The Encyclopedia of Life Support Systems (EOLSS). Oxford: Eolss Publishers.

Holland, J. H. (1992). Complex adaptive systems. *Daedalus, 121*(1), 17–30.

Holland, J. H. (2006). Studying complex adaptive systems. *Journal of Systems Science and Complexity, 19*(1), 1–8.

Holland, O., & Melhuish, C. (1999). Stigmergy, self-organization, and sorting in collective robotics. *Artificial Life, 5*(2), 173–202.

Hu, X., Zeigler, B., & Mittal, S. (2005). Variable structure in DEVS component-based modeling and simulation. *Transactions of SCS, 81*(2), 91–102.

Keil, D., & Goldin, D. (2003). Modeling indirect interaction in open computational systems. *12th IEEE Int'l Conference on Enabling Technologies: Infrastructure for Collaborative Enterprises*. Linz, Austria.

Miller, J. H., & Page, S. E. (2007). *Complex Adaptive Systems: An Introduction to Computational Models of Social Life.* Princeton, NJ: Princeton University Press.

Ministry of Defence Human Factors Integration Defence Technology Centre. (2008, July 15). *The Human View Handbook for MODAF.* Retrieved September 1, 2009, from www.hfidtc.com/MoDAF/HV%20Handbook%20First%20Issue.pdf.

Mittal, S. (2012). Emergence in stigmergic and complex adaptive systems: a formal discrete event systems perspective. *Cognitive Systems Research.* Retrieved from http://dx.doi.org/10.1016/j.cogsys.2012.06.003.

Mittal, S., Khargharia, B., & Zeigler, B. (2005). *Neuronal Path Formation Using Spiked Leaky Neuron Model.* Tucson, AZ: University of Arizona, ACIMS Lab, ECE 549 project.

Muller, J. (2003). Emergence of collective behavior and problem solving. *4th International Workshop on Engineering Societies in the Agents World.* London, UK.

NATO RTO HFM-155 Human View Workshop. (n.d.). *The NATO Human View Handbook.* National Defense Industrial Association. Retrieved September 1, 2009, from www.ndia.org/searchcenter/Pages/Results.aspx?k=human view.

Newman, M. (2001). Clustering and preferential attachment in growing networks. *Physical Review, E 64*(2), 025102.

Phillips, E. L. (2010). *The Development and Initial Evaluation of the Human Readiness Level Framework* (MS thesis).Monterey, CA: Naval Postgraduate School.

Pinker, S. (1997). *How the Mind Works.* New York, NY: W. W. Norton.

Shalizi, C. (2001). *Causal Architecture, Complexity and Self-Organization in Time Series and Cellular Automata* (PhD thesis). Madison: University of Wisconsin at Madison.

Uhrmacher, A. (2001). Dynamic structures in modeling and simulation—a reflective approach. *ACM Transactions on Modeling and Simulation, 11*(2), 206–232.

Uhrmacher, A., Ewald, R., John, M., Maus, C., Jeschke, M., & Biermann, S. (2007). Combining micro- and macro-modeling in DEVS computational Biology. *Winter Simulation Conference*. Washington, DC.

Uhrmacher, A., Himmelspach, J., & Ewald, R. (2011). Effective and efficient modeling and simulation with DEVS variants. In G. Wainer & P. Mostermann (Eds.), *Discrete Event Modeling and Simulation: Theory and Applications* (pp. 139–176). Boca Raton, FL: CRC Press.

Uhrmacher, A., Himmelspach, J., Rohl, M., & Ewald, R. (2006). Introducing variable ports and multi-couplings for cell biological modeling in DEVS. *Winter Simulation Conference.* Monterey, CA.

Uhrmacher, A., & Priami, C. (2005). Discrete event systems specification in systems biology—a discussion of stochastic pi calculus and DEVS. *Winter Simulation Conference.* Orlando, FL.

Uhrmacher, A., & Zeigler, B. P. (1996). Variable structure models in object-oriented simulation. *International Journal of General Systems, 24*(4), 359–375.

Watts, D. (1999). *Small Worlds.* Princeton, NJ: Princeton University Press.

Wolf, T. D., & Holvoet, T. (2005). *Emergence Versus Self-Organization: Different Concepts But Promising When Combined. Lecture Notes in Artificial Intelligence* (Vol. 3464, pp. 1–15). Springer-Verlag Berlin Hiedelberg.

Zeigler, B. P. (2004). Discrete event abstraction: an emerging paradigm for modeling complex adaptive systems. In L. Booker, S. Forrest, & M. Mitchell (Eds.), *Perspectives on Adaptation in Natural and Artificial Systems, Essays in Honor of John Holland.* Oxford: Oxford University Press.

Acronyms

ADU: Air Defense Unit
AFCAO: Air Force Chief Architect's Office
AFRL: Air Force Research Laboratory
AFSOC: Air Force Special Operations Command
ALSP: Aggregate Level Simulation Protocol
ALTBMD: Active Layered Theatre Ballistic Missile Defense
ATL: ATLAS Transformation Language
AV: All View/Viewpoint
AWACS: Air borne Warning and Control System
BPEL: Business Process Execution Language
BPMN: Business Process Modeling and Notation
BPS: Business Process Simulation
C2: command and control
C4ISR: command, control, computer, communication information surveillance reconnaissance
CADM: core architecture data model
CAOC: Combine Air Operations Center
CAS: complex adaptive systems
CBD: Chemical and Biological Defense
CDE: Collaborative Distributed Environment
CEP: complex event processing
CES: Core Enterprise Services
CIM: computation-independent model
CJCSI: Chairmain of the Joint Chief of Staff Instruction
CM: conceptual model
CMMI: Capability Maturity Model Integration
COTS: component off the shelf
CPN: Colored Petri Net
CSL: Constraint specification language
CS2F: Cognitive Systems Specification Framework
CV: Capability Viewpoint
DaaS: DEVS-as-a-Service
DAE: Differential Algebraic Equation
DAS-BOOT: Design and Specification-Based Object-Oriented Testing
DES: discrete event simulation
DESS: Differential Equation System Specification
DEVS: Discrete Event Systems Specification
DEVSML: DEVS Modeling Language
DEVSVM: DEVS virtual machine
DIS: Distributed Interactive Simulation
DoD: Department of Defense
DoDAF: Department of Defense Architecture Framework
DSDEVS: dynamic structure DEVS
DSL: domain-specific language
DSOL: Distributed Simulation Object Library

DTSS: Discrete Time System Specification
DUNIP: DEVS Unified Process
EBNF: Extended Backus Naur Form
EDA: Event-driven architecture
EFP: Experimental Frame-Processor
EMF: Eclipse Modeling Framework
EMOF: Essential Meta-Object Facility
ESB: Enterprise Service Bus
FTG: Formalism Transformation Graph
GEF: Graphical Editing Framework
GEMS: Generic Eclipse Modeling System
GENETSCOPE: Generic Network System Capable of Planned Expansion
GIG: Global Information Grid
GME: Generic Modeling Environment
GMF: Graphical Modeling Framework
GMT: Generative Modeling Technologies
GReAT: Graph Rewriting and Transformation Language
HFGCS: High-Frequency Global Communication System
HLA: High-Level Architecture
IDE: Integrated Development Environment
IDEF: Integrated definition method
JAX-WS: Java API for XML Web Services
JCA: Joint Capability Area
JCAS: Joint Close Air Support
JCIDS: Joint Capabilities Integration and Development System
JDAMs: Joint Direct Attack Munitions
JMT: Joint Mission Thread
JPEO: Joint Program Executive Office
JSOFT: Joint Special Operations Task Force
JSSEO: Joint SIAP Systems Engineering Organization
JTA: Joint Technical Architecture
JVM: Java Virtual Machine
KCGS: Knowledge-Based Contingency Driven Generative System
KIP: Key Interface Profile
LQA: Link Quality Assessment
M2DEVS: model-to-DEVS
M2DEVSML: model-to-DEVSML
M2M: model-to-model
MaaS: model-as-a-service
MBD: model-based design
MBE: model-based engineering
MBSE: model-based systems engineering
MDA: model-driven architecture
MDD: model-driven development
MDD4MS: model-driven development for modeling and simulation
MDE: model-driven engineering
MDSD: model-driven software development
MIC: model-integrated computing
MIPS: model-integrated program synthesis
MOF: Meta-Object Facility
MoE: Measures of Effectiveness

MoP: Measures of Performance
MSVC: Model Simulator View Controller
MVC: Model View Controller
NCES: Net-Centric Enterprise Services
NCOW/RM: Net-Centric Operational Warfare Reference Model
NFZ: nonflying zone
NIPRNET: Nonclassified internet protocol router network
NR-KPP: Net Ready Key Performance Parameters
NSW: Naval Special Warfare
NSWC: Naval Surface Warfare Center
OCL: Object Constraint Language
ODE: ordinary differential equation
OMG: Object Management Group
OOP: object-oriented programming
OpenUTF: Open Unified Technical Framework
OV: Operational view, Viewpoint
PaaS: Platform-as-a-Service
P-DEVS: parallel DEVS
PES: Pruned Entity Structure
PIM: platform-independent model
PISM: platform-independent simulation model
PSM: platform-specific model
PSSM: platform specific simulation model
QVT: Query/View/Transformation
RCP: Rich-client platform
RT-DEVS: real-time DEVS
RTI: Run-Time Infrastructure
RUP: Rational Unified Process
SAAM: Software Architecture Analysis Method
SaaS: Simulation-as-a-Service, Software-as-a-Service
SCOPE: Systems Capable of Planned Expansion
SES: System Entity Structure
SESM: Scalable Entity Structure Modeler
SIAP: Single Integrated Air Picture
SiMaaS: Simulation-as-a-Service
SIPRNET: secret internet protocol router network
SME: subject matter expert
SOA: Service-Oriented Architecture
SOAP: Simple Object Access Protocol
SOF: Special Operations Force
SOLE: Special Operations Liaison Element
SoS: system of systems
STE: software testing environments
SUT: system under test
SV: systems view, Viewpoint
SysML: Systems Modeling Language
T&E: testing and evaluation
TENA: test and training enabling architecture
TIS: test instrumentation system
TV: Technical View
UAV: unmanned aerial vehicle

UDDI: Universal Description Discovery and Integration
UML: Unified Modeling Language
WP: Way Point
WSDL: Web Service Description Language
WWW: World Wide Web
XMI: XML Metadata Interchange
XML: eXtensible Markup Language
XSLT: eXtensible Stylesheet Language Transformer

Index

D

E

F

I

J

K

L

M

N

O

P

S

V

W

X

Z

Printed and bound by CPI Group (UK) Ltd, Croydon, CR0 4YY
18/10/2024
01776261-0019